2023 IEEE 41st VLSI Test Symposium (VTS 2023)

San Diego, California, USA
24-26 April 2023

IEEE Catalog Number: CFP23029-POD
ISBN: 979-8-3503-4631-2

Copyright © 2023 by the Institute of Electrical and Electronics Engineers, Inc.
All Rights Reserved

Copyright and Reprint Permissions: Abstracting is permitted with credit to the source. Libraries are permitted to photocopy beyond the limit of U.S. copyright law for private use of patrons those articles in this volume that carry a code at the bottom of the first page, provided the per-copy fee indicated in the code is paid through Copyright Clearance Center, 222 Rosewood Drive, Danvers, MA 01923.

For other copying, reprint or republication permission, write to IEEE Copyrights Manager, IEEE Service Center, 445 Hoes Lane, Piscataway, NJ 08854. All rights reserved.

*** This is a print representation of what appears in the IEEE Digital Library. Some format issues inherent in the e-media version may also appear in this print version.

IEEE Catalog Number:	CFP23029-POD
ISBN (Print-On-Demand):	979-8-3503-4631-2
ISBN (Online):	979-8-3503-4630-5
ISSN:	1093-0167

Additional Copies of This Publication Are Available From:

Curran Associates, Inc
57 Morehouse Lane
Red Hook, NY 12571 USA
Phone: (845) 758-0400
Fax: (845) 758-2633
E-mail: curran@proceedings.com
Web: www.proceedings.com

2023 IEEE 41st VLSI Test Symposium (VTS 2023)

San Diego, California, USA
24-26 April 2023

IEEE Catalog Number: CFP23029-POD
ISBN: 979-8-3503-4631-2

Table of Contents

Outlier Detection for Analog Tests Using Deep Learning Techniques..........1

    Chin-Kuan LIN, Cheng-Che LU, Shuo-Wen CHANG, Ying-Hua CHU, Kai-Chiang WU, Mango CHAO

Machine Learning-Based Adaptive Outlier Detection for Underkill Reduction in Analog/RF IC Testing..........8

    Vineeth NIRANJAN, Deepika NEETHIRAJAN, Dallas WEBSTER, Amit NAHAR, Constantinos XANTHOPOULOS, Yiorgos MAKRIS

Architectural Radiation Hardening of CMOS Power Management Circuits through Bias Tuning..........15

    Gauri KOLI, Liam NGUYEN, Kitchen JENNIFER

Gerabaldi: A Temporal Simulator for Probabilistic IC Degradation and Failure Processes..........23

    Ian HILL, Andre IVANOV

Diagnosis of Quantum Circuits in the NISQ Era..........30

    Yu-Min LI, Cheng-Yun HSIEH, Yen-Wei LI, Chien-Mo LI

Targeted Custom High-Voltage Stress Patterns on Automotive Designs..........37

    Saidapet RAMESH, Jaiswal AKSHAY, Marchese ROBERT, Thota SUNNY, Dickson KRISTOFOR

Reliable Brain-inspired AI Accelerators using Classical and Emerging Memories..........41

    Mikail YAYLA, Simon THOMANN, Md Mazharul ISLAM, Ming-Liang WEI, Shu-Yin HO, Ahmedullah AZIZ, Chia-Lin YANG, Jian-Jia CHEN, Hussam AMROUCH

Innovation Practices Track: VLSI Functional Safety..........51

    Fei SU, Meirav NITZAN, Ankush SETHI, Vaibhav KUMAR, Dan ALEXANDRESCU

Special Session: Security Verification & Testing for SR-Latch TRNGs..........52

    Javad BAHRAMI, Mohammad EBRAHIMABADI, Jean-Luc DANGER, Sylvain GUILLEY, Naghmeh KARIMI

Silent Data Errors: Sources, Detection, and Modeling..........62

    Adit SINGH, Sreejit CHAKRAVARTY, George PAPADIMITRIOU, Dimitris GIZOPOULOS

A Novel LBIST Signature Computation Method for Automotive Microcontrollers using a Digital Twin..........74

    Daniel TILLE, Leon KLIMASCH, Sebastian HUHN

Vmin Prediction Using Nondestructive Stress Test..........80

    Chun CHEN, Jeng-Yu LIAO, Chien-Mo LI, Harry CHEN, Eric Jia-Wei

Predicting the Silent Data Error Prone Devices Using Machine Learning..........87

    Mohammad Ershad SHAIK, Abhishek Kumar

Functional Test Generation for AI Accelerators using Bayesian Optimization..........91

    Arjun CHAUDHURI, Ching-Yuan CHEN, Jonti TALUKDAR, Krishnendu CHAKRABARTY

Expanding a Pool of Functional Test Sequences to Support Test Compaction..........97

    Irith POMERANZ

A guided debugger-based fault injection methodology for assessing functional test programs..........104

Francesco ANGIONE, Paolo BERNARDI, Nicola DI GRUTTOLA GIARDINO, davide APPELLO, claudia BERTANI, Vincenzo TANCORRE

Refreshing the JTAG Family..........111

Michele PORTOLAN, Martin KEIM, Jeff REARICK, Heiko EHRENBERG

Innovation Practices Track: Silicon Lifecycle Management Challenges and Opportunities..........118

Robert JIN, Nilanjan MUKHERJEE, Yervant ZORIAN

Special Session: Approximation and Fault Resiliency of DNN Accelerators..........119

Mohammad Hasan AHMADILIVANI, Mario BARBARESCHI, Salvatore BARONE, Alberto BOSIO, Masoud DANESHTALAB, Salvatore DELLA TORCA, Gabriele GAVARINI, Maksim JENIHHIN, Jaan RAIK, Annachiara RUOSPO, Ernesto SANCHEZ, Mahdi TAHERI

Fully Deterministic Storage Based Logic Built-In Self-Test..........129

Subashini GOPALSAMY, Irith POMERANZ

Test Generation for Defect-Based Faults of Scan Flip-Flops..........136

Yu-Teng NIEN, Chen-Hong LI, Pei-Yin WU, Yung-Jheng WANG, Kai-Chiang WU, Mango CHAO

Design for testability (DFT) for RSFQ circuits..........143

Mingye LI, Yunkun LIN, Sandeep GUPTA

CAPEC: A Cellular Automata Guided FSM-based IP Authentication Scheme..........150

Mridha Md Mashahedur RAHMAN, Mohammad RAHMAN, Rasheed KIBRIA, Mike BORZA, Bandi REDDY, Adam CRON, Fahim RAHMAN, Mark TEHRANIPOOR, Farimah FARAHMANDI

A Low Overhead Checksum Technique for Error Correction in Memristive Crossbar for Deep Learning Applications..........158

Surendra HEMARAM, Soyed Tuhin AHMED, Mahta MAYAHINIA, Christopher MÜNCH, Mehdi TAHOORI

Thwarting Reverse Engineering Attacks through Keyless Logic Obfuscation..........165

Leon LI, Alex ORAILOGLU

Graph Neural Networks for Hardware Vulnerability Analysis - Can you Trust your GNN?.......171

Lilas ALRAHIS, Ozgur SINANOGLU

Special Session: Using Graph Neural Networks for Tier-Level Fault Localization in Monolithic 3D Ics..........175

Shao-Chun HUNG, Arjun CHAUDHURI, Sanmitra BANERJEE, Krishnendu CHAKRABARTY

An Efficient External Memory Test Solution: Case Study for HPC Application..........179

Keqing OUYANG, Minqiang PENG, Yunnong ZHU, Kang QI, Grigor TSHAGHARYAN, Arun KUMAR, Gurgen HARUTYUNYAN, Isaac WANG

Allocating Physically Aware Embedded Memory Test & Repair Processor using Floorplan Info at the RTL Design Level..........183

Vinay KUMAR, Bhrugurajsinh CHUDASAMA, Bin BW WANG, Manish ARORA, Bharath SHANKARANARAYANAN

Overcoming Embedded Memory Test & Repair Challenges in the Gate-All-Around Era..........187

Artur GHUKASYAN, Grigor TSHAGHARYAN, Gurgen HARUTYUNYAN, Yervant ZORIAN

Hybrid Binary Neural Networks: A Tutorial Review.........191

Ahmet Enis CETIN, Hongyi PAN

Innovation Practices Track: Innovation on Telemetry Monitoring..........203

Fei SU, Marc HUNTER, Chen HE, Sashi OBILISETTY

Kernel Smoothing Technique Based on Multiple-Coordinate System for Screening Potential Failures in NAND Flash Memory..........204

Gooyoung KIM, Youngseon MOON, Jongmin KIM

Pre and post silicon server platform transient performance using trans-inductor voltage regulator..........211

Judy AMANOR-BOADU, Rishik BAZAZ, Ritchie RICE, Azizi SHUMA, Horthense TAMDEM

Auxiliary State Machine Controlled Autonomous Design Verification Framework..........216

Gurumurti AVHAD

Special Session: Neuromorphic hardware design and reliability from traditional CMOS to emerging technologies..........221

Fabio PAVANELLO, Elena Ioana VATAJELU, Alberto BOSIO, Thomas VAN VAERENBERGH, Peter BIENSTMAN, Benoit CHARBONNIER, Alessio CARPEGNA, Stefano DI CARLO, Alessandro SAVINO

IP Session on Chiplet: Design, Assembly and Test..........231

Bapi VINNAKOTA, Jaber DERAKHSHANDEH, Eric BEYNE, Erik Jan MARINISSEN, Sreejit CHAKRAVARTY

Effective and Efficient Testing of Large Numbers of Inter-Die Interconnects in Chiplet-Based Multi-Die Packages..........232

Po-Yao CHUANG, Francesco LORENZELLI, SREEJIT CHAKRAVARTY, Cheng-Wen WU, Georges GIELEN, Erik Jan

An Exploration of ATPG Methods for Redacted IP and Reconfigurable Hardware..........238

Greg STITT, Naren Vikram Raj MASNA, Venkata KALLURU, Swarup BHUNIA, Nij DORAIRAJ, David KEHLET

Compact Set of Functional Broadside Tests with Fault Detection on Primary Outputs..........245

Irith POMERANZ

Innovation Practices Track: Testability and Dependability of AI Hardware and Autonomous Systems..........252

Fei SU, Eric ZHANG, Arjun CHAUDHURI, Michael PAULITSCH

Special Session: CAD for Hardware Security - Promising Directions for Automation of Security Assurance..........253

Sohrab AFTABJAHANI, Mark TEHRANIPOOR, Farimah FARAHMANDI, Bulbul AHMED, Ryan KASTNER, Francesco RESTUCCIA, Andres MEZA, Kaki RYAN, Nicole FERN, Jasper VAN WOUDENBERG, Rajesh VELEGALATI, Cees-Bart BREUNESSE, Cynthia STURTON, Calvin DEUTSCHBEIN

Unified Approaches for Silicon Debug..........263

Mike RICCHETTI, Sankaran MENON, David AKSELROD

PROCEEDINGS

2023 IEEE 41st VLSI Test Symposium (VTS)

—— VTS 2023 ——

April 24th – 26th 2023

San Diego, CA, USA

VTS 2023 Foreword

We welcome you all to attend the 41st IEEE VLSI Test Symposium (VTS 2023). This is the first time in person since 2020 we get together to communicate and discuss the challenges and solutions of our technical fileds and have a pleasure of meeting you all face to face while still keeping a virtual participation option. We were very excited to work with the Program Committee, the Organizing Committee, and the Steering Committee, and of course with all authors and presenters to put together the VTS 2023 program. We hope everyone will enjoy this event. VTS 2023 program is honored to have three keynote addresses. On the first day, we first have Subi Kengeri, Vice President, AI Systems Solutions at Applied Materials, to share "Heterogeneous Integration in the AI Era". The second address is given by Professor Madhavan Swaminathan of Penn State University to present "The Future of Semiconductor Packaging and Heterogeneous Integration is Now!". On the second day, we have Lorie Burmood, Director of Operations, NXP Semiconductors to address "The Heart of the Feedback Loop: MEMS Development and Test".

The rest of the technical program of VTS 2023 addresses many trends and test challenges in the semiconductor design and manufacturing process. Each day will also have a variety of regular paper sessions, industrial practice (IP) sessions, and academic-focused special sessions.

There are 24 regular papers in 8 sessions covering a range of topics: Analog Test, Failure Analysis, Reliability, Functional Test, BIST & DFT, Hardware Security, Post-Silicon Validation, Memory Test, and Test Generation. VTS also hosts the E.J. McCluskey Doctoral Thesis Competition to showcase the exciting student research spanning all of the above topics.

VTS 2023 also continues its tradition of drawing leading test practitioners and researchers in both industry and academia to contribute to the innovative practices (IP) and special sessions, enabling it to be a venue for debating future technology trends and test challenges, sharing test practices, and charting a new test research roadmap. We have 6 IP and 9 special sessions. The IP sessions include the topics of VLSI Functional Safety and RAS, Silicon Lifecycle Management Challenges and Opportunities, Innovation on Telemetry Monitoring, Chiplet Design & Assembly, & Test, Testability & Dependability of GPU, HPC, & Autonomous Systems, and Unified Approaches to Silicon Debug.

The more academic-focused special sessions include: Reliable Brain-Inspired Hardware using FeFET and Memristive Devices, Security Verification & Testing for SR Latches and TRNGs, Silent Error Corruption: Causes, Detection, and Diagnosis, JTAG: A Family Legacy, Approximation and Fault Resiliency of DNN Accelerators, Graph Neural Networks for Design, Security, & Trust, Tuning Memory Test Challenges & Solutions for Advanced Tech Nodes, Neuromorphic Hardware Design & Reliability from Traditional CMOS to Emerging Technologies, and CAD for Security.

This year's conference will also host a Monday Evening Wine & Cheese Panel covering Test and DFT in the era of the CHIPS and Science Act. There is also a panel on Wednesday focusing on test challenges. Finally, there are two embedded tutorials covering Hardware Design and Reliability Mitigation of Binary Bayesian Reasoning, and Hardware Efficient Binary Neural Network Architectures.

Because networking is such an important part of the VTS community, in additional to opportunities to talk to and meet colleagues over lunch and in breaks between sessions, there will also be a social program on Tuesday consisting of visiting Old Town in San Diego and a banquet at the conference site - Hyatt Regency Mission Bay Spa and Marina hotel.

VTS 2023 is a result of the work of many dedicated volunteers: the reviewers, the best paper award judges, the Program Committee, the Organizing Committee, and the Steering Committee. We wholeheartedly thank them all. We want to express our deep gratitude to all the authors who submitted papers. We are looking forward to seeing them present them in person at VTS 2023. We also thank all of the presenters at the special sessions and industrial practice sessions who contribute so much good information and discussion to the conference. Finally, we thank the IEEE Computer Society, the IEEE Philadelphia Section and the

IEEE Computer Society Test Technology Technical Council (TTTC) for the continued technical sponsorship and support.

We thank you all for making VTS a success by actively participating in it, assisting in its organization, and letting us always know how we can improve the symposium experience and increase its value for its audience.

Thank you all for contributing!

General Co-Chairs
Mehdi Tahoori (Karlsruhe Institute of Technology)
Peilin Song (IBM T. J. Watson Research Center)

Program Co-Chairs
Sule Ozev (Arizona State University)
Jennifer Dworak (Southern Methodist University)

VTS 2023 ORGANIZING COMMITTEE

General Co-Chair

Peilin Song
IBM Research, USA

General Co-Chair

Mehdi Tahoori
KIT, DE

Program Co-Chair

Sule Ozev
Arizona State University, USA

Program Co-Chair

Jennifer Dworak
Southern Methodist University, USA

Past Chair

Stefano Di Carlo
Politecnico di Torino, IT

Vice General Chair

Suriyaprakash Natarajan
Intel, USA

Vice Program Co-Chair

Naghmeh Karimi,
UMBC, USA

Vice Program Co-Chair

Gurgen Harutyunyan
Synopsys

Innovative Practices Co-Chair

Arani Sinha
Intel, USA

Innovative Practices Co-Chair

Fei Su
Intel, USA

Special Sessions Chair

Erik Larsson
Lund University, SE

New Initiatives Co-Chair

Xiaowei Li
Chinese Academy of Sciences, CN

Corporate Support Chair

Martin Keim
Mentor, Siemens, DE

Finance Chair

Ke Huang
San Diego State University, USA

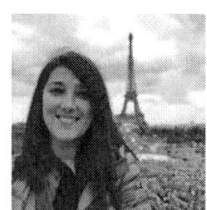

Registration Chair

Ioana Vatajelu
TIMA labs, FR

Publication and Media Chair

Marcello Traiola
Inria, IRISA, FR

Publicity Chair

Hussam Amrouch
University of Stuttgart, DE

Ex-Officio

Yervant Zorian
Synopsys, USA

Student Activities Chair

Ujjwal Guin
Auburn University, US

Japan Liaisons

Haruo Kobayashi
Gunma University, JP

Europe Liaisons

Haralampos Stratigopoulos
Sorbonne Univ., CNRS, LIP6, FR

India Liaisons

Rubin Parekhji
Texas Instruments, IN

China Liaisons

Xiaoxiao Wang
Beihang University, CN

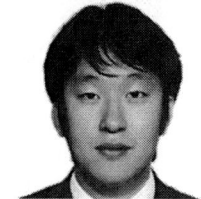

Korea Liaisons

Jaeyong Chung
Incheon National University, KR

Taiwan Liaisons

Mango Chao
National Chiao Tung University, TW

South America Liaisons

*Leticia Maria Bolzani Poehls
RWTH Aachen University, DE*

VTS 2023 STEERING AND PROGRAM COMMITTEES

Steering Committee

M. Abadir – Abadir & Associates
J. Figueras – Universidad Polytecnic de Catalunya
A. Ivanov – University of British Columbia
M. Nicolaidis – TIMA Laboratory
P. Prinetto – Politecnico di Torino
A. Singh – Auburn University
P. Varma – Veda Design Systems, Inc.
Y. Zorian – Synopsys

Program Committee

Members

V. Agrawal – Auburn University
L. Anghel – University Grenoble-Alpes
L. Balasubramanian – Ti
S. Banerjee – Intel Corp.
M. Barragan – Tima
K. Basu – Ut Dallas
D. Bhatta – Zeku
J. Blain Christen – Asu
R. Blanton – Cmu
L. Bolzani Poehls – Rwth Aachen University
A. Bosio – Inl ? Ecole Centrale De Lyon
K. Chakrabarty – Duke University
T. Chakraborty – Qualcomm Inc.
M. Chandrasekar – Synopsys
M. Chao – National Chiao Tung University
A. Chatterjee – Git
D. Chen – Iowa State University
K. Chung – Qualcomm
S. Di Carlo – Politecnico Di Torino
W. Dobbelaere – Onsemi
G. Fey – Tu Hamburg
P. Girard – Lirmm
D. Gizopoulos – University Athens
U. Guin – Auburn University
S. Gupta – Usc
G. Harutyunyan – Synopsys
S. Hellebrand – University Paderborn
K. Huang – San Diego State University
A. Ivanov – University British Columbia
K. Jennifer – Asu
N. Karimi – Umbc
C. Kavousianos – Universityioannina
M. Keim – Siemens
H. Kobayashi – Gunma University
C. Kumar – Ti India
E. Larsson – Lund University
X. Li – Cas
H. Li – Chinese Ac. Of Sciences

J. Li –
Y. Li – National Central University
T. Mak –
Y. Makris – Utd
S. Mir – Tima
A. Nahar – Ti
S. Natarajan – Intel Corp.
Z. Navabi – Worcester P.I.
A. Orailoglu – Ucsd
S. Ozev – Asu
R. Parekhji – Ti
I. Polian – University Stuttgart
I. Pomeranz – Purdue
J. Rajendran – Texas A&M University
J. Rajski – Mentor
E. Sanchez – Politecnico Di Torino
F. Saqib – Uncc
A. Sathaye – Intel
A. Savino – Politecnico Di Torino
S. Shoukourian – Synopsys
A. Singh – Auburn University
A. Sinha – Intel
P. Song – Ibm
H. Stratigopoulos – Sorbonne University
F. Su – Intel
S. Sunter – Siemens Eda
M. Tahoori – Karlsruhe
M. Tehranipoor – Ufl
C. Thibeault – E. Tech. Sup. Montreal
M. Traiola – Inria
P. Varma – Veda Design Systems
X. Wang – Beihang University
L. Wang – Ucsb
H. Wunderlich – University Stuttgart
E. Yilmaz – Cirrus Logic
R. Zhang – Cadence
Y. Zorian – Synopsys

Fast RF Mismatch Calibration Using Built-In Detectors

Muslum Emir Avci and Sule Ozev (ASU) and Chethan Kumar Y. B. (TI)

The VTS 2022 Best Paper Award selection committee is listed below. VTS extends special thanks to these individuals for reviewing the papers and offering invaluable comments.

Kanad Basu (UTD)
Manuel Barragan (TIMA)
Debesh Bhatta (Zeku)
Martin Keim (Siemens EDA)

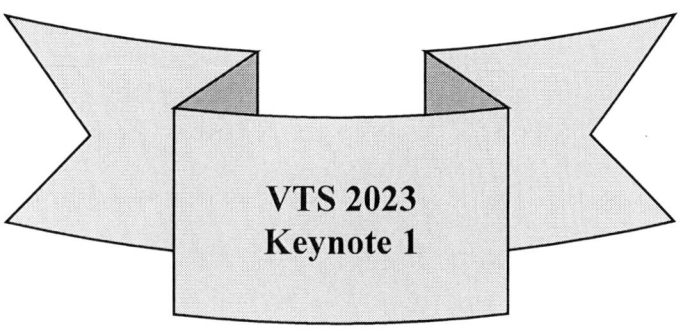

**VTS 2023
Keynote 1**

Heterogeneous Integration in the AI Era

Subramani (Subi) Kengeri
Vice President, AI Systems Solutions

Speaker bio: Subi Kengeri is the Vice President of AI Systems Solutions at Applied Materials. His team is chartered with the goal of architecting next generation AI Systems leveraging Applied's fundamental innovations.

Prior to joining Applied, Subi was the CTO and vice president of world-wide client solutions at Globalfoundries, responsible for enabling differentiated SoC and systems solutions. Subi joined Globalfoundries in 2009 as the vice president of global design solutions responsible for world-wide design engineering and semiconductor eco-system development. He was responsible for determining technology feasibility, competitiveness, and manufacturability of technology platform through cross-functional collaboration of customers, R&D and eco-system. In the role of vice president of CMOS Platforms Business Unit, Subi was responsible for business results.

Subi started his SoC design engineering career at Texas Instruments in 1991 and prior to joining Globalfoundries, he was the senior director of design-technology platform and head of North America Design center, at TSMC. Subi has been granted 50+ U.S. design engineering patents and has given over 100 invited talks and press interviews.

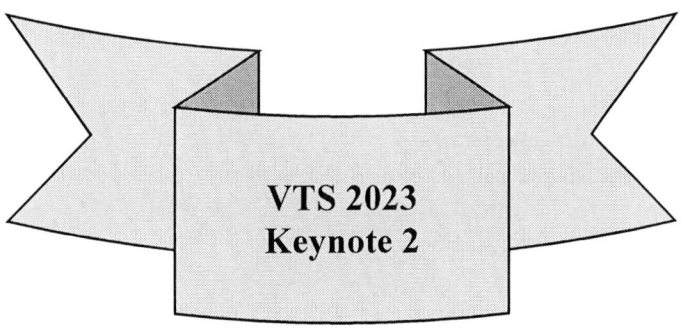

**VTS 2023
Keynote 2**

The Future of Semiconductor Packaging and Heterogeneous Integration is Now

Madhavan Swaminathan
Department Head of Electrical Engineering & William E. Leonhard Endowed Chair
Director, Center for Heterogeneous Integration of Micro Electronic Systems (CHIMES)
an SRC JUMP 2.0 Center
The Pennsylvania State University,

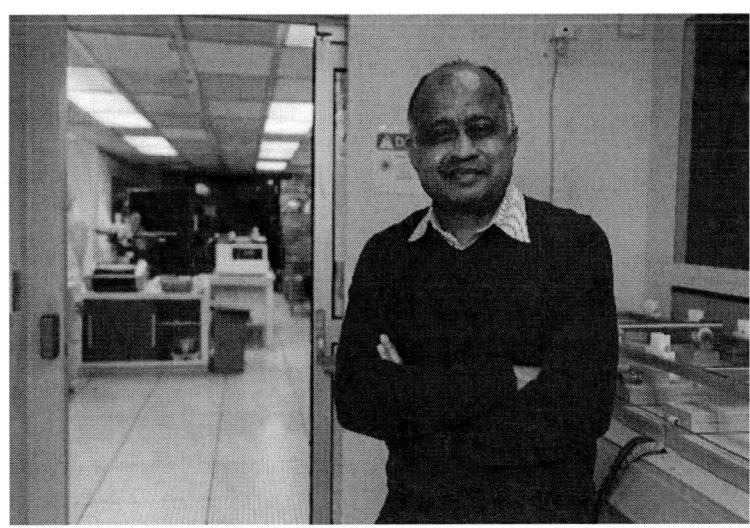

Abstract: The global semiconductor industry is projected to become a trillion-dollar industry by 2030. This is historic considering that it took the industry 55 years to reach half a trillion dollars in size and will take just another 10 years to double in size to a trillion dollars. Advanced packaging is expected to play an important role in making this happen.

So, what are the key drivers, where are the challenges and what are the innovations necessary in advanced packaging over the next decade and beyond to be able to support heterogeneous integration? Why is semiconductor packaging becoming so important? These questions will be addressed in the context of emerging applications.

Speaker bio: Madhavan Swaminathan is the Department Head of Electrical Engineering and is the William E. Leonhard Endowed Chair at Penn State University. He also serves as the Director for the Center for Heterogeneous Integration of Micro Electronic Systems (CHIMES), an SRC JUMP 2.0 Center.

Prior to joining Penn State University, he was the John Pippin Chair in Microsystems Packaging & Electromagnetics in the School of Electrical and Computer Engineering (ECE), Professor in ECE with a joint appointment in the School of Materials Science and Engineering (MSE), and Director of the 3D Systems Packaging Research Center (PRC), Georgia Tech (GT). Prior to GT, he was with IBM working on packaging for supercomputers.

He is the author of 550+ refereed technical publications and holds 31 patents. He is the primary author and co-editor of 3 books and 5 book chapters, founder and co-founder of two start-up companies, and founder of the IEEE Conference on Electrical Design of Advanced Packaging and Systems (EDAPS), a premier conference sponsored by the IEEE Electronics Packaging Society (EPS). He is a Fellow of IEEE, Fellow of the National Academy of Inventors (NAI), and has served as the Distinguished Lecturer for the IEEE Electromagnetic Compatibility (EMC) society.

He received his MS and PhD degrees in Electrical Engineering from Syracuse University in 1989 and 1991, respectively.

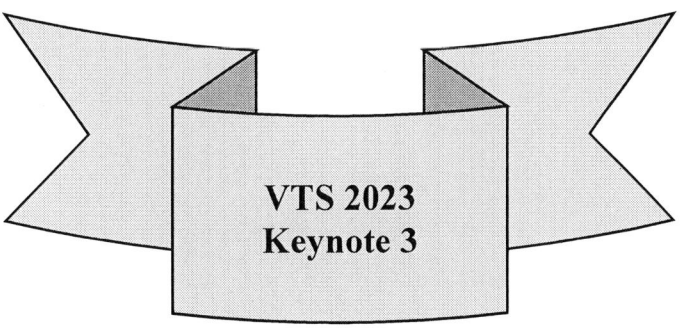

**VTS 2023
Keynote 3**

The Heart of the Feedback Loop: MEMS Development and Test

Lorie Burmood
Director of Operations, NXP Semiconductors, USA

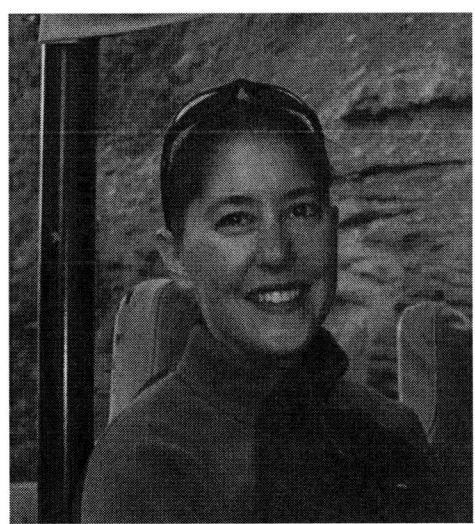

Speaker bio: Lorie Burmood is Director of Operations for Motions Sensors, NXP Semiconductors in Chandler Arizona. Her team is responsible for product and test engineering development of next generation MEMS technologies.

Lorie has been with Motorola/Freescale/NXP since 1998. She joined Sensors in 2004 on the Tire Pressure Monitor System development team, leading the effort to create a media compatible solution for pressure sensors. In 2014 she was the Product Engineering Manager for all Motion Sensor NPI. As Director of Operations, she is now responsible for R&D Test Development, NPI Product Engineering as well as site selections for new products. She oversees the various manufacturing sites (fab, assembly, and test), enabling new product introductions into high volume manufacturing.

Lorie has a double B.S. from Case Western Reserve University in Chemical Engineering and Materials Science Engineering. In 2009, she received her M.S.E. degree from Arizona State University, with emphasis in Analog and Digital VLSI Design. She has received a Six Sigma Black Belt Certificate from GE Healthcare.

TTTC's E. J. McCluskey Best Doctoral Thesis 2023 Award Contest @ VTS2023

List of Presentations

Applicant	Advisor	University	Title
Anuj Dubey	Aydin Aysu	NCSU	A Full-Stack Approach For Side-Channel Secure Ml Hardware
Arjun Chaudhuri	Krishnendu Chakrabarty	Duke	Fault Modeling, Design-For-Test, And Fault Tolerance For Machine Learning Hardware
Avinash Ayalasmoayajula	Farimah Farahamandi	UF	Automated Security Verification Of Pre-Silicon System-On-Chips
Md Rafid Muttaki	Farimah Farahamandi	UF	Mitigating Supply-Chain Threats Through Obfuscation And On-Chip Hardware Solutions
Muslum Emir Avci	Sule Ozev	Arizona State University	Low Overhead Rf Impedance Measurement By Using Periodic Structures
Rabin Yu Acharya	Domenic Forte	UF	Design And Application Of Evolutionary Algorithms For Hardware Security
Sanmitra Banerjee	Krishnendu Chakrabarty	Duke	Modeling And Optimization Of Emerging Technology-Based Artificial Intelligence Accelerators Under Imperfections
Shiva Shankar Thiagarajan	Yiorgos Makris	University Of Texas At Dallas	A Defect Tolerance Framework: Improving Manufacturing Yield Of Integrated Circuits
Sylwester Milewski	Jerzy Tyszer	Poznań University Of Technology	Hypercompression Of Test Data
Tao Zhang	Farimah Farimandi And Mark Tehranipoor	UF	Side Channel-Oriented Vulnerability Assessment Of Soc Security
Mikail Yayla	Jian-Jia Chen	TT Dortmund University, Germany	Robust And Efficient Machine Learning For Emerging Resource-Constrained Embedded Systems
Shamik Kundu	Kanad Basu	UT Dallas	Towards Functionally Safe Deep Learning Hardware Accelerators
Ayush Arunachalam	Kanad Basu	UT Dallas	Towards Synergizing Ai And Hardware

VTS 2023 SPONSORS

ADVANTEST®

SIEMENS

SYNOPSYS®
Silicon to Software™

Outlier Detection for Analog Tests Using Deep Learning Techniques

Chin-Kuan Lin*, Cheng-Che Lu[†], Shuo-Wen Chang[‡], Ying-Hua Chu[‡],
Kai-Chiang Wu[†] and Mango Chia-Tso Chao*

*Institute of Electronics, National Yang Ming Chiao Tung University, Hsinchu, Taiwan
[†]Department of Computer Science, National Yang Ming Chiao Tung University, Hsinchu, Taiwan
[‡]Qualcomm Semiconductor Limited, Hsinchu, Taiwan

Abstract—**With the increasing demand for high reliability of products, how to prevent potential defective devices from shipping to customers is a serious issue about which more and more companies are concerned. Toward this end, many test methods have been developed to screen out outliers. However, basic statistical paradigm may not be enough to handle the shrinking transistor size and increasingly complex circuit design. In this paper, we propose to use the concept of Z-score derived from our proposed neural network, called single density network (SDN), to define level of abnormality. We also define new metrics called self-excluded fail rate (SE fail rate) and normalized area under curve (AUC) to be our criteria to quantify and further visualize the outcome. To filter out spatially-correlated outliers, we make use of specific information of neighboring dice and encode them into our input features for the proposed SDN. A series of experimental results on industrial data reveal the effectiveness of our methodology and the better ability to identify defective outliers than existing conventional statistical approaches for a variety of analog tests.**

I. INTRODUCTION

Reducing potential customer returns is always a critical task for an IC design house, especially for products requiring high reliability such as automobile electronics. A general direction to reduce customer returns is to discard some abnormal parts even though they test good under every applied test item. Along this direction, one type of methods uses the pass/fail information within a targeted part's neighborhood [1]–[3], also called the GDBN-based (good-die-in-bad-neighborhood) methods. Another type uses the measurements of analog test items in the same lot, wafer or neighborhood to determine the level of abnormality of the targeted part [4]–[17], also called the analog outlier detection and the area in which this paper falls.

The methods for analog outlier detection can be divided into two main categories: (1) single-feature methods [4]–[10] and (2) multi-feature methods [11]–[17], depending on the number of analog test items (also called analog features in our later discussion) being considered to determine outliers. The most iconic single-feature method for outlier detection is DPAT (dynamic part average test) [4], which used a per-wafer-based sigma (σ, i.e., standard deviation) bound of a single analog feature to define the level of abnormality of a die as suggested in AEC-Q100 and AEC-Q101 when the distribution of the targeted analog feature is normal. For calculating this sigma bound, the initial version of DPAT used the mean plus k sigma. Its later version used the median plus k robust sigma instead to limit the impact of outliers themselves on other normal dice. When the distribution of the targeted analog feature is not normal, another version of DPAT, called AEC DPAT, used the 99th percentile minus median and the 1st percentile minus median to calculate the upper and lower limits, respectively, to handle the skewness of the feature distribution. [5] proposed another variant of DPAT, called RDPAT, which used Johnson transformation to

convert a non-normal distribution into a normal one before applying DPAT.

Instead of using the entire distribution of an analog feature on a wafer to determine a die's abnormality like DPAT, other single-feature methods [6]–[8] used only the feature values on the neighborhood of the targeted die to determine its abnormality. The idea behind is to utilize the locality of the process variation on a wafer, meaning that the feature values of physically closer dice are more correlated to the targeted feature value and in turn more suitable for reflecting its abnormality. The single-feature methods proposed by [6] and [7] are called NNR (nearest neighbor residual) and LA (location averaging), respectively. NNR and LA use the median and mean of the neighbored feature values, respectively, as the expected value of a targeted feature value, and the abnormality of a targeted feature value is defined as the absolute difference between its expected value and itself. NNR and LA also applied different strategies to select the neighbored feature values. [8] trains a stochastic regression model to predict the mean and sigma of a targeted feature value based on the other feature values within a fixed-size window, and then uses the Z-score of the targeted feature value outputted by the model to define its abnormality. This method is called SR (stochastic regression) in this paper and [8] applied CNN-based (convolutional-neural-network-based) MDN (mixture density network) [18] to train the model.

The above three single-feature methods, NNR, LA and SR, were initially designed to detect outliers for IDDQ test items. Later, [9] applied LA to general analog test items and helped to identify candidates that can skip burn-in test. [10] applied different single-feature methods, including NNR, LA, DPAT, AEC DPAT to identify outliers on general analog test items and compared their effectiveness based on different types of data.

Among the multi-feature methods for outlier detection [11]–[17], the former three [11]–[13] applied the technique called correlation testing, which selected two highly correlated features and used the absolute difference between the two normalized feature values of a die to define the die's level of abnormality. The two paired features can result from two separate test items or the same test item applied under different temperature or voltage conditions. On the other hand, [14]–[16] used more than two features to identify outliers. [14] applied unsupervised one-class SVM to identify outliers based on the relevant features reported by a two-class SVM model trained with known fail parts. [15] first used the sigma bound on the top-ranking principal components after PCA for outlier detection. After customer returns were collected, [15] further applied PCA plus one-class SVM on the top failing-related features for outlier detection. [16] applied multiple reversible lossless transformations to all features individually and trained a neural-network-based model to classify which transformation the received feature values of a targeted die result from. When the received feature values of a targeted die cannot be accurately classified, the die is considered as abnormal. [17] proposed a defect filter which uses PCA and an estimate of the joint probability to filter out outliers, with the main purpose to stabilize the model fitting of subsequent

979-8-3503-4631-2/23 $31.00 © 2023 IEEE

alternate tests.

In this paper, we follow the concept of using Z-score of a stochastic regression (SR) model to define the level of abnormality of a die based on a single feature as in [8], and further apply it to general analog test items instead of just IDDQ. The type of neural network for training the SR model in this paper is called single density network (SDN), which utilizes the mean and sigma of only one normal distribution for predicting the target feature and is a simplified version of MDN, which mixes the mean and sigma of multiple normal distributions. The baseline neural network architecture is MLP (multi-layer perceptron), which requires shorter training time. We also propose a new metric called self-excluded fail rate (SE fail rate) to evaluate the effectiveness of a single-feature outlier-detection methodology at the initial stage of production when no customer returns are collected. The concept of this SE fail rate is to examine the ratio of selected outlier parts that fail on a test item other than the one used for determining outliers. This SE fail rate can help us to select effectiveness test items for outlier detection as well. The experiments in this paper are conducted based on an early-stage industrial test data of an advanced IC product, including 2.25M parts with 561 analog test items.

The experimental results demonstrate that our SDN can outperform other conventional single-feature methods, such as DPAT, NNR and LA, in terms of SE fail rate. A series of experiments have also been conducted to verify the effectiveness of our SDN. We have further found that the advantage of our SDN over other conventional methods becomes more significant on test items that may generate site patterns on a wafer.

II. BACKGROUND

In this section, we will give an overview of principal background used throughout the paper. In Section II-A, we will show a general form for defining the level of abnormality, by which DPAT, NNR and LA can be distinguished. In Section II-B, particular metrics for quantitatively evaluating the performance of outlier detection will be presented.

A. Level of abnormality used in previous works

With respect to a specific analog test, the level of abnormality (LoA) for each die D can be generally defined as:

$$LoA(D) = \frac{t_{actual}(D) - t_{ref}(D)}{f_{norm}} \quad (1)$$

where t_{actual} is the actual measurement of die D w.r.t. the specific test item, t_{ref} is the reference value of die D w.r.t. the specific test item, and f_{norm} is the normalization factor.

DPAT (dynamic part average test): DPAT relies on per-wafer statistics to calculate LoA; t_{ref} for DPAT is the median of all dice's measurements across a wafer, and f_{norm} is the robust sigma of those measurements. An industrial convention for DPAT-based outlier detection is to check whether the calculated LoA is greater than +6 (upper specification limit, USL) or less than -6 (lower specification limit, LSL). In other words, a die will be considered as an outlier if its measurement w.r.t. a specific analog test is deviated from the reference point (t_{ref}) by $\pm6\sigma_{robust}$ (robust sigma). It can be noted that the ±6 criterion is extremely stringent and usually recognizes "extreme" outliers only. The problem of only recognizing extreme outliers can be relaxed by adopting a looser criterion (e.g., ±3 or ±4, instead of ±6). However, the result of outlier detection/recognition is fully based on cross-wafer statistics; spatial correlations in the surroundings of the die under test are not addressed.

NNR (nearest neighbor residual): NNR improves aforementioned conventional DPAT by taking into account the spatial correlations among surrounding dice. More explicitly, NNR calculates t_{ref} as the median of eight nearest neighbors' measurements (with f_{norm}

equal to 1); the eight nearest neighbors of a die are usually the eight immediate outer dice in its surroundings. NNR can be regarded as the most intuitive strategy for addressing spatial correlations.

LA (location averaging): LA further improves NNR by relaxing the assumption that the neighbors are those geographically close to each other. Instead, LA determines its t_{ref} by ranking a special "predictive capability" index within a window of 7x7 or 9x9 dice. Typically, 50% of the dice in the window are selected, and the average of selected dice's measurements is t_{ref} for LA (with f_{norm} equal to 1 as well). LA can be regarded as a successor of NNR which adopts better strategy for defining representative neighbors, other than intuitive strategy based on simple geographical proximity.

B. Performance metrics for outlier detection

We define *self-excluded fail rate*, abbreviated as *SE fail rate* and denoted by $r_{SEF(\bullet)}$, to be the foundation of our performance metric. The proposed outlier detection methodology is compared against other approaches by using this metric and its extended integral form. The bottom line behind $r_{SEF(\bullet)}$ is that, w.r.t. a specific analog test, the devices (dice) failing the test because of failure to satisfy a certain threshold are not the target of our outlier detection methodology. Therefore, these devices are excluded from the calculation of $r_{SEF(\bullet)}$; it turns out that "SE fail rate" focuses on the dice not identified by applying a simple cut-off threshold in the specific selected test, and is more precisely associated with the outliers of our concern. By excluding dice failing at a specific test item, SE fail rate potentially exhibits the ability to find dice failing at *other* test item(s). The higher SE fail rate w.r.t. the specific test item selected, the higher likelihood of dice failing (at other test items) and the more confident we are in finding outliers with sufficiently high degree of defectiveness.

Formally, with respect to a specific test item, $r_{SEF(i)}$ is the SE fail rate across a total of i dice, which can be derived by:

$$r_{SEF(i)} = \frac{SEF(i)}{i} \quad (2)$$

where $SEF(i)$ is the number of failed dice among top-ranked i potential outliers, excluding dice failing at the specific test item.

We can pick an i value for $r_{SEF(i)}$ such that the SE fail rate across top-ranked i potential outliers can be depicted. Note that the outliers here in the context are "potential" since they are not guaranteed to be real defective dice. To address this, we propose to depict $r_{SEF(i)}$ with i ranging from 1 to N, conceptually similar to integrating $r_{SEF(i)}$ from 1 to N. The numerical result is the area under curve of $r_{SEF(i)}$ with i from 1 to N. The normalized area under curve (AUC) quantifies how confident we are in identifying real outliers, that is, the confidence of real defective dice being identified from those (top-ranked i) "potential" outliers. The AUC of $r_{SEF(i)}$ normalized with respect to N is written as:

$$AUC(N) = \frac{\sum_{i=1}^{N} r_{SEF(i)}}{N} \quad (3)$$

A value of normalized AUC equal to 1 implies highest confidence, meaning that all of the identified potential outliers are real defective dice; a value of normalized AUC equal to 0 implies lowest confidence, meaning that none of them is defective from the perspective of other test items.

Fig. 1(a) illustrates our conception by using a simple example, where two outlier detection methods both capture 9 real defective dice out of 10 potential outliers. The upper red curve (better case, with a better order of top 10 outliers) successfully identifies defective dice and does not miss any one until the 10th-ranked outlier, while the lower blue curve (worse case, with a worse order of top 10 outliers) misidentifies the 1st-ranked outlier (actually a good die) as a bad die. The difference can be visualized by the gap between two curves and

979-8-3503-4631-2/23 $31.00 © 2023 IEEE

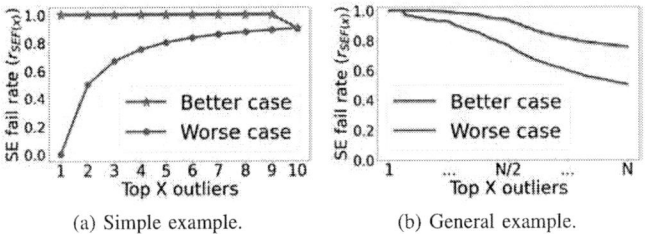

(a) Simple example.   (b) General example.

Fig. 1: Introductory example of SE fail rate and normalized AUC.

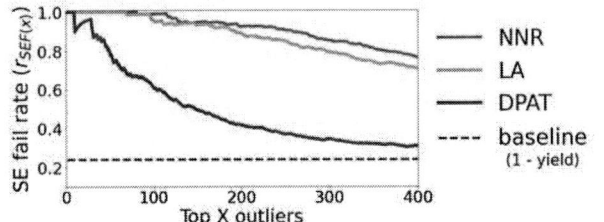

Fig. 2: Comparison among traditional methods in terms of SE fail rate and normalized AUC.

be quantified as eq. (2)[†‡] and eq. (3). In Fig. 1(a), the red curve has a normalized AUC of 0.99, and the blue curve, 0.707. This example shows a rough comparison between a very good scenario (red curve) and a not-so-good scenario (blue curve). The reason for the difference is the order of top X outliers of concern. We will later explain how to determine a good order of top X outliers such that the normalized AUC can be enlarged.

The plot of SE fail rate can be generalized as shown in Fig. 1(b), which is more like a real situation. We can easily tell the difference between two curves because the red one is all above the blue one. The ingenious about AUC is that it reveals the importance of evaluating the quality of outlier detection by observing $r_{SEF(i)}$ for a running span of i from 1 to N, instead of simply observing $r_{SEF(i)}$ at a specific value of i.

Fig. 2 shows a real demonstration of three curves of SE fail rate vs. top X outliers, where three curves are obtained by using NNR, LA and DPAT. One can note that NNR is comparable to LA, while NNR and LA both outperform DPAT especially when X gets larger. The horizontal dotted line reveals the overall fail rate across the entire set of dice, which equals 1 subtracted from the natural yield. Even with random guess, the curve of SE fail rate should be in line with the dotted line; on the other hand, it can be concluded that NNR, LA and DPAT are effective in identifying real defective dice from potential outliers. We in this work aim to propose a better methodology, as presented in the next section.

III. OUR DEEP-LEARNING OUTLIER DETECTOR

A. Overall Concept

In [8], CNN-based stochastic regression was employed for IDDQ outlier identification; in this work, we extend the use of stochastic regression to deal with the problem of outlier detection for a wide variety of analog tests. For our objective of outlier detection, we propose a stochastic regression model based on mixture density network (MDN) [18]. A MDN is a special class of neural network combined with a statistical model of mixture (probability) density functions. The output of a MDN can parameterize a mix of Gaussian density functions/distributions by a triplet of three vectors: $<\boldsymbol{\mu}, \boldsymbol{\sigma}, \boldsymbol{\alpha}>$ where

---

[†] $r_{SEF(i=1:10)}$ of red curve: $\{\frac{1}{1}, \frac{2}{2}, \frac{3}{3}, \frac{4}{4}, \frac{5}{5}, \frac{6}{6}, \frac{7}{7}, \frac{8}{8}, \frac{9}{9}, \frac{9}{10}\}$

[‡] $r_{SEF(i=1:10)}$ of blue curve: $\{\frac{0}{1}, \frac{1}{2}, \frac{2}{3}, \frac{3}{4}, \frac{4}{5}, \frac{5}{6}, \frac{6}{7}, \frac{7}{8}, \frac{8}{9}, \frac{9}{10}\}$

μ_i and σ_i ($i = 1, 2, ..., n$) are respectively the mean and standard deviation of n Gaussian distributions, and α_i ($i = 1, 2, ..., n$) is the weight for linear combination/mixture of the n Gaussian distributions. A MDN can be simplified to single density network (SDN) when only one (rather than multiple) Gaussian distribution is involved. In this work, as demonstrated later in Section IV, SDN is not outperformed by MDN and thus, for the sake of simplicity, SDN is extensively used as our stochastic regression model for outlier detection.

Input: 6 channels for each enumerative window of 13x13 dice in every wafer and every lot; that is, there is an input map of size 6 (channels) x 13 (dice on X-axis) x 13 (dice on Y-axis) for each die centered in the 13x13 window. For example, given a "train" data set of P lots, Q wafers per lot, R dice per wafer, there will be a total of $PxQxR$ input maps of size 6x13x13. A one-time preprocessing procedure needs to be done for extracting these many input maps from a whole batch of raw data.

Output: A duplet $<\mu, \sigma>$ where μ and σ are respectively the mean and standard deviation of a Gaussian distribution, i.e., prediction result of the stochastic regression model.

Objective: A stochastic regression model, given a "test" input map of 6x13x13, to output duplet $<\mu, \sigma>$ for the die under test (DUT), i.e., the die centered in the 13x13 map/window. Based on eq. (1), the LoA of the DUT is derived by letting t_{actual} be the DUT's actual measurement w.r.t. a specific analog test, t_{ref} be μ and f_{norm} be σ.

It is worth noting that the LoA as derived above is very much equivalent to the Z-score of the DUT where the mean and standard deviation of the population are those predicted by our stochastic regression model, but not purely based on statistics. After the LoA of each DUT is calculated, we sort all DUTs in decreasing order of LoA; according to the definition of LoA (level of abnormality), the more top-ranked a DUT, the more abnormal it is. Therefore, top-ranked (abnormal) DUTs are considered potential outliers and will be selected for evaluating the effectiveness of outlier detection based on examining AUC as defined in eq. (3). A typical number of DUTs selected for evaluation is 0.1% of the entire test data set.

B. Deep-Learning Model

In this subsection, we will demonstrate how and why we encode the specific information into our input features (see Fig. 3). The left part of Fig. 3 shows the wafer view of a real analog test and its corresponding legend. From this wafer view, we found that the variation of this test measurement has some spatial correlation. The magnitude distribution of the test measurement forms a donut shape, which has larger values around the middle circular belt, and smaller values near the wafer edge. This motivates us to include Channel 2, Channel 5 and Channel 6 in our feature encoding for better indicate locational information. Besides, we believe that the pass/fail log of dice needs to be clearly indicated because it directly influences whether the measurement value is of reference. Thus we can encode the pass/fail information when they are observed to be defective (early failure before the selected wafer-sort test items or even not proceeding to wafer-sort stage) into Channel 3 and Channel 4. For these failure/missing parts, their values at Channel 1 will be replaced with the average value of Channel 1 across the wafer. The middle part of Fig. 3 is the zoomed-in view of a window of 13x13 dice; it shows the target die and its corresponding 13x13 neighborhood which are used to judge whether the target is a outlier or not. Then we can encode those aforementioned pieces of information into our 6x13x13 input features as the right part of Fig. 3. Each of the six channels will be described in detail below.

Channel 1 (Measurement) Value of measurement normalized by RobustScaler: analog test items pre-selected by NNR and LA to help us quickly pick out a reduced set of test items which have the characteristic of spatially-correlated distribution of measurement values.

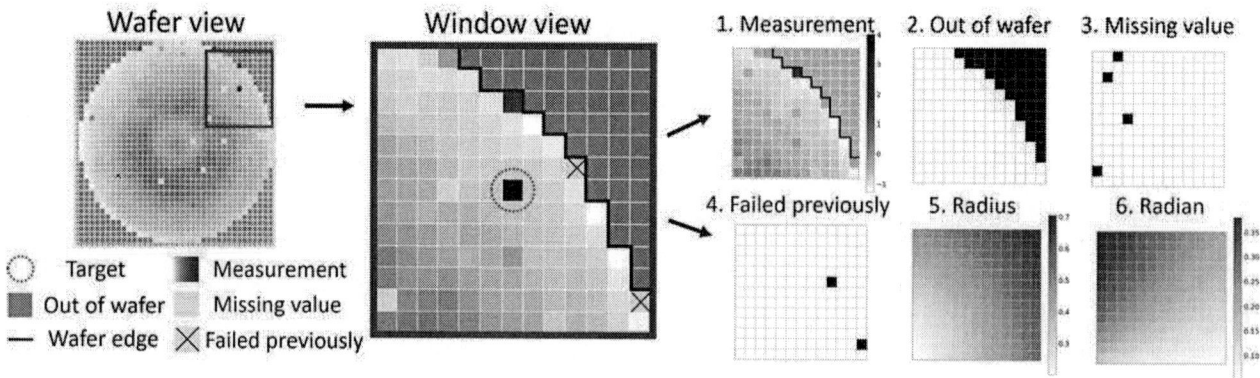

Fig. 3: Example of channel encoding.

Channel 2 (Out of wafer or not): As explained earlier, this channel shows if the neighboring dice are near the wafer edge. This information is useful for model to make more accurate prediction since the measurement values near the wafer edge typically have larger variations.

Channel 3 (Missing value or not): The test program implements a stop-on-fail policy. The dice sent to wafer-sort test may fail at some different stages so early termination is enforced to save testing cost and leave the subsequent measurements blank. The number of wafer-sort test items available for each die may vary due to this reason.

Channel 4 (Failed previously or not): Different from the above, the termination time defining this channel (failed previously) is earlier than Channel 3. This channel indicates the dice terminated without even being sent to the wafer-sort stage. The defect level associated with this channel is higher than the channel of missing value.

Channel 5 (Radius): The distance from the target to the center of the wafer is normalized to [0, 1].

Channel 6 (Radian): The center angle from 0 degree to 180 degrees is normalized to [0, 1], and -180 degrees to 0 degree is normalized to [-1, 0].

C. Model Architecture

TABLE I: **Architectural overview of our proposed SR models**

Input dimension (feature size): 6 x 13 x 13 (C x H x W)					
MLP		CNN		ATTN	
Layer	Dimension	Layer	Dimension	Layer	Dimension
Reshape	1 x 1014	Conv1	12 x 13 x 13	Reshape	6 x 169
FC1	1 x 1024	Conv2	16 x 13 x 13	Encoder1	6 x 169
FC2	1 x 1024	Conv3	16 x 13 x 13	Encoder2	6 x 169
FC3	1 x 512	FC4	1 x 512	FC3	1 x 512
FC4	1 x 256	FC5	1 x 256	FC4	1 x 256
Output dimension: 1 x 2 (mean and sigma)					

Typically, SDN (or MDN) can be realized by a multilayer perceptron (MLP), consisting of multiple fully-connected layers. In addition to MLP-based SDN, we also customize SDN by incorporating convolutional layers or attention modules. Based on MLP, convolution and attention, three different model architectures of SDN are outlined in Table I. The input dimension is 6 (channels) x 13 (dice on X-axis) x 13 (dice on Y-axis) as described earlier; the output is a vector of two elements: one is the predicted mean and the other is the predicted sigma (standard deviation).

Our MLP-based model has five fully-connected layers, designed as conventional expanding-shrinking architecture. The first layer is to reshape (or flatten) the input to 1x1014 where 1014 is equal to 6x13x13; the number of neurons expands from 1014 to 1024 and then

shrinks to 512 and further to 256 before being fed to the output layer. For CNN-based model, the first convolutional layer has 12 sets of 6 kernels/filters (a total of 6x12 filters), and every 6 filters correspond to the 6 channels in the input data. The convolution results of 6 channels are bundled and thus each input map of size 6x13x13 is transformed to a feature map of size 12x13x13 after the convolution. The second and third convolutional layers have 12x16 and 16x16 filters respectively, and both generate feature maps of size 16x13x13. The last two layers are fully-connected layers of the same dimensions as the MLP-based model. For attention-based model, after reshaping 6x13x13 to 6x169 (by flattening the latter two dimensions), two identical attention modules, Encoder1 and Encoder2, are employed and followed by two fully-connected layers of the same dimensions.

IV. EXPERIMENTAL RESULTS

The experiments in this work are conducted on an industrial product. We collected a total of 2,258,298 dice from 2,515 wafers, along with the test results of 561 analog test items in the wafer-sort test. The whole data set is divided into 8:1:1 (training:testing:validation) in chronological order of testing time (old to new). To narrow down the set of analog test items to be considered, we pick the best 100 from NNR and best 100 from LA; the union of these 200 test items is the candidate set of features selected for our outlier detection methodology. The resulting candidate set consists of 120 features. If not specifically mentioned, the default neural network setup for our SDN is MLP, and results of using different architectures such as CNN and attention will be discussed in Section IV-E.

A. Result of our SDN method

This subsection compares our single-variate SDN method against other existing location-based outlier detection methods such as NNA and LA. The results are demonstrated in Fig. 4 and Fig. 5. We select the top 10 analog test items based on $AUC(N = 200)$ results as our target features. With the same 10 features, we also implement NNR and LA as the references for comparison. Fig. 4(a) plots the curves of average $r_{SEF(\bullet)}$ across 10 selected features, where the bolder curve indicates which method is used for feature selection and the * mark denotes the value of $r_{SEF(200)}$. It can be clearly seen that our SDN-based method has the highest $r_{SEF(\bullet)}$ among three methods. As mentioned in Section II-B, the curve is better if it is closer to 1, which means that the outliers identified by the method are real defective dice. For a fair comparison, we can also select features using NNR and LA, as shown in Fig. 4(b) and Fig. 4(c). It is apparent that our SDN (red curve) outperforms other two methods regardless of how the features are selected.

Fig. 5(a) shows the best scenario of each of the three methods by gathering the border curves from Fig. 4(a), Fig. 4(b) and Fig. 4(c). To extend the experiment, we consider more features to ensure that

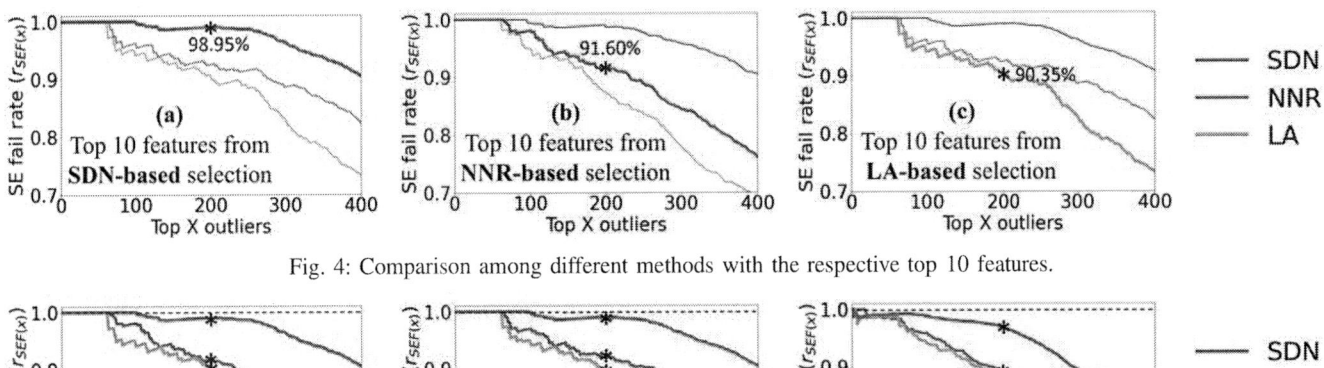

Fig. 4: Comparison among different methods with the respective top 10 features.

Fig. 5: Comparison among different methods with their best 10 / 20 / 40 features.

our SDN has the advantage of generalization. Toward this end, we average $r_{SEF(\bullet)}$ across the top 20 / 40 features selected by three respective methods, and show the results in Fig. 5(b) / Fig. 5(c). We can observe that, in Fig. 5(b), $r_{SEF(N=200)}$ of the red curve is at the same level as that in Fig. 5(a). However, in Fig. 5(c), $r_{SEF(N=200)}$ of the red curve has a perceptible drop, and the red curve declines more significantly than the cases in Fig. 5(a) and Fig. 5(b), and so do the blue and yellow curves in Fig. 5(c). These observations imply that the remaining features may not be perfect to apply on such location-based outlier detectors; topmost 10 to 20 representative features (analog test items) seems experimentally adequate for this product.

Table II tabulates the regression metrics corresponding to Fig. 5(a). The result shows that SDN not only has higher accuracy in predicting the target value but also has better ability to identify defective outliers.

TABLE II:
Regression metrics for single-variate works

	R square	MSE	MAE	RMSE
SDN	0.9840	0.0114	0.0568	0.1065
NNR	0.9473	0.0371	0.1017	0.1927
LA	0.9483	0.0364	0.0880	0.1908

B. Using residual of regression model to define abnormality

Among these location-based outlier detection methods, to precisely predict the value of the DUT is the basic requirement for defining LoA, and thus LoA can be calculated by subtracting this reference value from the actual value as shown in eq. (1). Our comparisons now are regression-only methods which use residual of regression model to define abnormality. Next, we will show that having higher ability in predicting the value of DUT does not mean that it is better at identifying outliers. We implement two regression model with two different NN architectures which are multilayer perceptron (MLP) and convolutional neural network (CNN) and also use the output mean of our SDN as regression task.

Fig. 6 shows our SDN is better than other regression-only methods though the MLP regression model has slightly better performance in R square. With the information of sigma output from SDN, the improvement of SDN over the methods that use residual to define abnormality is obvious. Our SDN can efficiently capture more failed dice that regression-only models missed among first N outliers.

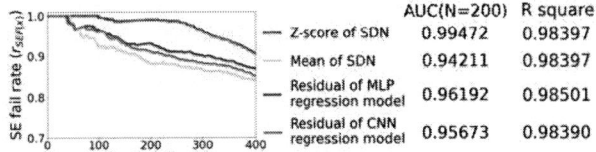

Fig. 6: Comparison among methods using different LoA settings.

C. SDN vs. MDN

In practice, the optimal number of kernels are determined experimentally and empirically, but must be set in advance as a hyper-parameter of our expected density function. Therefore, we set the number of kernels from one to three, and select the appropriate number of normal distributions according to the experimental results. Results in terms of $r_{SEF(\bullet)}$ and $AUC(\bullet)$ show that the simplest hypothesis leads to the best performance: SDN outperforms MDN, as demonstrated in Fig. 7 and Table III.

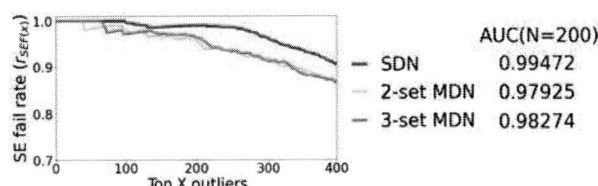

Fig. 7: Comparison between SDN and MDN.

TABLE III:
Regression metrics for SDN and MDN

	R square	MSE	MAE	RMSE
SDN	0.9840	0.0114	0.0568	0.1065
2-set MDN	0.9785	0.0153	0.0722	0.1235
3-set MDN	0.9787	0.0150	0.0713	0.1225

D. Using different loss functions

We also conduct some experiments on loss function. We find that to minimize the negative logarithm of probability instead of minimizing reciprocal of probability is more preferable. From the aspect of SE fail rate shown in Fig. 8, the performance does not have a big difference but the brown one is the testing result after training more than 500 epochs which is much longer than red one that only needs to train for less than 100 epochs. Besides, the red one is more stable when $X>200$ than the brown one. Thus the main difference is the convergence efficiency that can also be noted in Fig. 9. As Fig. 9(a) shown, the red curve quickly converges to a plateau, while another one needs more time to reach the same level. Fig. 9(b) also illustrates the same trend by showing the Z-score distribution, where the distribution is more ideal if it is more like a normal distribution (dash line). In Fig. 9(b), the purple distribution is the result of reciprocal of probability after training 100 epochs. The bell shape of purple one is sharper than red one which is the result of negative logarithm of probability after training 100 epochs. As training goes on to 500 epochs, the distribution becomes brown one which is more and more fitted to the normal distribution. From above experiment, we can get this conclusion that the convergence direction of these two loss functions are the same, but the efficiency is a vast difference.

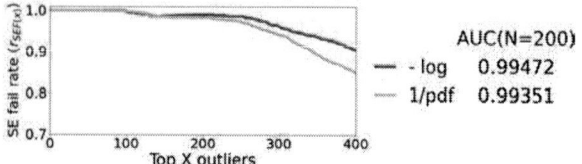

Fig. 8: Comparison between different loss functions.

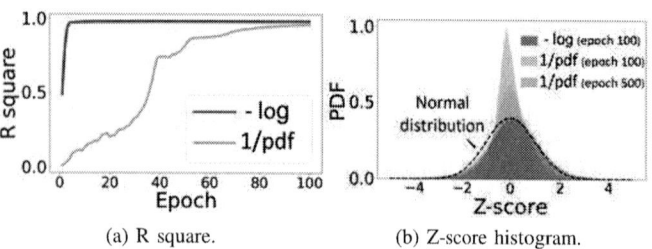

(a) R square. (b) Z-score histogram.

Fig. 9: Other comparisons between different loss functions.

E. Using different NN architectures

In this subsection, we compare various SR methods with different model architectures such as MLP, CNN, attention and Bayesian linear regression. As shown in Fig. 10, the upper three curves (all NN-based SR) are mostly overlapping and the difference in $r_{SEF(200)}$ among them are small enough to be ignored. Besides, we can also find that our proposed NN-based SR method is much better than traditional Bayesian linear regression in both $r_{SEF(200)}$ and regression statistics. Table IV shows the regression statistics of the aforementioned SR methods and also their respective training times. The training time for MLP is about one hour, which is the least among three NN-based SR methods. Although Bayesian linear regression requires only 6 minutes, the performance of $r_{SEF(200)}$ is much worse than NN-based SR methods. Therefore, this is the reason why we chose to use MLP as our default experimental setting.

TABLE IV:
Regression metrics for different SR models

	R square	RMSE	Training time
SDN (MLP)	0.9840	0.1065	4031s
SDN (CNN)	0.9814	0.1146	8355s
SDN (ATTN)	0.9810	0.1163	21300s
Bayesian	0.9592	0.1699	367s

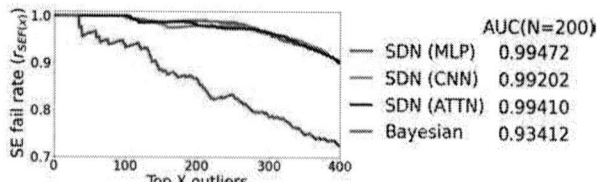

Fig. 10: Comparison among different SR models.

F. Site-pattern feature

When dealing with special wafer map patterns, we found that sometimes the values of test measurement are dominated by the regular pattern of site-to-site probing, as shown in Fig. 11, as known as site pattern. In order to handle such pattern, we encode the information of site pattern into our input features, by adding a new channel which marks each die by its corresponding site order/index of the probe card. To demonstrate the advantage of SDN and without loss of generalization, we randomly select 10 site patterns and plot the integrated result in Fig. 12. By observing the uppermost two curves (brown and red), we can infer that the model performs better with the help of site order. Moreover, the experimental results show that both of our SDN (with or without site encoding) have significant improvement over existing traditional methods when dealing with site patterns.

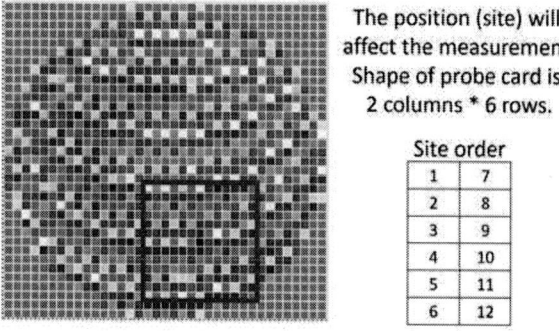

Fig. 11: Wafer view sample of site pattern.

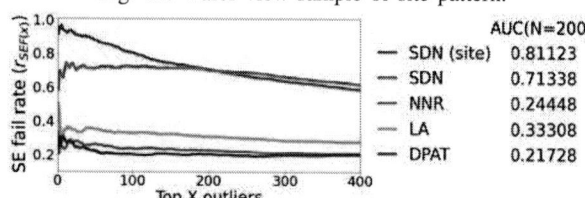

Fig. 12: Comparison on site patterns.

V. CONCLUSION

In this paper, to screen out potential defective dice, we present a deep-learning outlier detector and further apply it to general analog test items. Considering spatial correlations which may reveal the levels of abnormality (or suspiciousness) of dice, we utilize specific information of neighboring dice and encode them as inputs for our outlier detection methodology. Besides, by using our defined metrics, self-excluded fail rate ($r_{SEF(\bullet)}$) and normalized area under curve ($AUC(\bullet)$), we are able to quantify and visualize the difference in the quality of outlier detection. Finally, a series of experimental results demonstrate the effectiveness of our methodology and the better ability to identify defective outliers than existing conventional single-variate testing methods such as DPAT, NNR and LA on general analog test items. Superior performance in terms of the identification of real defective dice can be observed.

REFERENCES

[1] R. B. Miller and W. C. Riordan, "Unit level predicted yield: a method of identifying high defect density die at wafer sort," in *Proc. of Int'l Test Conf.* (ITC), pp. 1118–1127, Oct. 2001.

[2] C. Xanthopoulos, A. Neckermann, P. List, K.-P. Tschernay, P. Sarson, and Y. Makris, "Automated die inking," *IEEE Trans. on Device and Materials Reliability*, vol. 20, no. 2, pp. 295–307, June 2020.

[3] C. -H. Yang *et al.*, "Identifying good-dice-in-bad-neighborhoods using Artificial Neural Networks," in *Proc. of VLSI Test Symp.* (VTS), pp. 1–7, April 2021.

[4] T. Haifley *et al.*, "Guidelines for part average testing," *Automotive Electronics Council*, pp. 1–9, Dec. 2011.

[5] M. J. Moreno-Lizaranzu and F. Cuesta, "Improving Electronic Sensor Reliability by Robust Outlier Screening," *Sensors*, vol. 13, no. 10, pp. 13521–13542, 2013.

[6] W. R. Daasch, J. McNames, D. Bockelman, and K. Cota, "Variance reduction using wafer patterns in IddQ data," in *Proc. of Int'l Test Conf.* (ITC), pp. 189–198, Oct. 2000.

[7] W. R. Daasch, K. Cota, J. McNames, and R. Madge, "Neighbor selection for variance reduction in IDDQ and other parametric data," in *Proc. of Int'l Test Conf.* (ITC), pp. 1240–1248, Oct. 2001.

[8] C. -T. Chen *et al.*, "CNN-based stochastic regression for IDDQ outlier identification," in *Proc. of VLSI Test Symp.* (VTS), pp. 1–6, April. 2020.

[9] A. Nahar et al., "Quality improvement and cost reduction using statistical outlier methods," in *IEEE International Conference on Computer Design (ICCD)*, pp. 64–69, Oct. 2009.

[10] L. -C. Wang, S. Siatkowski, C. Shan, M. Nero, N. Sumikawa, and L. Winemberg, "Some considerations on choosing an outlier method for automotive product lines," in *Proc. of Int'l Test Conf.* (ITC), pp. 1–10, Oct. 2017.

[11] A. Keshavarzi, K. Roy, M. Sachdev, C. F. Hawkins, K. Soumyanath and V. De, "Multiple-parameter CMOS IC testing with increased sensitivity for I/sub DDQ/," in *Proc. of Int'l Test Conf.* (ITC), pp. 1051–1059, Oct. 2000.

[12] L. Fang, M. Lemnawar and Y. Xing, 2006. "Cost effective outliers screening with moving limits and correlation testing for analogue ICs," in *Proc. of Int'l Test Conf.* (ITC), pp. 1–10, Oct. 2006.

[13] Tadashi Sakamoto, Kazunori Yofu and Takashi Kyuho, "New method of screening out outlier; expanded PAT during package level test" *IEEE Trans. on Semiconductor Manufacturing*, vol. 30, no. 4, pp. 351–356, Nov. 2017.

[14] D. Drmanac, N. Sumikawa, L. Winemberg, L.-C. Wang, and M. S. Abadir, "Multidimensional parametric test set optimization of wafer probe data for predicting in field failures and setting tighter test limits," in *2011 Design, Automation & Test in Europe* (DATE), pp. 1–6, March 2011.

[15] N. Sumikawa, J. Tikkanen, L.-C. Wang, L. Winemberg, and M. S. Abadir, "Screening customer returns with multivariate test analysis," in *Proc. of Int'l Test Conf.* (ITC), pp. 1–10, Nov. 2012.

[16] H. Hu, N. Nguyen, C. He and P. Li, "Advanced Outlier Detection Using Unsupervised Learning for Screening Potential Customer Returns" in *Proc. of Int'l Test Conf.* (ITC), pp.1–10, Nov. 2020.

[17] H. -G. Stratigopoulos, S. Mir, E. Acar and S. Ozev, "Defect Filter for Alternate RF Test," in *Proc. of IEEE European Test Symp.* (ETS), pp. 101–106, May, 2009.

[18] C. M. Bishop, "Mixture density networks," *Neural Computing Research Group Report: NCRG/94/004*, pp. 1–26, Feb. 1994.

Machine Learning-Based Adaptive Outlier Detection for Underkill Reduction in Analog/RF IC Testing

V.A Niranjan*, D. Neethirajan*, C. Xanthopoulos*, D. Webster[†], A. Nahar[†] and Y. Makris*

*Department of Electrical and Computer Engineering, The University of Texas at Dallas, Richardson, TX USA, 75080
[†]Texas Instruments Inc., 12500 TI Boulevard, MS 8741, Dallas, TX 75243

Abstract—We present a solution for reducing the number of defective analog/RF integrated circuits (ICs) that escape detection during manufacturing testing. Also known as underkill, these ICs may fail when deployed in their target application and eventually become customer returns, casting doubt on the effectiveness of the employed test solution and affecting the bottom line. To ameliorate this problem, we introduce an adaptive outlier detection solution that identifies ICs which are suspect of becoming customer returns and proactively bins them as failing. The outlier detection boundary used by our method is dynamically computed based on the performance distribution of devices on each wafer and the underlying model is updated when new ICs are returned from customers and failure analysis confirms that they are indeed defective devices. The effectiveness of our method in reducing underkill while minimizing the incurred yield loss is evaluated using an industrial dataset from Texas Instruments.

I. INTRODUCTION

As the complexity of semiconductor devices has been growing exponentially over the years, their increasingly complex functionality has necessitated comprehensive post-silicon testing. Current expectations with respect to quality are measured at the rate of Defective Parts Per Billion (DPPB), with highly sensitive and competitive markets, such as the automotive sector, posing stringent requirements for near zero customer returns and near zero yield loss to be viable and profitable. A customer return occurs when a device passes all post-manufacturing tests but fails on-site. Several factors, such as overly optimistic test thresholds, insufficient test coverage, test environment factors, packaging issues, etc., can lead to the occurrence of these customer returns. We refer to these test escapes that occur during High Volume Manufacturing (HVM) as *underkill*. Figure 1 shows the binning of devices based on their measured performances. The two red quadrants include devices whose true functional status diverges from their actual test outcomes. Defective devices which are identified as passing based on their test outcome but end up failing on-site are shown in the bottom right red quadrant.

Underkill, leading to customer returns, is a rare but very unfavorable test outcome, as it not only damages the reputation of semiconductor manufacturers but also requires extensive root cause analysis and expensive retooling to modify the manufacturing and testing processes. A major challenge while developing solutions to reduce such test escapes is that the device failures which happen at the customer site are hard to be reproduced by the manufacturer. Most solutions that have been proposed earlier to address the issue of test escapes have used statistical techniques and univariate or, at best, bi-variate methods. In our work, we do not investigate the external factors influencing the occurrence of underkill; rather, we operate under the assumption that most semiconductor manufacturers preemptively implement advanced process control techniques to reduce the effects of process variation and environmental factors that exacerbate underkill.

The solution proposed herein explores the multidimensional test space, taking advantage of all available test measurements, in order to identify a subspace wherein the acceptable device behavior can be suitably captured and differentiated from any anomalous behavior pertaining to customer returns through multivariate statistics.

The remainder of the paper is structured as follows. In Section II, we provide an overview of the prior works on outlier detection in Analog/RF IC testing. In Section III, we present our proposed adaptive outlier detection for Underkill reduction. In Section IV, the experimental results from our industrial dataset are presented, followed by the concluding remarks in Section V.

II. RELATED WORK

In the past, several research efforts have been carried out that have specifically targeted test escape reduction. In safety-critical chips, such as automotive ICs, failure on-site can have a catastrophic impact. Ideally, semiconductor manufacturers want to achieve zero customer returns in order to cater to the strict demands of supplying automotive customers. To this end, most automotive device manufacturers include Dynamic Part Average Testing (DPAT) in their test programs. DPAT is derived from the concept of six sigma testing, where a given device is labeled as an outlier if the test measurements of the device are more than six standard deviations away from

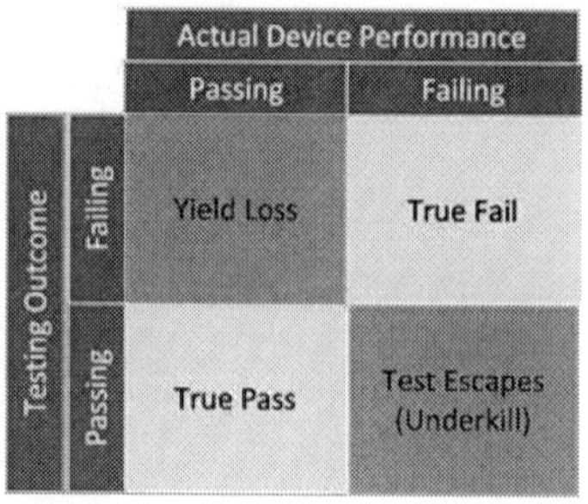

Figure 1: Device Classification Based on Test Outcome

979-8-3503-4631-2/23 $31.00 © 2023 IEEE

the mean test measurements, even if the device abides by the specification limits [1]. Prior works on outlier detection are categorized based on the usage of univariate or multivariate models to identify outliers. In univariate models, outlier detection boundaries are built for each individual test measurement or a specific test measurement to catch outliers. Univariate models such as DPAT or I_{DDQ} tests [2] [3] reject devices based on statistical thresholds for all test measurements or a specific test measurement that is known to correlate strongly with failures. The authors in [4] expand on DPAT by integrating information from visual inspection. Other univariate approaches include the use of temperature gradient testing to identify abnormal device behavior at different temperature insertions at package level testing, as described in [5], [6].

Similarly, multivariate models are used to build multi-dimensional outlier detection boundaries to screen outliers. For example, the authors of [7], [8] proposed a method where outliers are screened in a multi-dimensional space consisting of composite dimensions derived from the test measurements through principal component analysis (PCA). Other approaches include the use of a decision tree [9] to build a predictive model to identify failing chips using test data from multiple test insertions, along with models that identify signatures from a subset of tests that highly correlates with customer returns to identify failure-prone devices in [10], [11]. Alternatively, other approaches for reducing underkill include the use of die-inking locations on a wafer with a higher probability of failure based on neighborhood test outcomes, as proposed by authors in [12], and the use of adaptive test flow in mixed-signal circuits to achieve robust outlier detection in [13]. Lastly, the authors in [14] propose an unsupervised learning model that performs self-labeling of the training data via a transformation and a neural network-based supervised classifier to screen for potential customer returns.

In this work, we focus our efforts on selecting an optimal feature space using test measurements from different test insertions which help us isolate failure-prone devices, and we perform adaptive multivariate outlier detection using unsupervised clustering. The key novelty of our approach is in identifying a feature space without the loss of any information and in performing outlier detection in an adaptive manner by accounting for device performance variations seen across wafers, while maintaining low computational overhead.

III. PROPOSED APPROACH

To reduce the instances of underkill and effectively catch failure-prone devices prior to shipping them, our approach sets out to achieve the following, as shown in Figure 2:

- Identify and define a test measurement space that captures the device performance from the existing set of test measurements recorded at different temperature insertions.
- Cluster devices that exhibit similar performance in the previously selected test measurement space.
- Identify and isolate devices that exhibit anomalous behavior with regard to the performance within the cluster of devices that a device belongs to.

A. Feature Space Selection

Our approach takes as input test measurements taken at two different temperature insertions, namely *hot* and *room* temperature insertion. Individually, the outcome of each of these tests when evaluated against specification limits is not sufficient to capture failure-prone devices (customer returns). Hence, it is essential to select a set of features that allows our model to isolate and distinguish failure-prone devices from good devices. To this end, we chose the test measurement differential across the two insertions i.e., the Δ between the measurements obtained from the hot and room temperature test insertions. The motivation for performing our analysis in this feature space is to infer defectivity from abnormal variations in device performance across different temperature conditions. Indeed, based on well-documented prior research [5], [6], it is known that if a device exhibits abnormal variation in test measurements when tested at high temperatures the device has a higher probability of failing on-site. Assuming that the normalized test measurement vector for a given temperature insertion is T_{temp}, the feature space selected in our approach is given by:

$$\Delta\ Test\ Measurement\ Space = T_{Hot} - T_{Room} \quad (1)$$

where T_{Hot} and T_{Room} are normalized test measurement vectors for hot and room temperature test insertions, respectively.

B. Unsupervised Selection Using Gaussian Mixture Model

The second step in our approach is to perform unsupervised clustering of devices in the Δ test measurement space defined in Equation 1. Specifically, we use Gaussian Mixture Model (GMM) [15] to perform our clustering, which works under the assumption that there are multiple Gaussian distributions in the dataset and each of these distributions forms a separate cluster. A specific data point's association with a cluster is expressed as a percentage of mean, variance, and cluster density. In this work, clustering is performed at the wafer level and the number of clusters for any given wafer is identified by a silhouette score. This silhouette score helps us determine the optimal number of clusters in each wafer, ensuring cohesion among devices belonging to a cluster while maintaining sufficient separation between clusters. The silhouette score of our choice is the Bayesian Information Criterion (BIC). The BIC score identifies the optimal number of clusters and measures the goodness of fit of clusters by taking into account the number of features, data points, and the distance of each data point to the cluster centroid.

The objective of this unsupervised clustering is to identify devices that have similar behavior in the Δ test measurement space. Grouping the devices in our feature space implies comparable variation across test measurements and across the two different temperature test insertions. This helps in identifying devices that have abnormal variations. If the variations for a given device are abnormal or anomalous then they do not belong to their true clusters and are outliers, implying that these devices may be failure-prone and, thus, underkill suspects.

979-8-3503-4631-2/23 $31.00 © 2023 IEEE

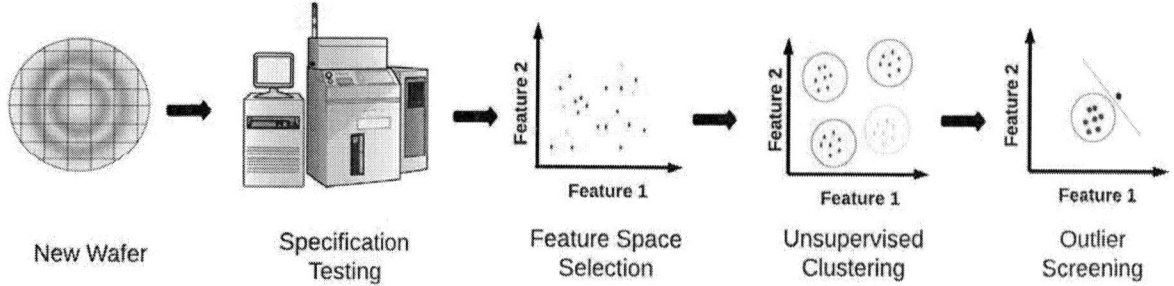

Figure 2: Experimental Flow of Proposed Machine Learning-based Adaptive Outlier Detection

C. Outlier Screening Algorithm

The next step in our approach is to identify devices that are outlying with respect to their cluster. To identify the outlying devices we use a modified formulation of the Cluster-Based Local Outlier Algorithm [16]. This multivariate outlier detection algorithm assigns each device in the cluster an outlier factor (CBLOF) i.e., a measure of the degree of anomaly. The CBLOF scores are computed based on the algorithm described in Algorithm 1. As per the algorithm, clusters are first separated into small and large clusters based on cluster size. For devices in a large cluster, the outlier factor is calculated as the product of cluster size and the distance of the device to its center. The outlier factor for devices in a small cluster is the product of cluster size and the distance of the device to its neighboring clusters. The outlier factor is computed for all devices in each cluster and the devices with the highest CBLOF scores are classified as outliers. To determine outliers, a threshold score is computed to draw the decision boundary and devices outside the decision boundary are proactively binned. The threshold score is calculated as $\mu + k * \sigma$, where μ is the mean CBLOF score, σ is the standard deviation of CBLOF score distribution and 'k' is a positive rational number. The value of 'k' is a crucial choice in correctly identifying customer returns (outliers) while maintaining low yield loss. There are several ways in which the value of 'k' can be chosen to achieve outlier detection. One such method is by picking a fixed, static value for all the wafers based on the performance metric the test engineers are targeting. If the test engineers want to achieve maximum coverage of customer returns i.e., to catch the most failure-prone devices, then the choice of 'k' will have to be made conservatively, albeit at the expense of increased collateral yield loss. While the proposed feature selection and unsupervised clustering methods are sufficient for effectively isolating failure-prone devices, in the next step of our method we seek to maximize coverage of customer returns while minimizing yield loss through an adaptive formulation of the outlier detection boundary.

D. Adaptive 'k' Selection

By analyzing and learning from the distribution of CBLOF and the optimal 'k' from prior wafers, the value of 'k' for successive wafers can be better inferred. By leveraging additional knowledge from previous wafers, the choice of 'k' for a given wafer can be adapted, resulting in better coverage

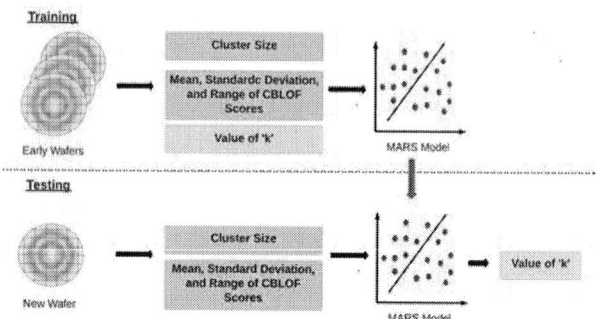

Figure 3: Dynamic Selection of 'k'

of customer returns while incurring less yield loss.

Static 'k' Selection: Static selection of 'k' explores two rule-based choices. Specifically, we choose either the statistical average or the minimum value of 'k' from previous wafers.

Dynamic 'k' Selection: The choice of 'k' in the equation $\mu + k * \sigma$ is dependent on the CBLOF score distribution. The mean, variance, and spread of the CBLOF score vary from wafer to wafer. To account for these variations and robustly pick 'k', we hypothesize the value of 'k' for each wafer is a function of cluster size, CBLOF score mean, standard deviation, and range given by the following Equation 2:

$$k \rightarrow f(Cluster\ Size, \mu(A_{score}), \sigma(A_{score}), Range(A_{score})) \tag{2}$$

To learn this function, a Multivariate Adaptive Regression Splines (MARS) [17] model is used. The MARS model is trained with a dataset from previous wafers and the trained model is used to make predictions on new wafers as illustrated in Figure 3.

IV. EXPERIMENTAL RESULTS

A. Dataset Overview

Our dataset consists of High Volume Manufacturing (HVM) devices and their corresponding test measurements from 19 wafers. For each device on the wafer, the dataset consists of probe test measurements that have been recorded at two different temperature insertions for all the devices along with their X and Y coordinates. There are a total of 728 probe test measurements for each device. Currently, all devices in the dataset are labeled as 'passing' based on the actual test outcome, as their performance passes the specification limits set by the manufacturer. However, there are 19 actual customer return devices in this dataset. These 19 devices passed the

979-8-3503-4631-2/23 $31.00 © 2023 IEEE

```
1  def
   ComputeCBLOF (C(C₁,C₂..,Cₙ),Tᵥₑ꜀(D₁,D₂,...Dₘ)) :
       Input:
           C: clusters membership of devices post GMM
       clustering,
           T_vec: test vectors of devices in Δ test
       measurement space
       Result: CBLOF Score for all devices
2      LC , SC = [ ];
3      // Set of large cluster and small cluster
4      CBLOF = [ ];
5      // CBLOF score for each device
6      for i in C :
7          if size(i) >= median(Size of cluster in C) :
8          LC.append(i);
9          else:
10         SC.append(i);
11     for t in T_vec :
12         if t belongs to SC,Cᵢ :
               CBLOF.append(length(Cᵢ)*minimum
               distance from t to its neighboring clusters'
               center)
13         elif t belongs to LC,Cᵢ :
               CBLOF.append(length(Cᵢ)* distance of t to
               its cluster center)
14     return CBLOF
       Algorithm 1: Cluster Based Local Outlier Factor
```

manufacturer's specification limits, yet failed on-site and were returned to the manufacturer, where they were subjected to failure analysis to reproduce and confirm that these are indeed failing devices. These 19 devices are spread evenly across our dataset, with each wafer containing exactly one customer return.

B. Unsupervised Clustering

As mentioned in Section III, the first step towards preemptively identifying a customer return and reducing instances of underkill is to select a feature space wherein we can identify and isolate failure-prone devices from good devices. To this end, we leverage the test measurement differential from hot and room temperature insertions to obtain our feature space. In this feature space, unsupervised clustering is performed using the Gaussian Mixture Model (GMM).

1) Feature Space Selection: Prior to performing unsupervised Clustering using GMM, we perform feature space selection. As per the proposed approach, outlined in Section III, the feature space selected is the Δ of all the test measurements across the two temperature insertions (hot/room). The effectiveness of our feature selection is measured by its ability to isolate failure-prone devices from good devices. To assess its effectiveness, we plot the device performances in the aforementioned feature space, as shown in Figure 4. In an effort to project the plot in 3-dimensional space for better visualization, we chose Principle Component Analysis

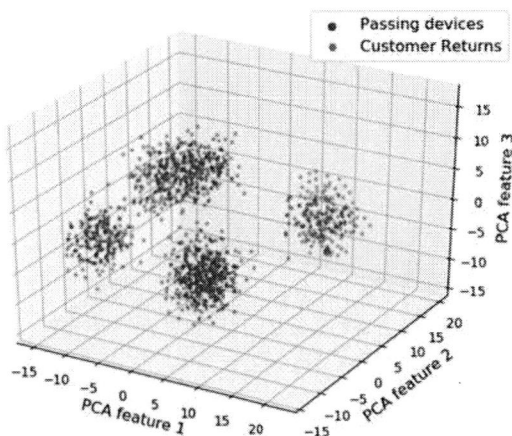

Figure 4: Device Performance in Δ Test Measurement Space

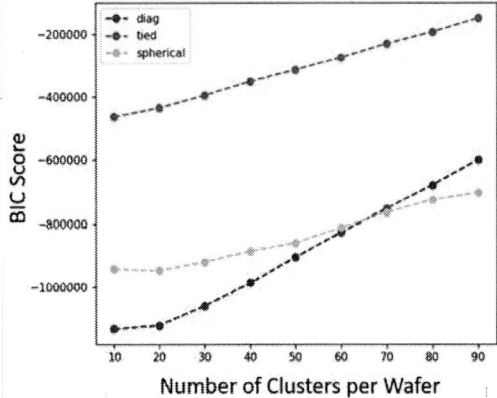

Figure 5: Bayesian Information Criterion vs. Cluster Size

(PCA) [18] for dimensionality reduction. We generated the plots for all 19 wafers and in Figure 4 we show the feature space for one randomly chosen wafer. Upon visual inspection of Figure 4, we observe that devices with similar device performance characteristics are clustered together whereas the failure-prone customer return device is located at the edges of these clusters. This confirms qualitatively that the space of Δ test measurements can effectively separate and identify failure-prone devices from good devices. Similar observations were made across all 19 wafers. Whilst the degree of isolation varies from wafer to wafer based on their test signatures, the overall level of separation between good and failure-prone devices was maintained.

2) Unsupervised Clustering using Gaussian Mixture Model: Once we establish the feature space for the devices under test, our next step is to implement the Gaussian Mixture Model (GMM) to perform unsupervised clustering of devices for each wafer. To discern the number of clusters ('n') for a given wafer, we use a silhouette score. The silhouette score is a measure of the similarity of a device to its own cluster, as compared to other clusters. Specifically, we use the Bayesian Information Criterion (BIC) score, with a lower score indicating a better fit. We sweep the value of 'n' and compute the BIC scores. The 'n' corresponding to the least BIC score is chosen as the optimal number of clusters per wafer. In our experiment, we swept the value of 'n' from 0 up to 100, and we recorded

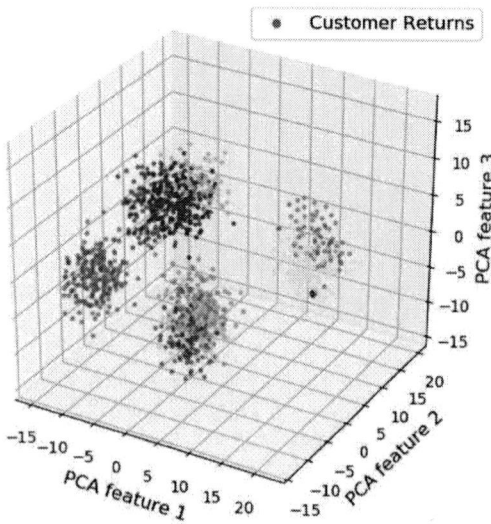

Figure 6: Gaussian Mixture Model Clustering

Value of 'k'	Customer Returns Identified	Yield Loss
2	19	10.35%
2.25	17	9.43%
2.5	15	8.66%
2.75	11	7.2%
3	9	5.4%
3.25	9	4.76%
3.5	3	1.99%
3.75	1	0.82%
4	0	0.51%

Table I: Customer Returns Identified and Yield Loss Incurred for Different Values of 'k'

the corresponding BIC score, as shown in Figure 5 for the chosen random wafer in Section IV-B1. For this wafer, the optimal number of clusters is $n = 10$. Upon performing the unsupervised clustering using GMM, we record the cluster membership of all the devices in a given wafer and we plot it in the 3-dimensional PCA space as explained previously in Section IV-B1. We observe that in Figure 6 the failure-prone device is at the periphery of the cluster that it belongs to. Indeed, when we visually inspect Figure 6, the customer return belongs to the cluster colored in yellow. The customer return device is at the edge of the yellow cluster and is located further away from the cluster center. Furthermore, we deduce that the failure-prone device is outlying with respect to its cluster. Similar trends were observed across all 19 wafers.

C. Outlier screening

Having established that the failure-prone devices are outlying with respect to their cluster, we proceed to evaluate the effectiveness in screening outliers using different methods with minimal yield loss and maximal coverage. First, the Cluster-Based Local Outlier Factor (CBLOF) scores are computed for all the devices in a wafer using the algorithm discussed in Section III-C. Then, the threshold is calculated as $\mu + k * \sigma$. Different methods to choose 'k' and identify the outliers are compared in the following subsection. Based on the choice of 'k' and the subsequent thresholding and binning, we record the yield loss incurred and coverage in identifying the customer returns for each method. We perform qualitative analysis for each method and explore the trade-off between identifying customer returns and incurring yield loss. First, we sweep the value of 'k' in $\mu + k * \sigma$ from k=4 down to k=2 in steps of 0.25 and we report the number of customer returns identified along with the overall yield loss incurred for a given choice of 'k' across all 19 wafers. The results are summarized in Table I and the following observations are made:

- As the value of 'k' increases, the number of customer returns identified decreases.

- As the value of 'k' decreases, the overall yield loss incurred due to thresholding and binning also increases.

The leniency of the threshold increases with the value of 'k' and the threshold becomes increasingly optimistic. An optimistic threshold may fail to identify failure-prone devices that are at the periphery of the cluster. Hence, a lower 'k' value is required to screen out such devices. For example, we observe that in Table I when k=3.75 the threshold identifies correctly only 1 out of the 19 customer returns (and even 0 customer returns at k=4), resulting in very low coverage. As the 'k' decreases to k=3 to implement a relatively stricter threshold, we observe that the number of correctly identified suspect devices increases to nine out of the nineteen customer returns. Furthermore, as the value of 'k' is decreased to k=2, the threshold identifies all 19 customer returns correctly. Conversely, when the value of 'k' decreases, the overall yield loss incurred due to thresholding and binning increases. This is expected, as the distribution of devices further from the mean CBLOF score is sparse. We know that as 'k' increases so does our threshold value. Hence, a large 'k' results in a very low percentage of yield loss. We observe that, in Table I, for k=2 the yield loss incurred is highest at 10.35%; as the value of 'k' increases to k=3, the yield loss decreases by almost half to a value of 5.4%, while for k=3.75 the yield loss goes down to 0.82% which is far more tolerable.

From the above observations, we can deduce that a lower value of 'k' results in better outlier detection and, thus, in better coverage of customer returns. However, this also results in increased yield loss, which is not desirable. Hence, there is an inverse relationship between the choice of 'k' and the coverage of customer returns and yield loss metrics. There is a direct trade-off between the coverage of customer returns and yield loss when performing an unsupervised selection of 'k'. While these results are important proof of concept that customer returns can be averted by our method at the expense of yield loss, identifying an appropriate value for 'k' needs to be performed intelligently and potentially adapted per wafer so that the best trade-off point between customer return prevention and yield loss can be reached.

D. Adaptive Selection of 'k'

To make an informed choice of 'k', we learn its value from previously seen customer returns. We split our dataset into a training and a test set, so that early wafers with known customer returns (i.e., training set) are used to learn the best

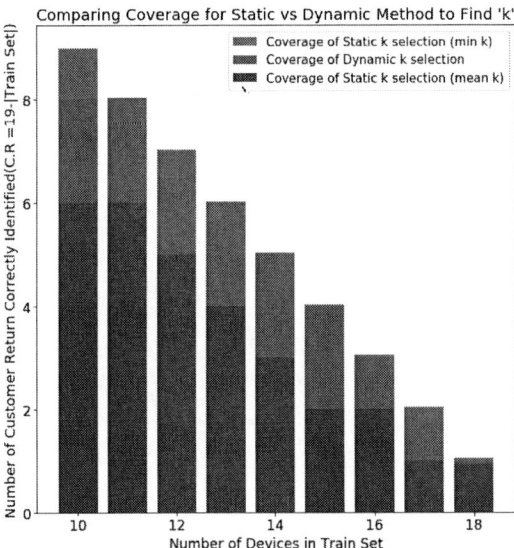

Figure 7: Comparison of Customer Returns Coverage using Static vs. Dynamic Selection of 'k'

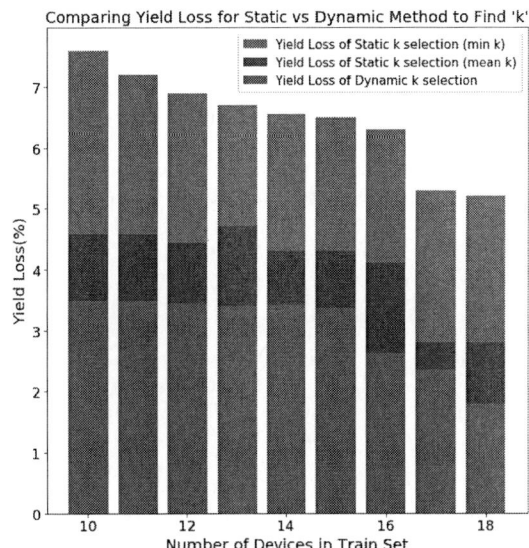

Figure 8: Comparison of Yield Loss for Static vs Dynamic Selection 'k'

value of 'k', which we then evaluate for coverage of customer returns and yield loss on the following wafers (i.e., test set). We evaluate the effectiveness of both static and dynamic selection of 'k', as described in Section III-D. We begin with a training set consisting of 10 customer returns from early wafers to learn the value of 'k' using static & dynamic methods. The value of 'k' is used to perform outlier detection on the remaining 9 wafers in the test set. We progressively increase the training set size, learning the value of 'k' from 10 up to 18 wafers (adding one wafer at a time to the train set) and evaluating the learned 'k' value on the remaining wafers. The coverage of customer returns and the yield loss incurred for both the static and dynamic 'k' selection methods are summarized and compared in Figures 7 and 8, respectively.

Static Selection of 'k': The value of 'k' is chosen as the minimum value of 'k' across the train set. We observe that the coverage of customer returns is 100% i.e., all customer returns in the test set are correctly identified as seen in Figure 7. The trade-off is that the incurred yield loss is significant. In Figure 8, we observe that the yield loss when selecting the minimum value of 'k' from the training set is the largest when compared to other methods. This is observed across different sizes of the training set: when the training set size is 10 the yield loss observed is 7.6%, and as we progress the yield loss reduces from 7.6% down to 5.2% for a training set size of 18. Alternatively, the choice of mean 'k' performs better in terms of yield loss than the minimum choice of 'k'. For training set of size 10 the yield loss is 4.8%; when the training set size increases to 12 the yield loss reduces to 4.44% and finally for training set size 18 it goes down to 2.8%. The yield loss incurred is lower when compared to the minimum choice of 'k', but the coverage of customer returns is also lower. For example, while the minimum choice of 'k' has 100% coverage, the mean choice of 'k' can correctly identify 6 out of the 9 customer returns in the test set when the training set size is 10. As we progress, the coverage of customer returns is 6 out of the 7 customer returns in the test set for a training set size

of 12 and, finally, 1 out of 2 customer returns in the test set for a training set size of 17.

Dynamic Selection of 'k': Lastly, we evaluate the effectiveness of a dynamic selection of 'k'. The value of 'k' is produced through a function of clustering details and CBLOF score distribution. In Figure 8 we observe that for the dynamic selection of 'k', the incurred yield loss is lower in comparison to that of the two static 'k' selection methods. For a training set size of 10, the incurred loss is 3.48%, as we progress the yield loss reduces to 1.8% for a training set size of 18. The dynamic 'k' selection also achieves better coverage of customer returns. In Figure 7, we see that the dynamic method can correctly identify all customer returns in almost all cases. The exception is when the training set size is 10, where it fails to catch 1 out of the 9 customer returns in the test set. As the training set size progresses, the coverage of customer returns becomes 100%.

Overall, dynamic selection of 'k' and adaptation per wafer is the best among the discussed options. The value 'k' predicted through this method is neither too optimistic to catch customer returns nor too conservative to incur large yield loss.

V. Conclusion

Towards reducing the instances of underkill in analog/RF IC testing and its repercussions in customer returned devices, we introduced a machine learning-based adaptive outlier detection methodology. Through a feature space selection process, followed by unsupervised clustering and an adaptive multivariate definition of the outlier boundary in the clusters of a wafer, devices that are failure prone are distinguished among the passing population and proactively binned as failing to prevent customer returns. As demonstrated on an industrial dataset from a contemporary analog/RF IC, the proposed methodology is able to correctly identify truly defective devices, which would have otherwise failed in the field, while minimizing the incurred yield loss.

REFERENCES

[1] "Guidelines for part average testing," *Automotive Electronic Council*, vol. AEC-Q001 Rev-D, 2011.

[2] S. S. Sabade and D. M. Walker, "Evaluation of effectiveness of median of absolute deviations outlier rejection-based IDDQ testing for burn-in reduction," in *IEEE VLSI Test Symposium (VTS)*, 2002, pp. 81–86.

[3] T. W. Williams, R. H. Dennard, R. Kapur, M. R. Mercer, and M. Maly, "IDDQ test: sensitivity analysis of scaling," in *IEEE International Test Conference (ITC)*, 1996, pp. 786–792.

[4] A. Coyette, W. Dobbelaere, R. Vanhooren, N. Xama, J. Gomez, and G. Gielen, "Latent defect screening with visually-enhanced dynamic part average testing," in *IEEE European Test Symposium (ETS)*, 2020, pp. 1–6.

[5] Q. Wang, T. Kyuho, H. Saihara, T. Kitamura, K. Yonemura, and H. Kariyazono, "Reliable screening for zero-defect quality improvement by temperature gradient testing," in *e-Manufacturing Design Collaboration Symposium (eMDC)*, 2013, pp. 1–4.

[6] T. Sakamoto, K. Yofu, and T. Kyuho, "New method of screening out outlier; expanded part average testing during package level test," *IEEE Transactions on Semiconductor Manufacturing*, vol. 30, no. 4, pp. 351–356, 2017.

[7] P. M. O'Neill, "Production multivariate outlier detection using principal components," in *IEEE International Test Conference (ITC)*, 2008, pp. 1–10.

[8] J. Yu, "Fault detection using principal components-based gaussian mixture model for semiconductor manufacturing processes," *IEEE Transactions on Semiconductor Manufacturing*, vol. 24, no. 3, pp. 432–444, 2011.

[9] S. Biswas and R. D. Blanton, "Statistical test compaction using binary decision trees," *IEEE Design & Test of Computers*, vol. 23, no. 6, pp. 452–462, 2006.

[10] A. E. Gattiker and W. Maly, "Current signatures," in *IEEE VLSI Test Symposium (VTS)*, 1996, pp. 112–117.

[11] N. Xama, J. Raymaekers, M. Andraud, J. Gomez, W. Dobbelaere, R. Vanhooren, A. Coyette, and G. Gielen, "Avoiding mixed-signal field returns by outlier detection of hard-to-detect defects based on multivariate statistics," in *IEEE European Test Symposium (ETS)*, 2020, pp. 1–6.

[12] C. Xanthopoulos, P. Sarson, H. Reiter, and Y. Makris, "Automated die inking: A pattern recognition-based approach," in *IEEE International Test Conference (ITC)*, 2017, pp. 1–6.

[13] H.-G. Stratigopoulos and C. Streitwieser, "Adaptive test with test escape estimation for mixed-signal ICs," *IEEE Transactions on Computer-Aided Design of Integrated Circuits and Systems*, vol. 37, no. 10, pp. 2125–2138, 2018.

[14] H. Hu, N. Nguyen, C. He, and P. Li, "Advanced outlier detection using unsupervised learning for screening potential customer returns," in *IEEE International Test Conference (ITC)*, 2020, pp. 1–10.

[15] Y. Li, M. Dong, and J. Hua, "A Gaussian Mixture Model to Detect Clusters Embedded in Feature Subspace," *Communications in Information Systems*, vol. 7, no. 4, pp. 337–352, 2007.

[16] Z. He, X. Xu, and S. Deng, "Discovering cluster-based local outliers," *Pattern Recognition Letters*, vol. 24, no. 9, pp. 1641–1650, 2003.

[17] J. H. Friedman, "Multivariate Adaptive Regression Splines," *The Annals of Statistics*, 1991.

[18] S. Wold, K. Esbensen, and P. Geladi, "Principal component analysis," *Chemometrics and Intelligent Laboratory Systems*, vol. 2, no. 1, pp. 37–52, 1987.

Architectural Radiation Hardening of CMOS Power Management Circuits through Bias Tuning

Gauri Koli, Liam Nguyen, Jennifer Kitchen

School of Electrical, Computer and Energy Engineering, Arizona State University, Tempe, AZ, USA.

Abstract—**Within the space electronics industry, several strategies have been implemented to mitigate heavy ion, neutron, and proton induced radiation effects in CMOS processes, including radiation through process technology alterations and through circuit design, layout, and architecture. In this work, a new method is proposed that adaptively calibrates integrated analog circuits in CMOS bulk technology through bias tuning to create a radiation hardened system. In this technique, the device parameters that vary due to total ionizing dose radiation are monitored using built-in-self-test circuitry. These monitored parameters are used to tune and calibrate circuit level performance parameters. This work analyzes the TID radiation induced performance shifts in three critical power management circuits and uses bias tune in each circuit to recover circuit performance. The three circuits include: a ring oscillator, a bandgap voltage reference, and a non-overlap (dead-time) clock generator.**

Index Terms—**Calibration, CMOS, analog, power management, radiation hardened, space electronics**

I. INTRODUCTION

Recent years have seen significant growth in the space industry, hence an increasing need for space electronics [1]. Inefficient and unreliable power systems continue to be the bottlenecks in electronics for space missions that span anywhere form a few days to multiple decades. These power systems oftentimes comprise of several point-of-load converters (PoL) that convert and distribute power from the central spacecraft bus to numerous spacecraft functions like the engine control, communications, steering control, and exploratory sensors and systems. It is desirable for these PoL converters to maintain performance and efficiencies above 90% over the mission's lifetime. Furthermore, space missions require continuously decreasing design cycle times and increasingly aggressive performance specifications, and would therefore benefit from using commercial circuit design methodologies. Unfortunately, very few state-of-the-art commercial platforms have been deployed in space due to their high rate of expected failures under radiation and temperature fluctuations.

A. Radiation Effects in Electronics

Radiation effects in electronics can be categorized into three types [2]: total ionizing dose effects (TID), single event effects (SEEs) and displacement damage (DD). DDs may also be referred to as non-ionizing dose effects. TID is caused by heavy ions, protons, and gamma rays. TID has long term and cumulative effects that gradually degrade CMOS device parameters, such as transistor threshold voltage shifts,

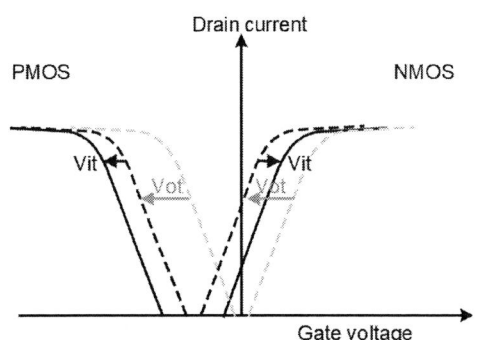

Fig. 1. TID induced threshold voltage shifts in MOSFETs

subthreshold currents, parasitic currents, and degradation in electron mobility and transconductance (g_m). SEEs can cause bit-flips within memory and current latch-up, which may result in device damage such as a single event burnout or gate rupture in power devices.

B. Radiation Hardening Techniques in Circuits

Several methods have been proposed to improve radiation performance in CMOS technologies [3], including reduction/elimination of mismatch causing elements, offset cancellation using averaging techniques in amplifiers [4], employing a combination of JFET-CMOS and BJT-CMOS structures [5] to overcome the mismatch in current mirrors due to TID, reduction of oxide thickness through process modification to mitigate TID, and triple modular redundancy [6] or error correcting codes [7] to mitigate SEE. For inaccessible electronics in harsh space environments, short repair times (if any) are preferred to reduce the cost of maintenance, which can only be achieved with highly reliable radiation-tolerant and self-calibrating electronics [8]. The most popular methods for ensuring reliable CMOS-based electronics pre-silicon fabrication is the prediction of radiation effects in semiconductors using time dependent analytical models [9] for deposition of charge due to ionizing radiation, and circuit synthesis using 6-sigma deviation with Monte-Carlo models provided by the foundry to estimate shifts due to radiation.

A great advantage of in-field integrated circuit parameter monitoring is that the cumulative radiation effects can be observed and recorded in continuous time. The cumulative radiation effects in CMOS circuits are observed through built in self test (BIST) circuits, and the relative changes in the parameters affected within the BIST circuits due to radiation are used to calibrate the PoL functional circuits. The adaptive

979-8-3503-4631-2/23 $31.00 © 2023 IEEE

calibration of circuits is a three-step process: the first being the prediction of radiation effects at design time, the second is in-field monitoring of circuit/system parameters without affecting the circuit/system's normal operation, and lastly, the tuning of the functional circuit/system device parameters to recover circuit performance. This work analyzes the performance parameter shifts for three critical circuits of a typical power management system: a ring oscillator, a bandgap voltage reference circuit, and a non-overlapping clock (dead-time) generator. This paper also proposes low overhead monitoring circuits to track performance shifts due to TID.

Section II gives a brief overview of the TID induced MOSFET device parameter shifts in CMOS technology. These shifts are modeled in this work to study performance shifts in critical power management circuits. Section III gives detailed analysis of the performance shifts in a ring oscillator, band-gap reference, and dead-time generator due to TID. Bias tuning options are proposed in each circuit for recovering circuit performance. Additionally, two low-overhead BIST circuits are proposed in Section IV for in-field performance monitoring, and a complete system for bias tuning is presented in Section V.

II. TID INDUCED DEVICE PARAMETER SHIFTS IN CMOS TECHNOLOGY

Ionizing radiation on a MOSFET deposits energy creating electron-hole pairs, which leads to non-recombined trapped charges in the oxide and at the interface of the oxide and silicon [2] and causes parametric effects on the MOSFET like threshold voltage (V_{TH}) shift. The V_{TH} shift due to traps is negative when the charge is positive, as seen in Fig. 1. Interface traps in NMOS are negatively charged, leading to a positive shift in V_{TH} (designated Vit in Fig. 1), and interface traps in PMOS are positively charged, leading to a negative V_{TH} shift. PMOS V_{TH} shift is higher than NMOS V_{TH} shift over time [10], and the V_{TH} shifts are bias dependent.

Interface build-up is a slower phenomenon compared to oxide trapping, which means that a change in Vot (shift due to oxide trapping) is observed earlier than a change in Vit. This explains why n-channel threshold voltage first decreases as a function of tunneling and annealing time and later begins to increase as interface traps increase. This is called the 'rebound' effect. These shifts are commonly observed in today's larger than 110 nm node CMOS bulk technologies. Due to the power requirements of most power management integrated circuits, thicker oxide technologies like 180 nm, 350 nm and 0.5 μm are still being used. Hence, TID still poses an issue in current CMOS technologies used for power electronics in space missions.

PMOS and NMOS V_{TH} shifts due to TID have been modeled in prior works [2], [10] using simple gate voltage offsets in PMOS and NMOS devices, where the offset voltage trends are determined by the TID levels. For 0.18 μm CMOS bulk technology, TID induced gate offsets have been observed (measured) with gate voltage offsets up to 400mV in PMOS devices and 8mV in NMOS devices. This process technology is chosen for design because it is a popular technology node for implementing integrated power management solutions for space applications. This work uses the voltage offset models derived in [10] to analyze the performance via simulation of three critical power management circuits with total dose up to 1 Mrad (SiO_2).

III. ANALYSIS AND BIAS TUNING OF PERFORMANCE PARAMETERS IN PM CIRCUITS UNDER TID

Three of the most critical circuits used in a PoL controller are a ring oscillator, a bandgap voltage reference, and a non-overlapping clock generator. This section analyzes the performance parameters of these three circuits under TID and proposes methods to recover performance under TID through bias tuning.

A. Ring Oscillator

A ring oscillator is generally used in a PoL system to set the switching frequency of the power converter. In some cases, it is also used to generate a clock for digital controllers in a switching power converter. A typical ring oscillator is shown in Fig. 2, whose frequency is set by the propagation delays of the inverters caused by the charging and discharging times of the equivalent capacitance at each node. This ring oscillator frequency can change due to TID induced threshold voltage shifts in the MOSFETs, which is an undesired phenomenon. Fig. 3(a) plots the shift in frequency with increasing TID for a 17-stage ring oscillator designed for a 20 MHz pre-radiation oscillation frequency in a 0.18 μm BCD-CMOS process technology with a 5.0 V supply voltage. The MOSFETs are modeled over TID using the gate offset models derived in [10] for this specific CMOS process.

In order to allow for flexibility in the oscillator's frequency, a 17-stage current starved ring oscillator topology shown in Fig. 4 is also implemented in the same process technology [11]. The current starved ring oscillator allows for bias tuning through $V_{control}$, which can be used to minimize undesired frequency shifts in the ring oscillator due to TID. $V_{control}$ changes the current bias (I_{bias}) values, where I_{bias} is set by the PMOS rail devices and determines the oscillation frequency. The simulated relative frequency shift for the current starved ring oscillator over TID is shown in Fig. 3(b) when the bias is kept constant over TID. The shift is plotted as the percentage frequency deviation from the pre-radiation oscillation frequency. As seen in Fig. 3(b), the frequency shifts by 3% at higher TID levels. When the control voltage ($V_{control}$) is ideally dynamically adjusted to tune the oscillator frequency with varying TID, the oscillator frequency can be maintained constant at the pre-radiation value. The required ideal voltage for $V_{control}$ required to maintain a constant (pre-radiation) oscillator output frequency with TID is also given in Fig. 3(b).

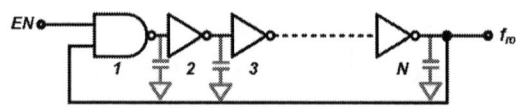

Fig. 2. A basic ring oscillator, where N=17 for the presented design

979-8-3503-4631-2/23 $31.00 © 2023 IEEE   16

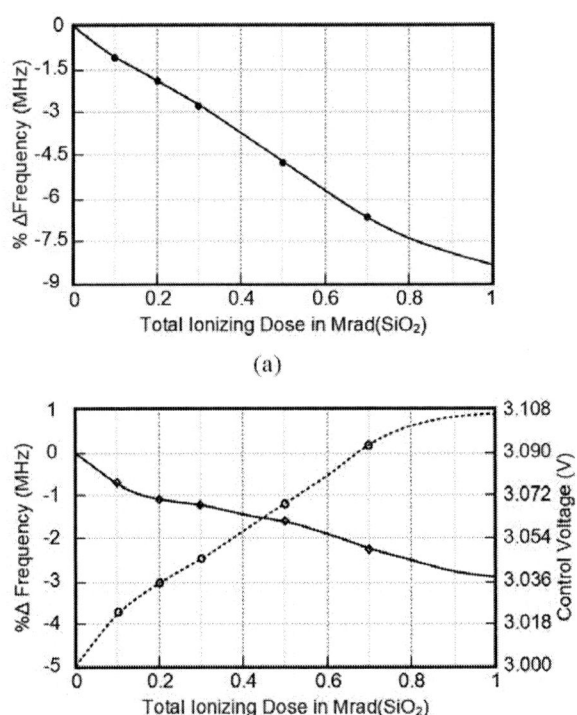

(a)

(b)

Fig. 3. (a) Percentage frequency shift in a 17-stage traditional ring oscillator versus TID, and (b) Percentage frequency shift in a 17-stage current starved ring oscillator with a fixed control voltage (solid line) and the control voltage value of $V_{control}$ required to tune the output frequency to the pre-radiation value (dotted line) versus TID

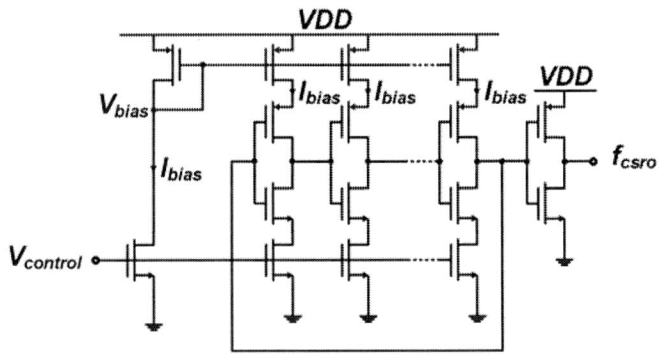

Fig. 4. Current starved ring oscillator with with bias tuning through $V_{control}$

The $V_{control}$ values are well within an acceptable range and resolution to support a simple DAC realization for system-level autonomous dynamic adjustment. Therefore, the $V_{control}$ bias tuning knob can be autonomously varied by employing in-field monitoring at the system level. Due to the reduced supply headroom, the frequency range for a 17-stage current starved ring oscillator is an order magnitude less than that of a 17-stage traditional ring oscillator and for this reason, the percentage change in oscillation frequency has been reported for both cases in order to clearly compare them. The NMOS rebound effect previously discussed in Section II can be seen at approximately 200 kRad (SiO_2) with the slight flattening of the curve, but this effect is not large since the NMOS

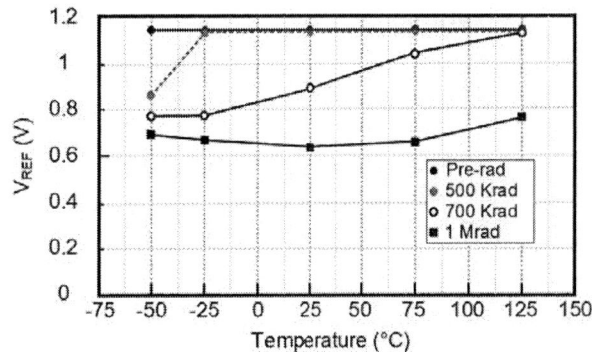

Fig. 5. Bandgap output reference voltage versus temperature at different TID levels

shifts are small compared to the PMOS V_{TH} shifts for device widths $10\mu m$ and larger [10]. The tuning of the current starved ring oscillator does not cause significant increase in power consumption, as the additional current for each inverter is only $3\mu A$ in the worst case.

B. Bandgap Voltage Reference Circuit

Bandgap voltage references (BGRs) provide an output reference voltage that is largely independent of process, voltage supply, and temperature (PVT) variations. For space electronics, the BGR should also provide a constant reference voltage under TID. If the reference voltage from the BGR shifts, this may compromise the performance of subsequent circuit blocks like error amplifiers in controller feedback loops, current biasing circuits, etc.

The expression in (1) describes the output reference voltage for the traditional bandgap that corresponds to Fig. 6, where V_{EB2} is the emitter-base voltage of Q_2, V_T is the thermal voltage of the pn junction, A_1 is the area of PNP Q_1, and A_2 is the area of PNP Q_2. This equation assumes that the currents I_1 and I_2 through the PNP devices are equal.

$$V_{REF} = V_{EB2} + \frac{R_2}{R_1} * V_T * ln(\frac{A_1}{A_2}) \qquad (1)$$

In reality, the precision of the output voltage of the BGR is highly dependent on the accuracy of the current mirrors used in the bandgap core (seen as I_1 and I_2 in Fig. 6). TID can affect mismatch between I_1 and I_2, which affects the BGR's output voltage precision [12], [13]. The threshold voltage shift due to TID may affect the threshold voltage of each MOSFET in the circuit differently. For the presented analysis, the threshold is modeled for the BGR as a uniform effect on all of the PMOSFETs or NMOSFETs. Under TID, the PMOS V_{TH} (threshold voltage) will shift, resulting in the variation of I_1 and I_2, which causes an offset in the bandgap output voltage. The expression in (2) describes the output reference voltage for the bandgap that corresponds to Fig. 6 as a function of current and V_{EB2}, where M_1 and M_2 threshold voltages are assumed to shift simultaneously (and have the same (W/L) and therefore the same nominal currents), such that I_1 and I_2 are equal before and during irradiation. This means I_1 can replace I_2 in (2). If I_2 decreases, the V_{REF} will also decrease. As the

979-8-3503-4631-2/23 $31.00 © 2023 IEEE   17

V_{TH} of the PMOS devices increase, they will lose current and the V_{REF} will subsequently decrease. The relationship between V_{TH} and I_2 is given in (4) as the standard saturation current equation.

$$V_{REF} = I_2 * R_2 + V_{EB2} \qquad (2)$$

V_{EB2} is expressed as a function of I_B (base current of the PNP) in (3), where G_E is the emitter Gummel number, A_E is the area of the emitter, n_i is the electron concentration and the hole concentration in undoped semiconductor material, and q is the electronic charge.

$$V_{EB2} = V_T * ln(\frac{I_B}{A_E} * \frac{G_E}{q * n_i^2} + 1) \qquad (3)$$

$$I_{1,2} = \frac{\mu_P * C_{OX}}{2} * \frac{W}{L} * (V_{GS} - V_{TH})^2 \qquad (4)$$

Fig. 5 plots the simulated V_{REF} versus temperature under varying TID radiation doses. The BGR is simulated in a 180 nm technology that offers vertical PNPs, and the BGR is designed with 20 μm^2 area PNPs. The operational amplifier is a simple 5-transistor OTA architecture. The simulated BGR under TID only includes the threshold shifts in the MOSFETs that are modeled using data from [10]. It is important to note that there are other device parameters that shift over TID and significantly affect BGR performance, namely the PNP diodes and their effective emitter area and the input offset voltage of the operational amplifier. A big design uncertainty are the PNP bipolar devices, which are used for their pn junctions that act as diodes and create a CTAT (Complementary To Absolute Temperature) voltage. The effects of radiation on diodes can shift the output voltage of a bandgap output voltage by a few hundred millivolts [14]. Furthermore, the voltages at the amplifier's inputs should be equal to provide a more precise output voltage. Across TID, the input pair of the amplifier will have V_{TH} shift, which can be mitigated by using offset cancellation circuitry such as chopper stabilization techniques. This is a radiation hardening by circuit design technique to improve the output voltage precision. Although this work focuses on analyzing MOSFET threshold voltage shifts under TID, the concept of using bias tune to recover performance can also accommodate performance shifts due to these other device parameter shifts over TID.

Fig. 6 shows the simplified schematic of the BGR with added PMOS current source (trim) devices in red that may be switched in/out and used to dynamically tune the bias current of the bandgap under TID in order to mitigate BGR output voltage shift. The pre-radiation nominal currents can be independently reduced or increased in the bandgap core through this current tuning, thus counteracting much of the V_{TH} shift due to TID. Simply increasing the width of the nominal PMOS current sources would not provide hardening against TID since the threshold shift occurs dynamically and oftentimes differently amongst the MOSFETs. Thus, increasing the BGR's PMOS current source device widths would yield

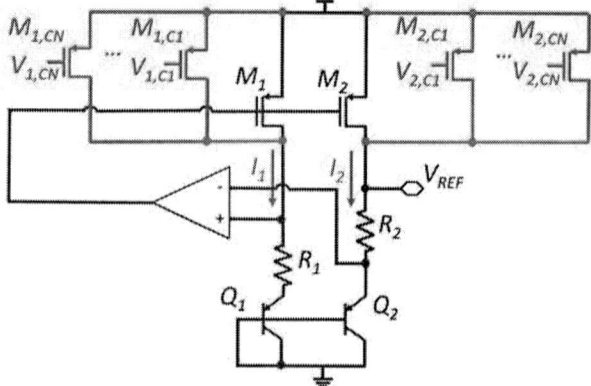

Fig. 6. Simplified schematic of the bandgap voltage reference core

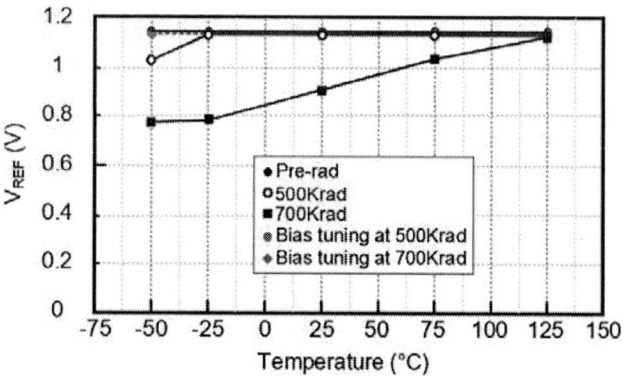

Fig. 7. Simulated V_{REF} versus temperature for 500 kRad and 700 kRad TID with and without bias tune

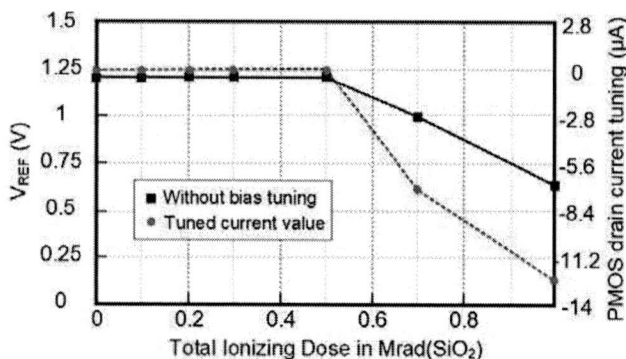

Fig. 8. Simulated Bandgap reference output voltage at 27 °C without tuning (black-solid) and drain current ($I_{1,2}$) of each BGR core branch after tuning (red-dashed) versus TID

the same result as an optimized BGR design with a non-width-increased current source, and bias tuning could again be used for hardening.

Fig. 7 plots V_{REF} versus temperature pre-radiation and under TID doses of 500 kRad (SiO_2) and 700 kRad (SiO_2). From Fig. 7, it is observed that ideally tuning the bias current by increasing the effective PMOS widths (M_1 and M_2) reduces the deviation of V_{REF} over temperature from the pre-radiation performance. This simulation shows that the 500 kRad (SiO_2) dose recovers well at -25 °C, while the 700 kRad (SiO_2) dose shows recovery throughout the temperature range of -25 °C to +125 °C.

Fig. 8 plots V_{REF} versus TID before bias tuning at a fixed

BGR temperature of 27 °C. V_{REF} varies from 1.207 V to 0.643 V over TID without bias tuning. When tuning the bias by varying the PMOS device widths (with ideally infinite width resolution) at each TID level, V_{REF} can be maintained at 1.207 V over TID. Fig. 8 also plots the PMOS drain (bias) current shift with respect to TID that is required to provide the V_{REF} performance with bias tuning. At 1 MRad (SiO_2), the current varies by 12.84 μA from its pre-radiation value, which is well within the range and resolution of the simple current bias trim illustrated in Fig. 6.

C. Non-overlapping Clock Generator

A non-overlapping clock is used to avoid shoot-through in the power stage of a switching power converter. This non-overlap period can vary due to parametric shifts from TID and must be recovered by employing tuning methods. A typical digital non-overlapping clock (also called a dead-time generator) is made up of digital NAND and NOT gates as shown in Fig. 9(a). The generated non-overlapping clock signals are ϕ_1 and ϕ_2. A simulation of the implemented non-overlapping clock generator in the 0.18 μm BCD process shows that the dead-time varies by 15% over the entire TID range for the untuned case in Fig. 9(a). This dead-time variation over TID is plotted in Fig. 10.

Several methods, such as that proposed in [15], and capacitor network tuning as illustrated in Fig. 9(b) have been explored in the past to tune the non-overlap period. Due to the on-chip area constraints from the capacitor network, the capacitor network tuning method poses limitations on the maximum tuning capability. In this work, the lowest overhead method of bias tuning is used to tune the non-overlapping clock. The bias tune for tuning the non-overlap employs the same technique for each inverter as the current starved ring oscillator discussed previously in Section III-A. The bias current is perfectly (ideally) tuned to maintain a constant pre-radiation dead-time (1.34 ns) over TID. The ideal control voltage ($V_{control}$) required to tune the bias current, and hence dead-time, to the pre-radiation value at each TID level is also plotted in Fig. 10. This control voltage varies significantly more than the tune for the ring oscillator because a 3-stage oscillator is used here. This control voltage range can be easily realized at a system level.

IV. BIST CIRCUITRY FOR LOW OVERHEAD IN-FIELD PERFORMANCE MONITORING UNDER TID

A. Ring Oscillator with an Output Charge Pump

In Section III, the performance shifts in a ring oscillator due to V_{TH} shifts over TID is analyzed, and a bias tuning method to recover the pre-radiation frequency is proposed. In this section, a similar ring oscillator *without tuning* is used to monitor frequency shifts due to TID. This is a low overhead monitoring circuit that can be easily integrated within the system.

A ring oscillator is oftentimes used to characterize the performance of a new technology and also used during wafer testing to observe manufacturing process variations (process

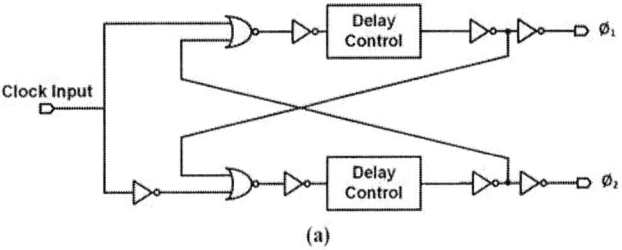

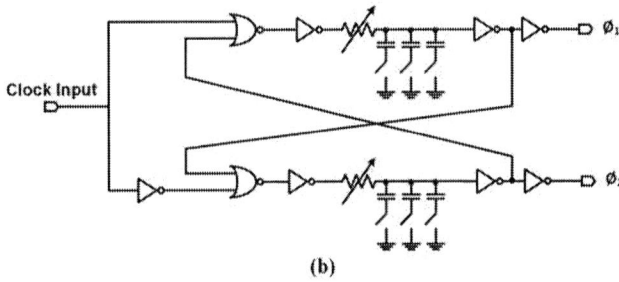

Fig. 9. Non-Overlapping clock generator with (a) delay control using basic bias tuning, and (b) digital implementation for tuning

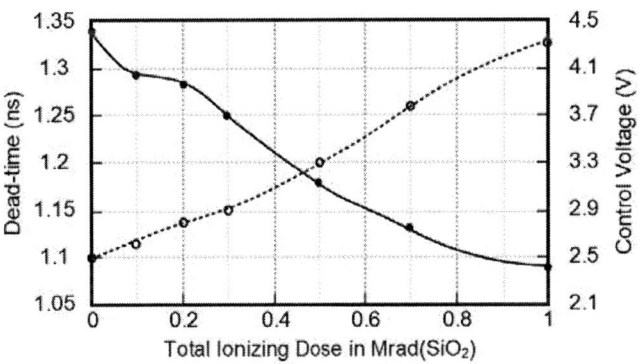

Fig. 10. Dead-time performance shift in a non-overlapping clock generator over TID (solid), and control voltage for tuning dead-time to the pre-radiation value over TID (dotted)

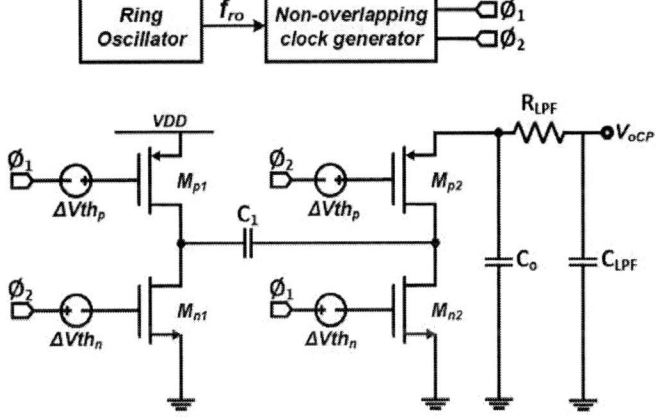

Fig. 11. Charge Pump used to convert dead-time to voltage for built-in performance monitoring over TID

control monitor). The ring oscillator provides a straightforward method to measure supply and temperature changes

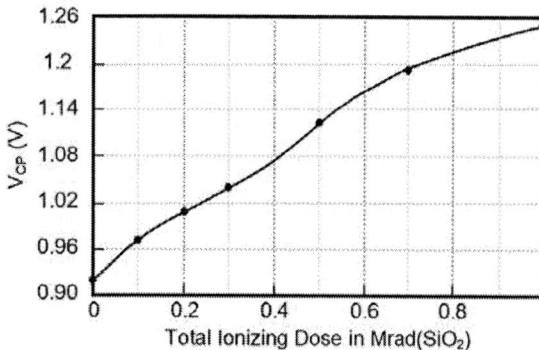

Fig. 12. Charge Pump DC output voltage versus TID

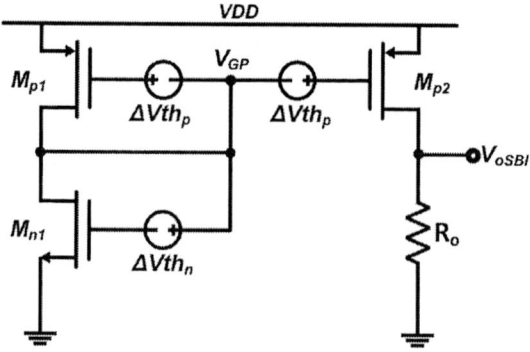

Fig. 13. Self-biased inverter BIST circuitry with modeled PMOS and NMOS V_{TH} shifts

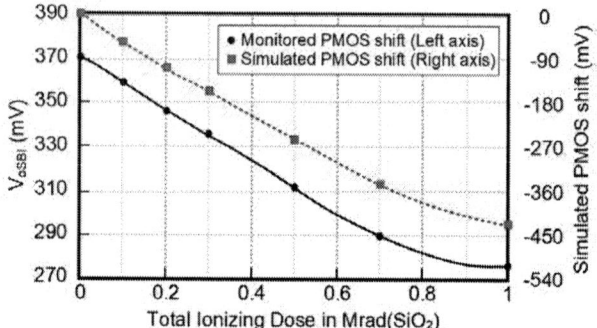

Fig. 14. Output voltage of the self-biased inverter that tracks V_{TH} shifts

on-chip. Oftentimes, it is costly to probe a high frequency signal (due to amplitude or bandwidth constraints) on chip. Thus, the oscillator's output signal is converted to a DC voltage on-chip. The DC voltage is then monitored off-chip. In this work, a charge pump is used to convert the oscillation frequency, which is typically in the higher MHz range for power management circuits, to a DC voltage that can be monitored off-chip. The simulation test bench for the proposed ring oscillator-based BIST circuit is shown in Fig. 11, where the ring oscillator is used to supply the input to the non-overlapping clock generator, which creates two non-overlapping clock signals $\phi1$ and $\phi2$. These signals are derived from the previously discussed non-overlapping clock generator circuit (also implemented on-chip). These clock signals are used as inputs to a charge pump that converts the frequency of the clock to a finite voltage V_{oCP}. A low pass filter is used to obtain a stable DC voltage that correlates to the ring oscillator's frequency. The TID-induced V_{TH} shifts in the NMOS and PMOS devices are modeled in simulation as a negative gate bias to observe a change in the frequency of the ring oscillator, as modeled by ΔVth_p and ΔVth_n in Fig. 11. All of the NMOS and PMOS devices are shifted simultaneously with TID. These V_{TH} shifts are modeled in all of the circuits, including the ring oscillator, non-overlapping clock generator and the charge pump, as the BIST circuitry's device parameters also shift with TID. The charge pump's output DC voltage curve correlating to parametric shifts in the frequency of the ring oscillator is plotted for varying TID level in Fig. 12.

This radiation monitoring circuitry can be used to control and tune critical functional PM circuits such as the current starved ring oscillator described in Section III-A.

B. Self-Biased Inverter

The self-biased inverter in Fig. 13 is made up of an inverter whose input is connected to the output. This configuration generates a gate voltage V_{GP} that is used to bias the PMOS M_{p2}. The drain current of M_{p2} is converted to a voltage V_{oSBI} with the help of resistor R_o. According to [6], if the width of the PMOS device is much greater than the width of the NMOS device in a self-biased inverter, then the gate voltage tracks V_{TH} shifts from PMOS devices only. Hence, the drain current derived from M_{p2} is a function of PMOS V_{TH} shifts alone. This is another low overhead method that can be easily integrated to monitor TID induced V_{TH} shifts in an integrated circuit.

TID induced threshold voltage shifts are modeled and added to all PMOS and NMOS devices of the self-biased inverter circuit, as shown in Fig. 13. The self-biased inverter's output voltage V_{oSBI} is plotted for various TID levels up to 1 Mrad (SiO_2) in Fig. 14. The PMOS V_{TH} shift is clearly tracked as shown by the graph, where the simulated PMOS shift is the actual PMOS V_{TH} shift modeled in the MOSFETs for each TID level. This self-biased inverter can be used in combination with other monitoring circuits to extract various parametric shifts.

The outputs of the ring oscillator with a charge pump and the self-biased inverter have different profiles over TID, thus allowing the designer flexibility to choose among the two radiation monitoring circuit implementations to match the bias tuning needs of the functional power management circuits.

V. COMPLETE SYSTEM FOR BIAS TUNING

To correlate the radiation induced performance shifts in monitoring circuits to the bias tuning that is required to tune the circuits to pre-radiation performance, a look up table is generated using CAD based simulations at design time from Section III. The output of the radiation monitoring circuit is then scaled to match the bias tuning control value from the look up table. The complete system is shown in Fig. 15. A single radiation monitoring circuit can be used to control multiple functional circuits in order to reduce area and power

979-8-3503-4631-2/23 $31.00 © 2023 IEEE

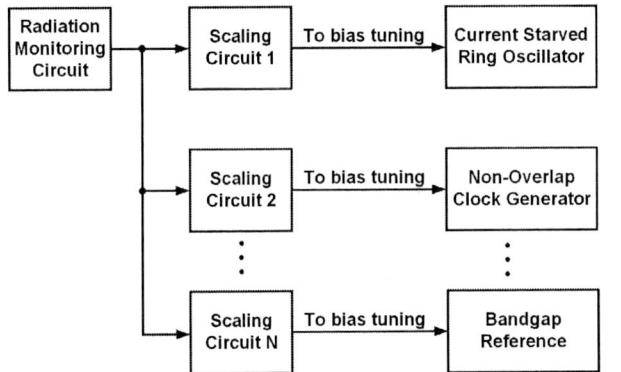

Fig. 15. A complete system for bias tuning using radiation monitoring circuits

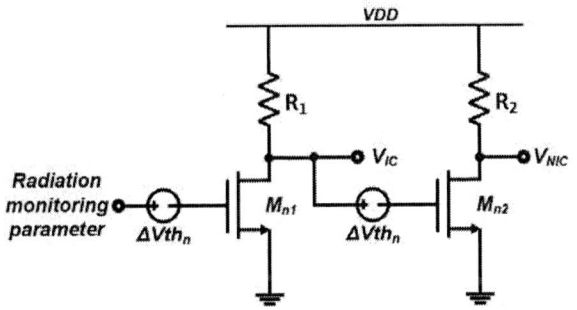

Fig. 16. An example of scaling circuit using minimum devices

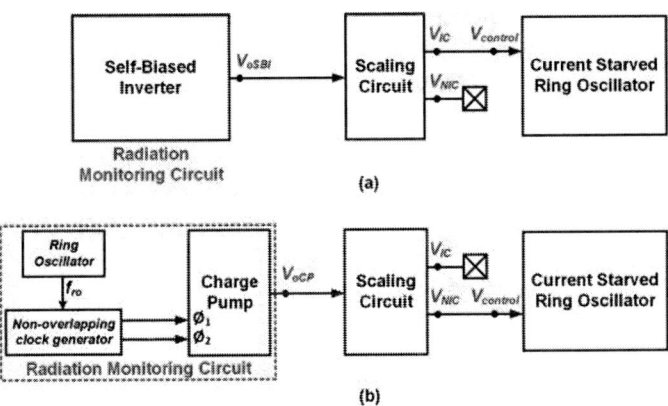

Fig. 17. Automatic bias-tuning system for TID induced parametric shifts with radiation monitoring circuit based on (a) self-biased inverter and (b) Ring oscillator with a charge pump

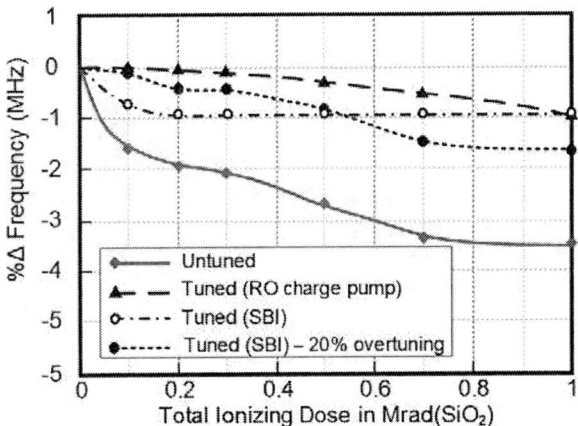

Fig. 18. Plot of un-tuned and bias-tuned ring oscillator frequency shift using the look up table method

ployed as the radiation monitoring circuit, the output of which is fed to the scaling circuit. The inverted output of the scaling circuit is used to tune the functional current starved ring oscillator. In the second method shown in Fig. 17(b), the ring oscillator based charge pump is used as the radiation monitoring circuit. The plots for the automatically tuned current starved ring oscillator frequency has been shown in Fig. 18 for both methods. The red solid curve shows the ring oscillator frequency without any bias tuning and the dotted curves show the output of the ring oscillators that take advantage of the autonomous bias tuning methods from Fig. 17(a) and Fig. 17(b). The plot also includes the tuned ring oscillator frequency using the Fig. 17(b) method, with 20% inaccuracy in the tuning value in order to demonstrate the robustness of the tuning method. Even with 20% inaccuracy in the tuning voltage, the shift in frequency is less than 1.8% of the pre-radiation value. Although 100% radiation hardening is not possible from this look-up table based method due to the small error margin from the non-linearity of the scaling and process variations, the presented methods of bias tuning are an essential step towards system radiation hardening by monitoring radiation induced parametric shifts in space technologies.

CONCLUSION

This paper analyzes the effects of MOSFET threshold voltage shifts due to TID on circuit performance parameters for critical analog PoL circuits. Three circuits are analyzed, and bias tuning is used in these three circuits to recover circuit performance over TID. Additionally, two radiation effects monitoring circuits are showcased that may be used to track TID-induced performance shifts in integrated power management circuits. Finally, two complete systems are demonstrated based on a look up table approach that automatically tunes a functional current starved oscillator to maintain performance over TID. One system uses the presented charge pump radiation monitoring circuit and the other system includes the self-biased inverter.

overhead. A separate scaling (interface) circuit is required for every PMIC circuit to match the bias tuning trend for that specific circuit design.

The scaling circuit can been realized by level shifting and tuning the slope of the radiation monitoring circuit's output. One such scaling circuit is illustrated in Fig. 16. The output V_{NIC} gives a level shifted voltage trend similar to Fig. 14 while the output V_{IC} gives an inverse voltage trend of the same plot. The performance of the scaling blocks is also affected by radiation induced threshold voltage shifts and these shifts have been taken into consideration for the simulations.

Two methods of automatically bias tuning the current starved ring oscillator have been implemented. In the first method shown in Fig. 17(a), the self-biased inverter is em-

REFERENCES

[1] T. Cook, A. Phillips, C. Siak, A. D. George and B. M. Grainger, "Evaluation of Point of Load Converters for Space Computational Loads," 2020 IEEE Aerospace Conference, 2020, pp. 1-12.

[2] T. R. Oldham and F. B. McLean, "Total ionizing dose effects in MOS oxides and devices," in IEEE Transactions on Nuclear Science, vol. 50, no. 3, pp. 483-499, June 2003.

[3] R. C. Lacoe, "Improving Integrated Circuit Performance Through the Application of Hardness-by-Design Methodology," in IEEE Transactions on Nuclear Science, vol. 55, no. 4, pp. 1903-1925, Aug. 2008.

[4] K. J. Shetler et al., "Radiation Hardening of Voltage References Using Chopper Stabilization," in IEEE Transactions on Nuclear Science, vol. 62, no. 6, pp. 3064-3071, Dec. 2015.

[5] N. Prokopenko, A. Bugakova, D. Denisenko, V. Chumakov and N. Butyrlagin, "Current Mirrors on Complementary Field-Effect Transistors with a Control PN Junction for Low-Temperature and Radiation-Hardened Analog ICs," 2021 IEEE East-West Design Test Symposium (EWDTS), 2021, pp. 1-6

[6] P. K. Samudrala, J. Ramos and S. Katkoori, "Selective triple Modular redundancy (STMR) based single-event upset (SEU) tolerant synthesis for FPGAs," in IEEE Transactions on Nuclear Science, vol. 51, no. 5, pp. 2957-2969, Oct. 2004.

[7] A. Dutta and N. A. Touba, "Multiple Bit Upset Tolerant Memory Using a Selective Cycle Avoidance Based SEC-DED-DAEC Code," 25th IEEE VLSI Test Symposium (VTS'07), 2007, pp. 349-354.

[8] S. Kumar and A. Mukherjee, "A Highly Robust and Low-Power Real-Time Double Node Upset Self-Healing Latch for Radiation-Prone Applications," in IEEE Transactions on Very Large Scale Integration (VLSI) Systems, vol. 29, no. 12, pp. 2076-2085, Dec. 2021, doi: 10.1109/TVLSI.2021.3110135.

[9] I. S. Esqueda and H. J. Barnaby, "Modeling the Non-Uniform Distribution of Radiation-Induced Interface Traps," in IEEE Transactions on Nuclear Science, vol. 59, no. 4, pp. 723-727, Aug. 2012.

[10] M. Gaillardin, V. Goiffon, S. Girard, M. Martinez, P. Magnan and P. Paillet, "Enhanced Radiation-Induced Narrow Channel Effects in Commercial 0.18 μ m Bulk Technology," in IEEE Transactions on Nuclear Science, vol. 58, no. 6, pp. 2807-2815, Dec. 2011.

[11] Jeffrey Prinzie, Michael Steyaert, Paul Leroux, " Radiation Hardened CMOS Integrated Circuits for Time-Based Signal Processing", Springer 2018.

[12] A. S. Cardoso et al., "Single-Event Transient and Total Dose Response of Precision Voltage Reference Circuits Designed in a 90-nm SiGe BiCMOS Technology," in IEEE Transactions on Nuclear Science, vol. 61, no. 6, pp. 3210-3217, Dec. 2014, doi: 10.1109/TNS.2014.2358078.

[13] Z. Chen, D. Ding, Y. Shan and Y. Dong, "Investigation of TID Effects on Subthreshold Bandgap Reference Circuits Fabricated in a SOI Process," 2018 International Conference on Radiation Effects of Electronic Devices (ICREED), Beijing, China, 2018, pp. 1-3, doi: 10.1109/ICREED.2018.8905099.

[14] H. Banba et al., "A CMOS bandgap reference circuit with sub-1-V operation," in IEEE Journal of Solid-State Circuits, vol. 34, no. 5, pp. 670-674, May 1999, doi: 10.1109/4.760378.

[15] A. Abuelnasr et al., "Self-Adjusting Deadtime Generator for High-Efficiency High-Voltage Switched-Mode Power Amplifiers," 2020 IEEE International Symposium on Circuits and Systems (ISCAS), 2020, pp. 1-5, doi: 10.1109/ISCAS45731.2020.9181166.

[16] L. Schirone and M. Macellari, "General purpose dynamic dead time generator," 2015 International Conference on Clean Electrical Power (ICCEP), 2015, pp. 529-533].

979-8-3503-4631-2/23 $31.00 © 2023 IEEE

Gerabaldi: A Temporal Simulator for Probabilistic IC Degradation and Failure Processes

Ian Hill
Department of Electrical and Computer Engineering
University of British Columbia
Vancouver, Canada
ianhill@ece.ubc.ca

André Ivanov
Department of Electrical and Computer Engineering
University of British Columbia
Vancouver, Canada
ivanov@ece.ubc.ca

Abstract—Wear-out reliability in integrated circuits is becoming an increasingly complex topic, with emerging high-reliability markets demanding stricter requirements, diverse workloads making stress characterization challenging, and sub-5nm device scaling aggravating variability in degradation processes. Efforts to tackle these complexities can benefit greatly from sophisticated techniques that effectively capture the variable and uncertain nature of semiconductor wear-out mechanisms. True-to-life stochastic modelling and computational Bayesian inference offer promising avenues in this pursuit but are difficult to leverage without a framework for specifying and evaluating wear-out models that capture this probabilistic information. We present a temporal wear-out simulator, *Gerabaldi*, as a foundation for enabling these statistical techniques for integrated circuit reliability engineering. The simulator introduces novel capabilities including layered stochastic parameter modelling, fully agnostic design enabling custom degradation model and stress test specifications, and wear-out model definition forms compatible with modern computational Bayesian inference frameworks. Here, we frame *Gerabaldi* within the context of existing wear-out analysis methods. We then present its key design features and two detailed example applications to illustrate the simulator's capabilities.

Index Terms—wear-out reliability, simulation, degradation physics, Bayesian inference, stochastic modelling

I. INTRODUCTION

The International Roadmap for Devices and Systems report for 2022 expresses continued concerns regarding potentially prohibitive costs of ensuring device reliability below 5nm [1]. At these scales, only a few atomic defects are needed to significantly affect device characteristics, increasing variability in degradation and failure mechanisms. Additionally, a recent survey we conducted identifies difficulties in meeting the increasingly diverse and strict sets of requirements needed for safety-critical functionality or extended operational lifetimes in emerging high-reliability markets [2]. Both of these concerns are statistical in nature, with incomplete physical understandings of atomic-level degradation mechanisms adding uncertainty challenges to the equation. A secondary identified impact of market diversification is the increasing variety of time-varying stress conditions that integrated circuits (ICs) are subjected to, such as brief exposures to extreme temperatures or shifting idle periods during weekly use cycles. These can be difficult to characterize and lead to situations where no

single degradation mechanism can be treated as dominant. This challenges assumptions of static stress test conditions and negligible degradation mechanisms widely used to simplify reliability prediction processes. Our research focuses on developing methods that can address these statistical challenges while not being subject to these assumptions, thereby allowing for wear-out qualification tests that can be customized to specific market requirements and allowing for lifespan estimates that explicitly account for predictive uncertainty.

We anticipate that computational Bayesian inference (CBI) techniques, now becoming feasible with modern computers, will be invaluable in tackling these challenges due to explicit consideration of the physical uncertainty and stochastic nature of IC wear-out processes. An early example of this potential is a recent work that used analytic Bayesian inference to address IC failure characterization [3]. CBI allows for similar solutions in the much broader class of analytically intractable problems. Unfortunately, these techniques are conceptually challenging and require changes in how we specify wear-out models. Our planned research in accelerated test design requires a framework that incorporates uncertainty/probabilistic information via wear-out models for CBI and allows us to simulate arbitrarily-specified accelerated wear-out tests; design exploration using physical testing is not practically feasible.

In this paper we present *Gerabaldi*, a temporal wear-out simulation framework for supporting sophisticated statistical exploration and analysis of IC reliability engineering problems. *Gerabaldi* is publicly available for use under an Apache 2.0 license [4] and introduces five novel features unavailable in existing wear-out simulators to the best of our knowledge:

1) Probabilistic wear-out and circuit model definitions for use in computational Bayesian inference frameworks;
2) Layered stochastic variability modelling of latent/hidden variables and stress conditions;
3) Temporal degradation simulation under time-varying stress conditions for a broader range of arbitrary model definitions;
4) Agnostic treatment of wear-out models, stress test specifications, and test environments;
5) Unbounded support for numerous stress conditions, competing failure mechanisms, degraded parameters, test devices, and test lots within a given simulation.

979-8-3503-4631-2/23 $31.00 © 2023 IEEE

TABLE I
FEATURE COMPARISON OF IC WEAR-OUT SIMULATORS

Simulator Type	Target Application	Mechanisms	Model Types	Simulation Type	Stress Support	Stochastic Injection
Gerabaldi	CBI-Based Test and Model Development	Degradation and Failure	Probabilistic and Deterministic	Temporal	Time-Varying	Layered Sources
Analog SPICE [5]–[11]	Reliability-Aware Analog Design	Degradation	Deterministic	Temporal	Time-Varying	Variability-Aware
Digital SPICE [12]–[14]	Post-Degradation Timing Analysis	Degradation	Deterministic	Temporal	Time-Varying	Variability-Aware
System MTTF [15]–[17]	Multi-Component Failure Analysis	Degradation and Failure	Deterministic	Direct to Failure Distributions	Constant	Model-Based

The remainder of this paper is organized as follows. Section II reviews related work, Section III discusses the design features of the simulator in detail, while Section IV presents example applications. Finally, Section V concludes and outlines plans for future work.

II. RELATED WORK

A. Existing Wear-Out Simulators

Gerabaldi's target purpose is to serve as a foundation for CBI techniques and as a tool for exploring accelerated wear-out test design, allowing for model and test comparisons in the context of Bayesian analysis. To be effective for these tasks, a simulator must support (i) probabilistic model forms across both degradation and hard failure mechanisms, (ii) arbitrary test definitions with potential for time-varying stress, and (iii) multi-source stochastic variation models. Many existing simulators that enable analysis of integrated circuit wear-out reliability are richly featured and backed by talented engineering teams. Thus, it is important to address why we propose a new solution based on these specific feature requirements. We surveyed existing simulators and simulation methods for IC wear-out reliability engineering and grouped them based on target application; a summary is shown in Table I.

The largest group consists of analog SPICE-based simulators primarily used to enable reliability-aware analog design techniques. Typically, an initial SPICE simulation is conducted for a target circuit, followed by a stress phase simulation which updates/degrades the circuit model. A second SPICE circuit simulation using the updated circuit model is then run to determine the degraded circuit performance. Mature simulators in this group incorporate detailed physical models, carefully avoid mathematical inaccuracies, and are performance-focused, helping designers identify and increase the maximum tolerable degradation before failure across process corners. Unfortunately, these applications rarely consider hard failure mechanisms and arbitrary test procedures, and the SPICE framework is not suitable for probabilistic modelling.

The second category is SPICE-based simulators for digital timing analysis. These are closely related to those for analog design, allowing designers to evaluate timing constraints as degradation proceeds. These simulators have similar feature sets to the analog SPICE-based simulators and are thus similarly unsuitable for developing CBI-compatible models.

A final category of simulators target system-level MTTF characterization. These are valuable analysis tools that can take in a single set of stress conditions and directly output expected failure times and distributions for complex systems with multiple components and mechanisms, thus avoiding computationally expensive temporal simulation and sampling. Discarding the temporal aspect makes time-varying stress simulations exceptionally difficult, however, and restricts the model space to empirical estimates of derived parameters.

B. Bayesian Inference for IC Reliability

Bayesian statistics and inference are exceptionally powerful methods for reasoning about uncertainty and for developing robust stochastic models of physical processes, but have historically been infeasible for most applications due to analytic intractability. Modern computing systems, however, have become sufficiently powerful in recent years to allow for computational methods of inference that circumnavigate the analytic intractability in similar reliability applications, such as fault identification within filter circuits [18]. Within IC reliability, a few recent papers have discussed the promise of, or developed specific applications of either analytic or computational Bayesian inference [3], [19], [20]. However, the field remains relatively unexplored. We hypothesize that this is due to both the conceptually challenging nature of CBI and the lack of tools compatible with Bayesian problem formulations, the latter of which *Gerabaldi* is designed to address.

Notably, recent works using standard machine learning techniques for wear-out prediction can be directly contrasted with CBI techniques by casting them to inference problems [21], [22]. Both techniques allow for latent/hidden variables (i.e., model parameters that cannot be directly measured) to be inferred from observed test data, thus allowing for wear-out predictions to be made. CBI offers two distinct benefits over other machine learning approaches, however. The first advantage is the explicit treatment of uncertainty, in that the resulting latent variables are probability distributions, allowing for direct analysis of the confidence that can be placed in a given prediction; and second, CBI conveniently allows for historical reliability knowledge to be leveraged through *informed priors* on the latent variables [19].

III. DESIGN

The core of *Gerabaldi* consists of three detailed data models for specifying degradable device models, physical environments, and stress/measurement specifications, along with a fully-featured program for executing tests.

A. Data Model and Execution Flow

A simplified structural diagram of the simulator data model is shown in Fig. 1. The blue arrows provide a rough representation of the execution flow between the three data models, and the purple-shaded items use the layered stochastic modelling strategy detailed in the following subsection. Although this diagram obscures the implementation complexity associated with the automatic management of latent variables, simulation state, and data handling, it is useful for conceptually understanding how the models interact.

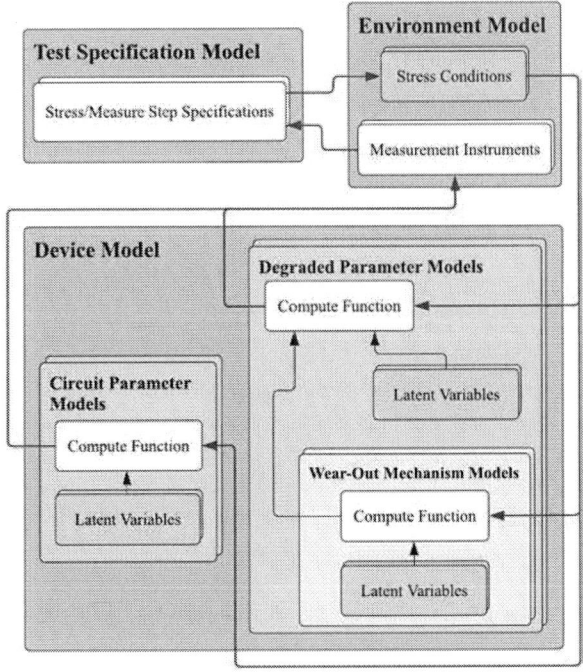

Fig. 1. Simplified data model structural diagram.

The *Test Specification Model* is centred around an arbitrarily-defined ordered sequence of steps, with each specifying either a stress phase under some stress conditions and duration or a measurement phase that defines the stress conditions to be applied during measurement along with the set of measurements to be collected. The *Environment Model* captures the non-ideal aspects of a physical test environment including an arbitrary set of stochastic models for different stress conditions (discussed in detail later) along with an arbitrary set of instrument models to simulate imperfect measurements of different parameters. Finally, the *Device Model* consists of an arbitrary set of degradable device parameters and any higher-level circuit parameters that are dependent on degraded parameters. The degradable parameters contain an arbitrary set of degradation and/or failure mechanism models along with information on the initial parameter values, the instantaneous stress dependence of the parameter, and how the mechanism models are combined to produce the output parameter value. All circuit, degradable parameter, and mechanism models use probabilistic latent variables. This allows for CBI techniques to be applied at any hierarchical level of

the *Device Model*, from reasoning about individual wear-out mechanisms and their effects, to reasoning about compound circuit parameters affected by numerous underlying degraded parameters affected by possibly many wear-out mechanisms. Note that the form of the "Compute Function" seen in Fig. 1 typically corresponds to the mathematical equation(s) seen in common semiconductor degradation mechanism models or solutions for circuit parameters. Additionally, however, these can be generic software functions of arbitrary complexity. This flexibility allows for expressive, programmatic models to be readily implemented (e.g., the gate oxide model in the second example application shown in Section IV).

Following the construction of the three described data models, wear-out tests can be simulated through a single program call. First, an initial physical state is generated from the *Device Model* based on the number of devices, chips, and lots required by the *Test Specification Model*. This test is then executed in sequence, with each step either carrying out a stress phase to degrade the device or conducting a set of device parameter measurements. Note that all physical simulation state information is stored/maintained externally to the data models from a software perspective, i.e., is compartmentalized within the described procedure call. This enables easy execution parallelism as multiple simultaneous test simulations can share device, test specification, and/or environment models. Additionally, input checking and error reporting tasks are performed at run time to aid in simulation debug.

B. Layered Stochastic Stress and Latent Variable Modelling

Two feature requirements informed the design of the stochastic model used for latent variables and stress conditions: (i) support for multi-source stochastic influences, and (ii) CBI compatibility. For the latter requirement, fitted/latent variable values within wear-out and parameter models must be described in terms of probability distributions as opposed to deterministic quantities with added stochastic variability. This allows device models to be compiled/translated for use in CBI software frameworks.

In semiconductor processes, physical parameters may vary due to minute differences across a single chip, as well as through imperfections in fabrication processes at the wafer and lot levels. Examples may include slight differences in chemical exposure times or equipment calibrations resulting in variations shared among all parameters on a given die or within a specific lot. Within the developed stochastic/probabilistic model, variability is divided into three layers: *Device-*, *Chip-*, and *Lot-Level*. The software class model implementing this layered variability approach and CBI compatibility is summarized in Fig. 2. To generate variable/stress values, samples from the distributions at each level are combined as a sum or product into a 3D array where each dimension represents one variability layer. A consequence of this approach is that when compiled for CBI, each latent variable is functionally represented as a combination of up to three probability distributions.

979-8-3503-4631-2/23 $31.00 © 2023 IEEE

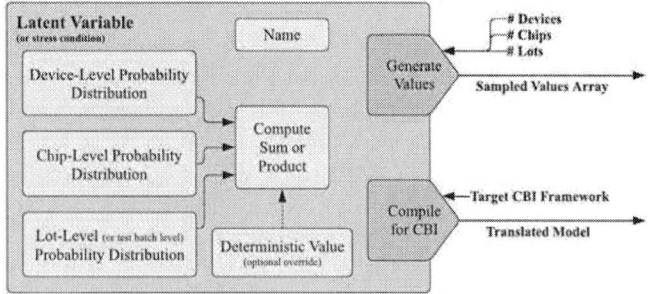

Fig. 2. Visual class diagram for the layered probabilistic model used by latent variables and stress conditions.

For simulations where *Chip-Level* and/or *Lot-Level* variability is not always known, users can leave these layers unspecified for any/all models and *Gerabaldi* will automatically adjust computations and CBI compilation based on which layers are provided. Finally, deterministic variable/stress definitions that override the layered stochastic model are supported for use cases where not all elements within a simulation need to be probabilistic. The only functional limitation of doing so is that error flags will be raised if a user tries to compile deterministic latent variables for CBI.

C. Time-Varying Stress Simulation

Gerabaldi supports both degradation and time-invariant failure mechanism models, with the latter being comparatively easy to implement for time-varying stress simulations. To simulate degradation processes under time-varying stress, we make the physical assumption that a system's future rate of degradation depends on the system's current physical state but *not* on how that state was arrived at. This assumption is implicitly used in standard wear-out lifetime extrapolation, testing and simulation practices. Here we have simply made it explicit. Under this assumption, we can leverage an equivalent time calculation method (also used by Lu *et al.* [23], Guo *et al.* [5], and Huang *et al.* [24] in specific application contexts) that permits the simulation of time-varying stress scenarios for arbitrary degradation models. A given model must obey three constraints, however: (i) that the degradation be strictly monotonic w.r.t. time for any used set of stress conditions; (ii) that for all used sets of stress conditions, the computed degradation values at time $t = 0$ must be identical; and (iii) that the computed values as $t \rightarrow \infty$ also all be identical.

We next walk through this method for the inductive base case where our set of stress conditions changes only once during the course of the test. We denote such a model by $D(t, X)$, where t is the elapsed duration of the stress and X is the set of values corresponding to some stress conditions. The objective is to compute the final degradation y_f as per the expression

$$D(t_f, X_v) = y_f \mid X_v = \begin{cases} X_0, t \leq t_{sw} \\ X_1, t > t_{sw} \end{cases} \quad (1)$$

where t_f is the total test duration and t_{sw} denotes the specific instant in time at which the (time-varying) stress conditions, X_v, switch/jump from values X_0 to X_1. The degradation y_f cannot be calculated directly, as our model is only defined for single values within X. Instead, we compute y_f indirectly by determining an equivalent *constant* stress test. To to so, first the intermediate degradation y_{int} is calculated directly:

$$D(t_{sw}, X_0) = y_{int} \quad (2)$$

Then, the equivalent time t_{equiv} needed to reach the degradation y_{int} under the second set of stress values X_1 in the equality

$$y_{int} = D(t_{equiv}, X_1) \quad (3)$$

is solved for using numerical optimization. Finally, the full degradation after the second interval of stress t_{sw} to t_f can be computed directly from the following:

$$D(t_{equiv} + (t_f - t_{sw}), X_1) = y_f = D(t_f, X_v) \quad (4)$$

This method generalizes inductively for multiple changes in stress by substituting the directly-calculated y_{int} in subsequent shifts/jumps with the degradation y_f from the base case or prior stress phases. Our earlier physical assumption ensures the equality in Equation 4, and the three model constraints ensure a valid solution exists for t_{equiv}. We extend this method to loosen the model constraints to non-strict monotonicity and non-identical degradation as $t \rightarrow \infty$ across the used stress condition space by defining the rate of degradation to be zero when the magnitude of degradation y_{int} can never be obtained under stress conditions X_1, and explicitly using the minimum valid t_{equiv} if multiple solutions exist. These small enhancements allow *Gerabaldi* to additionally simulate time-varying stress for the broad class of models that exhibit hard degradation saturation limits and/or stress threshold behaviours (e.g., wear-out mechanisms associated with phase changes, i.e., cases where no degradation occurs unless the temperature stress approaches a melting point).

D. Implementation Considerations

Gerabaldi is written in Python to improve compatibility with popular CBI frameworks such as Pyro and PyMC [25], [26]. The program interface is contained within the language and thus a basic level of Python programming experience is necessary to use the simulator. A graphical interface for non-programmers is slated as a potential item for development.

Simulation sizes in *Gerabaldi* can scale arbitrarily but are subject to tolerable run times; the time complexity is proportional to the product of the various scaled parameters (e.g., sample sizes, test complexity, model complexity). Future performance optimization via vectorized computation and parallel processing is feasible and could realize drastic performance increases due to the computational independence of simulating parameter degradation and measurement within each test step.

IV. EXAMPLE APPLICATIONS

We illustrate the unique capabilities of *Gerabaldi* by conducting illustrative reliability analyses that highlight the simulator's novel aspects. Adapting the following analyses for real process data and increased sampling sizes could readily be done; corresponding demonstration scripts within the simulator repository can be used to recreate the examples [4].

A. Pre-Silicon Degradation Variability Study

For our first example, we considered the pre-silicon preparation for an upcoming line of products in a newly-developed fabrication process. Specifically, we investigate the impact of this new process on an amplifier design subject to bias temperature instability (BTI) and hot carrier injection (HCI) degradation under temperature and voltage stress. In our scenario, we assume that foundry information indicates that in the new process, transistor threshold voltages should show twice the lot-to-lot variability in comparison to that found in the current process. In addition, we should expect wafer-to-wafer variability in the voltage dependence of BTI and HCI to double. The presumed goal in this illustrative use of *Gerabaldi* is to assess how the increased variability may impact accelerated wear-out test results.

Our *Device Model* consists of a basic small-signal common source amplifier gain equation with threshold voltages subject to BTI and HCI. We used probabilistic versions of a JEDEC BTI model [27] and an HCI model from Takeda and Suzuki with an added term for some negative temperature dependence [28]; both of these models conveniently have a voltage dependence proportional to a latent variable α. Distributions for the model latent variables have been inferred for the current process, and the new process is modelled by adjusting the *Chip-Level* variation distribution within the α latent variables in each mechanism model and the *Lot-Level* distribution in the initial value model for our threshold voltage. We constructed three *Test Specification Models* to see the increased variability impact on JEDEC's high temperature operating life (HTOL) and low temperature operating life (LTOL) tests [29], along with a custom stress-to-failure test that cycles 90 hours of stress with a 10 hour idle interval, ramping up the stress intensity at each cycle. Finally, we introduced minor variations in temperature and supply voltage during our tests through the application of *Gerabaldi*'s *Environment Model*.

The three tests were simulated with 1000 amplifiers sampled from devices spread across 10 lots with 10 wafers each for both the current and new processes. This required a total of 7.5 minutes to execute on a single core of an Intel© i5-12400 processor. The temporal simulation results for the amplifier gain as a function of time and stress conditions are shown in Fig. 3. The solid traces represent the mean degradation for each simulated test, while shaded regions show the range of values between the minimum and maximum observed samples at each step. The horizontal line indicates the gain at which an amplifier is deemed to have failed based on some arbitrary design specification.

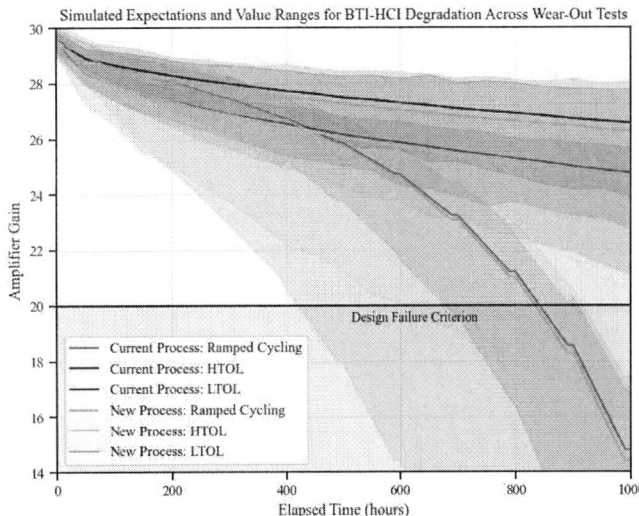

Fig. 3. Simulated amplifier gain degradation for different wear-out tests and fabrication processes.

In this example our amplifier minimum functional gain was set to be 20, against which we can see the increased potential for HTOL and LTOL qualification test failures in Fig. 4. Although the simulated mean degradation values for the qualification tests (the centre line of each rectangle) are similar for both processes and well above the failure threshold, increased outlier samples in the new process aggravate the non-linear amplifier dependence on the transistor threshold voltages. The result is 1% of our samples failing after HTOL stress. This type of simulation could provide an early indicator of potential qualification challenges and help in identifying pre-emptive actions to reduce the risk of test failures.

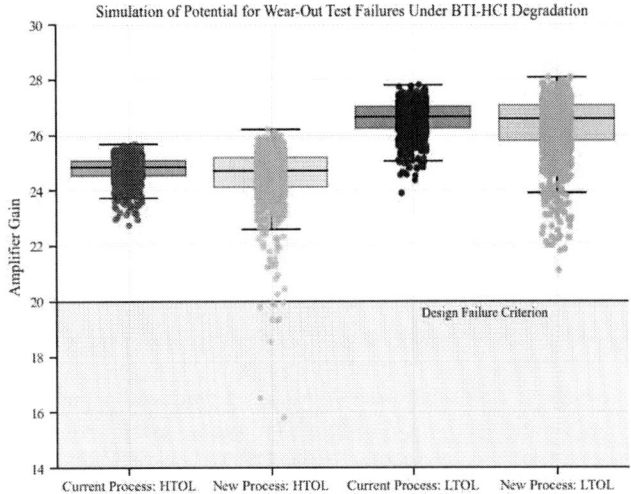

Fig. 4. Post-test gain distributions across fabrication processes.

B. Failure Model Basic Sensitivity Analysis

As a second example application, we considered the early use of a novel gate oxide time dependent dielectric breakdown (TDDB) model. We introduce a "toy" model for the 2-D

979-8-3503-4631-2/23 $31.00 © 2023 IEEE

transistor dielectric such that it be comprised of four columns, each three widths/rows thick, for a total 12 possible defect locations. We treat the transistor as failed once a defect forms in each width of the same column, representing a conductive path from gate to channel. This behaviour is implemented as a programmatic model that comprises the compute function (refer to Fig. 1) of our degraded gate oxide parameter model. Defect formation in each location for this toy model is governed by a time-independent, stochastic process in which the per-unit time probability of a defect forming is calculated based on the applied stress according to the following:

$$formation\ probability = k(\frac{T^\alpha}{\gamma})(\frac{V_{gs}}{\beta}) \qquad (5)$$

where T is temperature, V_{gs} is the applied gate-to-source voltage, and k, α, γ, and β are some arbitrary latent variables. The overall probability of a defect forming during a stress phase is then obtained by scaling the stress-dependent probability according to stress time, and random sampling is used to determine whether a defect formed or not. Note that this model is not proposed as an actual TDDB model for consideration. We designed it here to highlight the flexibility and dynamic stress compatibility of *Gerabaldi* for directly simulating TDDB failures in time. This is in contrast to simulators built for models that output derived quantities such as lifetime expectations.

Now, further suppose we have some experimental reliability data and have performed Bayesian inference to condition our model on this data. Say the width of the 95% likelihood region of the variable distribution for α were to be 1.05 to 1.15. Assuming this range of values would be of concern due to the strong mathematical influence on the defect formation probability, one could envisage needing to assess how the expected field-use TDDB failure rates would shift as a result.

We establish a scenario where we would have characterized user load cycles for a set of products, and have determined that typical users of such products follow an approximate weekly cycle consisting of 6 hours of moderate use each day, 2 hours of intensive use, and 16 hours of idle time, except for weekends which have 21 hours of idle and 3 hours of intensive use per day. We would also have characterized stochastic variation models for the voltage and temperature stress conditions experienced by each transistor.

Under the scenario above, we used *Gerabaldi* to simulate the three stress phases aggregated for each week across 20 years of our expected user load for 5 uniformly-spaced α values within our 95% likelihood region (fixing our other model variables), checking for TDDB failures once a week across 1000 gate oxide samples (requiring a total of 4.4 hours on five cores of an Intel© i5-12400 processor). The Python library "reliability" [30] was used to quickly fit the simulated gate oxide failures to Weibull distributions, shown in Fig. 5, and analyzed the worst-case 95% confidence interval estimate of expected time to 1% of transistors failing. Linearizing this expectation against shifts in α, we get a spread of ~2.4 years. This prediction could be concerning considering the extremely

poor overall TDDB reliability indicated by the conditioned model. For a realistic process where transistors rarely fail within a 10-year time span, simulation times would need to be increased significantly to obtain a reasonable Weibull fit.

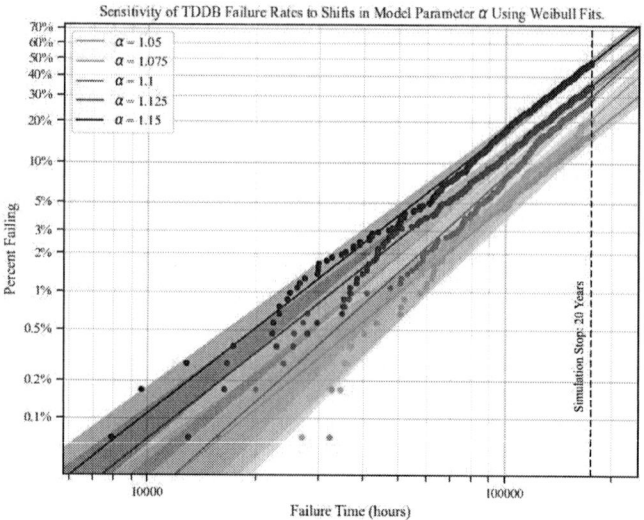

Fig. 5. Simulated TDDB failures for different values of latent variable α.

V. Conclusions and Future Work

We introduced a temporal wear-out simulator, *Gerabaldi*, that enables rich wear-out reliability analyses for a wide range of problems via novel features for generic probabilistic modelling with multiple sources of variability and time-varying stress simulation of arbitrary wear-out mechanisms. The illustrative simulations conducted would be infeasible without the simulator's unique ability to support custom non-algebraic models, independently specify chip and lot-level variability sources within semiconductor fabrication processes, and simulate arbitrarily complex wear-out stress tests.

Gerabaldi is now at a state sufficient for our planned research into CBI applications for semiconductor reliability. There are valuable additional features we plan to add in the future, however, for greater ease of use, decreased run times, and explicit handling for recovery processes. Further enhancements will be based potential user needs, and we encourage readers to contact us regarding desired improvements.

Acknowledgment

This work was supported in part by the Natural Sciences and Engineering Council of Canada (NSERC) through its Alliance Program and in part by the Huawei–UBC Joint Lab. The authors wish to thank Elmira Nezamfar and Mateo Rendon for their help with proofing.

References

[1] IRDS, "International roadmap for devices and systems™ 2022 edition: Metrology," IEEE, Tech. Rep., 2022.

[2] I. Hill, P. Chanawala, R. Singh, S. A. Sheikholeslam, and A. Ivanov, "CMOS reliability from past to future: A survey of requirements, trends, and prediction methods," *IEEE Transactions on Device and Materials Reliability*, vol. 22, no. 1, pp. 1–18, 2022.

979-8-3503-4631-2/23 $31.00 © 2023 IEEE

[3] W.-T. K. Chien, A. Chung, and W. Kuo, "Fast semiconductor reliability assessments using SPRT," *IEEE Transactions on Reliability*, vol. 68, no. 2, pp. 526–538, Jun. 2019.

[4] I. Hill, "Gerabaldi," GitHub repository, 2023. [Online]. Available: https://github.com/ianrmhill/gerabaldi

[5] S. Guo *et al.*, "Towards reliability-aware circuit design in nanoscale Fin-FET technology: — New-generation aging model and circuit reliability simulator," in *2017 IEEE/ACM International Conference on Computer-Aided Design (ICCAD)*, Nov. 2017, pp. 780–785, iSSN: 1558-2434.

[6] A. Schaldenbrand, "Analog reliability analysis for mission-critical applications," Cadence Design Systems, San Jose, CA, USA, Tech. Rep. 11038, Aug. 2019.

[7] B. Tudor, J. Wang, W. Liu, and H. Elhak, "MOS device aging analysis with HSPICE and CustomSim," Synopsys, Mountain View, CA, USA, Tech. Rep. 08/11.AP.CS753, Aug. 2011.

[8] M. Selim, E. Jeandeau, and C. Descleves, "Addressing IC reliability issues using Eldo," Siemens Digital Industries Software, Plano, TX, USA, Tech. Rep. 81656-C, 2016.

[9] E. Afacan, G. Berkol, G. Dündar, A. E. Pusane, and F. Başkaya, "A deterministic aging simulator and an analog circuit sizing tool robust to aging phenomena," in *2015 International Conference on Synthesis, Modeling, Analysis and Simulation Methods and Applications to Circuit Design (SMACD)*, Sep. 2015, pp. 1–4.

[10] S. Han, B.-S. Kim, and J. Kim, "Variation-aware aging analysis in digital ics," *IEEE Transactions on Very Large Scale Integration (VLSI) Systems*, vol. 21, no. 12, pp. 2214–2225, 2013.

[11] E. Maricau and G. Gielen, "Efficient variability-aware NBTI and hot carrier circuit reliability analysis," *IEEE Transactions on Computer-Aided Design of Integrated Circuits and Systems*, vol. 29, no. 12, pp. 1884–1893, 2010.

[12] T. Liu, C.-C. Chen, S. Cha, and L. Milor, "System-level variation-aware aging simulator using a unified novel gate-delay model for bias temperature instability, hot carrier injection, and gate oxide breakdown," *Microelectronics Reliability*, vol. 55, no. 9, pp. 1334–1340, 2015.

[13] J. B. Velamala, K. B. Sutaria, V. S. Ravi, and Y. Cao, "Failure analysis of asymmetric aging under NBTI," *IEEE Transactions on Device and Materials Reliability*, vol. 13, no. 2, pp. 340–349, 2013.

[14] A. Jafari, M. Raji, and B. Ghavami, "Impacts of process variations and aging on lifetime reliability of flip-flops: A comparative analysis," *IEEE Transactions on Device and Materials Reliability*, vol. 19, no. 3, pp. 551–562, 2019.

[15] J. Suan *et al.*, "RelSim: Computational framework for lifetime reliability assessment of heterogeneous accelerator systems," SSRN, Sep. 2022, preprint. [Online]. Available: https://papers.ssrn.com/abstract=4198756

[16] B. Wan, Y. Wang, Y. Su, and G. Fu, "Reliability evaluation of multi-mechanism failure for semiconductor devices using physics-of-failure technique and maximum entropy principle," *IEEE Access*, vol. 8, pp. 188 154–188 170, 2020.

[17] K. Yang, T. Liu, R. Zhang, and L. Milor, "A comprehensive time-dependent dielectric breakdown lifetime simulator for both traditional CMOS and FinFET technology," *IEEE Transactions on Very Large Scale Integration (VLSI) Systems*, vol. 26, no. 11, pp. 2470–2482, 2018.

[18] W. Harvey, A. Munk, A. G. Baydin, A. Bergholm, and F. Wood, "Attention for inference compilation," *CoRR*, vol. abs/1910.11961, 2019. [Online]. Available: http://arxiv.org/abs/1910.11961

[19] C. M. Tan, "Overview of reliability engineering," in *Theory and Practice of Quality and Reliability Engineering in Asia Industry*. Singapore: Springer, 2017, pp. 3–23.

[20] K. Date and Y. Tanaka, "Quality-oriented statistical process control utilizing bayesian modeling," *IEEE Transactions on Semiconductor Manufacturing*, vol. 34, no. 3, pp. 307–311, Aug. 2021.

[21] C. Yang and F. Feng, "Multi-step-ahead prediction for a CMOS low noise amplifier aging due to NBTI and HCI using neural networks," *Journal of Electronic Testing: Theory and Applications (JETTA)*, vol. 35, pp. 797–808, 12 2019.

[22] Y. Zhao and H. G. Kerkhoff, "A genetic algorithm based remaining lifetime prediction for a VLIW processor employing path delay and IDDX testing," in *2016 International Conference on Design and Technology of Integrated Systems in Nanoscale Era (DTIS)*, 2016, pp. 1–4.

[23] Y. Lu, L. Shang, H. Zhou, H. Zhu, F. Yang, and X. Zeng, "Statistical reliability analysis under process variation and aging effects," in *2009 46th ACM/IEEE Design Automation Conference*, 2009, pp. 514–519.

[24] K. Huang, X. Zhang, and N. Karimi, "Real-time prediction for IC aging based on machine learning," *IEEE Transactions on Instrumentation and Measurement*, vol. 68, no. 12, pp. 4756–4764, 2019.

[25] E. Bingham *et al.*, "Pyro: Deep universal probabilistic programming," *J. Mach. Learn. Res.*, vol. 20, pp. 28:1–28:6, 2019. [Online]. Available: http://jmlr.org/papers/v20/18-403.html

[26] J. Salvatier, T. V. Wiecki, and C. Fonnesbeck, "Probabilistic programming in Python using PyMC3," *PeerJ Computer Science*, vol. 2, p. e55, Apr. 2016. [Online]. Available: https://doi.org/10.7717/peerj-cs.55

[27] JC-14, "JEP122H: Failure mechanisms and models for semiconductor devices," JEDEC, Sep. 2016.

[28] E. Takeda and N. Suzuki, "An empirical model for device degradation due to hot-carrier injection," *IEEE Electron Device Letters*, vol. 4, no. 4, pp. 111–113, 1983.

[29] JC-14, "JESD22-A108F: Temperature, bias, and operating life," JEDEC, Jul. 2017.

[30] M. Reid, "Reliability: v0.8.6," Aug. 2022. [Online]. Available: https://doi.org/10.5281/zenodo.6969409

Diagnosis of Quantum Circuits in the NISQ Era

Yu-Min Li, Cheng-Yun Hsieh, Yen-Wei Li, and James Chien-Mo Li

Graduate Institute of Electronics Engineering, National Taiwan University, Taipei 10617, Taiwan

{r09943094, d08943012, r10943147, cmli}@ntu.edu.tw

Abstract—**Currently, noisy intermediate-scale quantum (NISQ) circuits may not always generate correct outputs due to noise and faults. In this work, we propose a diagnosis technique for NISQ circuits. The proposed technique contains static diagnosis and dynamic diagnosis. Static diagnosis uses a fault dictionary that contains output probability distribution for each fault. Dynamic diagnosis uses binary search to find the accurate fault locations of faulty quantum circuits. We demonstrate our technique using the Qiskit simulator with realistic noise models. We evaluate 15 benchmarks with unitary and non-unitary faults injected. Simulation results show that the average accuracy and resolution are 97.70% and 1.81. Experiments on the IBM Q devices have also been performed, and results show that our technique is feasible on real quantum circuit devices.**

Index Terms—**noisy intermediate-scale quantum, quantum circuit, fault diagnosis**

I. INTRODUCTION

Quantum circuits are becoming a promising computing technology. Quantum circuits can be very useful in various key applications, including machine learning, optimization, and quantum chemistry. Recently, quantum circuits have made significant development. There has been numerous superconducting hardware for quantum circuits developed, including Google's 53-qubit chip (*Sycamore* [1]), Intel's 49-qubit chip (*Tangle Lake* [2]), and IBM's 127-qubit chip (*Eagle* [3]). According to IBM's roadmap, a superconducting quantum computer with over 1,000 qubits could be available in 2023 [4].

Despite the fast progress in quantum circuit development, quantum circuits are vulnerable to *errors* in the *noisy intermediate-scale quantum* (NISQ) era [5]. Currently, quantum circuits are suffering from errors, and thus the output of quantum circuits may not be correct. There are mainly two sources of errors: *noise* and *faults*. In this paper, we distinguish them as follows. Noise is induced by the environment randomly, such as quantum decoherence. However, faults are induced by hardware defects or control imperfections. When executing a quantum circuit, noise changes with time but faults do not. Therefore, fault diagnosis is critical to fix root causes of systematic problems in quantum circuits. This diagnosis research aims to identify the fault location in quantum circuits. We list some applications of diagnosing quantum circuits. For the manufacturing aspect, diagnosis helps fix manufacturing defects. For platform providers such as IBM, diagnosis helps system maintenance and calibration. For users, they can identify the faulty sites or faulty gates and avoid using them. In this paper, we focus on diagnosing quantum circuits in the presence of faults, especially those systematically occur in quantum circuits.

There are some past research works related to diagnosis of quantum circuits [6] [7] [8]. However, these past research works did not consider noise, which is a serious problem in the NISQ era. The proposed technique in [8] requires input states for diagnosing quantum circuits. The input states require extra gates applied after the initial state, which may not be suitable in the NISQ circuits. In addition, they did not show effectiveness on *non-unitary faults*, which frequently occur in the NISQ era. Finally, they did not apply their work to real quantum circuit devices to demonstrate their effectiveness.

In this paper, we propose a diagnosis technique for quantum circuits, which can identify the fault locations of faulty quantum circuits. Our proposed technique contains two procedures: *static diagnosis* and *dynamic diagnosis*. Static diagnosis uses fault dictionary to find the similarity between outputs of *circuit under diagnosis* (CUD) and outputs of the fault dictionary. Dynamic diagnosis uses binary search to find the fault locations. Finally, our diagnosis technique ranks *suspects* in faulty quantum circuits according to their likelihood.

The advantages of our proposed diagnosis technique are described as follows: First, our static diagnosis is based on fault dictionary so that no iterative diagnosis is needed. We can perform diagnosis with the target model and report the ranked suspects quickly. Second, our dynamic diagnosis uses binary search to identify the fault location accurately, even when fault sizes are larger than our assumption. Those faults not in the dictionary, such as non-unitary ones, can also be diagnosed.

We evaluate 15 benchmarks with injected unitary and non-unitary faults. The diagnosis simulation results show that the average accuracy and resolution are 97.70% and 1.81. We also perform static diagnosis experiments on the IBM Q device, and the results show that our technique can be feasible on real quantum circuit devices.

The rest of this paper is organized as follows. Section II introduces quantum circuit concepts, fault modeling of quantum circuits, and past research for quantum circuits diagnosis. Section III explains our diagnosis flow in detail. Section IV shows the experimental results on simulation and real quantum circuit devices. Finally, Section V concludes this paper.

II. BACKGROUND

A. Quantum Circuit Concepts

The quantum state $|\psi\rangle$ of an n-qubit quantum circuit (QC) can be expressed as [9]:

$$|\psi\rangle = \sum_{i=0}^{2^n-1} c_i |i\rangle$$

979-8-3503-4631-2/23 $31.00 © 2023 IEEE

, where c_i is a complex number representing the amplitude of the i_{th} basis state $|i\rangle$. After measuring the QC, the quantum state will collapse into one of these 2^n states. The probability of observing $|i\rangle$ after measurement is $|c_i|^2$, and the summation of $|c_i|^2$ for all i equals 1.

A QC can be divided into *layers*. A layer contains a set of gates operated at the same time on disjoint qubits. To obtain the result of a QC, we need to execute the QC several repetitions. The number of repetitions is also referred to as the number of *shots*. Assumed we execute a 2-qubit QC by 1,000 shots. If the number of output 00, 01, 10, and 11 are respectively 488, 489, 12, and 11, then the *output probability distribution* (OPD) is (0.488, 0.489, 0.012, 0.011).

B. Fault Modeling of QC

The *scope* of a fault model determines how many gates are affected by a fault. In this work, we define two scopes: *qubit-specific faults* and *gate-specific faults*. A gate-specific fault only occurs on a single gate. A qubit-specific fault occurs on a specific qubit for all gates of a specific type. In previous work [10], the scope of fault models is qubit-specific, which is still not sufficient. The assumption of qubit-specific faults is based on superconducting QC, whose gates are applied by control signals. The assumption of gate-specific faults is based on photonic QC, whose gates are applied by physical elements. In this work, we consider both scopes: gate-specific faults and qubit-specific faults.

A *unitary fault* means that the fault behavior can be modeled by unitary matrices. The ratio faults cause the parameter to decay by a certain ratio. The bias faults cause the parameter with added bias. The truncation faults truncate the parameter with a certain value. For *CNOT* gates, one unitary fault model was proposed by [10]. A $U(\theta, \phi, \lambda)$ gate is injected before the control qubit of *CNOT*, and another $U(\theta', \phi', \lambda')$ gate is injected after the target qubit of *CNOT*. The parameters $[\theta, \phi, \lambda, \theta', \phi', \lambda']$ represents the fault size of *CNOT* fault.

For the purpose of diagnosis, unitary fault models are still insufficient. In this work, we also verify our diagnosis using non-unitary faults. A non-unitary fault means that unitary matrices cannot model the fault behavior. Please see Section IV for more details.

C. Past Research for QC Diagnosis

A fault diagnosis method based on *binary tomographic test* for QC is proposed in [7], which can detect and diagnose faults on phase gates. However, the method of slicing a QC may not be feasible. Inserting new gates in the QC under diagnosis may change the behavior of the original fault in QC. Another binary tomographic test method for diagnosing faulty QC has been proposed [8]. They determined appropriate input state vectors and measurement operators for diagnosing faulty QC. However, the input state of QC is initialized to the ground state. Moreover, those past research only consider noise-free QC, which is not practical for NISQ.

Randomized benchmarking (RB) is a helpful approach for calculating average error rates in QC [11]. RB analyzes the error rate of both single-qubit gates and two-qubit gates by applying random gates. RB is beneficial for identifying the errors on a certain qubit or a certain connection between two qubits. However, RB cannot identify that a fault occurs on a single gate or a specific gate type.

So far, there is still not enough research on fault diagnosis for QC in the NISQ era. In this work, we propose a diagnosis technique to find the accurate fault locations of faulty QC. Our proposed technique focuses on diagnosis of noisy QC in the NISQ era.

III. PROPOSED TECHNIQUE

Fig. 1 is the overall flow of our techniques. The overall flow contains two phases: the dictionary generation phase and the diagnosis phase. The dictionary generation phase generates dictionaries, which will be used in the diagnosis phase. Input to the dictionary generation phase is the CUD. The dictionary generation phase contains two procedures: *Fault Dictionary Generation* (A in Fig. 1), and *Partial QC Dictionary Generation* (C in Fig. 1). Outputs of the dictionary generation phase are the fault dictionary and the partial QC dictionary. The former is for faulty QC, and the latter is for fault-free QC.

The diagnosis phase reports the suspects of CUD. The diagnosis phase contains two procedures: *Static diagnosis* (B in Fig. 1), and *Dynamic diagnosis* (D in Fig. 1). We perform procedure (B) to identify the suspects of CUD. If the results of procedure (B) are acceptable, they are reported as suspects. Otherwise, we perform procedure (D), dynamic diagnosis. Static diagnosis takes less runtime than dynamic diagnosis, but the diagnosis resolution of the former can be worse than the latter. We perform procedure (D) to enhance the diagnosis resolution only when necessary. Outputs of the diagnosis phase are suspects identified by these two procedures. We introduce procedures A to D in the following subsections.

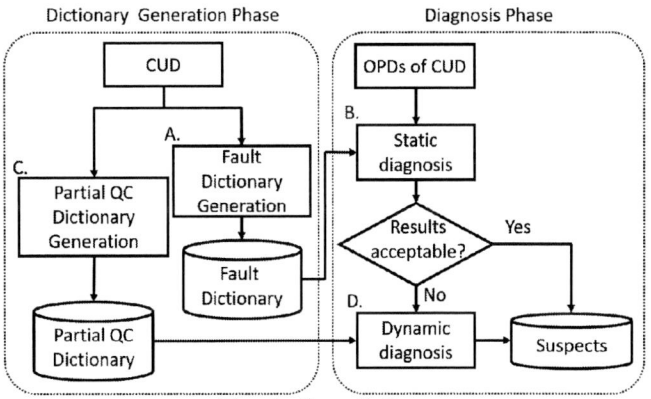

Fig. 1: Overall flow

A. Fault Dictionary Generation

In procedure (A), we generate the fault list and construct the fault dictionary. The fault dictionary contains many entries of OPD_f, which represents the OPD of a faulty QC with fault f. We generate a fault list based on the CUD and the user-specified fault model. For a given fault f, we perform many shots of fault simulations to obtain an OPD_f. A fault dictionary contains a complete OPD_f for all testable faults. Please note that

that we remove the untestable faults from the fault list because they cannot be distinguished from a good (fault-free) circuit. A fault is *untestable* if its quantum state is the same as that of the good circuit.

Consider a 2-qubit QC as an example. For demonstration purpose, we generate a small fault list containing only four faults: f_1, f_2, f_3 and f_4. Fig. 2 illustrates the fault dictionary of this example fault list, where OPD_{f1} to OPD_{f4} represents OPD_f of faults f_1 to f_4, respectively. Each OPD_f is expressed as four numbers in a pair of parenthesis. Four numbers in each pair of parenthesis are summed up to one.

Typically, the measurement of QC is performed on the z-axis only. However, the information of a QC is insufficient for diagnosis if we only measure on the z-axis. Therefore, in this work, we propose to measure the QC on the x-axis and y-axis as well. Each axis has its own fault dictionary, so there are three fault dictionaries constructed by this procedure. The generated fault dictionary will be used by the static diagnosis.

B. Static Diagnosis

Procedure (B) identifies the suspect faults based on a given fault dictionary. This is a *static* procedure because no iterative testing is needed. Let OPD_{CUD} denote the OPD of a CUD. Given the fault dictionary and the OPD_{CUD}, we calculate the distance between OPD_{CUD} and OPD_f of all faults in the dictionary. Fig. 2 demonstrates an example of the distance between OPD_{CUD} and OPD_f of fault f_3. The smaller the distance between two OPDs, the higher their similarity. Please note that there are three different distances in each fault because we measure a CUD on three axes, as mentioned in Section III-A. Three different distances are added up as the *total distance* for each fault. If the total distance between the CUD and a fault is smaller than a *threshold*, then the fault will be considered as a suspect.

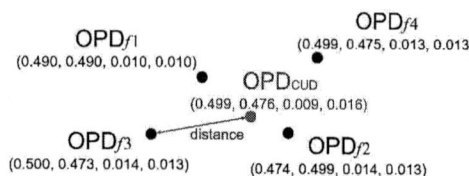

Fig. 2: Distance between OPD_{CUD} and OPD_{f3}

In this work, we use the *Hellinger distance* [12] as the distance metric. The Hellinger distance is defined as follows. P and Q are two discrete probability distributions where $P = (p_1, ..., p_n)$ and $Q = (q_1, ..., q_n)$. The Hellinger distance $HD(P, Q)$ between P and Q is shown as equation (1).

$$HD(P,Q) = \frac{1}{\sqrt{2}} \sqrt{\sum_{i=1}^{n} (\sqrt{p_i} - \sqrt{q_i})^2} \quad (1)$$

Fidelity $F(P, Q)$ of its corresponding Hellinger distance is shown as equation (2).

$$F(P,Q) = (1 - HD^2(P,Q))^2 \quad (2)$$

To obtain OPD_{CUD}, we require the number of test repetitions (shots) to execute a CUD. Fig. 3 shows the process of

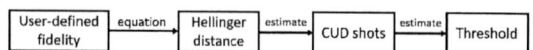

Fig. 3: Estimate CUD shots and threshold

estimating the required CUD shots and the threshold used in procedure (B). A user-defined fidelity determines how precise the static diagnosis is. We calculate the corresponding Hellinger distance of the user-defined fidelity by equation (2). This Hellinger distance represents the difference between ideal OPD and observed OPD. Then, we can estimate the corresponding shots of the Hellinger distance. Based on the concept of *standard error of the mean* [13], the difference between the ideal OPD and the observed OPD scales as $\frac{1}{\sqrt{shots}}$. If we execute a QC with a higher shots, the obtained OPD will be closer to the ideal OPD. We perform experiments and estimate the relationship between shots and Hellinger distance between ideal OPD and observed OPD. Finally, we can estimate the required CUD shots.

The threshold is used to determine whether a fault is considered as suspects or not. We estimate the threshold based on the required CUD shots in the previous step. We estimate the threshold with a guard band added. Adding a guard band to the threshold is necessary to avoid misclassifying the suspects of CUD. The guard band can be a confidence interval (e.g., two standard deviations), where the confidence interval is determined by the user. As mentioned in Section III-A, we also measure a CUD on three axes. Three different distances are added up as the total distance for each fault. If the total distance of a fault is smaller than three thresholds, the fault will be regarded as a suspect of CUD. We rank the suspects based on the total distance. If there is no suspect meet the threshold, we simply report the ten faults with the shortest total distances as suspects.

C. Partial QC Dictionary Generation

In procedure (C), we partition the CUD into partial QC and simulate their OPD. A *partial QC dictionary* is generated by many partial QC. We partition QC by layers such that a layer only contains a gate. Then, we generate partial QC based on the layers. Fig. 4 shows an example of partial QC by layer partitioning. Fig. 4 (a) is a complete QC, where C_g represents good QC. Before partitioning, C_g has five layers.

We denote $PC_{g,i}$ as the partial QC of C_g, containing layers from L_1 to L_i, where L_1 is the first layer and L_i is the the i_{th} layer. Fig. 4 (b) is a partial QC containing only the first three layers of Fig. 4 (a) ($PC_{g,3}$). Please note that all of the partial QC start from the first layer L_1. Our assumption is that a fault in a QC may be induced by executing previous gates, those gates on the left of the fault, so we must start from the first layer.

For each partial QC, we execute the partial QC several times with fixed shots. The output of each execution is an OPD, and thus we have several OPDs. We calculate the average OPD of these OPDs. The average OPD represents a high precision OPD. Then, we calculate the distance between the OPDs and the average OPD. Please note that we measure each partial QC on three axes as well. We use the total distance as mentioned

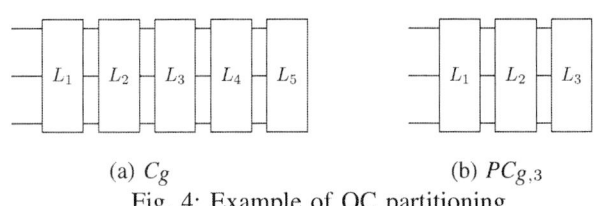

(a) C_g (b) $PC_{g,3}$

Fig. 4: Example of QC partitioning

in Section III-A. After that, we obtain several total distances. We construct a distribution based on these total distances. We use *normal distribution* to fit the distribution. The distribution can distinguish whether a partial QC is good or faulty. We store the average OPD and the distribution of all partial QC in the partial QC dictionary. The partial QC dictionary will be used by the dynamic diagnosis.

D. Dynamic Diagnosis

Dynamic Diagnosis identifies the faulty gate by executing the CUD iteratively. We use binary search in the iterative testing of CUD. First, we partition the CUD by the middle layer to generate a partial CUD. We generate the distance distribution of partial CUD to compare the partial CUD and the corresponding partial QC. If the partial CUD and the corresponding partial QC are matched, we call the partial CUD as good partial QC. Otherwise, we call the partial CUD as faulty partial QC. If the first half partial CUD is good, it means that the partial CUD does not contain any faulty gate in the first half layers. We continue to partition the CUD in the second half layers. On the contrary, if the first half partial CUD is faulty, it means that the faulty gate in the first half layers. The binary search process repeats until there is only one layer left. We call this layer as the *suspect layer*.

The output of binary search process is the suspect layer. After we find the suspect layer, the one gate g in the suspect layer is identified as the suspect gate. Please note that the fault list we assumed contains qubit-specific faults and gate-specific faults. Both faults are potential causes of faulty gates in CUD. We simply report both faults with the corresponding gate g as suspects in this work. After we perform static and dynamic diagnosis, we can get diagnosis results with more accurate and better diagnosis resolution.

To determine whether the partial CUD is good or faulty, we use the concept of *overlap coefficient* (OVL) [14]. The OVL is defined as follows. $f_1(x)$ and $f_2(x)$ are two continuous PDF. The OVL between $f_1(x)$ and $f_2(x)$ is shown as equation (3).

$$OVL = \int_{-\infty}^{\infty} min\{f_1(x), f_2(x)\}dx \quad (3)$$

Fig. 5 shows an example of classifying a partial CUD. The blue line represents the good distribution in the partial QC dictionary, while the red line represents the distribution of partial CUD. The shade represents the OVL between two distributions. A critical value of OVL is needed to classify whether a partial CUD is good or faulty. If the OVL between two distributions is greater than the critical value, we call the partial CUD a good partial QC. Otherwise, we call the partial CUD a faulty partial QC. In Fig. 5 (a), the OVL between two

distributions is smaller than the critical value. Thus we call the partial CUD faulty. In Fig. 5 (b), the OVL between two distributions is greater than the critical value. Thus we call the partial CUD good.

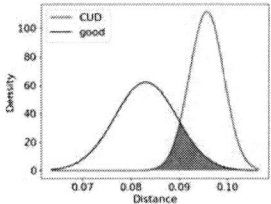 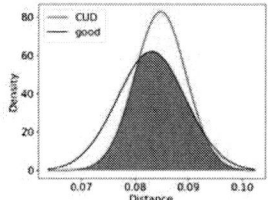

(a) Partial CUD is faulty (b) Partial CUD is good

Fig. 5: Example of classifying a partial CUD

IV. EXPERIMENTAL RESULTS

A. Experimental Setup

We now show the parameter setting of experimental results in this work. The fault dictionary is generated by a complete fault list, which considers two scopes: qubit-specific faults and gate-specific faults. Bias faults, ratio faults, truncation faults, and *CNOT* faults are considered in each scope. The simulation fault size represents the fault size we used in generating fault list. Simulation fault sizes of ratio faults, bias faults, and truncation faults are 0.2, 0.2π, and 0.25π, respectively. Simulation fault sizes of *CNOT* faults are the same as [10], but the parameters of U gate are 0.075π instead of 0.05π. The fault dictionary is generated by Monte Carlo method with 10^5 shots. The user-defined fidelity is 0.999, and the guard band of the threshold is two standard deviations. The shots for executing a CUD and the threshold used to determine the suspects are shown in Table I. Details of estimating the shots and the threshold are described in Section III-B. The critical value of OVL mentioned in Section III-D is set to 0.3.

TABLE I: CUD shots and Threshold

No. of qubits	CUD Shots	Threshold
3	1,595	0.106
4	3,805	0.092
5	7,733	0.084
6	16,021	0.080
7	33,560	0.074

We use *qasm_simulator* in *Qiskit* [15] as the QC simulator. For Section IV-B, we use the *depolarizing_error* as noise model of *gate error*. The error rate of single-qubit gates and two-qubit gates are 0.1% and 1%, respectively. The *readout error* rate is 1.5%, which represents the probability that output is inverted.

We use some benchmarks to evaluate our proposed technique. There are three types of functions, including Bernstein-Vazirani algorithm (BV) [16], Quantum Fourier Transform (QFT) [9] and Quantum Volume (QV) [17]. Each type of function contains five benchmarks with different numbers of qubits, from 3 to 7, so that we have 15 benchmarks in total. The benchmarks are *transpiled* to QC by Qiskit so they contain *basis gates* only. We choose the basis gates available in the IBM Q device, including $CNOT$, ID, R_z, $\sqrt{X}$, and X.

B. Diagnosis Results of Simulation

To evaluate our proposed diagnosis technique, we generate some faulty QCs as the CUDs and feed them into our proposed flow. We demonstrate the results of simulation with three scenarios. The first scenario is that the injected CUD fault size is equal to the simulation fault size. This scenario demonstrates a perfect situation where the faults of CUDs and the faults we assume are the same. The second scenario is that the injected CUD fault size differs from the simulation fault size. We demonstrate this scenario by injecting simulation fault size multiplied by some factors (shown as M in this paper). The last scenario is that the injected CUD faults are non-unitary. The fault models we used in fault simulation assume that all faults are unitary. However, non-unitary faults can also occur in a real faulty QC. We demonstrate this scenario by injecting non-unitary faults into CUDs using the Qiskit noise model.

Table II shows the diagnosis results of the first scenario. The first column shows the name of benchmarks. The second column shows the number of qubits in benchmarks. The third and fourth columns show the number of gates and faults in benchmarks. The fifth column shows the number of CUDs. Please note that the number of CUDs equals the number of faults because we injected each fault only once. The second to last column shows the average resolution, which is the average number of suspects reported by diagnosis flow. The last column shows the average accuracy, which is the average rate of diagnosing a fault successfully. Since the main objective of the diagnosis is to diagnose the fault location, if the reported suspects contain the injected fault location, the diagnosis is successful. For each benchmark, the average resolution varies from 1.52 to 2, and the average accuracy varies from 96.84% to 100%. In these 15 benchmarks, the average resolution and accuracy are 1.77 and 99.01% The results show that our diagnosis flow has very high accuracy and good resolution.

TABLE II: Diagnosis results (1st scenario)

Benchmark	No. of qubits	No. of gates	No. of faults	No. of CUDs	Res.	Acc. (%)
BV3	3	20	50	50	1.52	100.00
BV4	4	28	65	65	1.62	100.00
BV5	5	36	80	80	1.62	100.00
BV6	6	44	95	95	1.64	100.00
BV7	7	52	110	110	1.64	100.00
QFT3	3	26	142	142	1.70	99.30
QFT4	4	46	268	268	1.71	100.00
QFT5	5	68	391	391	1.83	98.21
QFT6	6	98	575	575	1.95	98.26
QFT7	7	130	756	756	1.97	97.09
QV3	3	41	139	139	1.77	99.28
QV4	4	79	274	274	1.69	99.27
QV5	5	80	274	274	1.89	98.54
QV6	6	119	411	411	1.97	98.30
QV7	7	120	411	411	2.00	96.84
Average					1.77	99.01

The diagnosis results of the second scenario are demonstrated in Table III. M represents the multiple of the injected CUD fault size with respect to the simulation fault size. In this experiment, the CUDs are injected with M=0.5, 1, 1.5, 2, and 2.5, where the results of M=1 are the same as in Table II. Table III shows the average accuracy and resolution of CUDs injected with these five fault sizes. For these five M, the average resolution varies from 1.77 to 1.86, and the average accuracy varies from 93.90% to 99.42%. For the same benchmark, CUD with larger M has more deviation, and hence results in better diagnosis accuracy. This is because the larger the injected fault size is, the more different between the good and faulty OPD. The result of M=0.5 has worst diagnosis accuracy than the other four due to the smallest fault size. The fault size is too small so that the fault effect is as small as the noise effect. For the diagnosis results of the second scenario, we conclude that even if the injected CUD fault sizes are different from our assumptions, our diagnosis flow still achieves high accuracy and good resolution.

TABLE III: Diagnosis results (2nd scenario)

Injected size	Res.	Acc. (%)
M=0.5	1.86	93.90
M=1	1.77	99.01
M=1.5	1.86	98.99
M=2	1.86	98.63
M=2.5	1.84	99.42

Table IV shows the results of the third scenario. We use noise models in Qiskit to simulate the behavior of non-unitary faults. The fault list contains four types of non-unitary faults: *depolarizing faults*, *reset faults*, *amplitude damping faults*, and *phase damping faults*. The parameters of noise models represent the fault size of these non-unitary faults. We set the fault sizes of depolarizing faults, reset faults, and amplitude damping faults to 0.3. We set the fault size of phase damping faults to 0.3π. We list the results of four non-unitary faults in Table IV. Because the injected faults differ significantly from the fault models we fault simulated, we rely most on dynamic diagnosis rather than static diagnosis. This experiment shows that our diagnosis results are still very good even if injected faults are different from our simulation fault models.

TABLE IV: Diagnosis results (3rd scenario)

Injected size	Res.	Acc. (%)
depolarizing fault	1.78	100.00
reset fault	1.85	96.00
amplitude damping fault	1.83	94.66
phase damping fault	1.80	94.74

Table V compares the effectiveness of static versus static-dynamic diagnosis. We compare the results of two conditions: (1) static diagnosis only, (2) static followed by dynamic diagnosis. We list average resolution and average accuracy in each condition. For the second condition, if the static diagnosis result is not acceptable, then we apply dynamic diagnosis. The last column shows the percentage of applying dynamic diagnosis. We list results in three rows representing the three scenarios mentioned above. The first row shows the results of unitary faults (1st scenario, M=1). The accuracy of static diagnosis is better than that of static+dynamic diagnosis in this scenario. Because the injected fault size is the same as the simulation fault size, it is easy to find the fault accurately by static diagnosis. The second row shows the results of unitary faults of the other four fault sizes, except $M=1$. In this row, the accuracy of static diagnosis decreases while the resolution increases, compared to $M=1$. The accuracy of static diagnosis is below 88%, and the resolution is over 8. In contrast, the results of static+dynamic diagnosis are not significantly worse than those of $M=1$. The third row shows the results of non-unitary faults (3rd scenario). The results of static diagnosis are much worse than those of unitary faults because the non-unitary faults are entirely different from the simulation fault models. In the last column of this table, if the injected faults are more different from the faults we simulated, there is a higher percentage that we apply dynamic diagnosis. We summarize that our proposed static+dynamic diagnosis flow reports good results in three scenarios.

TABLE V: Static versus static+dynamic diagnosis

	static		static + dynamic		
Injected fault	Res.	Acc. (%)	Res.	Acc. (%)	Applying dynamic (%)
unitary fault ($M=1$)	2.36	99.68	1.77	99.01	81.50
unitary fault (other)	8.51	87.18	1.86	97.43	97.78
non-unitary fault	9.03	46.69	1.82	96.35	99.24
Average	6.64	77.84	1.81	97.70	92.94

Last, we compare our method with other diagnosis method in the past, which are [7] and [8] where me mentioned in Section II-C. For the sake of fairness, we apply their missing gate fault model and diagnose two benchmark circuits on simulators. Table VI shows diagnosis resolution of all benchmark circuits. All benchmark circuits are successfully diagnosed with accuracy 1. From the results, we can see that our static+dynamic diagnosis method can precisely identify the fault location with the resolution for both benchmarks to be 1 even under the impact of noise.

TABLE VI: Average resolution of different diagnosis methods

Benchmark	Qubits	Depth	No. of CUDs	Resolution			
				[7]	[8]	static	static+dynamic
3qubit	3	5	6	1.83	1.00	1.33	1.00
simple-grover3	3	9	9	1.44	1.11	3.22	1.00

C. Diagnosis Results on IBM Q

To show the feasibility of our diagnosis flow on a real quantum device, we perform simulation with the IBM Q backend noise model [18]. Unlike Section IV-B, which used the depolarizing noise model, in this section, we use the *backend noise model* of real device provided by IBM Q. We choose the backend noise model of the 7-qubit device, *ibm_perth*.

Table VII shows the simulation results with *ibm_perth* backend noise model. The CUDs are benchmarks BV and QV with fault sizes from $M=0.5$ to 2.5. We show the average resolution and accuracy of two conditions: static diagnosis and static+dynamic diagnosis. In Table VII, the result of static+dynamic diagnosis has average resolution 1.92 and average accuracy 98.55%. The result of static+dynamic diagnosis

is much better than that of static diagnosis only. Also, the result of static+dynamic diagnosis has good average resolution and acceptable average accuracy. We conclude that our diagnosis is still successful in the presence of backend noise model.

TABLE VII: Diagnosis results (*ibm_perth*'s noise model)

	static		static + dynamic	
Benchmark	Res.	Acc. (%)	Res.	Acc. (%)
BV7 ($M=0.5\sim2.5$)	7.17	96.55	1.86	100.00
QV7 ($M=0.5\sim2.5$)	9.46	96.08	1.98	97.11
Average	8.32	96.32	1.92	98.55

Table VIII shows the results of experiments executed on the 7-qubit IBM Q device *ibm_perth*. We perform experiments on BV7 and QV7 with $M=1$, and we run static diagnosis only due to limits on IBM Q device. Dynamic diagnosis requires executing QC interactively, which is not allowed under the current queuing usage rule of IBM Q device. The CUDs in QV7 only contain *CNOT* faults because *CNOT* gates are more vulnerable to faults than single-qubit gates. The experimental results of resolution on BV7 and QV7 are 9.09 and 10.00, respectively. Both benchmarks have 100% accuracy. Although we only perform static diagnosis, we can infer that dynamic diagnosis would also be feasible on real quantum device. We conclude that our diagnosis flow can be feasible on real quantum device.

TABLE VIII: Diagnosis results (IBM Q device *ibm_perth*)

		static	
Benchmark	No. of CUDs	Res.	Acc. (%)
BV7 ($M=1$)	68	9.09	100.00
QV7 ($M=1$)	298	10.00	100.00

V. CONCLUSION

In this work, we proposed a diagnosis flow for QC in the NISQ era. The proposed technique contains static diagnosis and dynamic diagnosis. Static diagnosis uses a fault dictionary that contains output probability distribution for each fault. Dynamic diagnosis uses the concept of binary search to find the accurate fault locations of faulty QC. To demonstrate the effectiveness of our proposed technique, we evaluate 15 benchmarks with injected unitary and non-unitary faults. The diagnosis simulation results show that the average accuracy and resolution are 97.70% and 1.81. We also perform static diagnosis experiments on the IBM Q device, and the results show that our technique can be feasible on real QC devices.

REFERENCES

[1] F. Arute *et al.*, "Quantum supremacy using a programmable superconducting processor," *Nature*, vol. 574, no. 7779, pp. 505–510, 2019.

[2] J. Hsu, "Ces 2018: Intel's 49-qubit chip shoots for quantum supremacy," *IEEE Spectrum Tech Talk*, pp. 1–6, 2018.

[3] J. Chow, O. Dial, and J. Gambetta, "Ibm quantum breaks the 100-qubit processor barrier," *IBM Research Blog*, 2021.

[4] J. Gambetta, "Ibm's roadmap for scaling quantum technology," *IBM Research Blog [Internet]*, 2020.

[5] J. Preskill, "Quantum computing in the nisq era and beyond," *Quantum*, vol. 2, p. 79, 2018.

[6] A. Paler *et al.*, "Tomographic testing and validation of probabilistic circuits," in *2011 Sixteenth IEEE European Test Symposium*. IEEE, 2011, pp. 63–68.

[7] A. Paler, I. Polian, and J. P. Hayes, "Detection and diagnosis of faulty quantum circuits," in *17th Asia and South Pacific Design Automation Conference*. IEEE, 2012, pp. 181–186.

[8] D. Bera, "Detection and diagnosis of single faults in quantum circuits," *IEEE Transactions on Computer-Aided Design of Integrated Circuits and Systems*, vol. 37, no. 3, pp. 587–600, 2017.

[9] M. A. Nielsen and I. Chuang, "Quantum computation and quantum information," 2002.

[10] C.-H. Wu *et al.*, "qatg: Automatic test generation for quantum circuits," in *2020 IEEE International Test Conference (ITC)*. IEEE, 2020, pp. 1–10.

[11] E. Knill *et al.*, "Randomized benchmarking of quantum gates," *Physical Review A*, vol. 77, no. 1, p. 012307, 2008.

[12] E. Hellinger, "Neue begründung der theorie quadratischer formen von unendlichvielen veränderlichen." *Journal für die reine und angewandte Mathematik*, vol. 1909, no. 136, pp. 210–271, 1909.

[13] D. L. Streiner, "Maintaining standards: differences between the standard deviation and standard error, and when to use each," *The Canadian journal of psychiatry*, vol. 41, no. 8, pp. 498–502, 1996.

[14] H. F. Inman and E. L. Bradley Jr, "The overlapping coefficient as a measure of agreement between probability distributions and point estimation of the overlap of two normal densities," *Communications in Statistics-theory and Methods*, vol. 18, no. 10, pp. 3851–3874, 1989.

[15] M. S. Anis *et al.*, "Qiskit: An open-source framework for quantum computing," 2021.

[16] E. Bernstein and U. Vazirani, "Quantum complexity theory," *SIAM Journal on computing*, vol. 26, no. 5, pp. 1411–1473, 1997.

[17] A. W. Cross *et al.*, "Validating quantum computers using randomized model circuits," *Physical Review A*, vol. 100, no. 3, p. 032328, 2019.

[18] IBM Quantum, https://quantum-computing.ibm.com/, 2021.

(Industry Short Paper)

Targeted Custom High-Voltage Stress Patterns on Automotive Designs

Saidapet Ramesh[1], Kristofor Dickson[1], Akshay Jaiswal[2], Robert Marchese[1], Kiran Thota (Sunny) [1]

1. NXP Semiconductors, 6501 W. William Cannon Drive, Austin, TX 78735
2. NXP Semiconductors, 18 Film City Road, Sector 16A, Noida, Uttar Pradesh 201301, India

Abstract — Continuous analysis of multiple defective dies on a FinFET design indicated a failure mechanism with resistive defect on one library cell. All the units under study had passed the wafer probe test flow and only failed during internal quality and reliability study flows. In this paper, we present the details of developing and deploying targeted stress and test patterns using custom hand-written User Defined Fault Model (UDFM) files. This was essential to address a failure mechanism that was root-caused to not enough high-voltage stress applied in order to accelerate the resistive defect in wafer probe test flow. The defect location inside this cell created feedback in the circuit that limited the current through the defective path, making it more difficult to both accelerate and detect the existence of the problem in wafer probe test flow. A novel solution using the combination of targeted custom ATPG patterns to stress and accelerate the defect mechanism and custom ATPG patterns to also screen defects were generated using the custom UDFM file and deployed in the wafer probe test flow. Finally, we will present unique silicon fallout data from the targeted custom tests which confirmed effectiveness of targeted stress to accelerate and screen defective die early in wafer probe test flow.

Keywords — *Latent Defect, Stress, Focused Stress, Infant Mortality, ATPG, Zero Defect, UDFM, Fault Model, High Voltage Stress, Defect Oriented Test*

I. INTRODUCTION

With the ever shrinking process nodes in combination with increasingly complex and larger microprocessor and microcontroller designs, testing and screening defective units as early as possible in test flow to meet gross margin and 0 DPPB (Defective Parts Per Billion) requirements of automotive designs is becoming exponentially complex. Testing of manufactured die consists of two steps. The first step involves testing a set of manufactured die on a wafer before the good die are assembled in a package. The second step involves testing of final packaged product.

Potential defects in manufactured die fall into broad categories of hard and resistive defects. Hard defects include contact open, open metal lines, complete metal-via opens etc. Resistive defects include resistive metal-via contacts, weak bridges across two metal lines, weak metal-via bridges, etc. Resistive defects are very likely to become latent defects. If not stressed/accelerated correctly, latent defects escape production flow and end up as customer returns. Latent defects can be life threatening, if these CMOS devices are used for safety or critical applications like automotive, etc.

Identifying both types of defects on manufactured devices early in the entire testing cycle is very cost effective and guarantees higher quality. Hence, detecting defective die in the wafer probe test program flow becomes more critical. The testing process at wafer probe involves two steps. The first step is to create enough stress inside the stress flow of the test program to accelerate the potential latent defects into hard defects. And, the second step involves detection of the defects

using a combination of voltage and frequency test points. The authors of [1] introduce a high voltage test to put enough stress on a wafer. Continual efforts have been made by the same author by only putting enough stress on the die that is likely to be defective [2]. The static high voltage can screen out the early failing parts. The dynamic voltage stress can stress the chip at the real chip operation status [3]. Apart from high voltage, [4] has shown thermal shock is effective during wafer stress stage.

It is common practice to use a combination of traditional scan ATPG tests (generated primarily for screening defects inside wafer probe stress flows) to stress devices using three critical parameters - voltage, temperature and time. Using these three knobs, we are able to design a high voltage stress test (HVST) flow at wafer probe to screen early life failures (ELF). The usual consensus in the industry has been to ensure that such tests generate a certain degree of activity, as measured by toggle coverage (or internal wires reaching a Boolean 0 and a Boolean 1) during pre-silicon simulations. The industry practice so far has been to reuse the tests that have been originally created to "detect" defects for "accelerating" latent defects as well [5].

[5] only relies on stimuli based on two time frame based scan patterns while our custom stress methodology relies on stimuli based on one time-frame only. The other main difference is our custom stress strategy also uses a combination of static (one time frame) and IDDQ (quiescent current) based scan tests inside the HVST flows while [5] only used two time frame based ATPG tests inside their stress flow at wafer probe. Our final difference in strategy is we also use a combination of targeted exhaustive delay (two time frame) based ATPG patterns + static (one time frame) scan patterns for the targeted cell of interest in both pre-HVST and post-HVST flows.

This paper introduces the motivation using failure analysis findings from multiple defective dies of a FinFET design in Section II. Section III will cover the steps that were used to fully quantify the impact of the identified cell for all designs. Wafer-Probe flow will be discussed in detail in Section IV. The ATPG pattern generation using custom User Defined Fault Model (UDFM) files will be discussed in Section V. The unique silicon fallout will be analyzed using the new targeted test content deployed in wafer probe test flow in Section VI. On-going research and opportunities for expanding custom HVST stress patterns flow will be discussed in Section VII. Finally, the paper concludes in Section VIII.

II. MOTIVATION

Failure analysis findings from multiple defective dies indicated a physical defect at the exact same layout location inside a library cell instance spread across the design. The defective node had complete scan stuck-at, transition-delay, cell-aware, and toggle test coverage. All the units passed the wafer probe test flow and only failed during the internal

979-8-3503-4631-2/23 $31.00 © 2023 IEEE

quality and reliability study flows. This motivated us to deep-dive, find the root-cause of failure mechanism and deploy a custom stress and screening test solution.

A. Electrical Failure Analysis Findings

The failing units exhibited an elevated Vmin signature for transition delay scan patterns. Scan diagnosis reports from multiple quality and reliability study based failing units confirmed that the cell of interest was a multiplexor MXT2 standard library cell. Soft Defect Localization (SDL) [6], Mercury Cadmium Telluride (MCT) Photo-Emission (PEM) [7] and Laser Voltage Probe (LVP) [8] successfully identified a shorting defect within the cell as the prime cause of failure. A schematic representation of the short location is shown in Fig. 1.

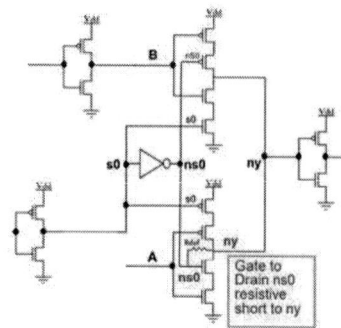

Fig. 1. Transistor schematic of MXT2 cell

B. Physical Failure Analysis Findings Deep Dive

Transmission Electron Microscopy (TEM) results revealed a tungsten bridging defect at Metal 0 between the metal gate connected to net ns0 and the drain tap connected to ny as shown in Fig. 2. All reject units were seen to fail for ns0 to ny shorts inside MXT2 cells at random die locations.

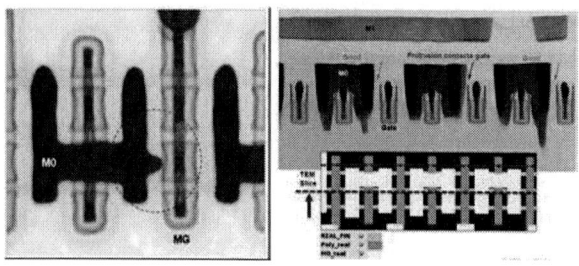

Fig. 2. TEM planar view and X-section view of bridging

In order to effectively accelerate time to failure, the two nets of interest "ns0" and "ny" must be held in opposite states during stress. This enforces the required current through the defect mechanism. When s0 is high (nS0 is low), input path B is selected. To stress the defect mechanism "Rdef", we need ny to be in contention, held in the opposite state to ns0. The issue with trying to stress the defect at the location inside the cell, is that under conditions when contention occurs between ny and ns0, the short between them will partially turn off the pull up or pull down stacks that need to be on to supply ny with current. The feedback in the circuit caused by the presence of the defect self-limits the stress that the defect is exposed to, making ageing the defect extremely difficult.

C. Modelling Cell Behavior

The 'Rdef' dependent cell behavior was modelled to measure output and the voltage drop across the defective short

versus the Vdd value applied during stress. Fig. 3 shows that the cell can function logically as expected, with 'Rdef' as low as 13kohm. Furthermore, the voltage across that resistive defect could be as low as 40% of the HVST Vdd applied.

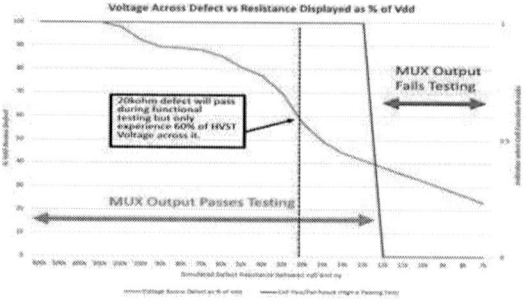

Fig. 3. Cell behavior model results of MXT2 cell with defect

In most cases HVST will fully expose shorting mechanisms so that they are successfully screened by testing. However, if the shorting occurs at the specific location in the MXT2 cell as shown, it is very difficult to screen and effectively accelerate during stress. For this reason, targeted User Defined Fault Model (UDFM) based patterns become crucial in order to successfully place the cell into the correct state, stress the failure mechanism as much as possible and then later screen for the problem effectively.

III. QUANTIFYING MXT2_GL1 CELL IMPACT

Once the root-cause of this issue was well understood and narrowed down to only one MXT2 cell with gate length 1 (GL1) combination, a thorough analysis on 3 different levels was performed to assess and quantify the MXT2_GL1 cell impact across all designs on this process node.

STEP - 1 (Gate length check): Fig. 4 shows the instance count of MXT2_GL1 instances comparison with the other MXT2_GL2, MXT2_GL3 & MXT2_GL4 instances with different gate length for a big New Product Introduction (NPI) Design A and a relatively small Design B.

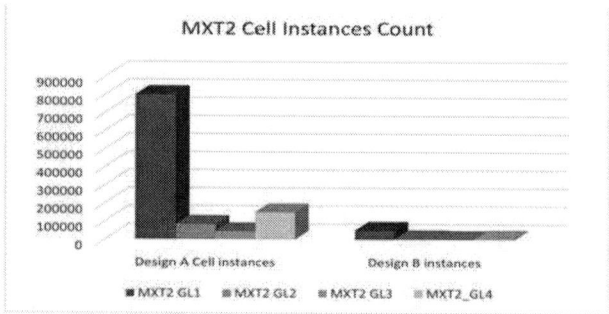

Fig. 4. MXT2_GL1/GL2/GL3 instances count comparison

Clearly, MXT2_GL1 cells were instantiated more by several folds in comparison to the other gate length combinations that were available in this process node. We could decipher that MXT2_GL1 cell had much higher probability of potential manufacturing defect compared to other MXT2 gate length flavors, once the minimum spacing between M0 layer and Gate poly layers were measured.

Fig. 5 explains the correlation between gate lengths and the minimum spacing of Metal 0 (M0) – Gate layers for the GL1 and GL4 combinations of MXT2 cells. As shown in Fig.

5, the gate length for GL1 was bigger than GL4 while the M0-Gate spacing was much tighter for GL1 cell.

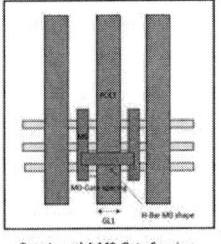 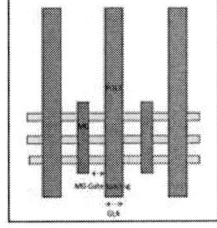

| Gate Length1 M0-Gate Spacing | Gate Length4 M0-Gate Spacing |

Fig. 5. Gate Length 1 (GL1) vs Gate Length 4 (GL4) representation

Additionally, it was checked and confirmed that the M0-Gate spacing verified in MXT2_GL1 cell had met the minimum spacing DRC rule for this process node. It was also observed that M0 H-bar layout shape was more common in GL1 cells than the other gate length cells. This layout analysis confirmed why the probability of a resistive shorting defect was higher in GL1 cells than the other gate length cells.

STEP – 2 (Look across designs): To quantify the cell instance count impact on all other NPI designs which used MXT2_GL1 cells from this process node, we had to compute the percentage of MXT2_GL1 instances with respect to the total cell instances for each design. Table. I shows the percentage of MXT2_GL1 instances with respect to the total cell instances for multiple designs.

TABLE I. MXT2_GL1 CELL INSTANCES % COMPARISON

FinFET design	# Cell Instances	% MXT2_GL1 Cell Instances
Design A	~ 24 million	3.47%
Design B	~ 9 million	2.45%
Design C	~ 4 million	1.44%

Data from Figures 4-5 and Table I confirmed that multiple designs were exposed to this HVST stress gap identified from the deep dive analysis of this issue on MXT2_GL1 cell.

STEP – 3 (Layout topology search): Once we established that the process marginality occurs due to H-bar topology in a standard library cell of this process node, search was expanded to other cells. List of potential impacted library cells, based on MXT2_GL1 symptoms were identified. Layout failure modes and effects analysis (FMEA) was performed on these cells to check that in the presence of a resistive defect, if those cells have the potential to create a feedback loop and prevent the latent defect from getting converted to a hard defect, thereby escaping test screens. However, no additional cells that had similar signature were found for this process node.

Team developed, verified and deployed a custom solution to enhance the stress and post-stress wafer probe test flows for MXT2_GL1 cells instances for all designs that were past tape-out stage.

IV. WAFER-PROBE STRESS

Typical ATE (automated test equipment) production screening consists of Wafer Probe (sorting wafers) and Final Test (on packaged parts). Probe does the heavy lifting of screening out majority of the process defects and outliers. High voltage stress techniques using three key knobs (voltage, temp and stress duration) are deployed at wafer probe.

During initial product development phase, post stress screening is a tool/metric to understand if the high voltage stress being applied is not causing an overkill or shipping out walking wounded units. Pre and post stress screening test content is exactly the same, to enable quicker decision making.

HVST includes two components: Dynamic Voltage Stress (DVS) and Elevated Voltage Stress (EVS) as shown in Fig. 6. Dynamic stress involves toggle based patterns executed at higher voltage for scan, memory and analog test modules. Usually dynamic voltage uses a voltage setting that is a few hundred mV higher than design specification high-voltage corner. While elevated stress is done at extremely higher voltage (~2-3x operating voltage), with the logic pre-conditioned using IDDQ (quiescent current) [9] patterns. Custom UDFM based scan patterns were added to the scan content being executed as part of both the dynamic and elevated stress.

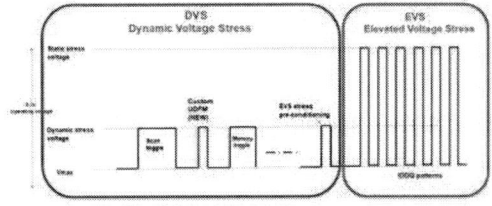

Fig. 6. High Voltage Dynamic + Elevated stress flow

As highlighted in [10], the circuit is stressed and de-stressed several times in EVS since dynamic voltage stress is approximately >4x as effective as static voltage stress in catching early silicon failures.

V. ATPG PATTERN GENERATION USING CUSTOM USER DEFINED FAULT MODEL (UDFM)

A. *User Defined Fault Model Generation*

We propose an innovative custom solution using a set of static and delay based custom hand-written UDFM files. These UDFM files were used to generate and deploy targeted ATPG test patterns.

This intra cell defect falls into the field of Cell-Aware test. Much research has been done on Cell-Aware test [11]. The test cost is really high with respect to time and design efforts. Although research has been done to reduce test cost as mentioned in [11] and [12], the cost of Cell-Aware test is considerably high since it requires analog SPICE simulations for each modeled library cell. Alternatively, the custom UDFM files have been developed without any analog SPICE simulations since it is purely logical values based only.

An example fault table that was used to write the custom static UDFM file is shown below in Table II. The identified test conditions to produce the maximum stress in a targeted manner were to drive opposite values on S0 and Y ports of this cell. To achieve this, we developed custom UDFM to set this specific stress condition using similar approaches in [13].

TABLE II. MUX2 STATIC FAULT TABLE

Fault Condition	A	B	S0	Y	
				Good Value	Bad Value
Stress Test 1	1	-	0	1	0
Stress Test 2	-	0	1	0	1

979-8-3503-4631-2/23 $31.00 © 2023 IEEE

B. Stress ATPG Pattern Generation

It is well known that, no single fault model covers all possible faults. For creating enough stress inside the wafer probe flow, a combination of ATPG static patterns and IDDQ patterns are used. The static UDFM was used to generate targeted ATPG static and IDDQ patterns which could then be used to create targeted stress inside the wafer probe HVST flow. The HVST flow consists of combination of DVS and EVS flows.

The targeted ATPG static patterns were used inside the DVS flow in addition to the existing traditional scan tests. Additionally, the targeted ATPG IDDQ patterns were used inside the EVS flow in addition to existing IDDQ tests to create maximum targeted voltage stress on all targeted cell instances in design.

C. Defect Testing ATPG Pattern Generation

The delay UDFM was used to generate targeted ATPG delay patterns to comprehensively test all MXT2_GL1 test instances and not inside the HVST stress flow. A combination of both the ATPG static patterns and ATPG delay patterns were deployed in both the pre-stress and post-stress flows to test and screen die with resistive defects from the targeted MXT2_GL1 cell instances. The delay MXT2_GL1 tests were also tested at very-low-voltage (VLV) corner in the post-HVST flow at a reduced speed to screen outlier resistive defective die efficiently.

VI. UNIQUE SILICON FALLOUT ANALYSIS

Once the targeted custom UDFM based stress and test patterns were deployed in high volume in the wafer probe test program, we started monitoring unique fallout for MXT2 only failures in the post-stress flow. A summary of unique fallout from design A is shown in Table III below. We were able to screen 2 units in the post stress flow which translates to potential projected 5.5 DPPM avoidance using the custom ATPG patterns. Additional fallout from at-speed TD patterns were also observed. Since the physical defect we are looking for is resistive in nature, only transition-delay patterns failing confirms effectiveness of the custom solution.

TABLE III. UNIQUE FALLOUT & DPPM AVOIDANCE

Custom ATPG Pattern Type	# Units	Post-Stress Fallout	% Fallout	Projected DPPM Avoidance
MXT2 Delay	366 K	2	0.0005	5.5
Total Count	366 K	2	0.0005	5.5

VII. EXPANDING CUSTOM HVST STRESS PATTERNS FLOW

The silicon results from this targeted MXT2_GL1 stress flow confirmed the effectiveness and gross-margin improvements of using custom UDFM based stress patterns to accelerate resistive defect mechanisms and screen defective die early in the wafer probe test flow. Multiple reject die analysis data confirmed resistive (latent) defects as the dominating mechanism from multiple FinFET designs.

Work is on-going to extend the custom UDFM based stress flow for all cell instances of a design. Research is also on-going to extend the custom UDFM based stress flows for Back-End-Of-Line (BEOL) layers to produce maximum current through the BEOL layers and potentially replace toggle test coverage metric for quantifying HVST stress in wafer probe test flow.

VIII. CONCLUSION

In this paper, we presented targeted custom UDFM based scan stress and scan tests generation and deployment flow. This new flow enabled additional targeted defect acceleration and detection and screening of defective die identified for a specific library cell for all NPI designs very early in test flow cycle at wafer probe test. High volume data analysis from one NPI design confirmed complete effectiveness of deployed solution. With the complex subtle failing defect mechanisms that come along with the ever shrinking and complex process nodes, the need and value for generating and deploying custom UDFM based scan patterns in addition to traditional ATPG patterns and cell-aware ATPG patterns will continue to become inevitable to meet the gross margin and 0 DPPB requirements for automotive designs.

ACKNOWLEDGMENT

The authors would like to multiple team members from NXP DFT design, backend design, failure analysis, and product and test engineering teams for their valued contributions in co-developing this custom solution.

REFERENCES

[1] C. He and Y. Yu, "Wafer Level Stress: Enabling Zero Defect Quality for Automotive Microcontrollers without Package Burn-in," in *2020 IEEE International Test Conference (ITC)*, 2020, pp. 1-10.

[2] C. He, P. Grosch, O. Anilturk, J. Witowski, C. Ford, and R. Kalyan, "Defect-Directed Stress Testing based on Inline Inspection results," in *2022 IEEE International Test Conference (ITC)*, 2022, pp. 398-406.

[3] Ghil-Geun Oh, Min-Hye Ho, Yeon-Jung Shin, Jae-Wook Choi, Ju-Youn Kim and Young-Dae Kim, "Dynamic Voltage Stress Sensing Circuits for Screening Out Early Device Reliability Issues in Advanced Technology Nodes," in *2021 IEEE Asian Solid-State Circuits Conference (A-SSCC)*, 2021, pp. 1-3.

[4] T. Vodenitcharova, L. C. Zhang, I. Zarudi, Y. Yin, H. Domyo, and T. Ho, "The Effect of Thermal Shocks on Stresses in a Sapphire Wafer," *IEEE Transactions on Semiconductor Manufacturing*, vol. 19, pp. 449-454, 2006.

[5] S. Natarajan, A. Sathaye, and O. Chaitali, "DEFCON: Defect Acceleration thorough Content Optimization," in *2022 IEEE International Test Conference (ITC)*, 2022, pp. 298-304.

[6] J. Rowlette and T.Eiles, "Critical timing analysis in microprocessors using near-ir laser assisted device alteration (lada)," in *International Test Conference (ITC) proceedings*, 20023, pp. 264-273.

[7] A. bahgat Shehata, F. Stellari, A. Weger, P. Song, H. Deslandes, T. Lundquist, and E. Ramsay, "Novel NIR Camera with Extended Sensitivity and Low Noise for Photon Emission Microscopy of VLSI Circuits," in *Proceedings of Internationsal Symposium for Testing and Failure Analysis (ISTFA)*, 2014, pp. 6-11.

[8] Keith A. Serrels, Ulrike Ganesh, "Laser Voltage Probing of Integrated Circuits: Implementation and Impact," *Microelectronics Failure Analysis: Desk Reference*, 7th ed., Edited By Tejinder Gandhi, ASM International, 2019, p 244–261.

[9] R. Kawahara, O. Nakayama, and T. Kurasawa, "The effectiveness of iddq and high voltage stress for burn-in elimination [cmos production]," in *Digest of Papers 1996 IEEE International Workshop on IDDQ Testing*, 1996, pp. 9–13.

[10] J. Zhang, A. Xu, D. Gitlin and D. Yeo, "Dynamic vs Static Burn-in for 16nm Production," 2020 IEEE International Reliability Physics Symposium (IRPS), Dallas, TX, USA, 2020, pp. 1-3.

[11] F. Hapke et al., "Cell-Aware Test," in *IEEE Transactions on Computer-Aided Design of Integrated Circuits and Systems*, vol. 33, no. 9, pp. 1396-1409, Sept. 2014.

[12] F. Hapke et al., "Defect-oriented cell-aware ATPG and fault simulation for industrial cell libraries and designs," *2009 International Test Conference*, 2009, pp. 1-10, doi: 10.1109/TEST.2009.5355741.

[13] S. Kundu, B. Gaurav, L. Endrinal, and L. Ranganathan, "Using custom fault models to improve understanding of silicon failures," *Internationsal Test Conference (ITC)*, 2022, *pp. 338-344*.

Reliable Brain-inspired AI Accelerators using Classical and Emerging Memories

Mikail Yayla[‡], Simon Thomann[§], Md Mazharul Islam[¶], Ming-Liang Wei[*†], Shu-Yin Ho[*†],
Ahmedullah Aziz[¶], Chia-Lin Yang[*], Jian-Jia Chen[‡], Hussam Amrouch[§]

[‡]Technical University of Dortmund, [§]University of Stuttgart, [¶]University of Tennessee Knoxville,
[*]National Taiwan University, [†]Macronix International Co., Ltd.
Corresponding author e-mail: amrouch@iti.uni-stuttgart.de

Abstract—By taking inspiration from the operation of biological brains, emerging brain-inspired hardware has the potential to revolutionize the way computations are performed. Brain-inspired computing can be realized using both classical CMOS and emerging beyond-CMOS technologies, whereas the latter holds the promise to provide substantial energy savings akin to the employment of non-volatile memories. One way to implement highly efficient brain-inspired AI applications is through analog computing schemes, such as Integrate-and-Fire (IF) Spiking Neural Networks (SNNs), which can be implemented using both CMOS and beyond-CMOS technologies as synaptic storage. However, managing the inherent degradation of computing accuracy in analog circuits and mitigating their effects on the predictive accuracy of AI systems remains a key challenge due to the inherent nature of analog computing.

In this paper, we discuss how the aforementioned challenges can be addressed. In the first part, we present our SPICE-Torch, a framework that connects low-level SPICE simulations of circuits and memories performing analog computations with high-level accuracy evaluations of NN models based on PyTorch. Furthermore, we present an example of neuromorphic optimization using classical CMOS technology. In the second part, we introduce memristors as an emerging beyond-CMOS technology that can retain their state without any outside influence and are well-suited for brain-inspired neuromorphic hardware. We demonstrate that brain-inspired hardware, realized using classical CMOS or beyond-CMOS technologies, has the potential to revolutionize the way we process information and solve complex computation problems. Nevertheless, to harness its full potential, reliability issues have to be managed carefully and HW/SW codesign is key. Our presented framework SPICE-Torch, which connects low-level SPICE simulations of circuits performing analog computations with high-level accuracy evaluations of NN models based on PyTorch is available as open-source in https://github.com/myay/SPICE-Torch.

I. INTRODUCTION

Emerging brain-inspired hardware is becoming increasingly important, as traditional computing architectures reach their limits to keep up with the demands of modern artificial intelligence (AI) applications. Inspired by the structure and function of the human brain, emerging brain-inspired HW uses massively parallel processing, near- or in-memory processing, while achieving low power consumption to achieve high performance. These novel HW architectures process and analyze data that is surrounding us in our everyday lives, particularly in the areas of object detection, natural language processing, speech recognition, and recommendation engines. Brain-inspired computing HW has the potential to revolutionize the way we process information and solve complex problems, and are an important area of research now and will be in the future, as AI applications increasingly pervade our daily lives. Brain-inspired computing can be realized using both classical CMOS and emerging beyond-CMOS technologies. Classical CMOS technology has been the backbone of the semiconductor industry for decades, providing the foundation for the development of modern digital circuits. However, as the limits of CMOS technology are reached, researchers are exploring alternative beyond-CMOS technologies such as emerging non-volatime memory (NVM) based technologies. Although state-of-the-art brain-inspired AI systems can

take only inspiration from biological systems, the difference can be lessened when beyond-CMOS NVM technologies are employed. This allows highly efficient computing, exploiting non-volatility and processing-in-memory, leading to increased energy efficiency and faster processing speeds. As such, both classical CMOS and emerging beyond-CMOS technologies hold promise for the development of brain-inspired computing. Despite using classical CMOS or emerging beyond-CMOS technology, reliability is always the key concern that designers need to carefully address due to many effects like temperature [1]–[7]. One efficient way of brain-inspired AI application is performing the computations in the analog domain. An analog computing scheme that exploits this for efficiency is Integrate-and-Fire (IF) Spiking Neural Networks (SNNs) [8]. Like in biological systems, in IF-SNNs, neural activity is event-driven. Input spike signals are processed together with stored information by the integration of voltage spikes over time, until a threshold reached which may cause output spike signals used in further processing. IF-SNNs can be combined together with classical or emerging memory technologies to realize the storage, while the current coming out of them when sensing can be used summing in the analog domain by Kirchhoff's circuit law [9]. However, managing the inherent degradation of computing accuracy in the analog circuits due to the impacts of nonidealities is a key challenge, since it can severely affect the accuracy of AI models. While high-level frameworks such as PyTorch or Tensorflow assume ideal conditions, accurate SPICE simulations become highly complex when the nonideal behaviour is modeled. A unified framework that connects accurate SPICE-level simulations with high-level accuracy evaluations would benefit hardware design across abstraction layers. **In the first part** of this special session, we present SPICE-Torch, a framework that connects low-level SPICE simulations of circuits performing analog computations with high-level accuracy evaluations of NN models based on PyTorch. Our framework is available as open-source in https://github.com/myay/SPICE-Torch.

In the second part of this special session, we focus on emerging memories. While the family of emerging memory technologies is vast, we limit our discussion to only memristive memory because of the recent surging interest in this technology [10]. A memristor is a resistor that can exhibit two or more resistance levels based on the history of the input electrical stimuli. Memristors can be either non-volatile or volatile [10]. The former is used to design memory arrays and resistive synapses [11], [12], while the latter can be used to make neuristors [13] or memcapacitive synapses [14], [15]. Mechanisms that facilitate memristive switching include but are not limited to correlated electrons, ionic drift, oxygen vacancy migration, electrochemical filament formation, and interface effects [16]–[21]. Memristors can be prime enablers of brain-inspired neuromorphic hardware due to their scalability, high endurance, inherent non-linearity, and CMOS compatibility [10].

979-8-3503-4631-2/23 $31.00 © 2023 IEEE

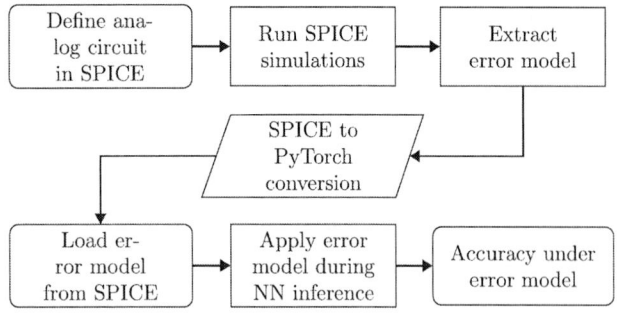

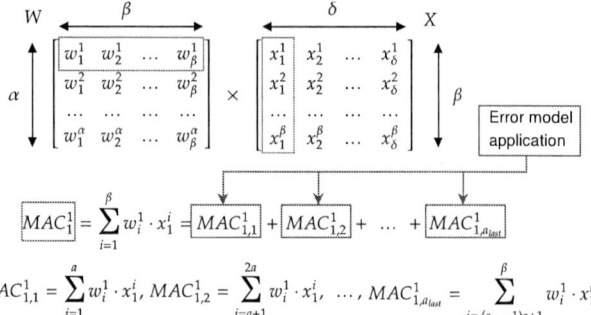

Fig. 1: Operational steps of our SPICE-Torch framework.

II. SPICE-TORCH: CONNECTING SPICE AND PYTORCH

We present SPICE-Torch, a framework that connects the low-level SPICE simulations of analog computing circuits with the high-level accuracy evaluations of NNs in PyTorch.

The section is organized as follows. In Sec. II, we first present our framework SPICE-Torch. Then, as an example of analog computing, we introduce IF-SNNs in Sec. III-A. The errors that may occur in IF-SNNs are presented in Sec. III-B, which motivates analyses using SPICE-Torch. We demonstrate the capabilities of SPICE-Torch by optimizing the IF-SNN circuits in Sec. III-C.

A. High-level Overview of SPICE-Torch

The general flow of our framework is illustrated in Fig. 1. There are mainly three steps in SPICE-Torch. (1) The user defines the analog circuit design in SPICE. Then, the SPICE simulation is run. Afterwards, the error models are extracted based on the data from SPICE. (2) The error models from SPICE are converted into a format that PyTorch can load. (3) The error models are loaded into PyTorch. This allows the application of the error models from the second step. Finally, the accuracy under the error models is returned. Is is also supported by our framework to run the SPICE simulation (to generate error models) and the NN accuracy evaluation (to evaluate accuracy under the mappings) separately. Our framework is available as open-source in https://github.com/myay/SPICE-Torch.

B. Description and Generation of Error Models

SPICE-Torch connects SPICE and PyTorch by mappings between sets of MAC values. Consider an analog circuit (e.g. the neuron circuit in Sec. III) in which the MAC values can only take values from a set $S_{\mathrm{MAC}} = \{q_0, q_1, \ldots, q_L\}$, where for each $q_j \in \mathbb{N}$, $q_j < q_{j+1}$. Assume that for each MAC value there exists one level in S_{MAC}. If there are no errors in the computations, each value q_j in S_{MAC} is also mapped to q_j, i.e. $m_j : q_j \rightarrow q_j$. Then the relation $S_{\mathrm{MAC}} \rightarrow S_{\mathrm{MAC}}$ is the identity function. However, with errors from analog computing, the identity property may not hold anymore. Consider a different set $S'_{\mathrm{MAC}} = \{q'_0, q'_1, \ldots, q'_L\}$, where $q'_k \in \mathbb{N}$ and $q'_k \leq q'_{k+1}$ (i.e. values can be repeated). In the case with errors, the relation is $S_{\mathrm{MAC}} \rightarrow S'_{\mathrm{MAC}}$. Values q_j of S_{MAC} may be mapped to different values q'_k of S'_{MAC}, causing different MAC values to be computed.

SPICE-Torch implements two variants of such mappings. In the first variant, a direct mapping has the form $m_j : q_j \rightarrow q'_k$, where the value q_j of S_{MAC} is deterministically mapped to q'_k of S'_{MAC}. In the second variant, a probability based mapping has the form $m_{j,k} : q_j \rightarrow q'_k$, where the value q_j of S_{MAC} can take any value in S'_{MAC} with a certain probability.

These mappings are extracted from SPICE simulations. The data from SPICE is first transformed into a csv-file, which represents the mapping. After that, the csv-file is transformed into a numpy array,

Fig. 2: Workload of NNs in matrix notation and error application at the MAC-level. α: Number of neurons, β: Number of weights, δ: Columns in unrolled input matrix, a: Array size.

so that it can be loaded into PyTorch. In the direct mapping case it is a 1d array and in the probabilistic mapping case a 2d array.

C. Application of Error Models during NN Inference

Our framework loads the described mappings, and applies them during the MAC computations in PyTorch. In general, when the computing array and neuron circuit in Fig. 3 is used for computations, the MAC computations are separated into sub-MAC computations (see Fig. 2). This is due to the technological limitations for the number of components connected in an array. MAC_1^1 is the result of a vector product between the first row of the weight matrix $\mathbf{W}$ and the first column of the input matrix $\mathbf{X}$. When it is computed with one computing array, the result of MAC_1^1 is summed up by sub-MAC results, i.e. $\mathrm{MAC}_{1,1}^1$, $\mathrm{MAC}_{1,2}^1$, etc. To compute with β weights and inputs, one array of size a needs to be invoked $a_{last} = \lceil \frac{\beta}{a} \rceil$ times, and the results need to be summed, so that the entire value of MAC_1^1 can be computed.

In our framework, the error is applied at the level of these sub-MAC results. In standard deep learning libraries (such as PyTorch), the sub-MAC results are not accessible and likely not computed. PyTorch for example uses a MAC engine that employs CUDA-based function calls for matrix multiplications, which are not available as open source. Therefore, it is impossible to apply error models on the sub-MAC results. To still enable to application of error models on the sub-MAC results, it is necessary to replace the closed source MAC engine with an own custom MAC engine.

We implemented our own custom MAC engine based on GPU CUDA kernel extensions for PyTorch in our framework. Our custom CUDA-based MAC engine is called instead of the standard MAC engine of PyTorch. With the full control over our custom MAC-engine, we equip it with functionality that enables error model application with any array size. With that, arbitrary error models in the form illustrated above can be applied on the sub-MAC results. Furthermore, applying the error models during the training procedure (i.e. retraining) is also supported. To achieve this, during execution of the NN, the correct MAC results are replaced with the MAC results computed by the application of error models. In this way, the parameters of the NNs are optimized based on the error model, without the involvement of our custom operations, because the custom MAC operations are always detached from the computation graphs after execution.

III. BRAIN-INSPIRED ANALOG COMPUTING WITH IF-SNNs AND NEUROMORPHIC OPTIMIZATION WITH SPICE-TORCH

A. Basic Operation of IF-SNNs

The circuit of IF-SNNs is shown in Fig. 3. The design is based on [22]. In the computing array, a is the array size, $x_{s,i}$ the inputs, and

979-8-3503-4631-2/23 $31.00 © 2023 IEEE

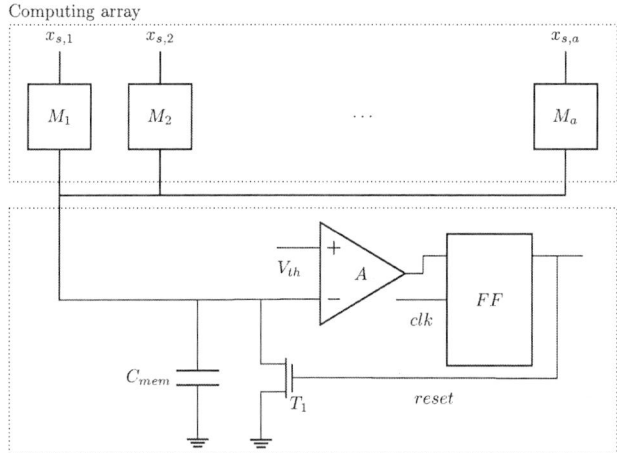

Computing array

Neuron circuit

Fig. 3: IF-SNN circuit. Top: Computing array with multipliers M_1 to M_a. Bottom: Neuron circuit.

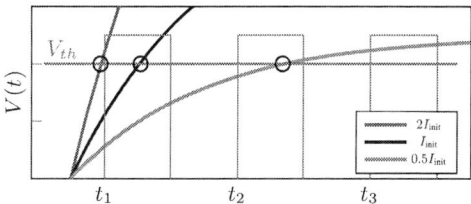

Fig. 4: Voltage across the capacitor over time. t_1, t_2, and t_3 are the spike times recorded by the clock of the FF. Gray curve: clock of the FF (latches at rising edges). Points in the circles: Ideal spike times.

M_i the multipliers. To realize multiplication, different technologies can be used [23]. The neuron circuit consists of a membrane capacitor C_{mem} with capacitance C, an analog comparator A, and a flip flop (FF).

For the input spikes, state-of-the-art SNN implementations use stochastic coding [24]. This exploits the noise tolerance capability of NN models for computing efficiency. To perform inference with stochastic encoding, random variables are sampled. They are then used to conduct Bernoulli experiments, where the occurrence of a spike ("1") or no occurrence ("0") is determined. The spike value is denoted as $x_{s,i}$ (also shown in Fig. 3). A spike occurs with the probability $\frac{x_i}{x_{max}}$. The value x_i is the actual input value and x_{max} the maximum input value. For the binary input case, x_{max} is 1, and p_i is either 1 or 0 without a need for Bernoulli experiments. Note that in the non-binary case the random sampling and the subsequent computations can be repeated multiple times to achieve better estimates of the MAC results. The steps of computations of the MAC results in SNNs are as follows:

(1) The multipliers are first loaded with the correct weights or are assumed to be already loaded. Then the input spikes $x_{s,i}$ are provided to all multipliers in parallel. The multiplications are all computed in parallel as well. Finally, the output currents of all multipliers are accumulated by Kirchhoff's law and the resulting current is transmitted to the neuron circuit.

(2) In the neuron circuit, the transmitted current from the computing array charges C_{mem}. Once the voltage across C_{mem} reaches the threshold voltage V_{th}, an output spike is generated (see the analog comparator in Fig. 3). The spike time t_{fire} is acquired by a digital counter (not shown here) that tracks the clock cycles until the FF latches the spike signal.

(3) The MAC-value is determined by

$$\frac{v}{t_{fire}} = \sum_i w_i x_i, \qquad (1)$$

where $v = x_{max}\frac{CV_{th}}{I_{ON}}$ and I_{ON} is the on-state current of the multiplier. To perform the conversion from spike time to MAC value, a relation is used. Consider the set of spike times $S_{t_{fire}} = \{t_1, t_2, \ldots, t_L\}$, where t_L is the largest firing time, and $t_j < t_{j+1}$. Consider also the set of MAC-values $S_{\text{MAC}} = \{q_1, q_2, \ldots, q_L\}$, where $q_j \leq q_{j+1}$. In the relation $S_{t_{fire}} \rightarrow S_{\text{MAC}}$, the values are mapped using $m_j : t_j \rightarrow q_{L-j+1}$ due to the reciprocal. For this, a reciprocal unit (not shown

here) can be used to realize the mapping, which transforms t_{fire} to a MAC value. After completing the calculations, the neuron is reset by the transistor T_1 and can perform subsequent computations.

Because of the limited computing array size, a large vector product (i.e. with dimension higher than a) is separated into multiple smaller vector products. Due to this, digital adders are necessary for accumulation. These digital components are not shown here (adder, reciprocal unit, counter, etc.). They follow conventional designs and are not further discussed.

B. Errors in Analog Computing based IF-SNNs

A key aspect of the IF-SNNs is the spike time t_{fire}. It encodes the MAC value (see Eq. (1)). The time t_{fire} is a specific spike time in the set of ideal firing times $S_{t_{fire}}$, indexed with j. We drop j in some cases for readability. Consider the following definitions of t_{fire}. The *ideal spike time* (t_{fire}^{ideal}) is derived by comparing the voltage across C_{mem} directly with V_{th} (circles in Fig. 4). The *spike time recorded by the analog comparator* (t_{fire}^{acomp}) is the time at which the comparator returns a spike signal. The *spike time recorded by the clock* (t_{fire}^{clk}) is the time at which the clock of the FF produces a rising edge (t_i in Fig. 4). We use these definitions to reason about the errors in the analog computations of IF-SNNs.

Summed currents: In the computing array, the output currents of the multipliers are summed. The single or summed currents can have variations, which affect the correctness of the MAC value. For example, the technology used in the circuit is prone to process variation, or external factors such as temperature may affect the operation. Due to such effects the accumulated current changes. This shifts the firing time, i.e. t_{fire}^{ideal} becomes $t_{fire}^{ideal} + t_{var}$, whereas the sign of the value of t_{var} depends on the variation characteristics.

Capacitor: A capacitor has a certain capacitance C and charges when a voltage V_0 is applied. The curves for different applied voltages are shown in Fig. 4. The charge increases rapidly first, but slows down and eventually stops at the maximum capacitor charge i.e. when charged to $Q = CV_0$. The charging of a capacitor by application of V_0 is described by $V(t) = V_0(1 - e^{(-\frac{t}{\tau})})$, where t is the time, $\tau = RC$ is the time constant, and R is the internal resistance of the capacitor. Since τ is in the denominator, smaller C lead to larger absolute values in the exponent, which in turn causes faster capacitor charging. On the other hand, a larger capacitor leads to slower charging. Considering the neuron circuit in Sec. III the firing occurs when the charge in C_{mem} reaches the threshold charge by accumulating the current I_a with $Q_{fire} = CV_{th} = \int_0^{t_{fire}} I_j dt$, where index j determines the MAC value (see Sec. III, step (3)). This means the firing occurs ideally when $V(t) = V_{th}$. Moreover, the MAC value q_j has the current $I_j = \frac{q_j \cdot I_{on}}{x_{max}}$. From above integral over I_j, we get $CV_{th}=I_j t_{fire}$ and the relation between C and t_{fire} in case of a MAC value q_j: $\frac{C}{q_j} \propto t_{fire}$. In the integral equation, C is proportional to the firing time t_{L-j+1}. As the result, when C is too small, the voltage across it rises quickly. In this case, the frequency of *clk* in the FF (called f_{clk}, and T_{clk} is

979-8-3503-4631-2/23 $31.00 © 2023 IEEE   43

TABLE I
Datasets used for experiments.

Name	# Train	# Test	# Dim	# classes
FashionMNIST	60000	10000	(1,28,28)	10
KuzushijiMNIST	60000	10000	(1,28,28)	10
SVHN	73257	26032	(3,32,32)	10
CIFAR10	50000	10000	(3,32,32)	10

the clock period) cannot provide sufficient resolution to distinguish the t_{fire} with MAC values larger than q_j. For example, consider $S_{MAC} = \{q_1, q_2, q_3\}$ and $S_{t_{fire}} = \{t_1, t_2, t_3\}$. When $t_1 < T_{clk}$ and $t_2 < T_{clk}$, t_1 and t_2 are both latched at same rising edge, because $\lceil \frac{t_1}{T_{clk}} \rceil = \lceil \frac{t_2}{T_{clk}} \rceil$. This causes t_1 and t_2 both to be mapped to the same value q_3.

Analog comparator: If the voltage at the terminal $+$ is larger than the voltage at the terminal $-$, the output will be high and otherwise low. A real, nonideal comparator requires a certain absolute difference between the two inputs to return a high. This means, the comparator does not immediately return high when the voltage across the capacitor exceeds V_{th}, it only does so when there is a difference of Δ, i.e. when $V(t) + \Delta \geq V_{th}$. The size of Δ depends on the comparator implementation. This effect is further exacerbated by the logarithmically increasing function of the capacitor voltage over time. Due to its slow increase, the comparator may sense the spike signal with delays, i.e. $t_{fire}^{ideal} \leq t_{fire}^{acomp}$.

Flip-Flop (FF): It latches the output from the analog comparator. It also resets the charge in the capacitor to zero by transistor T_1. The clock signal clk determines the sensing frequency of the FF, as discussed above. As the signal from the analog comparator can only be latched at a rising edge of the clock, the spike time registered by the clock is delayed by at most one clock cycle (see Fig. 4), where $t_{fire}^{acomp} \leq t_{fire}^{clk}$.

In summary, we define the error caused by these nonidealities as the difference between the spike time registered by the clock and the ideal spike time, i.e. $\epsilon_{NI} = t_{fire}^{clk} - t_{fire}^{ideal}$. Inserting this error into Eq. (1) affects the MAC values:

$$\frac{v}{t_{fire} + \boxed{\epsilon_{NI}}} = \sum_i w_i x_i + \boxed{\epsilon_{MAC}}. \quad (2)$$

To extract the spike timing or MAC value deviations, SPICE simulations need to be performed and the error models need to be then extracted. To assess the impact of the error models on the accuracy of NN models, the error model has to be applied during operation. In the following, we use our tool SPICE-Torch to generate error models of IF-SNN circuits and assess their impact on NN accuracy, allowing us to perform neuromorphic optimization.

C. Neuromorphic Optimization: Experiment Setups and Results

To demonstrate the capabilities of framework SPICE-Torch, we use the following experiment setting. We employ Binarized Neural Networks (BNNs), which are executed as IF-SNNs using the hardware configuration shown in Fig. 3. By this, the multiplication components become XNOR gates, for which we use SRAM-based cells. We focus on the optimization of the capacitor size of C_{mem} and the resulting errors from it (see Eq. (2)), i.e. the reduction of the precision of the MAC result under different device settings. The capacitor size is a major bottleneck in the circuit design of IF-SNNs, as it also leads to high energy, area, and latency cost [25], [26]. SPICE-Torch is not limited to BNNs or the optimization of capacitor size. Any Quantized NN (QNN) can be used. Furthermore, any other component in the neuron circuit can be replaced by alternative designs or swept for optimization.

Setup for PyTorch: We evaluate following BNNs. A VGG3-based BNN, with FashionMNIST and KuzujishiMNIST, and a VGG7-based

TABLE II
BNN architectures. Layer types are fully connected (FC), convolutional (C), and maxpool (MP). Each convolutional layer is followed by a batch normalization layer, except the output layer.

Name	Architecture
VGG3	In $\rightarrow$ C64 $\rightarrow$ MP2 $\rightarrow$ C64 $\rightarrow$ MP2 $\rightarrow$ FC2048 $\rightarrow$ FC10
VGG7	In $\rightarrow$ C128 $\rightarrow$ C128 $\rightarrow$ MP2 $\rightarrow$ C256 $\rightarrow$ C256 $\rightarrow$ MP2 $\rightarrow$ C512 $\rightarrow$ C512 $\rightarrow$ MP2 $\rightarrow$ FC1024 $\rightarrow$ FC10

BNN with SVHN and CIFAR10. VGG3 and VGG7 are modified versions of the VGG-architectures [27], adapted for the above datasets. The details of the datasets and BNN architectures are in Tab. I and Tab. II. The BNNs use convolutional (C) layers with size 3×3, fully connected (FC), maxpool (MP) with size 2×2, and batch normalization (BN). We use Adam for optimizing BNNs and the modified hinge loss (MHL) with the hyperparameter $b = 128$ [28]. The batch size is 256 and the initial learning rate (LR) is 10^{-3} in all cases. We halve the LR every 10th epoch for Fashion, Kuzujishi, SVHN, and halve it every 50th epoch for CIFAR10. For each model we train 100 epochs for Fashion, Kuzujishi, SVHN, and 200 epochs for CIFAR10. When retraining, i.e. applying the error model during training, we load the training state of the BNN models trained without any error model and continue the training with application of the error model. We use additional five epochs for Fashion and additional ten for CIFAR10 with an LR of 10^{-4} in all cases.

Setup for SPICE As the underlying technology in the computing array (Fig. 3), we use Static Random Access Memory (SRAM)-based XNOR cells with 14 nm Fully-Depleted Silicon-on-Insulator (FD-SOI) technology. For this, we reproduce industry measurements of 14 nm FD-SOI technology with a ultra-thin body and buried oxide (BOX) design [29], for both, p-type and n-type. The transistor model-card parameters for the industry-standard compact model of FD-SOI (BSIM-IMG) are carefully tuned until they are in excellent agreement with the measurements. Additionally, the model has been calibrated to device-to-device variation measurements. For a comprehensive variability representation all important sources of process variation (gate work function, channel dimension, BOX and channel thickness) are considered. Through Simulation with Integrated Circuit Emphasis (SPICE) Monte-Carlo simulations based on the calibrated compact model, the standard deviation for each model parameter has been tuned to match the observed variation in the measurements. We use an array of $a = 32$ XNOR cells to realize the computing array. Each XNOR cell connects V_{DD} to the shared Match Line (ML) and forms a conducting path if the weight does *not* match the respective multiplication result, realizing the XNOR operation. Through the shared ML, Kirchhoff's law accumulates the individual results. The resulting current is proportional to the MAC level. This current is used to excite the neuron circuit; hence ML is connected to the membrane capacitor which charges it over time. To reduce SPICE simulation time, we use ideal verilog-a implementations of the remaining components (comparator and FF with 2 GHz). Opposite to the conventional XNOR cell design [30], which connects ML to GND (to *dis*charge), in this work, we need to charge ML, wherefore we connect the ML to V_{DD}. Consequently, the transistors along the charge path are substituted with p-type transistors.

Capacitor Size Optimization without Input Variation To optimize the capacitor size, we sweep C_{mem}. An ideal mapping of spike times to MAC values is found when each spike time recorded by the clock (t_{fire}^{clk}) has a unique MAC value, in which case the mapping has maximum precision. For efficiency, the smallest capacitor size with maximum mapping precision should be founf. However, C_{mem} can be further reduced at the cost of precision to represent the MAC values, as

979-8-3503-4631-2/23 $31.00 © 2023 IEEE

Nr. of MAC levels	30	20	10
FashionMNIST	85.99, (0.23, 0.33)	82.73, (0.17, 0.16)	62.14, (0.91, 0.82)
KuzushijiMNIST	91.36, (0.36, 0.25)	88.88, (0.13, 0.13)	51.96, (0.65, 0.75)
SVHN	71.21, (0.17, 0.14)	43.92, (0.28, 0.39)	12.70, (0.14, 0.17)
CIFAR10	50.55, (0.06, 0.07)	33.97, (0.19, 0.17)	10.1, (0.16, 0.11)

TABLE III

Avg. and (max.-avg., avg.-min.) accuracy for the case with variation in Fig. 5.

Case	Baseline	Direct mapping	Distr.-based mapping
VGG3	0.68, (0.07, 0.04)	3.23, (0.03, 0.03)	22.60, (0.18, 0.15)
VGG7	3.39 (0.09, 0.11)	122.59, (0.17, 0.13)	1001.82, (0.14, 0.16)

TABLE IV

Avg. and (max.-avg., avg.-min.) experiment run times in seconds for ten runs, 10^5 inputs in each run. Hardware: Intel Core i7-8700K 3.70 Ghz, 32 GB RAM, GeForce GTX 1080 8 GB.

explained in Sec. III-B (and Eq. (2)). In Fig. 5, we show the trade-off between the number of MAC levels (which determines the capacitor size) and the inference accuracy, using direct mappings (see Sec. II-B). The range of capacitor size is from $135.2\,\mathrm{pF}$ (highest nr. of MAC-levels, i.e. 32 levels) to $1\,\mathrm{pF}$ (lowest nr. of MAC-levels, i.e. 2 levels). For Fashion and Kuzujishi, 16 levels ($32.36\,\mathrm{pF}$) retains high inference accuracy. For a smaller number of levels, the accuracy collapses. For SVHN and CIFAR10, more levels (a larger capacitor) is needed, since high accuracy is retained only for down to 20 levels ($52.40\,\mathrm{pF}$). When retraining (purple plots), i.e. applying the error model during training, the capacitor size can be reduced even further (to $1.74\,\mathrm{pF}$), and only drops completely for less than four MAC values for all datasets.

Capacitor Size Optimization under Input Variation When process variation is considered for the transistors in the XNOR array, the ON current (the current that each XNOR cell releases to return a "1") is subject to variation. This varies the charge speed of C_{mem}. Consequently, the spike times registered by the clock (or the MAC levels) may change, leading to the computation of wrong MAC values. To extract the error model for the case with process variation, we use a Monte-Carlo approach and 1000 samples per MAC level. For each MAC level, a bucket is formed by placing decision boundaries midway between the spike times on either side. The Monte-Carlo samples of a given MAC level are sorted into these buckets, counted, and the result normalized. This yields the probability to map from a MAC value $q_i \in S_{\mathrm{MAC}}$ to all possible values of a different set S'_{MAC} (see Sec. II-B). Repeating this for all MAC levels in S_{MAC} constructs the entire mapping $S_{\mathrm{MAC}} \to S'_{\mathrm{MAC}}$. The errors from process variation are injected during the inference of NNs and the average test accuracy is reported in Fig. 5. For Fashion/Kuzujishi five experiment repetitions are conducted and for SVHN/CIFAR10 three. The average and min. and max. values observed are shown in Tab. III. We observe that for Fashion and Kuzujishi, the accuracy under process variation is lower than with no variation. For these two datasets, the accuracy can be up to $15\,\%$ lower compared to the original accuracy, for up to 15 MAC levels. For SVHN and CIFAR10, the accuracy drop is even larger and drops by around $50\,\%$. These drops are expected, since the MAC mappings are not deterministic. Because of this there are multiple possibilities for wrong MAC values. We conclude that if process variation is considered, the design of analog computation hardware needs methods to mitigate the effects of process variation.

Execution time: We report the run times for the experiments of Fig. 5. Extracting direct mappings from SPICE takes two hours and for the probability based mappings 20 hours. The run times for the PyTorch evaluations are in Tab. IV. Applying error models with direct mapping takes less time than the probability based mapping. Further, applying the error models for VGG7 takes more time than for VGG3.

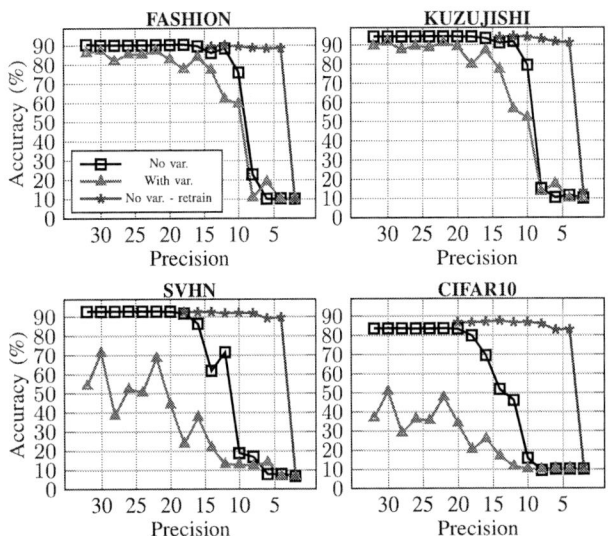

Fig. 5: Accuracy over number of MAC levels, which correspond to capacitor sizes. Capacitor size range: From $135.2\,\mathrm{pF}$ (highest nr. of MAC-levels, i.e. 32 levels) to $1\,\mathrm{pF}$ (lowest nr. of MAC-levels, i.e. 2 levels). Case with no variation (additionally with retraining) and no variation is shown. For case with variation, deviations are in Tab. III.

IV. MEMRISTOR BASED BRAIN INSPIRED HARDWARE

A. Overview of Memristor Technology

Memristor is a two-terminal device which can regulate the flow of current based on the history of the applied voltage across it [31]. A wide variety of material stacks with both volatile and non-volatile characteristics have been reported in the literature [10]. These materials exhibit resistive switching due to diverse switching mechanisms including ionic drift [32], electrochemical filament formation [21], [33], oxygen vacancy migration [20], and interface effects [34], etc. There are various advantages and disadvantages associated with each memristive switching mechanism, which are summarized in Table I. Memristors can also be differentiated based on their volatility, with volatile memristors requiring a continuous power supply to maintain their resistive state. In contrast, nonvolatile memristors retain their resistive state even when no power is applied, making them suitable for non-volatile memory applications like solid-state drives (SSDs) [21], [35]. Volatile memristors are typically used in dynamic random-access memory (DRAM) applications where stored data is not required when the power is off [36]. Furthermore, they are highly scalable and suitable for brain-inspired neuromorphic hardware, thanks to their low operating voltage [10].

B. Memristor-based Neurons

In neuroscience, the neuron is recognized as the fundamental unit of computation. As neuromorphic computing seeks to replicate the neurobiological computational primitive, the neuron remains the central building block of such systems [10]. Neurons in neuromorphic computing are typically categorized into two main types: bio-plausible and bio-inspired models [37], [38]. Bio-plausible models attempt to mimic the behavior of biological neurons, while bio-inspired models leverage some of the features of biological neurons [37], [38]. To achieve design simplicity and energy efficiency, memristors are employed in memristor-based neuron circuits to replace some of the

Table I: Commonly Observed Switching Mechanisms in Memristors.

Switching Mechanism	Advantages	Disadvantages
Ionic Drift Mechanism	Fast Operation. Ease of tunability.	Low retention. High sensitivity to variation
Electrochemical Filament formation.	High Endurance Can be easily adjusted through voltage.	Slow Operation High sensitivity to variation
Oxygen-vacancy migration	High efficiency	Prone to environmental factors
Interface Effects	High tunability	Low reliability

traditional CMOS components Fig. 6(a-c). A variety of memristor-based bio-realistic neuron implementations have been proposed, including the Hodgkin-Huxley model [39], Morris and Lecar model [40], FitzHugh-Naguma model [41], and Hindmarsh-Rose model [42]. Memristors have also been used in developing bioinspired neuron models that incorporate dynamics such as neuron stochasticity [43], integration, and firing [44]. For example, Lashkare et al. proposed an integrate and fire-spiking neuron that utilized memristors [45]. Memristors with both unipolar and bipolar switching have been modeled and simulated to design novel neuron topologies [46], [47]. Al-Shedivat demonstrated how the stochasticity in the switching of memristors can be used to create a stochastic neuron [48]. In a recent study, Pantazi et al. proposed an all-memristor architecture-level implementation of a neuromorphic network that utilized the integrate and fire functionality in the neuron and plasticity in the synapse component [49].

The use of memristors in neuron circuits brings about significant advantages such as design simplicity and reduced area. However, it is highly challenging to design bio-realistic neurons based on memristors only. While CMOS technology can replicate the dynamics of memristors, the reverse is not true, as memristors are unable to fully emulate all the dynamics offered by CMOS [50], [51]. As a result, memristors are incorporated into neuron designs to replace some of the CMOS components [52]. Despite this, most proposed neuromorphic networks are implemented on hybrid platforms as memristors are primarily suitable for synaptic characteristics [53]. In the upcoming section, we will delve further into the unique features of memristors that make them ideal for synaptic functions.

C. Memristor-based Synapses

A synapse is a key component in neuromorphic computing. It resides between the neurons and controls the transfer of spikes by varying the connectivity strength [54]. The primary features of the memristor-based synapses are (i) Linearity, (ii) Non-volatility, (iii) symmetry, (iv) plasticity, and (v) low power consumption. Due to material-level challenges, all of these features have not been achieved simultaneously. However, innovation at the material and device level has resulted in the development of a wide range of memristors that exhibit a diverse range of operating voltages and features (Table II). Typically, memristors are connected between neurons in a crossbar fashion and their weights are updated based on the timing history of the pre-neuron and post-neuron known as synaptic time-dependent plasticity (STDP) Fig. 6(e-g). This leads to non-linear values for the synaptic weight, which is typically used for unsupervised training/learning with the winner-take-all (WTA) algorithm in the machine-learning- based classification task.

In deep learning, the synaptic weight's linearity is of great importance as it is highly required for efficient vector-matrix multiplication (VMM). A memristor used for VMM processing must exhibit a linear increase and decrease in their potentiation and depression of conductance. Moreover, the memristor's symmetry between potentiation and depression is crucial for effective learning in the neural network. Some of the proposed memristor-based synapse and neuron concepts reported in the literature are summarized in Table II [34], [55]–[69]. Memristors with both unipolar and bipolar switching is conceptualized as the synapse. However, bipolar memristors offer a significant advantage over unipolar memristors by providing multistate capabilities in both potentiation and depression [70]. One of the most critical features of memristor-based synapses is their non-volatile nature, allowing them to retain synaptic weights without requiring power. Modulating non-volatile memristors with the STDP algorithm on-chip is relatively easy to accomplish. Even if the switching of the memristor is volatile, it is possible to achieve linear potentiation by careful optimization of the pulse [71]. to maintain high network accuracy, it may be necessary to precisely modulate the device conductance over a wide dynamic range while ensuring linearity. In this type of synapse, the synaptic weight can be represented by the combined conductance of multiple cells [72]. In addition to STDP learning, conventional neural network learning has also made use of the memristor-based synapses [73].

D. Memristor-based learning

In neuromorphic computing, on-chip learning mainly utilizes synaptic time-dependent plasticity (STDP), which is a learning technique that depends on the timing of pre-neuron and post-neuron activity to cause either potentiation or depression [74]. Literature has documented the use of both supervised and unsupervised STDP learning techniques [75], [76]. Pedretti et al. first demonstrated the use of memristor-based STDP learning for tasks such as data clustering and anomaly detection [77]. Stochastic memristors have also been used to perform an STDP-based visual pattern recognition task in a later work [78]. To implement supervised STDP learning, the polarization property of ferroelectric memristors was utilized. Here, the bi-directional polarization is utilized for both forward propagation and backpropagation algorithm. During the back-propagation scheme, the weight is adjusted in every iteration of training. To implement back-propagation, synapses need to be accessible from both forward and backward directions for epoch and updating data [79]. IBM has developed a back-propagation circuit, shown in Fig. 6(h), which utilizes two select transistors to access and update the memristor within the synapse structure [80].

Off-chip learning of memristive synapse has also been proposed. When computational speed and power are crucial in hardware, off-chip learning can be performed separately and later mapped [81]. Normally, on-chip STDP learning is more power-efficient than off-chip STDP learning because it eliminates the need for data transfer between the chip and external memory, which consumes considerable power. However, if the network's size is very large, the power consumption of on-chip STDP learning may increase, making off-chip learning a more energy-efficient alternative.

E. Memristor-based neuromorphic system

A complete neuromorphic architecture commonly utilizes the hybrid analog/digital circuit for mimicking brain functionality as its computational primitive [82], [83]. The existing implementation involves CMOS technology for mixed-signal computational primitive as a hardware accelerator Fig. 6(i). While these implementations show impressive accuracy, they still suffer from the limitation of separate memory and computation block, known as the von Neumann bottleneck [84]. Memristors offer a promising solution to this challenge as

979-8-3503-4631-2/23 $31.00 © 2023 IEEE

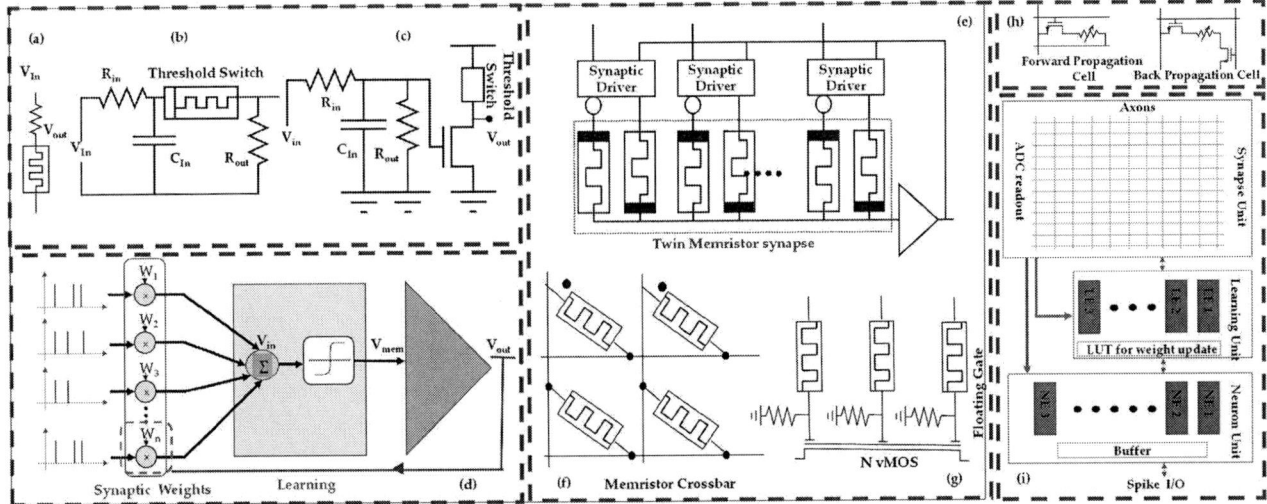

Fig. 6: (a)-(c): Different Neuron Topologies [67],[18]. (d) Electronic model of a neuromorphic system showing the integration of weighted spikes (e) Twin memristor synapse arrays connected with a neuron [76]. (f) A memristor crossbar synapse [93]. (g) A memristor synapse with a floating gate transistor [93]. (h) Memrsitor-based forward and back-propagation cell [80](i) A complete memristor-based neuromorphic architecture as shown in [83].

they can be used for storage and computation simultaneously owing to their non-volatility and scalability.

Neuromorphic systems rely heavily on vector-matrix multiplication for their computation, in which the input vector is multiplied with a stored vector. Neurons with the highest multiplication results transmit their spikes to subsequent stages of the network. The values of the stored vector are directly mapped onto the memristor crossbar arrays as synaptic weights. The input vector is encoded in the amplitude of the input voltage spike and the output is represented by the current through the memristor computed naturally by Ohm's law. Several recent efforts have incorporated memristor-based crossbar arrays for vector-matrix multiplication and dot-product operation [85], [86]. Successful demonstrations of the memristor-based neuromorphic system include object classification, pattern recognition, sparse encoding, etc [87]. Hewlett Packard Labs has shown the initial Table II: Different Memristor stacks reported in the literature [88].demonstration of a memristor-based dot-product engine for neuromorphic application which encouraged several other works of dot-product engine-based classification tasks. Both on-chip learning and off-chip learning-based tasks have been extensively demonstrated by researchers in the neuromorphic community [89]. By far, off-chip learning shows promising performance in the most computationally intensive tasks as on-chip learning is still challenging to implement because of non-linearity and interconnect issues.

F. Challenges and prospects of Memristor technologies

Memristors suffer from several materials and device-level challenges which hamper the implementation of large-scale efficient neuromorphic hardware [90]. Linear and incremental synaptic weight update features are crucial for maintaining sufficient accuracy in most machine learning algorithms, but this is primarily governed by the materials and device structure, posing a limitation. This limitation can be addressed by using memristors with fewer levels and employing multiple memristors for a single weight [72]. However, it causes the degradation of synaptic weight density and additional area overhead.

Device level variation is another challenge of memristors in a synaptic network as it causes inconsistent memristor behavior altering the computed output [91]. However, device variability may not be a major concern for on-chip learning due to the adaptable nature of learning. Here, the impact of device level variation can be mitigated during the learning process [92]. Some of the recent works even leveraged the device level variation to achieve a faster reconstruction and convergence. Since the read operation of memristors dominates the vector-matrix multiplication operation, and the network training phase requires only a few set/reset cycles, device degradation due to frequent data overwrite is not a significant concern in most cases [93]. The Memristor crossbars present a viable platform to execute vector-matrix operations necessary for neuromorphic computing and other data-intensive tasks in a highly efficient manner on a vast scale [93].

While there is a potential for the widespread application of large-scale neuromorphic systems with the progress of device technology and architecture, accomplishing these objectives will necessitate further research involving materials, devices, architecture, and algorithm development.

V. CONCLUSION

In the first part of this special session, we present SPICE-Torch, a framework that connects the low-level SPICE simulations of analog computing circuits with the high-level accuracy evaluations of NNs in PyTorch. We demonstrate the capabilities of our framework by optimizing the capacitor size of an analog IF-SNN circuit. When considering no process variation, the capacitor size can be optimized with retention of high inference accuracy. However, with process variation, the accuracy drops are very large. This shows that our framework can be used for neuromorphic optimization and can identify problems in the designs early. Users of our framework can replace any component in the SPICE simulation or design custom analog computing circuits. They can also use other quantized NNs for evaluating the inference accuracy under the circuit configurations. We have used SPICE-Torch or early variants of it to perform the experiments regarding brain-inspired hardware using emerging memories in the following works [28], [94]. Our framework is available as open-source in https://github.com/myay/SPICE-Torch. With SPICE-Torch, we envision great benefits for hardware design across the abstraction layers. We believe that our tool will benefit the research and industry

979-8-3503-4631-2/23 $31.00 © 2023 IEEE

Table II: A collection of memristor stacks reported in the literature that are promising for neuromorphic applications

Materials/ Stack	Switching Mechanism	# of Synaptic States	Neuron Type	R_{ON} / R_{Off}	Endurance, Retention	V_{set} , V_{reset}	Ref.
Ag/Pd/SiGe	Electrochemical Filament	100	N/A	100	> 10^9 cycles, 48 h	4 V , -3 V	[55]
Ag/AgInSbTe	Electrochemical Filament	50	N/A	2	- , >10 min	0.5 V , -1 V	[56]
Ag/Si	Electrochemical Filament	100	N/A	10	1.5×10^8 cycles, ~ 5 years	3.2 V, -2.8 V	[57]
HfO_2 / AlO_x	Oxygen Vacancy Filament	40	N/A	3	- , 3000s	0.9 V , -1.0 V	[58]
Ta/TaO_x/TiO_2/Ti	Interface Effects	50	N/A	2	5000 cycles, ~3-5 ms	5 V, - 6V	[60]
Mo/PCMO	Interface Effects	32	N/A	15	100 cycles, >30s	-1.5 to -3.5 V, 3V	[61]
Al/Mo/PCMO	Interface Effects	100	N/A	100	50 cycles, 3×10^4 s	-3 V , 2 V	[34]
Mo/TiO_x	Interface Effects	64	N/A	20			[62]
TiO_x/TiO_y	Interface Effects	100	N/A	10	>50 cycles, -	15 V, -15 V	[63]
BTO/LSMO	Ferroelectric Tunneling	100	N/A	10			[64]
Ag/HfO_2/Pt	Oxygen Vacanccy Filament	-	Sigmoid	10^4	>100 cycles, 2000 s	0.21 V, -0.1 V	[66]
Ag/SiO_2/Pt	Electrochemical Filament	8	Stochastic	10^4-10^7	10^{12} cycles, -	-1.6 V, -1.5 V	[65]
TiN/$SiO_{1.3}$/TiN	Oxygen Vacanccy Filament	9	I & F	150	>150 cycles, 10^4 s	3.5 V-4 V, 2.5V	[67]
TiN/PCMO/Pt	Interface Effects	~10	I & F	10^3	-	4 V,-3 V	[68]
Ag/Si_3N_4/TiN	Electrochemical Filament	-	I & F	10^5	-	~1 V, ~ -1 V	[69]

for evaluating the feasibility of edge AI systems designs that use analog and digital computations.

In the second part of this special session, we discussed the prospect of using memristive devices to design brain-inspired computing hardware. Memristors have proven to be a promising alternative to traditional CMOS components in neuromorphic computing. Memristors have been proven to be suitable for bio-plausible functions as well. However, it is challenging to design systems based solely on memristors. Most of the proposed neuromorphic networks are implemented on hybrid platforms. Nonetheless, memristors show great potential for developing energy-efficient and highly parallel hardware. The ongoing research in memristive devices is expected to pave the way for the development of next-generation computing systems which attempt to reach the efficiency and robustness of the human brain. In future work, we will perform neuromorphic exploration with SPICE-Torch regarding the design of robust memristor-based AI hardware.

ACKNOWLEDGMENT

This paper has been supported by Deutsche Forschungsgemeinschaft (DFG) by project ACCROSS (428566201), project OneMemory (405422836), by the Collaborative Research Center SFB 876 "Providing Information by Resource-Constrained Analysis" (project number 124020371), subproject A1 (http://sfb876.tu-dortmund.de) and by the Federal Ministry of Education and Research of Germany and the state of NRW as part of the Lamarr-Institute for ML and AI, LAMARR22B. This work was also supported in part by the Air Force Research Laboratory (USA) under Agreement FA8750-21-1-1018. The U.S. Government is authorized to reproduce and distribute reprints for Governmental purposes notwithstanding any copyright notation

thereon. The views and conclusions contained herein are those of the authors and should not be interpreted as necessarily representing the official policies or endorsements, either expressed or implied, of Air Force Research Laboratory or the U.S. Government.

REFERENCES

[1] A. Gupta, K. Ni, O. Prakash, X. S. Hu, and H. Amrouch, "Temperature Dependence and Temperature-Aware Sensing in Ferroelectric FET," in *2020 IEEE International Reliability Physics Symposium (IRPS)*, 2020, pp. 1–5.

[2] H. Amrouch, S. Salamin, G. Pahwa, A. D. Gaidhane, J. Henkel *et al.*, "Unveiling the Impact of IR-Drop on Performance Gain in NCFET-Based Processors," *IEEE Transactions on Electron Devices*, vol. 66, no. 7, pp. 3215–3223, 2019.

[3] O. Prakash, A. Gupta, G. Pahwa, J. Henkel, Y. S. Chauhan *et al.*, "Impact of Interface Traps on Negative Capacitance Transistor: Device and Circuit Reliability," *IEEE Journal of the Electron Devices Society*, vol. 8, pp. 1193–1201, 2020.

[4] M. Rapp, S. Salamin, H. Amrouch, G. Pahwa, Y. Chauhan *et al.*, "Performance, power and cooling trade-offs with NCFET-based many-cores," in *Proceedings of the 56th Annual Design Automation Conference 2019*, 2019, pp. 1–6.

[5] G. Paim, L. M. G. Rocha, H. Amrouch, E. A. C. da Costa, S. Bampi *et al.*, "A Cross-Layer Gate-Level-to-Application Co-Simulation for Design Space Exploration of Approximate Circuits in HEVC Video Encoders," *IEEE Transactions on Circuits and Systems for Video Technology*, vol. 30, no. 10, pp. 3814–3828, 2020.

[6] H. Amrouch and J. Henkel, "Lucid infrared thermography of thermally-constrained processors," in *2015 IEEE/ACM International Symposium on Low Power Electronics and Design (ISLPED)*, 2015, pp. 347–352.

[7] M. Rapp, H. Amrouch, Y. Lin, B. Yu, D. Z. Pan *et al.*, "MLCAD: A Survey of Research in Machine Learning for CAD Keynote Paper," *IEEE Transactions on Computer-Aided Design of Integrated Circuits and Systems*, vol. 41, no. 10, pp. 3162–3181, 2022.

979-8-3503-4631-2/23 $31.00 © 2023 IEEE

[8] M. Bouvier, A. Valentian, T. Mesquida, F. Rummens, M. Reyboz et al., "Spiking Neural Networks Hardware Implementations and Challenges: A Survey," *J. Emerg. Technol. Comput. Syst.*, vol. 15, no. 2, Apr. 2019. [Online]. Available: https://doi.org/10.1145/3304103

[9] A. Shafiee, A. Nag, N. Muralimanohar, R. Balasubramanian, J. P. Strachan et al., "ISAAC: A Convolutional Neural Network Accelerator with In-Situ Analog Arithmetic in Crossbars," in *International Symposium on Computer Architecture (ISCA)*, 2016.

[10] J. Zhu, T. Zhang, Y. Yang, and R. Huang, "A comprehensive review on emerging artificial neuromorphic devices," *Applied Physics Reviews*, vol. 7, no. 1, p. 011312, 2020.

[11] S. Alam, M. M. Islam, J. Hutchins, N. Cady, S. Gupta et al., "Threshold Switch Assisted Memristive Memory with Enhanced Read Distinguishability," in *2022 IEEE 22nd International Conference on Nanotechnology (NANO)*. IEEE, 2022, pp. 531–534.

[12] A. Aziz, N. Jao, S. Datta, and S. K. Gupta, "Analysis of functional oxide based selectors for cross-point memories," *IEEE Transactions on Circuits and Systems I: Regular Papers*, vol. 63, no. 12, pp. 2222–2235, 2016.

[13] S. Alam, M. M. Islam, A. Jaiswal, N. Cady, G. Rose et al., "Variation-aware Design Space Exploration of Mott Memristor-based Neuristors," in *2022 IEEE Computer Society Annual Symposium on VLSI (ISVLSI)*. IEEE, 2022, pp. 68–73.

[14] Z. Wang, M. Rao, J.-W. Han, J. Zhang, P. Lin et al., "Capacitive neural network with neuro-transistors," *Nature communications*, vol. 9, no. 1, p. 3208, 2018.

[15] K.-U. Demasius, A. Kirschen, and S. Parkin, "Energy-efficient memcapacitor devices for neuromorphic computing," *Nature Electronics*, vol. 4, no. 10, pp. 748–756, 2021.

[16] A. Zeumault, S. Alam, M. Omar Faruk, and A. Aziz, "Memristor compact model with oxygen vacancy concentrations as state variables," *Journal of Applied Physics*, vol. 131, no. 12, p. 124502, 2022.

[17] A. Zeumault, S. Alam, Z. Wood, R. J. Weiss, A. Aziz et al., "TCAD modeling of resistive-switching of HfO2 memristors: Efficient device-circuit co-design for neuromorphic systems," *Frontiers in Nanotechnology*, vol. 3, p. 734121, 2021.

[18] A. Aziz and K. Roy, "Insulator-Metal Transition Material Based Artificial Neurons: A Design Perspective," in *2020 21st International Symposium on Quality Electronic Design (ISQED)*. IEEE, 2020, pp. 444–451.

[19] A. Aziz, "Device-circuit co-design employing phase transition materials for low power electronics," Ph.D. dissertation, Purdue University Graduate School, 2019.

[20] H. G. Hwang, Y. Pyo, J. U. Woo, I. S. Kim, S. W. Kim et al., "Engineering synaptic plasticity through the control of oxygen vacancy concentration for the improvement of learning accuracy in a Ta2O5 memristor," *Journal of Alloys and Compounds*, 2022.

[21] J. H. Ryu, F. Hussain, C. Mahata, M. Ismail, Y. Abbas et al., "Filamentary and interface switching of CMOS-compatible Ta2O5 memristor for non-volatile memory and synaptic devices," *Applied Surface Science*, 2020.

[22] T. Tang, L. Xia, B. Li, R. Luo, Y. Chen et al., "Spiking neural network with rram: Can we use it for real-world application?" in *Design, Automation & Test in Europe Conference & Exhibition (DATE)*. IEEE, 2015.

[23] M.-L. Wei, H. Amrouch, C.-L. Sung, H.-T. Lue, C.-L. Yang et al., "Robust Brain-Inspired Computing: On the Reliability of Spiking Neural Network Using Emerging Non-Volatile Synapses," in *International Reliability Physics Symposium (IRPS)*, 2021.

[24] M. Koo, G. Srinivasan, Y. Shim, and K. Roy, "SBSNN: Stochastic-bits enabled binary spiking neural network with on-chip learning for energy efficient neuromorphic computing at the edge," *IEEE Transactions on Circuits and Systems I: Regular Papers*, vol. 67, no. 8, 2020.

[25] Y. Xiang, P. Huang, R. Han, C. Li, K. Wang et al., "Efficient and Robust Spike-Driven Deep Convolutional Neural Networks Based on NOR Flash Computing Array," *IEEE Transactions on Electron Devices*, vol. 67, no. 6, 2020.

[26] M.-L. Wei, M. Yayla, S.-Y. Ho, J.-J. Chen, C.-L. Yang et al., "Binarized SNNs: Efficient and Error-Resilient Spiking Neural Networks through Binarization," in *International Conference On Computer Aided Design (ICCAD)*, 2021.

[27] K. Simonyan and A. Zisserman, "Very Deep Convolutional Networks for Large-Scale Image Recognition," in *International Conference on Learning Representations, (ICLR)*, 2015.

[28] S. Buschjäger, J.-J. Chen, K.-H. Chen, M. Günzel, C. Hakert et al., "Margin-Maximization in Binarized Neural Networks for Optimizing Bit Error Tolerance," in *Design, Automation Test in Europe Conference Exhibition (DATE)*, 2021.

[29] Q. Liu, M. Vinet, J. Gimbert, N. Loubet, R. Wacquez et al., "High performance UTBB FDSOI devices featuring 20nm gate length for 14nm node and beyond," in *International Electron Devices Meeting*, 2013.

[30] M. Bennaser and C. A. Moritz, "Power and failure analysis of cam cells due to process variations," in *International Conference on Electronics, Circuits and Systems*, 2006.

[31] L. Chua, "Memristor-The missing circuit element," *IEEE Transactions on Circuit Theory*, vol. 18, no. 5, pp. 507–519, 1971.

[32] D. B. Strukov and R. S. Williams, "Exponential ionic drift: Fast switching and low volatility of thin-film memristors," *Applied Physics A: Materials Science and Processing*, 2009.

[33] C. Yakopcic, S. Wang, W. Wang, E. Shin, J. Boeckl et al., "Filament formation in lithium niobate memristors supports neuromorphic programming capability," *Neural Computing and Applications*, 2018.

[34] K. Moon, A. Fumarola, S. Sidler, J. Jang, P. Narayanan et al., "Bidirectional non-filamentary RRAM as an analog neuromorphic synapse, Part I: Al/Mo/Pr0.7Ca0.3MnO3 material improvements and device measurements," *IEEE Journal of the Electron Devices Society*, 2018.

[35] M. Jerry, P. Y. Chen, J. Zhang, P. Sharma, K. Ni et al., "Ferroelectric FET analog synapse for acceleration of deep neural network training," in *Technical Digest - International Electron Devices Meeting, IEDM*, 2018.

[36] S. Hamdioui, H. Aziza, and G. C. Sirakoulis, "Memristor based memories: Technology, design and test," in *Proceedings - 2014 9th IEEE International Conference on Design and Technology of Integrated Systems in Nanoscale Era, DTIS 2014*, 2014.

[37] A. K. Bhoi, P. K. Mallick, C.-M. Liu, and V. E. Balas, *Bio-inspired neurocomputing*. Springer, 2021, vol. 310.

[38] G. Qiao, S. Hu, J. Wang, C. Zhang, T. Chen et al., "A neuromorphic-hardware oriented bio-plausible online-learning spiking neural network model," *IEEE Access*, vol. 7, pp. 71730–71740, 2019.

[39] F. Corinto, A. Ascoli, and S. K. Sung-Mo, "Memristor-based neural circuits," in *2013 IEEE International Symposium on Circuits and Systems (ISCAS)*. IEEE, 2013, pp. 417–420.

[40] A. Amirsoleimani, M. Ahmadi, and A. Ahmadi, "STDP-based unsupervised learning of memristive spiking neural network by Morris-Lecar model," in *2017 International Joint Conference on Neural Networks (IJCNN)*. IEEE, 2017, pp. 3409–3414.

[41] J. Zhang and X. Liao, "Synchronization and chaos in coupled memristor-based FitzHugh-Nagumo circuits with memristor synapse," *Aeu-international journal of electronics and communications*, vol. 75, pp. 82–90, 2017.

[42] B. Bao, A. Hu, H. Bao, Q. Xu, M. Chen et al., "Three-dimensional memristive Hindmarsh–Rose neuron model with hidden coexisting asymmetric behaviors," *Complexity*, vol. 2018, pp. 1–11, 2018.

[43] T. Tuma, A. Pantazi, M. Le Gallo, A. Sebastian, and E. Eleftheriou, "Stochastic phase-change neurons," *Nature nanotechnology*, vol. 11, no. 8, pp. 693–699, 2016.

[44] D. Lee, M. Kwak, K. Moon, W. Choi, J. Park et al., "Various threshold switching devices for integrate and fire neuron applications," *Advanced Electronic Materials*, vol. 5, no. 9, p. 1800866, 2019.

[45] S. Lashkare, S. Chouhan, T. Chavan, A. Bhat, P. Kumbhare et al., "PCMO RRAM for integrate-and-fire neuron in spiking neural networks," *IEEE Electron Device Letters*, vol. 39, no. 4, pp. 484–487, 2018.

[46] D. Howard, L. Bull, and B. de Lacy Costello, "Evolving unipolar memristor spiking neural networks," *Connection Science*, vol. 27, no. 4, pp. 397–416, 2015.

[47] V. Ntinas, I. Vourkas, A. Abusleme, G. C. Sirakoulis, and A. Rubio, "Experimental study of artificial neural networks using a digital memristor simulator," *IEEE transactions on neural networks and learning systems*, vol. 29, no. 10, pp. 5098–5110, 2018.

[48] M. Al-Shedivat, R. Naous, G. Cauwenberghs, and K. N. Salama, "Memristors empower spiking neurons with stochasticity," *IEEE journal on Emerging and selected topics in circuits and systems*, vol. 5, no. 2, pp. 242–253, 2015.

[49] A. Pantazi, S. Woźniak, T. Tuma, and E. Eleftheriou, "All-memristive neuromorphic computing with level-tuned neurons," *Nanotechnology*, vol. 27, no. 35, p. 355205, 2016.

[50] V. Saxena, X. Wu, and K. Zhu, "Energy-efficient CMOS memristive synapses for mixed-signal neuromorphic system-on-a-chip," in *2018 IEEE International Symposium on Circuits and Systems (ISCAS)*. IEEE, 2018, pp. 1–5.

[51] Y. Babacan and F. Kaçar, "Memristor emulator with spike-timing-dependent-plasticity," *AEU-International Journal of Electronics and Communications*, vol. 73, pp. 16–22, 2017.

979-8-3503-4631-2/23 $31.00 © 2023 IEEE

[52] A. Mehonic and A. J. Kenyon, "Emulating the electrical activity of the neuron using a silicon oxide RRAM cell," *Frontiers in neuroscience*, vol. 10, p. 57, 2016.

[53] M. R. Azghadi, B. Linares-Barranco, D. Abbott, and P. H. Leong, "A hybrid CMOS-memristor neuromorphic synapse," *IEEE transactions on biomedical circuits and systems*, vol. 11, no. 2, pp. 434–445, 2016.

[54] Z. Lv, Y. Zhou, S.-T. Han, and V. Roy, "From biomaterial-based data storage to bio-inspired artificial synapse," *Materials today*, vol. 21, no. 5, pp. 537–552, 2018.

[55] S. Choi, S. H. Tan, Z. Li, Y. Kim, C. Choi *et al.*, "SiGe epitaxial memory for neuromorphic computing with reproducible high performance based on engineered dislocations," *Nature Materials*, 2018.

[56] J. J. Zhang, H. J. Sun, Y. Li, Q. Wang, X. H. Xu *et al.*, "AgInSbTe memristor with gradual resistance tuning," *Applied Physics Letters*, 2013.

[57] S. H. Jo, T. Chang, I. Ebong, B. B. Bhadviya, P. Mazumder *et al.*, "Nanoscale memristor device as synapse in neuromorphic systems," *Nano Letters*, 2010.

[58] J. Woo, K. Moon, J. Song, S. Lee, M. Kwak *et al.*, "Improved synaptic behavior under identical pulses using AlOx/HfO2 bilayer RRAM array for neuromorphic systems," *IEEE Electron Device Letters*, 2016.

[59] Z. Wang, M. Yin, T. Zhang, Y. Cai, Y. Wang *et al.*, "Engineering incremental resistive switching in TaO: X based memristors for brain-inspired computing," *Nanoscale*, 2016.

[60] I.-T. Wang, C.-C. Chang, L.-W. Chiu, T. Chou, and T.-H. Hou, "3D Ta/TaOx/TiO2/Ti synaptic array and linearity tuning of weight update for hardware neural network applications," *Nanotechnology*, vol. 27, no. 36, p. 365204, 2016.

[61] K. Moon, E. Cha, J. Park, S. Gi, M. Chu *et al.*, "Analog synapse device with 5-b MLC and improved data retention for neuromorphic system," *IEEE Electron Device Letters*, 2016.

[62] J. Park, M. Kwak, K. Moon, J. Woo, D. Lee *et al.*, "TiO x-based RRAM synapse with 64-levels of conductance and symmetric conductance change by adopting a hybrid pulse scheme for neuromorphic computing," *IEEE Electron Device Letters*, vol. 37, no. 12, pp. 1559–1562, 2016.

[63] K. Seo, I. Kim, S. Jung, M. Jo, S. Park *et al.*, "Analog memory and spike-timing-dependent plasticity characteristics of a nanoscale titanium oxide bilayer resistive switching device," *Nanotechnology*, 2011.

[64] A. Chanthbouala, V. Garcia, R. O. Cherifi, K. Bouzehouane, S. Fusil *et al.*, "A ferroelectric memristor," *Nature Materials*, 2012.

[65] P. Wijesinghe, A. Ankit, A. Sengupta, and K. Roy, "An all-memristor deep spiking neural computing system: A step toward realizing the low-power stochastic brain," *IEEE Transactions on Emerging Topics in Computational Intelligence*, vol. 2, no. 5, pp. 345–358, 2018.

[66] Q. Duan, L. Xu, J. Zhu, X. Sun, Y. Yang *et al.*, "Resistive switching and synaptic plasticity in HfO2-based memristors with single-layer and bilayer structures," in *China Semiconductor Technology International Conference 2018, CSTIC 2018*, 2018.

[67] A. Mehonic and A. J. Kenyon, "Emulating the electrical activity of the neuron using a silicon oxide RRAM cell," *Frontiers in Neuroscience*, 2016.

[68] K. Baek, S. Park, J. Park, Y.-M. Kim, H. Hwang *et al.*, "In situ TEM observation on the interface-type resistive switching by electrochemical redox reactions at a TiN/PCMO interface," *Nanoscale*, vol. 9, no. 2, pp. 582–593, 2017.

[69] M. W. Kwon, S. Kim, M. H. Kim, J. Park, H. Kim *et al.*, "Integrate-and-fire (IF) neuron circuit using resistive-switching random access memory (RRAM)," *Journal of Nanoscience and Nanotechnology*, 2017.

[70] F. Garcia-Redondo, R. P. Gowers, A. Crespo-Yepes, M. Lopez-Vallejo, and L. Jiang, "SPICE compact modeling of bipolar/unipolar memristor switching governed by electrical thresholds," *IEEE Transactions on Circuits and Systems I: Regular Papers*, vol. 63, no. 8, pp. 1255–1264, 2016.

[71] R. Yang, K. Terabe, Y. Yao, T. Tsuruoka, T. Hasegawa *et al.*, "Synaptic plasticity and memory functions achieved in a WO 3-x-based nanoionics device by using the principle of atomic switch operation," *Nanotechnology*, 2013.

[72] I. Boybat, M. Le Gallo, S. Nandakumar, T. Moraitis, T. Parnell *et al.*, "Neuromorphic computing with multi-memristive synapses," *Nature communications*, vol. 9, no. 1, p. 2514, 2018.

[73] H. An, M. A. Ehsan, Z. Zhou, F. Shen, and Y. Yi, "Monolithic 3D neuromorphic computing system with hybrid CMOS and memristor-based synapses and neurons," *Integration*, vol. 65, pp. 273–281, 2019.

[74] C. D. Schuman, T. E. Potok, R. M. Patton, J. D. Birdwell, M. E. Dean *et al.*, "A survey of neuromorphic computing and neural networks in hardware," *arXiv preprint arXiv:1705.06963*, 2017.

[75] A. Shrestha, K. Ahmed, Y. Wang, and Q. Qiu, "Stable spike-timing dependent plasticity rule for multilayer unsupervised and supervised learning," in *2017 international joint conference on neural networks (IJCNN)*. IEEE, 2017, pp. 1999–2006.

[76] M. M. Adnan, S. Sayyaparaju, G. S. Rose, C. D. Schuman, B. W. Ku *et al.*, "A twin memristor synapse for spike timing dependent learning in neuromorphic systems," in *2018 31st IEEE International System-on-Chip Conference (SOCC)*. IEEE, 2018, pp. 37–42.

[77] G. Pedretti, V. Milo, S. Ambrogio, R. Carboni, S. Bianchi *et al.*, "Memristive neural network for on-line learning and tracking with brain-inspired spike timing dependent plasticity," *Scientific reports*, vol. 7, no. 1, pp. 1–10, 2017.

[78] D. R. Ly, A. Grossi, T. Werner, T. Dalgaty, C. Fenouillet-Beranger *et al.*, "Role of synaptic variability in spike-based neuromorphic circuits with unsupervised learning," in *2018 IEEE International Symposium on Circuits and Systems (ISCAS)*. IEEE, 2018, pp. 1–5.

[79] L. Deng, D. Wang, Z. Zhang, P. Tang, G. Li *et al.*, "Energy consumption analysis for various memristive networks under different learning strategies," *Physics Letters A*, vol. 380, no. 7-8, pp. 903–909, 2016.

[80] IBM, "Hardware analog-digital neural networks," U.S. Patent 8 275 727.

[81] J. Shamsi, A. Amirsoleimani, S. Mirzakuchaki, and M. Ahmadi, "Modular neuron comprises of memristor-based synapse," *Neural Computing and Applications*, vol. 28, pp. 1–11, 2017.

[82] Y. Halawani, B. Mohammad, M. Al-Qutayri, and S. F. Al-Sarawi, "Memristor-based hardware accelerator for image compression," *IEEE Transactions on Very Large Scale Integration (VLSI) Systems*, vol. 26, no. 12, pp. 2749–2758, 2018.

[83] Q. Wang, Y. Kim, and P. Li, "Neuromorphic processors with memristive synapses: Synaptic interface and architectural exploration," *ACM Journal on Emerging Technologies in Computing Systems (JETC)*, vol. 12, no. 4, pp. 1–22, 2016.

[84] U. K. Agarwal, S. Makhija, V. Tripathi, and K. Singh, "An Investigation into Neuromorphic ICs using Memristor-CMOS Hybrid Circuits," *arXiv preprint arXiv:2210.15593*, 2022.

[85] M. Hu, J. P. Strachan, Z. Li, E. M. Grafals, N. Davila *et al.*, "Dot-product engine for neuromorphic computing: Programming 1T1M crossbar to accelerate matrix-vector multiplication," in *Proceedings of the 53rd annual design automation conference*, 2016, pp. 1–6.

[86] M. Nourazar, V. Rashtchi, A. Azarpeyvand, and F. Merrikh-Bayat, "Code acceleration using memristor-based approximate matrix multiplier: Application to convolutional neural networks," *IEEE Transactions on Very Large Scale Integration (VLSI) Systems*, vol. 26, no. 12, pp. 2684–2695, 2018.

[87] F. M. Bayat, M. Prezioso, B. Chakrabarti, I. Kataeva, and D. Strukov, "Memristor-based perceptron classifier: Increasing complexity and coping with imperfect hardware," in *2017 IEEE/ACM International Conference on Computer-Aided Design*. IEEE, 2017, pp. 549–554.

[88] M. Hu, C. E. Graves, C. Li, Y. Li, N. Ge *et al.*, "Memristor-based analog computation and neural network classification with a dot product engine," *Advanced Materials*, vol. 30, no. 9, p. 1705914, 2018.

[89] R. Hasan, T. M. Taha, and C. Yakopcic, "On-chip training of memristor based deep neural networks," in *2017 International Joint Conference on Neural Networks (IJCNN)*. IEEE, 2017, pp. 3527–3534.

[90] K. Yang, J. Joshua Yang, R. Huang, and Y. Yang, "Nonlinearity in memristors for neuromorphic dynamic systems," *Small Science*, vol. 2, no. 1, p. 2100049, 2022.

[91] D. Gaol, G. L. Zhang, X. Yin, B. Li, U. Schlichtmann *et al.*, "Reliable memristor-based neuromorphic design using variation-and defect-aware training," in *2021 IEEE/ACM International Conference On Computer Aided Design (ICCAD)*. IEEE, 2021, pp. 1–9.

[92] S. De, M. A. Baig, B.-H. Qiu, H.-H. Le, Y.-J. Lee *et al.*, "Neuromorphic computing with fe-finfets in the presence of variation," in *2022 international symposium on vlsi technology, systems and applications (VLSI-TSA)*. IEEE, 2022, pp. 1–2.

[93] C. Sung, H. Hwang, and I. K. Yoo, "Perspective: A review on memristive hardware for neuromorphic computation," *Journal of Applied Physics*, vol. 124, no. 15, p. 151903, 2018.

[94] M.-L. Wei, H. Amrouch, C.-L. Sung, H.-T. Lue, C.-L. Yang *et al.*, "Robust Brain-Inspired Computing: On the Reliability of Spiking Neural Network Using Emerging Non-Volatile Synapses," in *International Reliability Physics Symposium (IRPS)*, 2021.

Innovation Practices Track: VLSI Functional Safety

Fei Su	Meirav Nitzan	Ankush Sethi and Vaibhav Kumar	Dan Alexandrescu
Intel Corporation	Qualcomm	NXP Semiconductor	Synopsys
U.S.	U.S.	U.S.	U.S.
fei.su@intel.com	mnitzan@qti.qualcomm.com	ankush.sethi@nxp.com	alexand@synopsys.com

I. INTRODUCTION (FEI SU)

In safety-critical applications, e.g., automotive, how to avoid or alleviate risk due to hazards caused by silicon/IC malfunction is a key field, aka functional safety (FuSa). While there have been several standards established for FuSa, e.g., ISO 26262, there are still many technical challenges in this field, for example, tradeoff between cost and effectiveness, new safety mechanism for transient, intermittent or degrading faults. In this session we invited the industry experts to present the latest developments in these domains and share their insights.

II. SAFETY MECHANISMS FOR TRANSIENT FAULTS IN AUTOMOTIVE - COST VS. EFFECTIVENESS (MEIRAV NITZAN)

With ever growing design complexity in the Automotive domain, the semiconductor industry needs to address the Integrated Circuits Functional Safety with innovative solutions. ISO 26262, the Road vehicles standard, provides a method to compute the reduced failure rate of the random Single Point Faults after applying safety mechanisms with detect the presence of faults. It is called SPFM – Single Point Fault Metric. The ASIL (Automotive Safety Integrity Level) assigned to the design determines the SPFM ratio (from 90% for ASIL B to 99% to ASIL D).

Random faults in semiconductor can be either permanent fault, which are caused by manufacturing detects and component aging, and transient faults, caused by disturbance such as cosmic radiation or electrical noise. In this presentation we explore the techniques and mechanisms to reduce the transient failure rate, focusing on Radiation Hardened Flops (or High Resilience Flops). We examine if it should be considered only as a way to reduce the initial failure rate, or if it can be considered a safety mechanism, and how each approach affects the SPFM.

III. FAULT INJECTION FOR AUTOMOTIVE FUNCTIONAL SAFETY (ANKUSH SETHI)

With the increasing complexity of SoC design (Microcontroller) for automotive safety application, there are a lot of custom safety mechanisms which needs to be devised to cater to the new requirements. For standard safety mechanisms like ECC and Redundancy, the effectiveness of safety mechanism is easy to estimate whereas it is hard to judge it for custom safety mechanisms. So, we need to perform fault injection verification to measure their effectiveness. Also, for the ASIL-D development the ISO26262 recommends performing fault injection verification.

This fault injection verification is not intended primarily to verify or confirm the correct implementation of the design and its safety mechanisms i.e., the systematic integrity of the product. The fault injection verification is done to increase the confidence into quantitative safety analysis of random hardware failures. This fault injection verification helps establish the proof for effectiveness of the safety mechanisms. This helps give feedback if any design change is needed to increase the diagnostic coverage of the safety mechanisms.

Typically, the fault injection campaign is executed for the targeted design using a fault simulation tool. However, for a more efficient and productive fault campaign, there is a need to use additional techniques, like formal analysis, to improve the overall process. This presentation also talks about fault injection verification also talks about the RTL vs. Gate level fault injection techniques.

IV. MANAGING INTERMITTENT AND DEGRADING FAULTS THROUGH ADVANCED SENSORS, ANALYTICS AND PREDICTIVE MAINTENANCE (DAN ALEXANDRESCU)

Safety and reliability requirements need to be understood and managed over a wide set of operating conditions and increasingly longer product lifetimes. The unprecedented use of cutting-edge technological processes, advanced IPs and complex designs exposes automotive IC and solutions providers to risks caused by process variability, aging and degradation. We show how advanced test instruments and sensors can be upgraded through analytics to act as capable safety mechanisms against intermittent and degrading faults in the context of current and future revisions of ISO 26262. The proposed multi-stage approach uses information from advanced sensors tightly embedded in critical design blocks. Path margins monitoring, pre-error detections, ECC and BIST events, outlier and deviant conditions, subtle disagreements on measurements from the various sensors are evaluated cohesively into relevant risk metrics. Actionable insights are then issued to the system actuators to promptly correct safety and reliability threats. Deep silicon data is sent to edge and cloud platforms for fleet-level monitoring. Advanced analytics at all stages provide a quick and precise understanding of very infrequent events and phenomena, helping deployed products to be used safely and enabling higher quality for future designs. Production testing is poised to benefit from the same SLM infrastructure and analytics, decisively enabling automotive's impressive goal of 10 Defective Parts per Billion (10 DPPB).

979-8-3503-4631-2/23 $31.00 © 2023 IEEE

Special Session: Security Verification & Testing for SR-Latch TRNGs

Javad Bahrami*, Mohammad Ebrahimabadi*, Jean-Luc Danger[†], Sylvain Guilley[†‡], Naghmeh Karimi*

*University of Maryland Baltimore County, United States
[†]LTCI, Télécom Paris, Institut Polytechnique de Paris, France
[‡]Secure-IC S.A.S., France

Abstract—Secure chips implement cryptographic algorithms and protocols to ensure self-protection (e.g., firmware authenticity) as well as user data protection (e.g., encrypted data storage). In turn, cryptography needs to defer to incorruptible sources of entropy to implement their functions according to their mandatory usage guidance. Typically, keys, nonces, initialization vectors, tweaks, etc. shall not be guessed by attackers. In practice, True Random Number Generators (TRNGs) are in charge of producing such sensitive elements.

Fully aware of the central role of TRNGs in the proper implementation of security in chips, stakeholders have been formalizing the requirements recently. The methods to strengthen such requirements are manifold. In this paper, we discuss and apply three of them by targeting the Set-Reset Latch TRNG which is an alternative to Ring-Oscillator (RO) TRNGs as it provides faster throughputs. The first method concerns the confidence in the TRNG being random enough. It explores how the TRNG properties can be reliably predicted by simulation, compared to real silicon experiments. The second aspect dealt with in this paper is the assessment of the TRNG properties over time, i.e., considering the impact of aging in the TRNG properties. Such knowledge is important as secure chips are expected to be in service for a long period, and it would be detrimental to the service they render if the quality of the entropy they deliver would be declining over time. Eventually, the third aspect of this paper is the timely detection of unforeseen failures or malevolent attacks. The mitigation lies in leveraging "health tests" launched prior to using random numbers.

This paper focuses on a particular type of TRNG that is not prone to biasing by attackers: it is the so-called Set-Reset Latch (SR-latch) TRNG and exploits a race condition in an arbitration gate. Such kind of TRNG is of great practical interest as an alternative design compared to the mainstream "Ring Oscillator" TRNG, and it is also very amenable to analyses by various sorts of simulations aiming at properly characterizing its security in various operational environments.

Index Terms—True Random Number Generators (TRNGs), Set-Reset Latch TRNG, Security, Verification & Testing, Aging, Health tests, Compliance to standards.

I. INTRODUCTION

Secure chips require a source of uniformly distributed bit-strings. Such source allows to generate all the random parameters required to implement cryptographic protocols properly. Indeed, most, if not all, cryptographic algorithms implicitly rely on the availability of fresh random values. For instance, AES-CBC initialization vectors, HMAC keys, ECDSA nonces,

Crystals Kyber noise, etc. all need to be provided by "True Random Number Generators" (TRNGs). Those sources of entropy should be independent such that an attacker cannot get to know or influence them. For this reason, hardware TRNGs have been put forward.

Ring-Oscillator based TRNGs (RO-TRNG) have been out for a while, and are trustworthy. However, it is not enough to only rely on RO-TRNGs, as based on current security assurance requirements there is a need for an alternate TRNG design as well to avoid *single point of failures* (SPOFs). There are various TRNGs based on other rationales. However, there is an obviously advantage for designs which are integrable in CMOS logic, and another bonus for PPA-efficient[1] structures. Accordingly, in this paper, we tackle one such TRNG, i.e., Set-Reset latch (SR-latch) based TRNG. The main concern here is thus to analyze the randomness of such security primitive not only when these primitives are new but also when they have been used for a while.

II. IN SILICON VALIDATION OF SR-LATCH TRNG

A. Principle of SR-latch TRNG

A TRNG designed in digital devices exploits either the noise coming from clock jitters, the noise around metastable states, or both. The exploitation of the phase jitter noise is carried out by means of oscillators, like ROs, and represents a robust way to generate randomness. The use of metastable state is more tricky as it requires analog and custom cell [10] but it is much faster than the oscillator-based TRNG. A proposal to build such TRNG in full digital environment is to use latches like in [6], [11] which exploit both the jitter and metastable state. The use of NOR-based SR-latch, represented in Figure 1, allows to capture the noise around a metastable state.

The Set (S) and Reset (R) inputs are derived by the same signal SR. If the two NOR gates are perfectly balanced, when the SR input goes to zero, the latch would be in a metastable state at around $V_{dd}/2$ voltage. However, the environmental noise at the input creates a convergence towards a stable state (either V_{dd}, interpreted as logic '1', or 0 V, interpreted as logic '0'). This is depicted in the simulations shown in Figure 2, showing the values of Q (output of the SR-latch) over 100

---

[1]PPA stands for "Performance, Power and Area", three important figures of merit (aside Safety & Security) in hardware designs.

979-8-3503-4631-2/23 $31.00 © 2023 IEEE

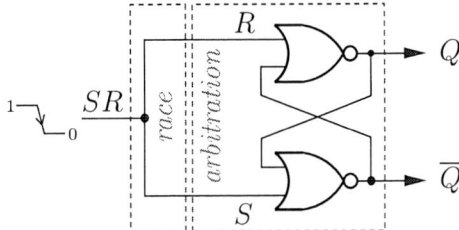

Fig. 1. An SR-latch realized via cross-coupled NOR gates.

repetitions. In this figure, the value of $V_{dd} = 1.2$ V. It can be seen that it takes a certain time for the final value to converge and its entropy depends on the amount of the physical noise.

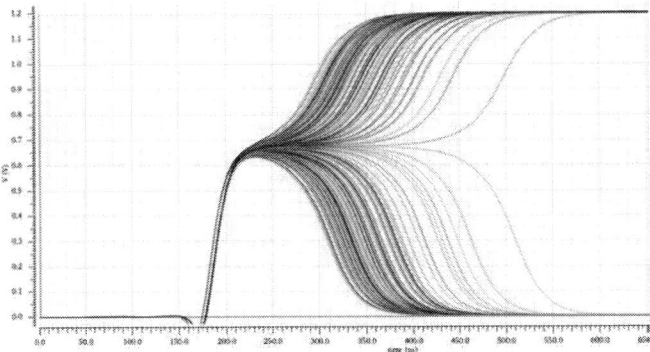

Fig. 2. Transition from a metastable to a stable state due to input noise.

The metastability, shown to several hundreds of picoseconds, can be resolved by adding a synchronizing flip-flop, clocked by SR) at the output Q or $\bar{Q}$ of the SR-latch. Henceforth, the SR-latch TRNG can be operated with a periodic signal SR which can be the system clock. Thus one random bit is gotten at every clock period, which makes the entropy source at the same time fast and compact (hence economic in terms of PPA; referring to Power-Performance-Area).

As discussed, a latch can be used as a **TRNG**, as its outcome depends on the dynamic noise. Note that if the two NOR gates are strongly unbalanced, due to *locally correlated* process mismatch, the latch will always converge to the same stable state. In this case the latch behaves like a "Physically Unclonable Function" (also known as **PUF** [12]), as its value depends on the fabrication noise which has become static. To enhance the probability of getting a true random output, it is necessary to consider a set of latches; instead of only one latch. Figure 3 illustrates the architecture where the embedded latches' outputs are XORed together to realize a TRNG.

B. Stochastic model of the SR-latch TRNG

A stochastic model can be built to assess the entropy of the SR-latch TRNG according to the number of latches and the process mismatch. Indeed, the XOR combination of latches can give an entropic random number if there are a minimum number of latches around a metastable state and if the process mismatch is not too high. A first approach has been

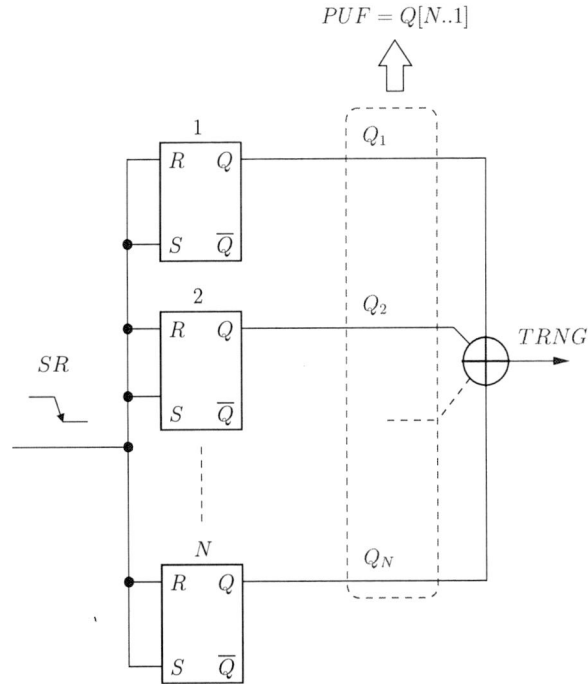

Fig. 3. The PUF-TRNG circuitry composed of a set of SR-latches.

studied in [4] but did not propose a closed form expression to assess the entropy and the number of latches according to the mismatch.

Every latch i outputs a random variable Q_i which can be considered as a Bernoulli random variable associated with a probability $\mathbb{P}[Q_i = 1] = p_i$. The XOR between these variables can be obtained by changing the variable $p_i \in [0, 1]$ by $q_i \in [-1, 1]$, such that:

$$p_i = \frac{1 + q_i}{2}, \qquad q_i = 2p_i - 1.$$

The probability of the XOR between two latches Q_i and Q_j is thus:

$$\mathbb{P}[Q_i \oplus Q_j = 0] = p_i \cdot p_j + (1 - p_i)(1 - p_j) = \frac{1 + q_i q_j}{2}.$$

By induction (refer to the *piling-up lemma* [14]), one gets that the resulting probability of the node $TRNG$ with the XOR of N latches is equal to:

$$P_0 = \mathbb{P}[TRNG = 0] = \frac{1 + \prod_{i=1}^{N}(2p_i - 1)}{2} , \quad P_1 = 1 - P_0. \tag{1}$$

The probability p_i highly depends on the imbalance between the S and R input. This bias comes from mismatches in the process variation and the routing. It is equivalent to a delay offset Δ_M which has a Gaussian distribution: $\Delta_M \sim \mathcal{N}(0, \Sigma^2)$. With respect to the noise, Z being also considered Gaussian $Z \sim \mathcal{N}(0, \sigma^2)$, we define the "Mismatch to Noise Ratio" MNR as being:

$$\mathrm{MNR} = \frac{\Sigma}{\sigma} . \tag{2}$$

The mean probability of p_i according to MNR is given in Equation 3, and it can be deduced from the section III.A of the article from Schaub et al. [19]. This study considers the SR-latch as a PUF and thus gives the Bit Error Rate (BER) probability which is fully equivalent to the p_i probability when the SR-latch is used as a TRNG.

$$\widehat{p_i} = \frac{1}{\pi} \arctan\left(\frac{1}{\text{MNR}}\right). \tag{3}$$

If we use N SR-latches in parallel, Equation 1 applies. If we consider the noise independent from one latch to another, The mean probability is given by Equation 4:

$$\widehat{P_0} = \frac{1 + (2\widehat{p_i} - 1)^N}{2} = \frac{1 + (\frac{2}{\pi}\arctan(\text{MNR}))^N}{2} \tag{4}$$

and, by complementarity,

$$\widehat{P_1} = 1 - \widehat{P_0}. \tag{}$$

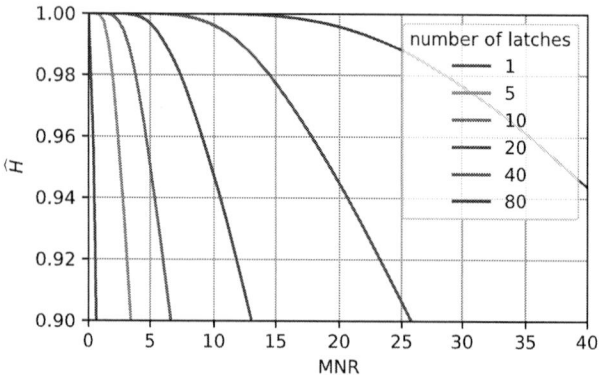

Fig. 4. Mean Entropy $\widehat{H}$ according to Mismatch to Noise Ratio (MNR) for different number of latches.

The mean Shannon entropy is given by Equation (5):

$$\widehat{H} = -\sum_{i \in \{0,1\}} \widehat{P_i} \log(\widehat{P_i}). \tag{5}$$

It is illustrated in Figure 4 according to the number of latches and the MNR parameter.

This figure shows that it is necessary to have a significant number of SR-latches to ensure a good entropy, especially if the technology has a high process variance. The study presented in the next section III gives some information about the MNR quantity in the current technology.

III. TESTS ON AN ASIC VERSION OF THE SR-LATCH TRNG

A. Architecture

A design with 1024 SR-latches has been implemented in a testchip fabricated in 28nm FD-SOI process and presented in [7]. As a primary goal, it has been built to demonstrate the impact of the body-bias voltage on the properties of a PUF-TRNG architecture. In order to drive the latches from outside

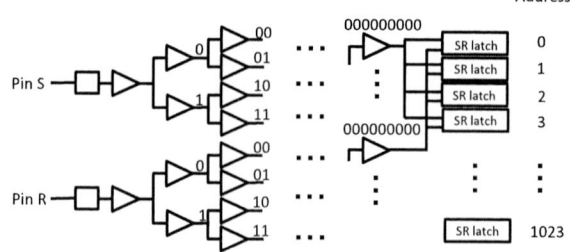

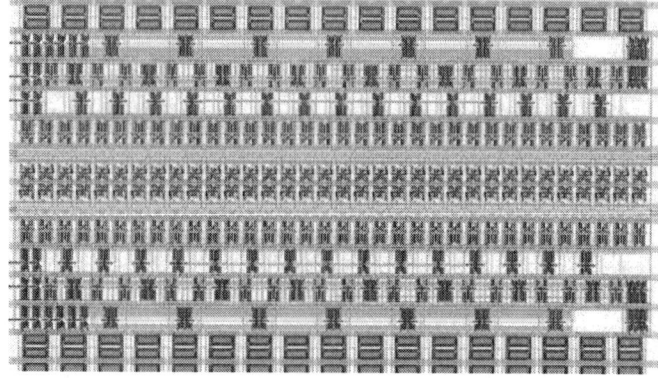

Fig. 5. Structure of the two buffer trees of the PUF-TRNG testchip.

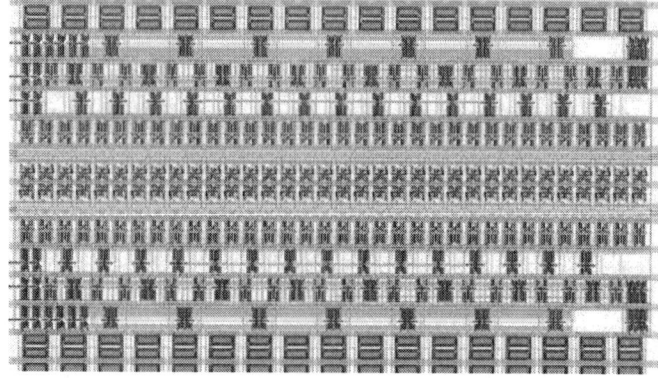

Fig. 6. Partial layout of the PUF-TRNG circuit.

of the chip, a buffer tree is added on each S and R input which have respective input pins, as represented in Figure 5.

The trees have been carefully balanced to guarantee the simultaneous arrival time of the S and R signals to all the latches. But as the buffer trees introduce a higher sensitivity to process mismatches between the S and R inputs, the architecture is in favor of PUFs (i.e. stable outputs of latches). On the opposite, the imbalance generated by the buffer trees could decrease the number of metastable latches necessary for the TRNG.

Figure 6 shows a part of the layout. The latches are placed in the middle rows whereas the buffers are placed in a balanced manner at the top and the bottom of the latches

The NOR gates are built using standard regular threshold voltage V_{th} transistors, but the PMOS transistors of the top NOR and bottom NOR have different and controllable reverse body bias, denoted $VB1$ and $VB2$ respectively. It is shown in [7] that the range of $VB1, VB2$ is quite large to get a high number of SR-latches either stable for PUF, or unstable for latches for TRNG. Another takeaway result is that the optimal pair $(VB1, VB2)$ where there is a maximum of latches stable for PUF happens to be exactly the same as the optimal pair $(VB1, VB2)$ where there is a maximum of latches unstable for TRNG.

B. Evaluation of the Mismatch to Noise Ratio

The delay Δt between the S and R input is swept to measure the number of stable and unstable latches, more precisely to get the probability p_i of every latch. It has to be noted that

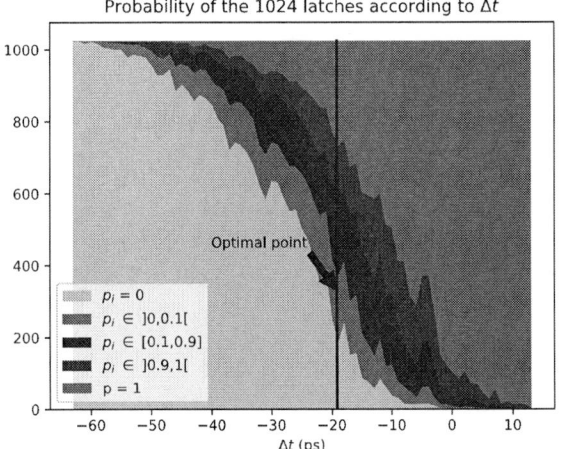

Fig. 7. Distribution of the 1024 latches output value probability (p_i) vs. Δt.

Δt is relatively important (a few ps) as there is an imbalance between the paths of the S and R signals. This imbalance results from the placement and routing of the pins and the stages of the buffer tree. In Figure 7 we can see the proportion of latches stable with p_i at 0 or 1, and unstable latches with $p_i > 0$ and $p_i < 1$.

At Δt around -20 ps, the mean probability $\widehat{P_1}$ is assessed at 0.045, leading to an MNR estimated at around 7 according to Equation 3. As there is a high mismatch coming from the buffer tree, the real MNR of SR-latches having S and R connected together as in Figure 1 is certainly less than 7.

With $MNR = 7$, if one wants to get an entropy equal to 0.997 bit (as mandated by AIS31 standard [17]), thirty (30) SR-latches would be necessary according to equation 5. Figure 7 also shows that if Δt is not exactly at -20ps, the MNR value stays almost the same. That means that there is a tolerance if the internal delay offset is biased, meaning that Δ_M introduced in section II-B may not be necessarily centered.

IV. PREDICTING THE EFFECT OF AGING ON TRNG

As the transistors' specification change over time due to the so-called device aging, we need to investigate how the SR-latch TRNG behavior evolves over time. Accordingly, in this section, we first give a brief review of device aging, its causes and impacts, and then we present our experimental results showing how aging affects the TRNG outcome over time.

Indeed in this study, the open question is that due to aging which of the following cases may happen?

- *Less Randomness* because of independent aging of the two cross-coupled NOR (more unbalancedness); or
- *More/less aging* on the PMOS of the NOR gate that was faster when the device was new (not aged);

If the second case applies in a negative sense, is it possible to take advantage of the differential aging of the 2 NOR gates

to make up for aging by recovering? Indeed, by setting some values in the SR-latch at rest, can a bias be compensated, as in the case of PUF (see [8], [15])?

A. Background on Device Aging

Aging mechanisms result in performance degradation and eventual failure of digital circuits over time. In CMOS technology, the two leading factors of aging are Bias Temperature Instability (BTI) and Hot Carrier Injection (HCI) [2]. Both mechanisms result in increasing switching voltage and path delays. NBTI (one class of BTI) [2] affects PMOS transistors, while PBTI (another class of BTI) as well as HCI affects NMOS devices.

BTI Aging: A PMOS (NMOS) transistor goes under two phases of NBTI (PBTI) depending on its operating condition [2]. The first phase, i.e., *stress*, occurs when the related transistor is "ON". Here, charges are trapped at the Si-SiO$_2$ interface and lead to an increase in the threshold voltage. The second phase, *recovery*, occurs when the transistor is off. In this phase, the charges trapped in the stress phase are partially removed, and thus the threshold voltage (V_{th}) drift that occurred during the stress phase partially recovers. The impact of BTI depends on the supply voltage, temperature, physical parameters of the transistor under stress, and stress time. Fig. 8 depicts the V_{th} drift of a PMOS transistor when it is continuously under stress for 6 months versus the case that it experiences stress and recovery phases every other month. In this figure, the values on the Y-axis are not shown intentionally to make the figure generic and technology independent.

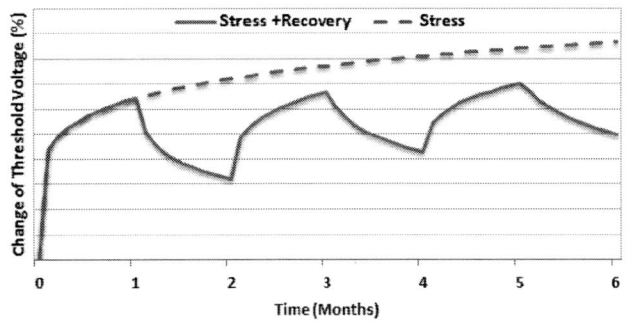

Fig. 8. NBTI-induced V_{th} drift of a PMOS transistor that is always under the stress, and a transistor that is under the stress and recovery alternatively.

HCI Aging: HCI happens in an NMOS when hot carriers are injected into the gate dielectric during transistor switching and remain there. HCI is a function of switching activity; degrading the circuit by shifting the threshold voltage and drain current of stressed transistors. The threshold voltage drift induced by HCI depends on the activity factor of the transistor under stress, its temperature, clock frequency, and usage duration [16].

B. Experimental Setup

To answer the above question and study the impact of aging on the SR-Latch-based TRNGs, we conducted extensive Spice

979-8-3503-4631-2/23 $31.00 © 2023 IEEE

simulations. We implemented our TRNG at the transistor level using a 45nm open-source NANGATE library [1]. Our netlist includes 20 SR-latches in parallel (following Fig. 3). To mimic the real-silicon behavior, we considered process variation (PV) through Monte-Carlo simulations with Gaussian distributions: transistor gate length L: $3\sigma = 10\%$, threshold voltage V_{TH}: $3\sigma = 30\%$, and gate-oxide thickness t_{OX}: $3\sigma = 3\%$. This allows us to derive the delay offset of the mismatch Δ_M as defined in section II-B.

In our implementation of the SR-latch TRNG, while the $\overline{Q}$ outputs of all latches are left floating, the Q outputs are XORed together to build the final single bit of randomness. We benefited from Synopsys HSpice for the transistor-level simulations, and the HSpice built-in MOSRA Level 3 model [20] to extract NBTI, PBTI, and HCI aging effects. The effect of aging was evaluated for 7 years of device operation in time steps of 6 months. The simulations were conducted for temperature of 85°C, $V_{dd} = 1.2$ V, and the operating frequency of 500 MHz.

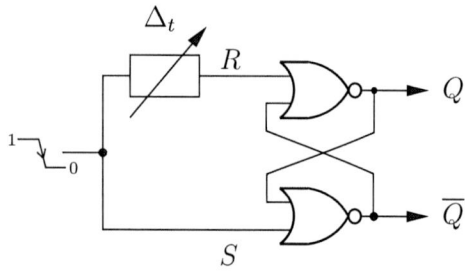

Fig. 9. SR-latch with a controlled Δ_t to study the mismatch and noise impact.

In a perfect netlist, the R and S signals arrive simultaneously, hence the latches get metastable. However, as there is a mismatch in processes and wiring, the outputs of latches might go to a known state and the entire circuit behaves as a PUF. For instance, PMOS transistors in one of the NOR gates residing in the latch might have a lower threshold voltage V_{th} and get evaluated faster than the other ones (PV). Bulk and flicker noise (specifically in low frequencies) are other important sources of unwanted behavior in CMOS technology that might cause the circuit to behave predictably.

As modeling all these effects separately is not trivial, we abstract them away and add a controlled delay Δt to the R signal of each embedded latch while S signals are assumed to be clean as shown in Figure 9.

C. Analyzing the Process Variation Induced Delay

To assess the impact of process variation in our latches, we extract the time Δ_M at which the output of each of the embedded latches (shown in Fig. 3) toggles. To do so, we sweep the delay Δt shown in Figure 9 in the R signal path towards the S signal with the step size of 80 fs. With such sweeping, the latch goes first to the metastable state (red area in Figure 10) and then the output is toggled with more sweeping. The exact moment of Δ_M is when our latch toggles with the accuracy of 80 fs similar to the sweeping step size.

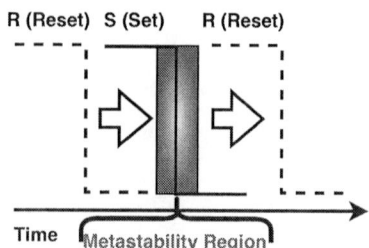

Fig. 10. Sweeping Δt of the R signal with step size of 80 fs.

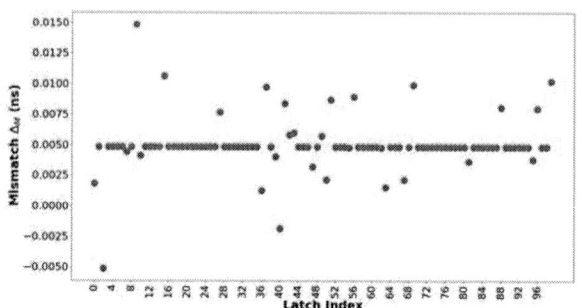

Fig. 11. Mismatch distribution (ns) of the circuit with 100 latches.

We collect this data for each of the embedded latches and use this data for the statistical analysis discussed below. Such information provides a precise insight into the statistical behavior of SR-latch-based TRNGs and helps us to answer the following questions:

- What is the value of Δ_M for every latch?
- What is the mean and variance of Δ_M distribution (denoted as μ and Σ^2, respectively)?

Figure 11 shows the results of the analysis for a TRNG realized via 100 parallel latches. The takeaway points from this set of results include:

- The output of the majority of the embedded latches flips while the S and R signals are 5 ps apart from each other ($\mu = 5$ ps) due to the architectural bias we have in our implementation. Hence, Δ_M is not centered, $\Delta_M \sim \mathcal{N}(\mu, \Sigma^2)$, and the stochastic model proposed in section II-B may be optimistic even if the ASIC results in section III show that this bias is not problematic in the circuit designed in FD-SOI.
- Figure 12 shows the distribution of Δ_M which should be near a Gaussian distribution. The estimated standard deviation is $\Sigma = 2$ ps.

We repeated the experiments with 20 latches, this time by adding a random Gaussian delay on Δ_t of the R signals with $\sigma = 0.25$ ps and $\sigma = 3$ ps. With the knowledge of the delay offset Δ_M coming from the mismatch, the noise standard deviation $\sigma \in \{0.25, 3\}$ ps corresponds to the mismatch to noise ratio MNR of $\{8, 2/3\}$, respectively. Thus in case of considering the XOR of 20 latches, we can deduce from

979-8-3503-4631-2/23 $31.00 © 2023 IEEE

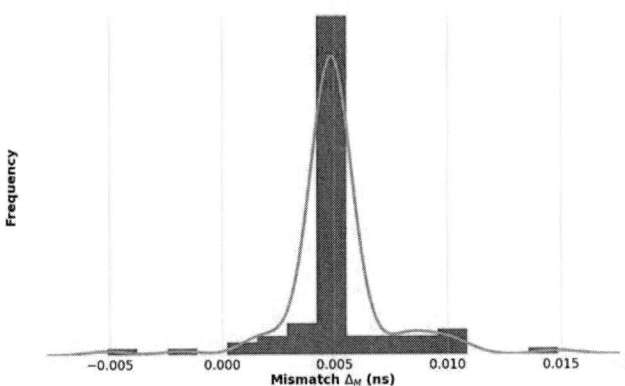

Fig. 12. Mismatch distribution histogram (ns) of the circuit with 100 latches.

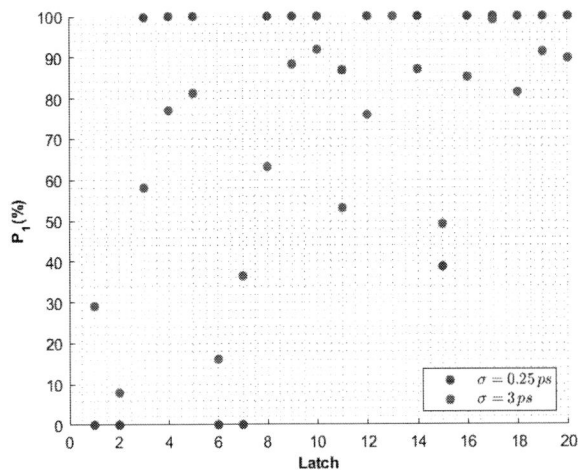

Fig. 13. The randomness of all 20 SR-latch TRNGs under different noise levels ($\sigma = 0.25$ ps and $\sigma = 3$ ps).

Equation 5 and Fig. 4 that we should get an entropy greater than 0.96 bit for $\sigma = 0.25$ ps (as its MNR is 8) and an entropy of 1 for $\sigma = 3$ ps. However, the Δ_M distribution is not centered and has a mean of $\mu = 5$ ps, which should deteriorate the entropy.

As expected and also shown in Fig. 13, we got the best randomness in our latches for the case of $\sigma = 3$ ps and the worst for $\sigma = 0.25$ ps. By the best we mean the majority of latches before being XORed have an acceptable randomness (distribution of '1' vs. '0') and by the worst we mean the circuit mainly behaves as a PUF (constant output). Hereby, we picked these two extreme cases for the aging analysis to analyze how device aging can affect the randomness of the TRNGs in these two cases. Please note that by randomness we mean the distribution of '1' and '0' values which is also referred to as uniformity in literature. More precisely, by randomness, we refer to the number of '1's generated by the TRNG over the total number of '1' and '0's.

D. Aging Induced Impacts on the SR-Latch Based TRNG's outcome

We used HSpice MOSRA to simulate the aging impacts on the targeted TRNG realized by 20 parallel latches for the course of 7 years with the steps of 6 months. The results are shown in Fig. 14 and Fig. 15. The former represents the best case mentioned earlier ($\sigma = 3$ ps) and the latter depicts the case in which ($\sigma = 0.25$ ps) which seemed to be the worst case among the ones considered in this study in terms of randomness. In each of these figures, for the sake of clarity, we only showed the randomness of two randomly selected (out of 20) latches, namely latches 11 and 15, along with the randomness of the whole TRNG realized by XORing the 20 latches. As shown, for the new device both latches (called Latch 11 and Latch 15) revealed highly acceptable randomness in the range of 50%±5%. The trend is the same during the course of aging for 7 years. Moreover, the main TRNG output realized by XORing all latches also exhibits a very similar (close to 50%) randomness. On the other hand as shown in Fig. 15, this trend is very different when the noise standard deviation (σ) is 0.25 ps. In this case, the selected latches exhibit very low randomness; having a randomness of 85% and 39% when new. This trend changes during the course of aging resulting in 29% and 66% randomness after 7 years.

Another important observation that can be made from Fig. 14 and Fig. 15 is that the randomness of the main TRNG (realized by XORing all 20 latches) becomes close to 50% in both cases of noise standard deviations equal to 3 ps and 0.25 ps during the course of usage even if for the latter the randomness was around 58% initially.

In sum based on the above discussion and observations we can conclude:

- Latch-based TRNGs may have a close to optimum behavior when the circuit timing (jitters) is accurately controlled. These optimizations might be achieved by explicitly using delay elements or an accurate placement and routing (PnR).
- On the other hand, for the circuit with a poor choice of standard deviations (noise not well-controlled), the value of randomness changes substantially. While the circuit gets close to the optimum point of randomness at some points (50%), not having a predictable behavior would give less confidence to the designer.
- Even for the circuit with poor control of randomness when new, one can achieve optimized randomness during the course of aging.

Figure 16 explains why the randomness of Latch 11 in Figure 15 varies substantially over time due to aging. As in this case ($\sigma = 0.25$ ps), the time difference between the falling edge of S and R signals is well below the Mismatch (Δ_M), and considering that the upper NOR gate operates much faster in the new device due to Mismatch (referring to the high value of the P_1 for Latch 11), the upper gate (generating the Q signal) gets evaluated before the other NOR gate. Under this condition ($Q = 1$ and $\overline{Q} = 0$) that puts 0 to the input of the

979-8-3503-4631-2/23 $31.00 © 2023 IEEE

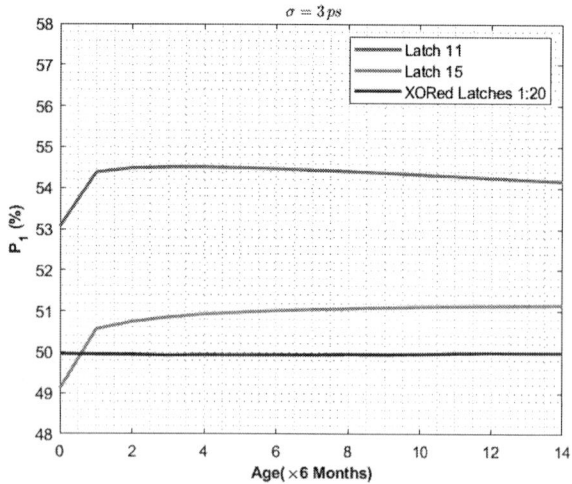

Fig. 14. Behavior of the circuit with noise standard deviation of 3 ps under 7 years of aging.

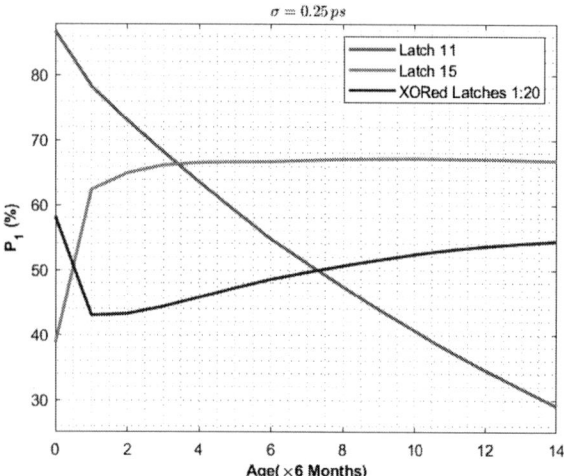

Fig. 15. Behavior of the circuit with noise standard deviation of 0.25 ps under 7 years of aging.

M3 transistor (it is assumed that these values are generated after the active edge of S and R signals goes from 1 to 0), PMOS transistor (M3) ages faster due to the NBTI impacts and the upper NOR gate gets slower while the other gate is not much stressed. It means the NOR gate at the bottom will be operating faster after some years than the upper one and $\overline{Q}$ path will be evaluated before the Q path resulting in more 0s at the output than 1s. Consequently, Latch 11 has a high P_i value in the beginning and gets very close to 30% at the end.

Figure 17 depicts the entropy value of the XORed 20 latches for two different value of $\sigma = 3$ ps, $\sigma = 0.25$ ps. As shown, the entropy for $\sigma = 3$ ps is stable at 0.5 for all ages of the circuit while this entropy for $\sigma = 0.25$ ps is increased at earlier ages of the circuit. This figure shows how a biased TRNG ($\sigma = 0.25$ ps) would reveal information to the potential attacker while the circuit with ($\sigma = 3$ ps) has good random

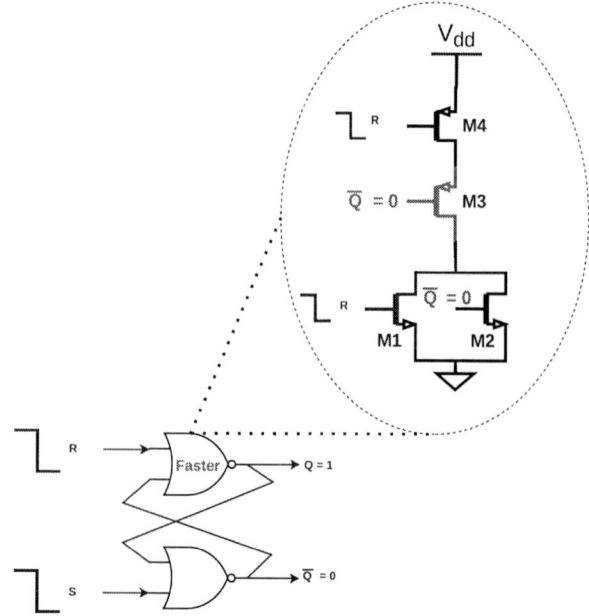

Fig. 16. Transistor level behavior of a faster NOR gate under aging.

TABLE I
NIST TEST RESULT RUNNING ON THE TRNG OUTPUT (AFTER XORING THE OUTPUT OF 20 INCLUDED LATCHES) BOTH FOR FRESH AND AGED CIRCUITS WITH $\sigma = 3$ PS.

Test	Age 0		Age 7-Year	
	Passed/ Total	P Value	Passed/ Total	P Value
Monobit	24/24	0.44	24/24	0.44
Frequency	24/24	0.47	24/24	0.45
Runs	24/24	0.50	24/24	0.54
Longest run	24/24	0.54	24/24	0.51
Binary matrix	6/8	0.51	7/8	0.39
DFT	23/24	0.44	23/24	0.42
Non overlapping	23/24	0.41	24/24	0.51
Overlapping	1/1	0.96	1/1	0.02
Universal	1/1	0.99	1/1	0.99
Linear Complexity	1/1	0.59	1/1	0.15
Serial	24/24	0.58	24/24	0.49
Approximate Entropy	24/24	0.62	24/24	0.51
Cumulative Sums	24/24	0.44	24/24	0.42
Random Excursion	1/1	0.42	1/1	0.43
Random Excursion Variant	1/1	0.45	1/1	0.43

characteristics both for fresh and aged devices.

Implemented TRNG (after XORing the 20 latches) was further evaluated using 15 statistical tests offered by NIST FIPS SP 800-22 [18] to assess their randomness within 1,200,000 cycles. The responses were divided into 24 blocks each including 50,000 random numbers, and we applied the NIST tests to each block. The results are shown in Table I and Table II when the circuit is new and aged for 7 years. Note that some of the tests (e.g., Universal) need larger blocks so we partitioned our responses accordingly. As shown in Table I, the TRNG passes almost all NIST tests when $\sigma = 3$ ps. On the other hand, when $\sigma = 0.25$ ps, the NIST tests are

979-8-3503-4631-2/23 $31.00 © 2023 IEEE

TABLE II
NIST TEST RESULT RUNNING ON THE TRNG OUTPUT (AFTER XORing THE OUTPUT OF 20 INCLUDED LATCHES) BOTH FOR FRESH AND AGED CIRCUITS WITH $\sigma = 0.25$ PS

Test	Age 0		Age 7-Year	
	Passed/ Total	P Value	Passed/ Total	P Value
Monobit	0/24	0	0/24	0
Frequency	0/24	0	0/24	0
Runs	0/24	0	0/24	0
Longest run	21/24	0.235528	24/24	0.468735
Binary matrix	7/8	0.364195	8/8	0.354716
DFT	0/24	0.000349	21/24	0.323806
Non overlapping	5/24	0.114004	11/24	0.218996
Overlapping	0/1	0	0/1	0
Universal	1/1	0.952023	1/1	0.98463
Linear Complexity	1/1	0.638685	1/1	0.345916
Serial	0/24	0	0/24	0.139295
Approximate Entropy	0/24	0	0/24	0
Cumulative Sums	0/24	0	0/24	0
Random Excursion	1/1	0.491643	0/1	0.63267
Random Excursion Variant	1/1	0.596831	1/1	0.632342

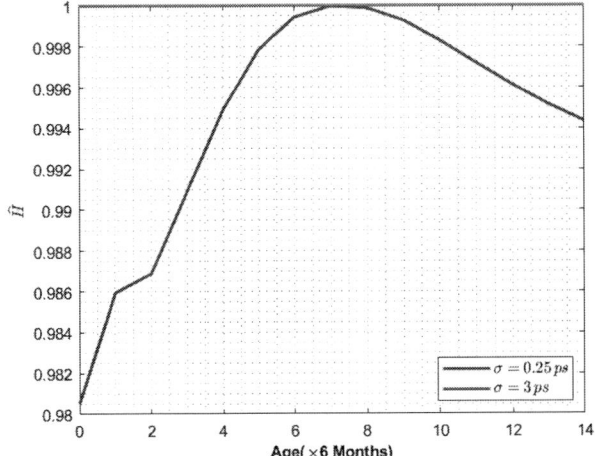

Fig. 17. Entropy of 20 latches (after XOR) for different σ values.

not successful (Table II). However, based on the information presented in these tables, it can be observed that the aging process not only does not lead to the degradation of the True Random Number Generator (TRNG) metrics, **but it also results in a minor improvement**, in some cases. For example, in Table I, the fresh device passes 6 out of 8 "Binary Matrix" tests while this number is 7 for the aged device.

In practice, it may not be possible to achieve the required randomness by only using a single latch as this mainly needs an analog/mixed-signal circuitry to change the jitter dynamically in order to achieve a randomness close 50%.

V. HEALTH TESTS IN VARIOUS CERTIFICATION SCHEMES (NIST SP 800 90B, AIS BSI 31, OSCCA GM/T 0078)

True Random Number Generators are amplifying some noise in order to create an unbiased and continuous flow of entropy. Even though the TRNG source has a sound design, multiple defects can cause it to malfunction. Let us list below a (non-exhaustive) list of reasons for a TRNG to fail:

- The TRNG can have a non-standard mode of operation, e.g., leveraging a free-running oscillator (FRO). The FRO is functioning at an impressive speed, and therefore likely cause local heat, which can result, in the long run, in some physical damage resulting from the thermal stress. In the case of a FRO-based TRNG, this can cause the loop to have such a breakdown that the loop gets opened. It is thus no longer oscillating hence no jitter (source of entropy) is accumulated. This dramatic situation is customarily denoted as "total failure".
- At a least extent, the TRNG source can become biased. Recall indeed that TRNGs are leveraging tiny noise and hence are fragile. For instance, the FRO can change speed, resulting in less entropy being collected, hence worth the quality of randomness when the circuit aged.
- Besides natural decay, a TRNG can also be manipulated by an attacker. Typically, the attacker would force the environment to couple to the FRO and therefore always resynchronize it, attenuating each effect of the jitter [3].
- Still more critically, the external source imposed by the attacker in the vicinity of the TRNG can be so strong that the FRO is completely controlled from the outside, yielding a mere cancellation of the noise due to jitter [13].

For all those reasons, it is reasonable not only to test the TRNG once at boot, but on a regular basis. A sane practice is to test the TRNG before each and every use. Such tests are termed *"health tests"* and are prescribed by multiple standards. Requirements for NIST, AIS, and OSCCA schemes are described next.

A. NIST SP 800 90B

The NIST is specifying in its publication FIPS 140-3 tests for cryptographic modules. An electronic device that is designed to pass all tests can be certified according to a FIPS 140-3 scheme, known as Cryptographic Module Validation Program (CMVP).

The 8th area of security requirements is termed "Sensitive Security Parameter Management", and includes the Random bit generators. In practice, the Entropy Source Validation (ESV) defers the evaluation to a Special Publication (SP) of the NIST, referred to as NIST SP 800 90B [21].

The ESV mandates independent health tests. They shall be run *at start-up* and *continuously*.

By default, two standard tests are indicated. Namely:

- Repetition Count Test (§4.4.1): it amounts to detect that some value is consecutively repeated many more times than expected, given the assessed entropy per sample of the source.
- Adaptive Proportion Test (§4.4.2): it aims at detecting that some value becomes much more common in the sequence of noise source outputs than expected, given the assessed entropy per sample of the source.

Those tests are reproduced in Alg. 1 and 2, where C is a given cutoff value. Obviously, some perfect sequences (meaning X_i are independent and identically distributed for all $0 \leq i < N$) will fail to pass the NIST canonical tests. Therefore the NIST prescribed that *false positives* occur with a limited rate between 2^{-20} and 2^{-40}.

Algorithm 1: NIST Repetition Count Test

static parameters	: C: cutoff value.
input	: Sequence of bits R_i ($0 \leq i < N$, where N can be equal to $+\infty$).
output	: Failure or Success (or no response if $N = \infty$ and sequence X_i is random).

1 $A \leftarrow R_0, B \leftarrow 1$ // Initialization
2 **for** $i \in \{1, \ldots, N-1\}$ **do**   // Continuous test
3     $X \leftarrow R_i$
4   **if** $X = A$ **then**
5        $B \leftarrow B + 1$
6   **if** $B \geq C$ **then**
7   **return** Failure
8   **else**
9        $A \leftarrow X, B \leftarrow 1$
10 **return** Success

Algorithm 2: NIST Adaptive Proportion Test

static parameters	: C: cutoff value; W: window size.
input	: Sequence of bits R_i ($0 \leq i < N$, where N is a multiple of W, and can be equal to $+\infty$).
output	: Failure or Success (or no response if $N = \infty$ and sequence X_i is random).

1 $i \leftarrow 0$ // Initialization
2 **for** $k \in \{0, \ldots, N/W - 1\}$ **do**   // Continuous test
3     $A \leftarrow R_{Wk}, B \leftarrow 1$
4   **for** $j \in \{1, \ldots, W-1\}$ **do**
5   **if** $A = R_{Wk+j}$ **then**
6           $B \leftarrow B + 1$
7   **if** $B \geq C$ **then**
8   **return** Failure
9 **return** Success

In order to deal with suspected failure modes that only the entropy source designers would be aware of, it is made possible by clause "Developer-Defined Alternatives to the Continuous Health Tests" (§4.5) to define custom tests. Those developer-defined continuous tests shall nevertheless detect failures timely.

The tests shall be run on at least 1,024 consecutive samples.

B. AIS BSI 31

The current applicable version is 2.0 [17]; notice that section §2.4 mentions other RNG standards. A new draft (version 2.35) is in preparation [9] and is open for public comments.

The requirements for PTG.2 and PTG.3 dictate a reliable online test (health testing) and a reliable total failure test that

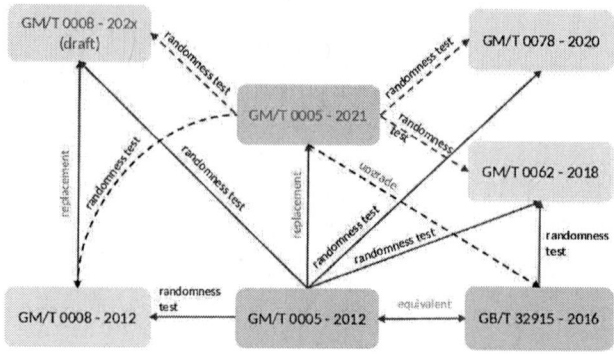

Fig. 18. Interplay between OSCCA standards for random number generation (Courtesy of Dr. Wei Cheng from Secure-IC certification team)

shall prevent undetected degradation of the entropy of the internal random numbers or an undetected total breakdown of the physical noise source.

The PTG.2 requirements (see §3.4.3 Functionality Class PTG.2) mostly focus on total failure tests and available entropy that can be extracted. There is currently no prescription regarding the health tests.

Notice however that the new version of the standard will require four novel online tests, namely:

- <u>Test T1:</u> monobit test
- <u>Test T2:</u> distance test
- <u>Test T3:</u> poker test
- <u>Test T4:</u> autocorrelation test on binary sequences

These will certainly have an impact on the complexity of health tests, since they are heterogeneous with other tests (such as that from NIST, namely Alg. 1 and 2). Thus a multi-compliance is still hard to achieve.

C. OSCCA GM/T 0078

The Office of State Commercial Cryptography Administration (abridged OSCCA) regulates the use and proper implementation of techniques involving cryptography in China. The OSCCA requirements regarding TRNGs have become stringent with the publication of GM/T 0078, which occurred in 2020 [5]. An overview of the OSCCA standards, already in place and to be launched, is given in Fig. 18.

An OSCCA-compliant module must implement four tests which are the same as that of NIST SP 800-22 T1, T2, T3, and T4, but on $2 \cdot 10^4$ bits. Notice that the test T3 of GM/T 0078-2020 (runs test) actually slightly differs from that of NIST SP 800-22. Besides, an OSCCA module must embed a total failure test, that reads 32 bits of data of the entropy source and returns an error defined as:

```
total_failure_error = (data == 0x00000000) \
                    | (data == 0xFFFFFFFF) ;
```

It is noteworthy that GM/T 0078 is dated from 2022, and then was **not** a requirement at the time GM/T 0008:2012 (or GM/T 0028:2017) came into force.

979-8-3503-4631-2/23 $31.00 © 2023 IEEE

VI. CONCLUSIONS

We presented in this paper different methods to evaluate and validate a TRNG. We focused on the SR-latch TRNG which has a different principle than the well-known Ring-Oscillator TRNG. It is important, for instance in OSCCA certification, to implement jointly two TRNGs with different rationales, as they do not have the same failure modes, making the dual construction more resistant to attacks.

First, we propose a stochastic model of the SR-latch TRNG in order to validate its principle and adapt the architecture to the targeted entropy. It is notably shown that the number of latches of the architecture should be chosen according to the knowledge of process mismatch and noise. An ASIC implementation of 1024 SR-latches TRNG in 28 nm FD-SOI technology has shown the feasibility of the SR-latch TRNG to get a 1-bit entropy by using a subset of latches.

We moved one step forward and assessed the impact of aging on this TRNG type using HSpice/MOSRA. This study concluded that aging contributes to enhancing entropy if the mismatch to noise ratio is not too high. It also shows that the mean of the process mismatch should be relatively small compared to the noise level to keep a minimum number of latches near a metastable state, thus providing a high level of entropy.

Finally, a review of health tests is presented. It allows the user to verify that TRNG delivers a minimum of entropy and ensures no failure or attacks compromise the TRNG's ability to deliver sufficient entropy.

ACKNOWLEDGMENT

This research has been supported in part by the National Science Foundation CAREER Award (NSF CNS-1943224).

REFERENCES

[1] "Nangate 45nm open cell library," "http://www.nangate.com".

[2] J. Bahrami, M. Ebrahimabadi, J.-L. Danger, S. Guilley, and N. Karimi, "Leakage Power Analysis in Different S-Box Masking Protection Schemes," in *2022 Design, Automation and Test in Europe Conference and Exhibition (DATE)*, 2022, pp. 1263–1268.

[3] P. Bayon, L. Bossuet, A. Aubert, V. Fischer, F. Poucheret, B. Robisson, and P. Maurine, "Contactless electromagnetic active attack on ring oscillator based true random number generator," in *Proceedings of the Third international conference on Constructive Side-Channel Analysis and Secure Design*, ser. COSADE'12. Berlin, Heidelberg: Springer-Verlag, 2012, pp. 151–166.

[4] M. Ben-Romdhane, T. Graba, and J.-L. Danger, "Stochastic model of a metastability-based true random number generator," in *Trust and Trustworthy Computing*. Springer Berlin Heidelberg, 2013, vol. 7904, pp. 92–105.

[5] Cryptography Industry Standard of the People's Republic of China, "GM/T 0078-2020: The Design Guidelines for Cryptographic Random Number Generation Module," December 28 2020.

[6] J.-L. Danger, S. Guilley, and P. Hoogvorst, "High speed true random number generator based on open loop structures in FPGAs," *Microelectronics journal*, vol. 40, no. 11, pp. 1650–1656, 2009.

[7] J.-L. Danger, R. Yashiro, T. Graba, Y. Mathieu, A. Si-Merabet, K. Sakiyama, N. Miura, M. Nagata, and S. Guilley, "Analysis of Mixed PUF-TRNG Circuit Based on SR-Latches in FD-SOI Technology," in *Euromicro Conference on Digital System Design (DSD)*, 2018, pp. 508–515.

[8] S. Kiamehr, M. S. Golanbari, and M. B. Tahoori, "Leveraging aging effect to improve SRAM-based true random number generators," in *Design, Automation & Test in Europe Conference & Exhibition (DATE)*, 2017, pp. 882–885.

[9] W. Killmann and W. Schindler, "A Proposal for Functionality Classes for Random Number Generators," September 18 2011, Version 2.35 - DRAFT: https://www.bsi.bund.de/SharedDocs/Downloads/DE/BSI/Zertifizierung/Interpretationen/AIS_31_Functionality_classes_for_random_number_generators_e.pdf?__blob=publicationFile&v=2.

[10] D. Kinniment and E. Chester, "Design of an on-chip random number generator using metastability," in *European Solid-State Circuits Conf.*, 2002, pp. 595–598.

[11] F. Lozach, M. Ben-Romdhane, T. Graba, and J.-L. Danger, "FPGA design of an open-loop true random number generator," in *Euromicro Conference on Digital System Design*, 2013, pp. 615–622.

[12] R. Maes, *Physically Unclonable Functions - Constructions, Properties and Applications*. Springer, 2013. [Online]. Available: https://doi.org/10.1007/978-3-642-41395-7

[13] A. T. Markettos and S. W. Moore, "The Frequency Injection Attack on Ring-Oscillator-Based True Random Number Generators," in *Cryptographic Hardware and Embedded Systems (CHES)*, C. Clavier and K. Gaj, Eds., vol. 5747, 2009, pp. 317–331.

[14] M. Matsui, "Linear Cryptanalysis Method for DES Cipher," in *EUROCRYPT*, ser. Lecture Notes in Computer Science, T. Helleseth, Ed., vol. 765. Springer, 1993, pp. 386–397.

[15] A. Muthukumar, N. Sivasankari, and K. Rampriya, "Anti-aging true random number generator for secured database storage," in *2017 4th International Conference on Advanced Computing and Communication Systems (ICACCS)*, 2017, pp. 1–7.

[16] F. Oboril and M. B. Tahoori, "Extratime: Modeling and analysis of wearout due to transistor aging at microarchitecture-level," in *IEEE DSN*, 2012, pp. 1–12.

[17] M. Peter and W. Schindler, "A Proposal for Functionality Classes for Random Number Generators," September 2 2022, Version 2.0 - https://www.bsi.bund.de/SharedDocs/Downloads/EN/BSI/Certification/Interpretations/AIS_31_Functionality_classes_for_random_number_generators_e.pdf?__blob=publicationFile&v=5.

[18] A. Rukhin, J. Soto, J. Nechvatal, M. Smid, E. Barker, S. Leigh, M. Levenson, M. Vangel, D. Banks, A. Heckert, J. Dray, and S. Vo, "A Statistical Test Suite for the Validation of Random Number Generators and Pseudo Random Number Generators for Cryptographic Applications," NIST, april 2010, https://nvlpubs.nist.gov/nistpubs/legacy/sp/nistspecialpublication800-22r1a.pdf.

[19] A. Schaub, J.-L. Danger, S. Guilley, and O. Rioul, "An Improved Analysis of Reliability and Entropy for Delay PUFs," in *Euromicro Conference on Digital System Design, (DSD)*, 2018, pp. 553–560.

[20] Synopsys, "HSPICE User Guide: Basic Simulation and Analysis," 2016.

[21] M. S. Turan, E. Barker, J. Kelsey, K. McKay, M. Baish, and M. Boyle, "SP 800-90B: Recommendation for the Entropy Sources Used for Random Bit Generation," January 2018, https://doi.org/10.6028/NIST.SP.800-90B.

979-8-3503-4631-2/23 $31.00 © 2023 IEEE

Silent Data Errors:
Sources, Detection, and Modeling

Adit Singh[*]   Sreejit Chakravarty[§]   George Papadimitriou[†]   Dimitris Gizopoulos[†]

[*]*Auburn University,* Auburn, AL, USA, singhad@auburn.edu
[§]*Intel,* Santa Clara, CA, USA, sreejit.chakravary@intel.com
[†]*University of Athens,* Greece, {georgepap | dgizop}@di.uoa.gr

Abstract—Chip manufacturers and hyperscalers are becoming increasingly aware of the problem posed by Silent Data Errors (SDE) and are taking steps to address it. Major computing facilities operators like Meta and Google have emphasized the critical role of SDEs in today's microprocessors. Numerous studies in the literature have highlighted the severity of this issue, especially in datacenter applications operating at large scales. These errors can lead to data loss and require a significant amount of time and effort to resolve through debugging engineering efforts, which can take months to complete. In this paper, we provide an overview of the issue of SDEs, including an explanation of the problem and the current methods used to address it, as well as gaps that still exist in addressing the issue. We also discuss the different sources of SDEs, including post-manufacturing testing failures, voltage and timing marginalities, and hard-to-detect faults. The paper emphasizes the impact of timing marginalities as a significant source of SDEs. Finally, our spotlight points to the architecture and system dimensions of the problem: we describe the challenges of measuring the true (still unknown) rates of SDE from CPUs, and emphasize on the role of detailed microarchitectural simulation models for this purpose. We present data on the severity of SDEs and their predicted rates under various operating conditions, sources of faults, and technology fabrication nodes.

Index Terms—Silent data corruptions, microprocessors, timing marginalities, voltage failures, microarchitectural simulation, microarchitectural modeling, fault injection, failure rates

I. INTRODUCTION

In modern computing systems, microprocessors are critical components that power a wide range of applications, from personal devices to large-scale data centers. Extreme performance levels are expected to be reached by future supercomputers by leveraging millions of microprocessor cores and specialized accelerators. Ensuring dependability is one of the most difficult barriers to overcome in achieving exascale computing [1]. Microprocessors employ the most aggressive design and manufacturing techniques for the above purposes and are, therefore, far from immune to errors, and in particular, the occurrence of silent data errors (SDEs) can have serious consequences for system reliability and computational accuracy [2]–[4]. Silent data errors are one of the most insidious and difficult-to-detect problems in modern computing. A silent data error occurs when data is corrupted in such a way that it still appears valid (no hardware or software level "alarm" is raised), but produces incorrect results when used in computation. In the

context of microprocessors, these errors can arise from a variety of sources, including cosmic rays, hardware defects, and hardware bugs. One of the known sources of silent data errors is cosmic rays. These high-energy particles can strike computer memory and cause bit flips, where a 0 becomes a 1 or vice versa. Other sources of silent data errors include hardware defects due to manufacturing and aging, hardware design bugs, and power supply fluctuations. Further, low voltage operation makes the chip more susceptible to silent data errors, since the critical charge to flip a bit is lower at reduced voltages [5]–[12].

Identifying and measuring silent data errors in microprocessors can be challenging because they often occur sporadically and are difficult to reproduce [13], [14]. To minimize the impact of on-chip memory errors, error correcting codes (ECC) are used to detect and correct such errors [15]. However, the use of ECC methods results in additional storage requirements and increased complexity, and cannot detect or correct all hardware-induced errors [16]. Although commonly used ECC methods can identify and correct some faults, they are limited in their ability to do so, with the most common method, single error correction, double error detection (SECDED), able to detect up to two flipped bits and correct only one flipped bit per 64 bits [15], [16]. In addition, multiple-bit faults are more common in on-chip memory structures in newer fabrication technologies [17]. While ECC can be beneficial in reducing failure rates in some on-chip memory structures, it is not always applicable to all functional, control, and memory blocks of the microprocessor. Even when using ECC methods, silent data corruptions are still possible, especially in large-scale datacenter infrastructures, which poses a significant threat to program integrity [13], [14], [18].

Silent data errors constitute a significant challenge for modern microprocessors and the computing systems they power. However, through the use of sophisticated error detection techniques, as well as the development of fault-tolerant and error-correcting models, researchers have made significant progress over the last decades in mitigating the rate and impact of silent data errors. As computing systems continue to become more complex and more critical to our daily lives, it is likely that this area of research will continue to be an important focus for the computing community. It is therefore critical to identify the sources of errors that are most likely to affect the program's execution silently, to propose novel ways for modeling and

---

979-8-3503-4631-2/23 $31.00 © 2023 IEEE

detection of silent data errors. One approach is to use fault tolerance techniques, such as redundancy or replication, to ensure that multiple copies of critical data are available. This can help ensure that even if a silent data error occurs, the system can continue to operate correctly. Another approach is to use error-correcting codes that can detect and correct errors automatically. These techniques can be especially important in safety-critical systems, where the consequences of a silent data error can be catastrophic.

In this paper, we summarize the importance of SDEs by first defining the SDE problem. We present the current approaches and the vital gaps in addressing the important problem of SDEs, and the potential sources of failure that could cause SDEs. We also discuss the impact of hard-to-detect faults that escape from post-manufacturing testing on SDEs, by investigating several types of manufacturing test, such as cell aware tests, scan timing tests, and system level tests. We stress another major source of potential SDEs, which is the sensitivity to voltage failures and timing marginalities, and discuss that timing marginalities being the source of a significant number of SDEs. Finally, we present the challenges in measuring the SDE rates on real microprocessors and in detailed low-level simulation models, and present the severity of SDE rates in different technology fabrication nodes and under several operating conditions using early microarchitecture level modeling and measurement of silent data errors.

II. SILENT DATA ERROR PROBLEM
Sreejit Chakravarty

After defining the SDE problem, the methodology used to address it is summarized. and gaps in the current approach are highlighted. Silicon error sources are shown in Fig. 1(b). Fig. 1(a) abstracts the silicon life cycle and depicts how silicon error sources are targeted. Logic bugs can also cause errors but are assumed to be eliminated prior to productization of the product and not considered. Silicon provider's high-volume manufacturing (HVM) flow consists of two screens to weed out faulty silicon: ATE tests and system level tests (SLTs). ATE tests are primarily structural tests like scan, memory BIST, etc. SLTs are application-based tests. Both screens target defects, circuit marginality issues and early life failures. HVM manufacturing screens are not perfect and there are residual faulty silicon escapes, referred to as HVM DPM escapes.

After shipment, silicon is assembled into the final system and undergo additional tests. These tests are more exhaustive system tests, than those used by silicon providers. The goal is to screen for HVM DPM escapes.

A. SDE Definition

Silent Data Error is often discussed in two different contexts. It is important to distinguish between the two since the required solutions are different.

1) **Low DPM SDE.** The first of this is a new name for "DPM escape". DPM escape from HVM testing is a fact of life. But, as shown in Fig. 2(a), it has a different impact on the end customer based on the size of the installation base. DPM1, which could be in the low 100s, is an adequate DPM level for installation base that are small, may be 10s of thousands. However, DPM1 is not acceptable for larger installation base since it leads to unacceptable levels of failure. For larger installation bases a much lower DPM level, DPM2 which could be in the low 10s is required. Achieving such low DPM level is difficult by itself but has been exacerbated due to the rise in the complexity of silicon designs. Fig. 2(b) shows the increase in design size resulting from complex designs incorporating added functionality on a single piece of silicon. This has prompted companies having large installation base to highlight the issue [13], [14]. The central problem, as pointed out in these papers, is the root causing the failures and finding an effective fix to plug the small DPM escapes from silicon providers.

2) **InField SDE.** In Fig. 1(a), InField refers to the silicon life cycle phase when the end user uses the silicon device. InField errors are radiation induced soft-errors or aging induced reliability hard failures. Protection mechanisms, like memory error-correcting codes (ECC), are added for InField errors. InField errors are said to be silent if the protection mechanism does not detect them. For example, an error in the memory that is not detected

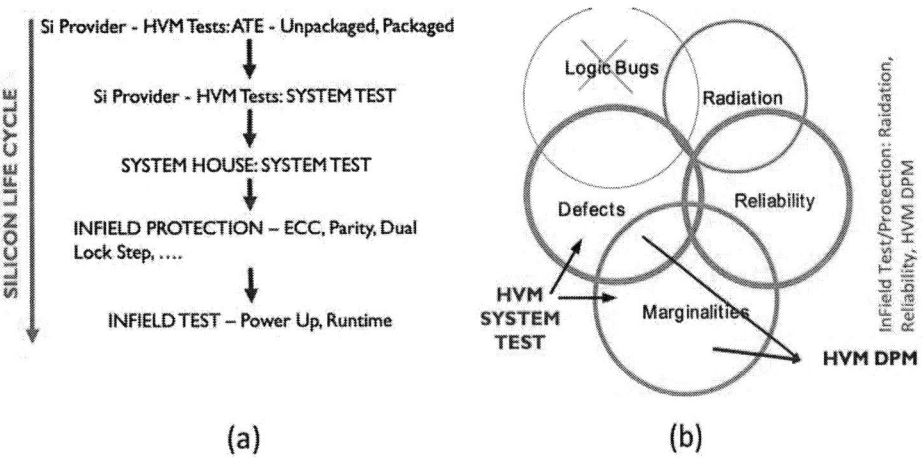

(a)   (b)

Fig. 1. Mapping of silicon error sources to silicon life cycle.

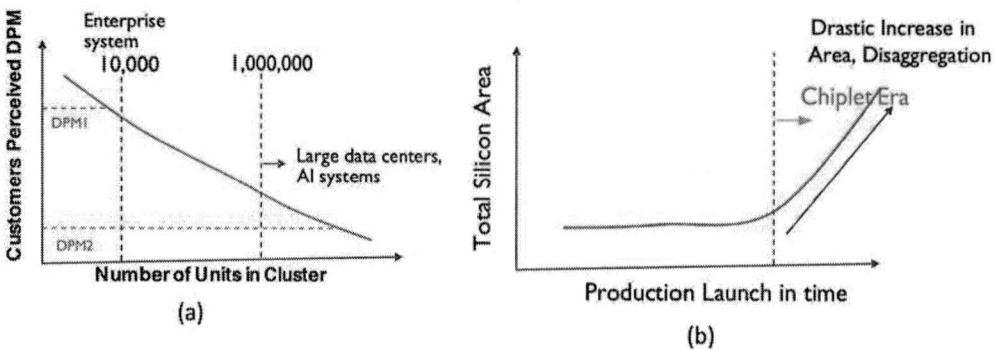

Fig. 2. DPM impact on customer installation.

and/or corrected by the ECC scheme is a silent error. It is silent in the sense that the silicon does not recognize or flag the error. The resulting error could be identified in other system components, like firmware, or not.

B. Current Approach and Gaps in Addressing SDE

1) Brick and Mortar Story of Low DPM SDE: Bick-and-mortar analogy is used to address the current methodology for Low DPM SDE problem. For ATE structural testing, the SoC is divided into small DFT domains like memories, logic within a clock domain, analog IPs, etc. Specialized DFT added to each DFT domain, shown as bricks in Fig. 3(a), is used to test it. Inter-domain structural tests, like hierarchical scan, are used to test the silicon interface between DFT domains, shown as mortars in Fig. 3(a). However, inadequate coverage or inadequacy of the target fault model result in DPM leaks, shown as brick test drips in Fig. 3(b). Inadequacy of inter-domain structural tests results in mortar drips of Fig. 3(b). "Mortar test sealants" (HVM-SLT) of Fig. 3(c), which are HVM SLTs, are added to patch these leaks. There are two major issues with this HVM test flow.

- Structural tests are based on fault models, and work well, unless very low outgoing DPM is a requirement. Ongoing effort to improve fault models, to reduce HVM DPM, targets only the "brick test drips" of Fig. 3(b).
- Filling "mortar drips" of Fig. 3(b) is getting more difficult due to the increase in design size and aggressive design trends. The inadequacy of fault model based structural tests and automation tool capacity are falling short in addressing this problem.
- HVM-SLTs, which are randomly generated tests with different instruction and data mixes, use a "shot gun"

approach to fill the brick-and-mortar drips. Increasing the number of HVM-SLTs improves its quality. However, there is no meaningful approach to identify effective HVM-SLTs or to measure an HVM-SLT's quality. This approach adds very significantly to test cost. They are therefore truncated, leaving a DPM gap.

- As shown in Fig. 3(d), system houses al run a barrage of tests before enabling their systems for end user use. These are a mix of frequently used applications, mixed in with some longer tests from the HVM system test flow. Such tests run for longer duration, and based on the usage model, run often enough prior to farming it out to the end user. Specific tests that fail are fed back to the silicon provider to plug their test holes.

Hence, we have reached an inflection point and new ideas are needed to fill the gap left by the adhoc HVM test screens.

2) InField SDE: InField error prevention relies on protection mechanisms which are assumed to provide good coverage. This is a fallacious assumption! Protection schemes were developed to protect against radiation induced soft errors and provide poor protection against aging and HVM DPM related hard failures, as illustrated using the example of Fig. 4. Fig. 4(a) shows a memory with data ECC and address parity protection. Fig. 4(b) shows a detailed view of the array. Note that sense amplifiers (SA) do not change state if its differential inputs are not driven.

Assume wordline W0 s@0. On a read from location 0, the cells are not selected, the SA's differential inputs are not driven, and memory output retains the old state. If the previous read was fault-free, then the ECC decoder will not detect this error. For the memory writes and reads sequence of Table I, the output after the second read should be BB. In the faulty

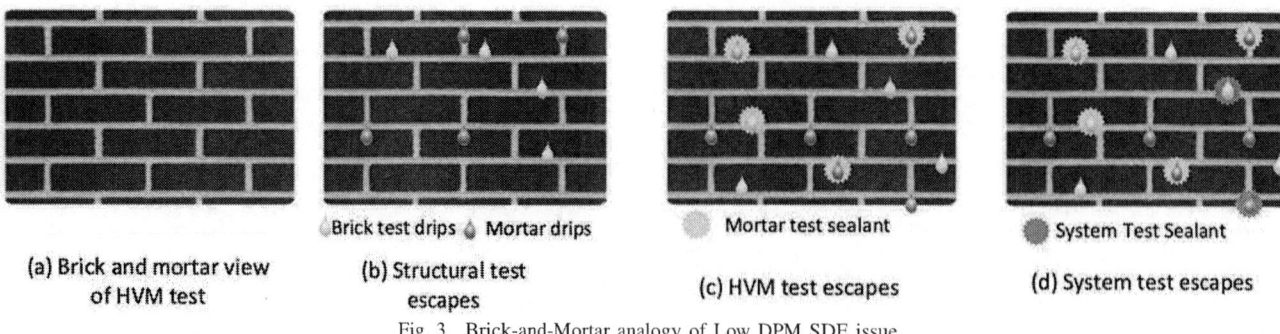

(a) Brick and mortar view of HVM test

(b) Structural test escapes · Brick test drips · Mortar drips

(c) HVM test escapes · Mortar test sealant

(d) System test escapes · System Test Sealant

Fig. 3. Brick-and-Mortar analogy of Low DPM SDE issue.

979-8-3503-4631-2/23 $31.00 © 2023 IEEE

TABLE I
A PERSPECTIVE ON SDE SOURCES.

Memory Operations			Good Data	Faulty Data
1st	Write	ADDR3, AA	N/A	N/A
2nd	Read	ADDR3	AA, ECC[AA]	AA, ECC[AA]
3rd	Write	ADDR0, BB	N/A	N/A
4th	Read	ADDR0	BB, ECC[BB]	AA, ECC[AA]

case, it is AA instead. However, since the ECC[AA] matches the faulty data AA the decoder will not flag this error. This is an example of an InField SDE. A similar observation can be made using reliability studies of ECC protected memories alluded to in Fig. 4(c). It shows the percentage of reliability induced errors escaping detection (and correction) by ECC decoders, at different voltage and operating corners.

The key question is: how is the gap left by protection schemes filled? Repeated or periodic InField testing is the predominant practice. Unfortunately, there is a major gap in InField testing.

- Scan/LBIST tests with 90% or more coverage is often considered to be very good logic test.
- March C- is assumed to be a good infield test for memories.

What is the basis for using the above tests for InField testing? It seems to be based on convenience rather than any sound engineering basis. In addition to the above structural tests, SLTs are often used as InField tests. Once again, based on our inability to measure the quality of SLTs there is no rationale basis for the use of such tests. They are more of a feel-good test! Questions that remain unanswered are: what is an appropriate InField test for logic and memories? How often should we apply these tests? How do we create measurably effective SLTs for InField Testing?

III. TESTING TIMING MARGINALITIES THAT CAUSE SDEs
Adit Singh

A. Understanding Escapes from Manufacturing Tests

The previous section has outlined the many varied sources of failure that can potentially cause silent data errors (SDEs) in computing systems. Any defective, marginal, or unstable IC or SOC that is not detected and screened out during post manufacturing tests can potentially cause failure during operational deployment. additionally, early life failures may

occur from latent defects of device aging. While many test escapes result in repeated and severe malfunction during operation, including system crashes, that are readily observed and detected, it is not uncommon for more subtle faults with a limited error impact to go unnoticed, particularly if they are infrequently activated. Observe that rare activation with minimal error impact are precisely the characteristics of the "hard to detect faults" that escape post manufacturing testing in the first place.

1) New Cell Aware Tests: The solution to minimizing test escapes that cause field failures is obviously better testing, with more complete test coverage of the actual defects and anomalies observed in manufacturing. This was discussed at some length in the previous section. Classical stuck-at (SA) and transition delay fault (TDF) test generation explicitly targets faults only at the circuit nodes, i.e., the interconnects between the standard cells; defects and faults within the standard cells of the design are not targeted during test generation, and are only serendipitously detected. Plugging this test coverage gap was the motivation behind the recent introduction of the Cell Aware (CA) test methodology [19] which also targets shorts and opens within the standard cells of the design during test generation. CA tests have been shown to significantly reduce test escapes in volume production. However, comprehensive CA test generation significantly increases test set size and test application times. Additionally, targeting resistive defects, beyond ideal shorts and opens, can make test costs prohibitive. Consequently, CA tests are generated for only a limited set of defect resistances in practice, resulting in many of the test escapes discussed earlier with the help of the Brick-and-Mortar analogy in Fig. 3. Nevertheless, in principle, effective test methodologies are available today to minimize field failures, including silent data errors, that are caused by classical permanent faults. In this area, the current focus of test development in advancing DFT (design-for-test) and ATPG (Automatic Test Pattern Generation) towards better detection of these faults is largely aimed at reducing the cost of test generation and application to facilitate high coverage CA tests.

2) Scan Timing Tests and Process Variations: Meanwhile, scan test methods that target timing failures appear to be less effective in screening failing state-of-the art circuits. Both tradition TDF, as well as two-cycle CA delay tests, only target

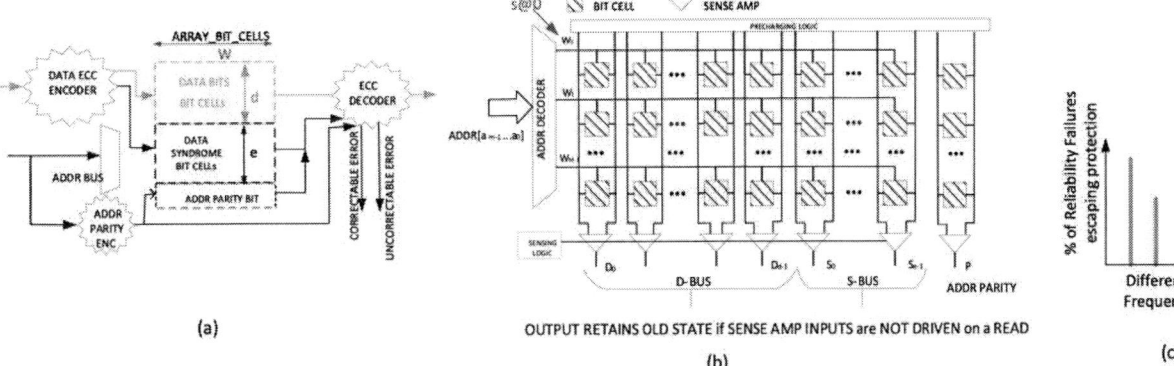

Fig. 4. Silent Data Error of Data and Address protected Memories.

979-8-3503-4631-2/23 $31.00 © 2023 IEEE

localized (lumped) gross delays. While these tests can be effective in screening out large delay faults caused by mechanisms such as an isolated resistive or open defect, they can miss timing failures caused by an accumulation of distributed delays in a circuit path caused by process variations. Such delay variability is presenting an increasing reliability challenge at advanced technology nodes. Even timing aware scan TDF tests [20], developed for small delay defects (SDDs), have not proven effective against random process variations that can impact every component in a circuit to some degree; each gate or standard cell can display a unique delay. Given the billions of paths in a processor, a significant number (due to pure statistical chance) can contain several gates significantly slower than nominal. Therefore, because the long paths in high performance circuits are carefully optimized during design to all have nearly equal delays to allow fast clock rates, each manufactured instance of a design can have a different slowest path in silicon due to random process variations. Consequently, as has long been recognized, reliable testing to ensure that a circuit meets timing in the presence of significant random gate delay variability requires a comprehensive path delay test. Unfortunately, for many reasons, including the increase in test time due to the very large number of paths that need to be tested, effective scan-based path delay testing [21] has so far not proven practical.

3) System Level Tests: In the absence of an effective scan-based path delay test capability industry, in recent years, is increasing relying on at-speed functional system level tests (SLTs) as a final test screen, primarily to eliminate undetected timing failures. At-speed functional tests have been the gold standard for testing circuit timing in an operation environment, but are expensive both to develop and apply. Also, the test coverage is unknown and can be limited. Therefore, in practice, low-cost scan-based structural tests, applying high coverage test content generated using SA, TDF, and increasingly also Cell Aware (CA) fault models are used during the wafer probe and post packaging test insertions. A comprehensive functional test is then performed at the final stage. This "component" system level test (SLT), that specifically targets only the DUT (Design Under Test), involves temporarily mounting the packaged part on a test board that accurately mimics the intended application hardware, including all its electrical characteristics. The part is then extensively tested in functional operation for as long as an hour or more, at full rated speeds, over a range of user applications and operating conditions. New highly parallelized SLT testers, that allow tight temperature control of the device during test to replicate actual operating conditions, have been developed to support such long test times with high test throughput. However, the coverage of functional tests is difficult to estimate and quantify; test escapes remain a challenge.

B. Voltage Sensitive Failures and Timing Marginalities

1) Industrial Data on Voltage Sensitive V_{min} Failures: The need for aggressive dynamic voltage frequency scaling (DVFS) for thermal management and temperature control in high performance processor SOCs appears to be significantly increasing the occurrence of a class of voltage sensitive timing failures, being referred to as V_{min} failures. These are parts that only fail close to the minimum specified operating voltage, while passing at higher voltages. Table II summarizes some data from a recent industrial test experiment [22] on Intel 14 nm FinFET processors that was aimed at studying the addition fallout observed from CA (cell aware) tests after prior testing with SA and TDF tests. The total additional fallout from CA tests, after screening by the traditional tests, was 4300 DPPM. Out of these, approximately 90% needed two-cycle CA-delay tests for detection. A significant number, 1600 DPPM, were only detected close to V_{min}. While the data clearly highlights the benefits of CA-tests, the large number of V_{min} only failures raise concern about the possibility of additional undetected voltage sensitive marginal parts in the population that may experience failure under less favorable circuit operating conditions. This is illustrated in Fig. 5.

2) Marginal Timing Failures: Fig. 5 plots the longest path delay for each instance of a hypothetical collection of manufactured ICs (Integrated Circuits) of the same design. Worst case path delays for individual instances are different because of manufacturing process variations. Also shown in the figure is the time period and clock edge corresponding to the clock frequency at which the circuit is expected to operate correctly, down to V_{min}. While more tightly clustered together at higher supply voltages, the critical path delays increase and spread out as the supply voltage is lowered towards V_{min}. This is because circuits slow down significantly as supply voltage is lowered. More importantly, the increase in delay of a slow, high threshold voltage, transistor can easily be 3 or 4 times greater close to V_{min}. Low voltage operation greatly accentuates delay variability due to process variations. The plot in Fig. 5 illustrates this spread for a few sample V_{DD} values. At $V_{DD} = V_{min}$, the spread of worst-case delays in the collection of circuits plotted exceeds the clock period for some parts. These are the V_{min} only failures.

TABLE II
DISTRIBUTION OF FALL-OUT FROM DIFFERENT CAT TESTS IN [22].

CAT-Static (Before Delay Tests)	CAT-Delay (Fails at V_{max} & V_{min})	CAT-Delay (V_{min} Only Fails)
400 DPPM	2500 DPPM	1400 DPPM

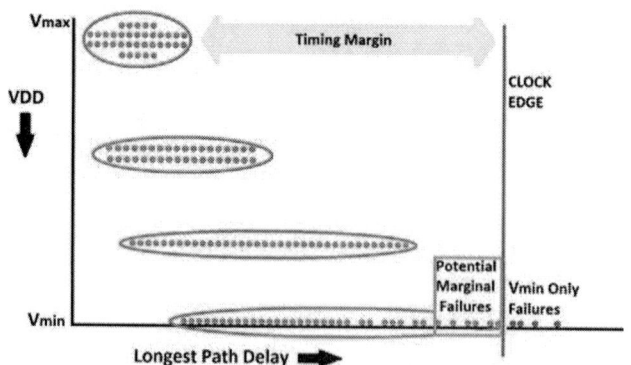

Fig. 5. Critical Path Delays in a Collection of ICs.

Observe in Fig. 5 that path delays are much smaller than the clock period at high values of V_{DD}, allowing a significant noise margin. Thus, even if an occasional IC may fail timing because of an open or short defect, the bulk of the path delays are well clear of the clock edge, and highly unlikely to fail due to circuit noise of other environment conditions. However, at V_{min}, even if all the parts failing the timing test are scrapped, there are many additional marginal parts near the clock edge that narrowly pass timing as shown. These may fail in operation under more unfavorable operating conditions, e.g., power supply noise, that can momentarily lower V_{DD} further. These are marginal parts that may malfunction and cause occasional, unpredictable errors in operation. Unfortunately, as explained below, the specified V_{min} cannot be conservatively raised to increase timing margins and eliminate such failures.

3) Timing Margins Reduced for Power Management: The throughput of high-performance processors has long been limited by the need to manage heat dissipation. Even though clock rates have not increased in two decades to hold down power dissipation, the exponential increase in transistor counts from technology scaling that continues to track Moore's Law means that all the cores in a package cannot always be operated at their maximum frequency. Active thermal management is an integral part of modern processors, where dynamic voltage frequency scaling (DVFS) is used to reduce supply voltage and clock frequency and cool off the die as needed to maintain acceptable operating temperatures. Modern designs offer the on-chip thermal management system a choice of several operating frequencies, each with a specified V_{min}. This is illustrated for a hypothetical processor in Table III. The V_{min} values provide guidance to thermal management on how far the supply voltage can be lowered to save power while operating at each frequency. Observe that since maximum operating voltages today are well below a volt, only modest voltage, and therefore power, reduction is available when reducing operating frequency by each step. For this reason, V_{min} cannot be conservatively chosen; it must be set to be as low as possible. In fact, to maximize power savings at each operating frequency, the V_{min} values are individually estimated and assigned (as shown in Table III) to each processor core. This is to avoid the need to be conservative and use high V_{min} values to accommodate systematic core-to core, and die-to-die parameter variations, if common V_{min} assignments are made to an entire batch of processors. Note, however, that each V_{min} is the best estimate, based on limited test measurements. Since accurate V_{min} measurement involves a search requiring repeating an accurate timing test multiple times at different candidate V_{min} voltages, the cost of obtaining a measured V_{min} value for each core at each target frequency is prohibitive. Unfortunately, since the estimated V_{min} values cannot to overly conservative, this leaves open the possibility of some timing failures at V_{min} for a few outlier parts with delays in the tail of the process variations distribution. These are the parts that fail at V_{min} in Fig. 5. Setting a higher, more conservative V_{min}, e.g., V_{DD} can minimize such failures, but since power consumption is a quadratic function of V_{DD}, this would significantly reduce the power savings for the many parts with near nominal delays. It is implicitly expected that the few parts failing at V_{min} will be screened out by testing.

4) Timing Marginalities and SDEs: Unfortunately, as shown in Fig. 5, when the tail of the worst-case path delay distribution approaches or crosses the clock period boundary, without any timing margin remaining to absorb electrical or environmental noise experienced by the circuit, a significant number of the marginal parts that barely pass the timing test can fail in operation under worse case operating conditions. Any such part that is not detected by subsequent functional systems tests and ends up in a system deployed in the field can exhibit unpredictable and intermittent errors and can potentially be a source of SDEs.

There are hints in the published work that point to timing marginalities being the source of many of such errors. Studies [23] have found that SDE rates are higher in very low frequency, low voltage operation. This is consistent with slow outlier process variability paths being the source of the failures because the delay of such paths is greatly accentuated in low voltage operation. It has also been found the rate of occurrence of SDEs increases with device age. Again, modern transistors and their circuits slow down a little over time due to aging mechanisms such as NBTI. It is, therefore, expected that the delay of marginal paths will increase with time, and the timing error rate will also go up.

C. Detecting Timing Marginalities in Processors

1) Key Characteristic of Slow Outlier Paths: Recent research [24] studying gate and path delay increases from process variations has suggested that because of the extremely large number of transistors in modern SOCs, a noticeable number of devices can be expected to be from the extreme tail of the variability distribution. Published work [25] has found that transistor threshold voltages (V_{th}) are normally distributed, at least out to ± 5 sigma (σ). Assuming that this distribution remains normal, since the probability of a 6σ transistor in about 1 in a billion, every SOC incorporating billions of transistors can be expected to contain many such extreme transistors. A population of a million such SOCs (used to measure DPM) will even contain thousands of 7σ transistors and beyond. Such extreme transistors, with greatly increased threshold voltages, can slow down dramatically when operating at the lowest V_{min} corresponding to the lowest operating frequency in a processor. This is because the transistor gate voltage is not far above V_{th} when the transistor is on, resulting in a very weak gate overdrive. For example, a logic gate containing an extreme 7σ transistors can have 10X or more

TABLE III
FREQUENCY AND V_{min} TABLE.

Freq	V_{min}
F1	$V_{min\_1} = 0.90V$
F2	$V_{min\_2} = 0.79V$
F3	$V_{min\_3} = 0.71V$
F4	$V_{min\_4} = 0.60V$
F5	$V_{min\_5} = 0.49V$

delay compared to a similar logic gate operating at the same low voltage with a nominal transistor. Such outlier transistors can alone contribute to slow outlier paths and significant DPM from timing failures.

The above discussion can be more formally understood based on modeling logic gate delays using an analytical gate delay model, such as the well validated Sakurai-Newton alpha-power law [26]. This approximates switching delay to be proportional to $1/(V_{DD} - V_{th})^\alpha$. The literature suggests that for advanced FinFET technologies, the best fit α appears to be 1.25. Observe that the switching delay blows up as $V_{DD} - V_{th}$ approaches zero. At very low power saving operating voltages around 0.5V, the gate overdrive $V_{DD} - V_{th}$ is typically only 150-200mV for nominal V_{th}. Statistically extreme transistors can raise V_{th} sufficiently high to reduce this overdrive to near zero, introducing very large delays in the circuit. Failure analysis has measured functional transistors with V_{th} increase due to variability of more than 200mV.

Given this dramatic non-linear increase in delay caused by the extreme shift in V_{th} in $7 - 8\sigma$ transistors, it has been speculated [24] that many, if not most, of the rare long paths that fail timing due to process variations get most of their increased delay from a single extreme V_{th} transistor. There is credible experimental evidence validating this conjecture. However, given the rare occurrence rates (1 in $10^9 - 10^{12}$) of 7σ and 8σ transistors, direct experimental validation appears impractical. Even meaningful Monte Carlo simulations of circuit path delays using SPICE are very time-consuming to perform. Nevertheless, early trends from results of a few hundred million simulations runs appear to confirm that the longest circuit paths due to process variations do contain an extreme outlier transistor. These are the paths that can fail at V_{min}, or are marginal paths with the potential to fail in operation.

2) Screening V_{min} Marginal Parts: Observe that if a target path to be delay tested contains a single extreme outlier transistor that contributes most of the increased delay in the path, then the path exhibits a large lumped delay at the output of the gate containing the slow transistor, which will likely be detected by two-cycle TDF or CA tests. (It is important to remember here that all tests are probabilistic; no test can guarantee that all defects in an IC are detected.) This observation is significant, because it suggests that path delay tests may not be essential to detect timing failures caused by large random process variations. It also explains the unexpected success of CA timing tests in detecting all V_{min} SLT failures in [22].

The two-cycle timing test targeting the extremely weak transistor in the slow path can be further sensitized by conducting the test at V_{DD} 10-25mV below V_{min}. A supply voltage 25mV below V_{min} will further reduce the $(V_{DD} - V_{th})$ term in the Sakurai-Newton equation above, causing an extreme ($> 7\sigma$) or ($> 8\sigma$) transistor in the path to add another 10+ nominal gate delays, significantly slowing down the path further and making it easier to detect. Meanwhile, delays in statistically less extreme transistors (the assumption is that there is only a single extreme transistor in the slow path to be detected) will increase much less. Any risk of incorrectly failing such paths can be mitigated by slowing the clock to increase the clock period by a couple of nominal gate delays to absorb the slowdown in good paths. The optimal supply voltage and clock period/frequency to maximize detection of slow and marginal paths while avoiding yield loss from over-testing will need to be worked out on the test floor from experience, as is common practice.

In summary, research is ongoing to validate and exploit the assumption that many of the slow paths with large delays due to process variations that cause timing failures in advanced processors experience most of their increased delay from a single statistically extreme outlier transistor. This should allow traditional lumped delay scan timing tests to remain effective in detecting such failures. Furthermore, overtesting somewhat below V_{min} with appropriate adjustment in clock frequency can further sensitize the slow path detection during both scan and SLT testing.

IV. THE CHALLENGE OF MEASURING SDE RATES
G. Papadimitriou, D. Gizopoulos

A. *The Challenging Task of Unveiling Errors at System-Level*

The main challenge of measuring SDE rates is that, per their definition, these errors are silent, which means that no hardware-based or software-based error handling mechanism can detect them. The SDE rates are often low, and strongly depend on the defective hardware structure and the executed software workload. This means that in order to measure SDE rates accurately, a large amount of data must be processed coming from non-negligible numbers of faulty chips. For example, in a typical datacenter, billions of bytes of data may be processed every second. This requires specialized equipment and techniques, such as hardware monitors or software-based profiling tools [27]. Another reason why measuring SDE rates is challenging is that these errors can be highly dependent on the system's configuration and workload. For example, SDE rates may be higher in systems that operate in harsh environments or experience frequent power fluctuations [7], [9], [10]. They may also be higher in systems that run complex and demanding applications. Therefore, in order to accurately measure SDE rates, it is necessary to perform large-scale experiments under a wide range of conditions, which is both time-consuming and expensive; it can practically be realized only by owners of extreme scale systems.

Enterprise and cloud data centers are installing increasingly complex System-on-Chip (SoC) devices in large numbers, which raises the likelihood of undetected faults that can cause unexpected crashes or silent data errors (SDEs). Although soft errors due to cosmic rays are widely recognized [28], [29], the scale of data center infrastructure means that SDEs caused by escaped manufacturing defects and in-field reliability mechanisms must also be taken into account [13], [14], [27]. Defects leading to SDEs are difficult to detect and screen due to the multiple conditions required for their occurrence, such as specific machine instruction sequence,

979-8-3503-4631-2/23 $31.00 © 2023 IEEE

operating voltage, frequency, and temperature conditions, and platform behavior like interrupts [14]. These factors result in limited repeatability of SDE detection tests and the need for extended test duration to identify failures, making it crucial to design test methods with this behavior in mind. One method is to execute SDE-targeting code multiple times during tests, while another is to use pseudo-random instruction and data sequences on every execution loop to increase the number of unique data sequences applied by the tests.

Numerous factors can cause faults in an SoC, such as radiation, electrical marginalities, and manufacturing defects. Even silicon defects that are not detected (or even exist) during manufacturing can result in faults [30], [31] in the field. The way these faults affect the operation of a workload depends on the circuit where the fault occurs [32], [33]. If the fault occurs in a circuit that includes error detection and correction, such as a cache or memory with error correction code, the hardware can correct the error.

Silent data errors go undetected, do not interrupt the machine operation, but instead result in data errors. Data errors are more likely to occur when faults occur in circuits that are not used for program control, such as the SoC's integer of floating-point units [35]. The effects of silent data errors are unpredictable and depend on various factors. While an incorrect calculation of a single pixel value may not be significant, a data error in a financial transaction calculation could require corrective action [27]. Since a single fault can manifest in different ways over time due to workload variations, managing faults that can cause SDE at scale is crucial, particularly when millions of processing cores are installed in a data center or a supercomputer. Lerner et al. in [27] presented that a datacenter of modest size (i.e., 100,000 SoCs) is likely to experience at least one SDE event per month with a rate of 10 failures in time (FIT)[1]. For larger installations, frequent SDE events are likely, even at 1 FIT. To this end, it is crucial to minimize the rate of SDE, for example, by periodically testing the datacenter infrastructure to identify defective hardware components that perform wrong calculations.

B. The Need for Billions of Real Machines or Billions of Years of Simulation

Given the challenges associated with measuring SDE rates, discussed in the previous section, it is not surprising that many researchers have turned to simulations as a way to study such errors: simulation-based analysis provides the opportunity to evaluate faulty chips even without having access to any faulty

[1] 1 FIT equals to one failure every 10^9 (one billion) hours of operation

physical chip. Simulations have their limitations. In particular, measuring SDE rates at the RTL (register-transfer level) provides very high detail and accuracy, which, unfortunately, is extremely computationally expensive. In fact, measuring real SDE rates at the RTL is practically impossible since it can take many years, even with the most powerful computers available today. Table IV shows the most common ways to evaluate the reliability (including the expected SDE rates) of computing devices, comparing the time and cost required to complete the study, how many of the available resources can be accessed (or are modeled), if the faults are induced by processes that are natural (i.e., realistic error rates) or synthetic (i.e., models chosen by the user), if the study can be performed in the early stages of the project or only on the final product, and how much information can be gathered on faults generation and propagation (observability). Alternatively, researchers can attempt to measure SDE rates using real machines [13], [14]. However, this approach is possible only for hyperscalers, i.e., owners of huge fleets of computing machines to study SDE rates accurately. For example, in order to precisely measure the SDE rates, billions of machines may be required [13], [14]. Even with the proliferation of cloud computing and big data platforms, it is hard to obtain access to such large numbers of machines. For microprocessors consisting of several million bits and programs consisting of several billions of cycles, determining the real probability of failure (or the FIT rate) is an extremely difficult, if at all possible, task. Specifically, there are two stages at which the FIT rate is measured.

Early stages: when both the design of the microprocessor hardware and the design of the software are under development and major modifications can be applied. In early stages, the FIT rate of a microprocessor, program pair is actually estimated or predicted and not really measured. This is because there are certain parts of the hardware structure of the microprocessor that are still unknown in detail or are deliberately removed from the abstraction to facilitate the design and simulation of the system. Analysis of the failure rates at early stages can be only implemented using architectural (ISA), or microarchitectural models of the system, both of which are available very early in the design flow. Architectural models do not include any hardware information, but only the ISA visible hardware locations (memory and registers) [36]. Microarchitectural models (also referred to as performance models) have a significant detail of the microprocessor hardware: they contain all major hardware storage components that occupy a very large part of the final silicon estate (registers, register files, buffers, queues, caches, predictors, etc.) but they model the combinational logic and the random sequential logic

TABLE IV

SILENT DATA ERROR RATE MEASUREMENT METHODOLOGIES [34].

Evaluation Method	Time Needed	Cost	Accessible Resources	Fault Source	Availability	Observability
Field, Lifetime data	months/years	very high	all	natural	final product	limited
Beam testing	hours	high	all	natural	final product	limited
Software-level fault injection	hours	low	limited	synthetic	early/final product	medium
Architecture-level fault injection	days	low	limited	synthetic	early	medium
Microarchitecture-level fault injection	**days/weeks**	**low**	**most**	**synthetic**	**early**	**very high**
RTL fault injection	years	low	all	synthetic	late	very high

(state machines in control) only functionally. Program executables (assembly/machine instructions) can run on both an architectural and a microarchitectural model. The architectural model is typically around three orders of magnitude faster to simulate than the more detailed microarchitectural one. Moreover, microarchitecture-level models can also be used for bug modeling during the CPU validation phase, e.g., [31], [37].

Late stages: when the microprocessor design, as well as the program design, are very close to completion and design changes (particularly to the hardware) are either impossible or extremely costly. At these stages, the failure rate of a microprocessor, program pair can be actually measured because almost all details of the hardware design are in place, unlike the early models. In late stages, measurement of the failure rates are mainly employed for validation purposes. Late stages measurements can be realized when the program runs on two setups: a gate-level (RT level) model of the microprocessor design, and an actually manufactured silicon chip. Unfortunately, the simulation speed of such fine-grained late stage models is prohibitive to run reasonably long programs. The simulation throughput of the gate level models is typically three or more orders of magnitude slower than the microarchitecture (performance) models. Therefore, the combined effect of the hardware design and software design on the failure rate of the system cannot be measured at the gate or the RTL.

Finally, measurement of the failure rate on actually manufactured chips is the only true physical experiment which runs at the true speed of silicon. The major drawback of this experiment is it requires expensive accelerating testing of the chips with dense beams of particles. Such particles (neutrons or others) are blindly hitting the chip when the program runs, and the failing executions (output corruptions or abnormal terminations) are recorded. There is no way to isolate the hardware spot that was affected and the bits that were flipped. However, the failure rates of such physical beaming experiments are employed by the industry to better emulate the actual physical conditions in an accelerated setup to reach statistically significant results for the failure rates.

Summary: Typically, RAS architects rely either on Statistical Fault Injection (SFI) [38] or on analytical methods, such as the Architecturally Correct Execution (ACE) analysis [39], to provide insights into the programs' resiliency toward transient faults, because both methods aim to report the cross-layer vulnerability. Unlike lower-level simulation models (e.g., gate and RTL), microarchitecture-level fault-injection based on performance models allows deterministic end-to-end execution of large workloads on top of an operating system, i.e., full system analysis, which is impossible at lower levels [33], [36]. Further, injection on RTL models [40] would marginally augment vulnerability analyses with combinational logic vulnerability, since logic has very low raw failure rates compared to storage elements.

We, therefore, employ GeFIN [6], which has been developed and extended on top of the gem5 simulator [41], which is a state-of-the-art microarchitecture-level simulator. Recent studies have shown that fault injection based on microarchitecture-level models in gem5 simulator can provide vulnerability results of the entire CPU during 18 days [33], in contrast to RTL fault injections, which could need several years (see Table IV). In the next subsections, we summarize a number of recent vulnerability studies using our gem5-based simulation and injection set of tools.

C. SDE Failures in Time Analysis

Failures in Time (FIT) rate of a device is the number of failures that can be expected in one billion (10^9) device-hours of operation. For each hardware structure in a microprocessor, a different FIT is computed using the formula below.

$$FIT_{struct} = AVF_{struct} \times rawFIT_{bit} \times \#Bits_{struct}$$

The FIT of the structure is determined by three components: the FIT_{BIT} (or raw FIT) rate, which is determined by the fabrication technology and expresses the fault rate of a single bit, the number of bits of the structure and the SDE AVF of the structure, which is affected by the microarchitecture and the running workload. The raw FIT rate expresses the number of SDEs that will be introduced in the component, while the AVF (architectural vulnerability factor) is the derating factor that quantifies how many of these errors will lead to a failure. The product equals the FIT rate of a component. The SDE FIT rate of the entire CPU is calculated by adding the individual SDE FITs of the individual hardware structures. In the following subsections, the AVF is determined using microarchitecture level fault injection on gem5 simulator.

Fig. 6 shows the SDE FIT rate for each technology node [17]. The red color indicates the percentage of SDE FIT due to multi-bit faults, which starts from 0% in 250 nm node and reaches a high 12% in 22 nm. We can also see that the SDE FIT for each technology node is increasing until the point of 130 nm. After that, the SDE FIT rate starts to decrease, reaching the lowest FIT values at 22 nm. These values correspond to the exact same microarchitecture with the exact same configuration. The differences observed are due to the much smaller area that the chip occupies in the higher density technologies, which results in a significantly smaller number of particles that will eventually strike the microprocessor.

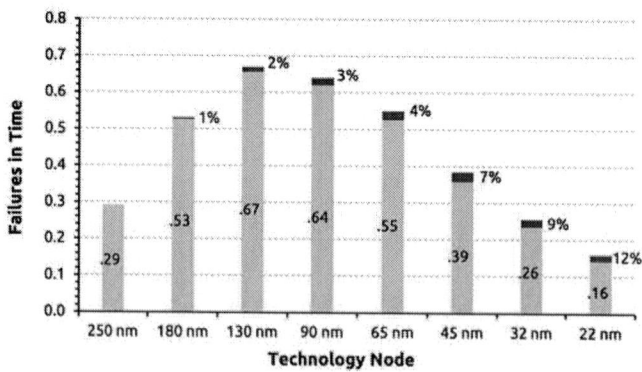

Fig. 6. SDE FIT for the entire CPU core for different technology nodes (numbers inside the green bars) due to transient faults. Red color areas correspond to the contribution of multi-bit upsets. The graph shows only the FIT rate for SDEs [17].

979-8-3503-4631-2/23 $31.00 © 2023 IEEE

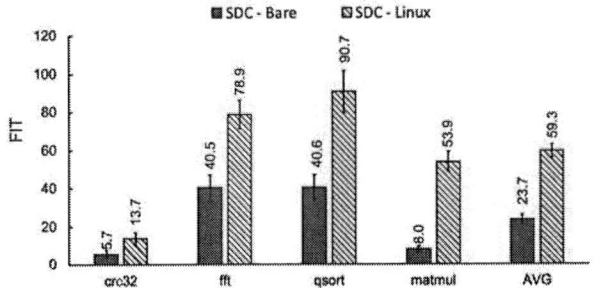

Fig. 7. Cortex A5 Bare-metal and Linux beam FIT rates for SDEs [34].

D. SDE Rates for Bare-metal versus OS Executions

In this section, we summarize a characterization study which is performed through physical beam experiments on an Arm Cortex-A5 microprocessor, to show the contribution of OS (Operating System) to the SDE rates. To this end, we show a comparison between SDE FIT rates of bare-metal executions and with Linux OS. In Fig. 7 we can observe that the average SDE rates for A5 is 23.7% for bare metal and 59.3% for Linux. It is also clear from Fig. 7 that the SDE rate is constantly higher when the applications run on top of Linux, in contrast to bare-metal execution. Specifically, we can see that the difference of the SDE rates between bare-metal and Linux OS can be as high as 6.7×. However, as we discussed earlier, during beam experiments it is very difficult to investigate the SDE rates in finer granularity. To this end, in the next subsection we show results from the state-of-the-art microarchitecture-level fault injection framework, named GeFIN [42] which is based on the gem5 simulator.

It is essential to note that there have been several recent attempts for validating the results of microarchitecture-level models using physical accelerated beam experiments [34], [43]. Fig. 8 shows the SDE FIT rates comparison between beam and GeFIN fault injection. If the FIT rate obtained with beam experiments is higher than the fault injection the value is represented as positive; negative otherwise. It is clear from Fig. 8 that the GeFIN SDE FIT rate prediction is very close to the one measured with beam experiments.

E. SDE Correlation to on-chip storage structures

In this section, we examine the relationship between SDEs and the major on-chip memory structures of modern microprocessors. It is essential to evaluate the SDE rates of individual hardware structures to understand their susceptibility. Fig. 9 illustrates the susceptibility of each structure to non-benign faults that are not masked at the hardware level. An SDE can occur if a hardware error eventually becomes available at the software and silently affects the execution of the program.

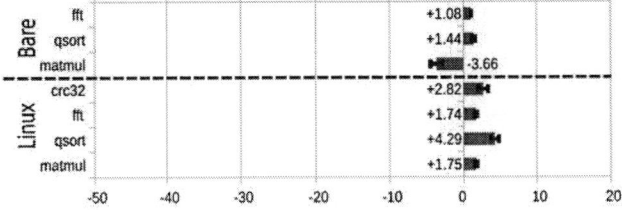

Fig. 8. Beam and fault injection SDE FIT rates comparison [34].

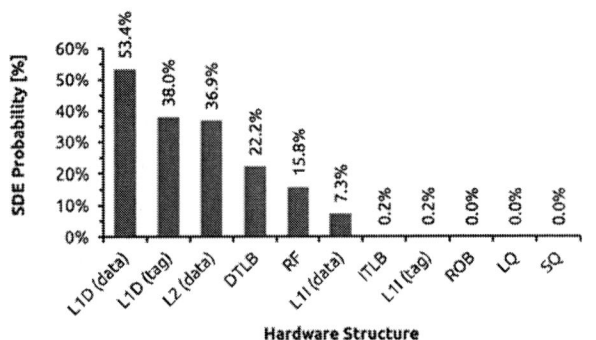

Fig. 9. The percentage of hardware corruptions at the software level (i.e., non-masked errors at hardware level) that eventually result in SDE.

Therefore, in Fig. 9 show the percentage of these errors to result in an SDE. Our first and most significant observation is that the Re-Order Buffer (ROB), Load Queue (LQ), and Store Queue (SQ) have a zero probability of experiencing SDEs. This is because any fault that occurs in these structures is not architecturally visible due to dependency graph checks that fail before the commit stage. Memory structures like the ROB, LQ, and SQ, which are deep in the microprocessor's pipeline, ensure proper instruction ordering when instructions are ready to commit. Any corruption in these structures may lead to dependency graph check failures before the commit stage and result in a crash.

V. CONCLUSION

The problem of silent data errors (SDEs) in today's microprocessors is a critical challenge that demands urgent attention. With the emergence of smaller feature sizes, complex computational structures, and specialized silicon features, the potential for temporary computational errors that go unnoticed during manufacturing tests is on the rise. SDEs may not be addressed by methods such as microcode updates, and can be linked to specific components within the microprocessor. Further, since the nature of these failures is silent, incorrect computation happens without any indication of an error. As such, it is crucial to develop effective methods for detecting, mitigating, and preventing SDEs in microprocessor chips to ensure reliable and accurate computing performance. We described the challenges of measuring SDE rates in both real microprocessors and simulation models (based on microarchitectural modeling on gem5). The study of SDEs is an ongoing area of research, and it is essential to continue to explore new approaches to address this critical issue in the design and implementation of microprocessors.

ACKNOWLEDGMENT

Adit Singh's research was supported in part by the US National Science Foundation under grant CCF-1910964. George Papadimitriou and Dimitris Gizopoulos' work has received funding in part from the EU Horizon Europe research and innovation programme under grant agreements No 101070238 (NEUROPULS), No 101093062 (Vitamin-V), and No 101097224 (REBECCA), and in part by research gifts from Meta and Intel.

979-8-3503-4631-2/23 $31.00 © 2023 IEEE

REFERENCES

[1] M. Snir, R. W. Wisniewski, J. A. Abraham, S. V. Adve, S. Bagchi, P. Balaji, J. Belak, P. Bose, F. Cappello, B. Carlson, A. A. Chien, P. Coteus, N. A. Debardeleben, P. C. Diniz, C. Engelmann, M. Erez, S. Fazzari, A. Geist, R. Gupta, F. Johnson, S. Krishnamoorthy, S. Leyffer, D. Liberty, S. Mitra, T. Munson, R. Schreiber, J. Stearley, and E. V. Hensbergen, "Addressing failures in exascale computing," *Int. J. High Perform. Comput. Appl.*, vol. 28, no. 2, p. 129–173, may 2014. [Online]. Available: https://doi.org/10.1177/1094342014522573

[2] C. Constantinescu, I. Parulkar, R. Harper, and S. Michalak, "Silent data corruption — myth or reality?" in *2008 IEEE International Conference on Dependable Systems and Networks With FTCS and DCC (DSN)*, 2008, pp. 108–109.

[3] R. Lucas, "Top ten exascale research challenges," in *DOE ASCAC Subcommittee Report*, 2014.

[4] E. Ibe, H. Taniguchi, Y. Yahagi, K.-i. Shimbo, and T. Toba, "Impact of scaling on neutron-induced soft error in srams from a 250 nm to a 22 nm design rule," *IEEE Transactions on Electron Devices*, vol. 57, no. 7, pp. 1527–1538, 2010.

[5] L. Bautista-Gomez, F. Zyulkyarov, O. Unsal, and S. McIntosh-Smith, "Unprotected computing: A large-scale study of dram raw error rate on a supercomputer," in *SC '16: Proceedings of the International Conference for High Performance Computing, Networking, Storage and Analysis*, 2016, pp. 645–655.

[6] A. Chatzidimitriou, G. Papadimitriou, D. Gizopoulos, S. Ganapathy, and J. Kalamatianos, "Assessing the effects of low voltage in branch prediction units," in *2019 IEEE International Symposium on Performance Analysis of Systems and Software (ISPASS)*, 2019, pp. 127–136.

[7] P. Koutsovasilis, C. D. Antonopoulos, N. Bellas, S. Lalis, G. Papadimitriou, A. Chatzidimitriou, and D. Gizopoulos, "The impact of cpu voltage margins on power-constrained execution," *IEEE Transactions on Sustainable Computing*, vol. 7, no. 1, pp. 221–234, 2022.

[8] A. Chatzidimitriou, G. Papadimitriou, D. Gizopoulos, S. Ganapathy, and J. Kalamatianos, "Analysis and characterization of ultra low power branch predictors," in *2018 IEEE 36th International Conference on Computer Design (ICCD)*, 2018, pp. 144–147.

[9] A. Chatzidimitriou, G. Papadimitriou, and D. Gizopoulos, "Healthlog monitor: A flexible system-monitoring linux service," in *2018 IEEE 24th International Symposium on On-Line Testing And Robust System Design (IOLTS)*, 2018, pp. 183–188.

[10] D. Gizopoulos, G. Papadimitriou, A. Chatzidimitriou, V. J. Reddi, B. Salami, O. S. Unsal, A. C. Kestelman, and J. Leng, "Modern hardware margins: Cpus, gpus, fpgas recent system-level studies," in *2019 IEEE 25th International Symposium on On-Line Testing and Robust System Design (IOLTS)*, 2019, pp. 129–134.

[11] A. Chatzidimitriou, G. Papadimitriou, and D. Gizopoulos, "Healthlog monitor: Errors, symptoms and reactions consolidated," *IEEE Transactions on Device and Materials Reliability*, vol. 19, no. 1, pp. 46–54, 2019.

[12] G. Papadimitriou, A. Chatzidimitriou, D. Gizopoulos, V. J. Reddi, J. Leng, B. Salami, O. S. Unsal, and A. C. Kestelman, "Exceeding conservative limits: A consolidated analysis on modern hardware margins," *IEEE Transactions on Device and Materials Reliability*, vol. 20, no. 2, pp. 341–350, 2020.

[13] H. D. Dixit, S. Pendharkar, M. Beadon, C. Mason, T. Chakravarthy, B. Muthiah, and S. Sankar, "Silent Data Corruptions at Scale," 2021. [Online]. Available: https://arxiv.org/abs/2102.11245

[14] P. H. Hochschild, P. Turner, J. C. Mogul, R. Govindaraju, P. Ranganathan, D. E. Culler, and A. Vahdat, "Cores That Don't Count," in *Proceedings of the Workshop on Hot Topics in Operating Systems*, ser. HotOS '21. New York, NY, USA: Association for Computing Machinery, 2021, p. 9–16. [Online]. Available: https://doi.org/10.1145/3458336.3465297

[15] R. W. Hamming, "Error detecting and error correcting codes," *The Bell System Technical Journal*, vol. 29, no. 2, pp. 147–160, 1950.

[16] Y. Luo, S. Govindan, B. Sharma, M. Santaniello, J. Meza, A. Kansal, J. Liu, B. Khessib, K. Vaid, and O. Mutlu, "Characterizing application memory error vulnerability to optimize datacenter cost via heterogeneous-reliability memory," in *2014 44th Annual IEEE/IFIP International Conference on Dependable Systems and Networks*, 2014, pp. 467–478.

[17] A. Chatzidimitriou, G. Papadimitriou, C. Gavanas, G. Katsoridas, and D. Gizopoulos, "Multi-bit upsets vulnerability analysis of modern microprocessors," in *2019 IEEE International Symposium on Workload Characterization (IISWC)*, 2019, pp. 119–130.

[18] J. Meza, Q. Wu, S. Kumar, and O. Mutlu, "Revisiting memory errors in large-scale production data centers: Analysis and modeling of new trends from the field," in *2015 45th Annual IEEE/IFIP International Conference on Dependable Systems and Networks*, 2015, pp. 415–426.

[19] F. Hapke, W. Redemund, A. Glowatz, J. Rajski, M. Reese, M. Hustava, M. Keim, J. Schloeffel, and A. Fast, "Cell-aware test," *IEEE Transactions on Computer-Aided Design of Integrated Circuits and Systems*, vol. 33, no. 9, pp. 1396–1409, 2014.

[20] X. Lin, K.-h. Tsai, C. Wang, M. Kassab, J. Rajski, T. Kobayashi, R. Klingenberg, Y. Sato, S. Hamada, and T. Aikyo, "Timing-aware atpg for high quality at-speed testing of small delay defects," in *2006 15th Asian Test Symposium*, 2006, pp. 139–146.

[21] C. J. Lin and S. Reddy, "On delay fault testing in logic circuits," *IEEE Transactions on Computer-Aided Design of Integrated Circuits and Systems*, vol. 6, no. 5, pp. 694–703, 1987.

[22] W. Howell, F. Hapke, E. Brazil, S. Venkataraman, R. Datta, A. Glowatz, W. Redemund, J. Schmerberg, A. Fast, and J. Rajski, "Dppm reduction methods and new defect oriented test methods applied to advanced finfet technologies," in *2018 IEEE International Test Conference (ITC)*, 2018, pp. 1–10.

[23] T. Claburn. (2021) Fyi: Today's computer chips are so advanced, they are more 'mercurial' than precise – and here's the proof. Accessed: March 2023. [Online]. Available: https://www.theregister.com/2021/06/04/google_chip_flaws/

[24] A. D. Singh, "Understanding vmin failures for improved testing of timing marginalities," in *2022 IEEE International Test Conference (ITC)*, 2022, pp. 372–381.

[25] M. D. Giles, N. Arkali Radhakrishna, D. Becher, A. Kornfeld, K. Maurice, S. Mudanai, S. Natarajan, P. Newman, P. Packan, and T. Rakshit, "High sigma measurement of random threshold voltage variation in 14nm logic finfet technology," in *2015 Symposium on VLSI Technology (VLSI Technology)*, 2015, pp. T150–T151.

[26] T. Sakurai and A. R. Newton, "Alpha-power law mosfet model and its applications to cmos inverter delay and other formulas," *IEEE Journal of Solid-state Circuits*, vol. 25, pp. 584–594, 1990.

[27] D. P. Lerner, B. Inkley, S. H. Sahasrabudhe, E. Hansen, L. D. R. Munoz, and A. v. de Ven, "Optimization of tests for managing silicon defects in data centers," in *2022 IEEE International Test Conference (ITC)*, 2022, pp. 578–582.

[28] T. C. May and M. H. Woods, "A new physical mechanism for soft errors in dynamic memories," in *16th International Reliability Physics Symposium*, 1978, pp. 33–40.

[29] R. Baumann, "Radiation-induced soft errors in advanced semiconductor technologies," *IEEE Transactions on Device and Materials Reliability*, vol. 5, no. 3, pp. 305–316, 2005.

[30] M. D. McCluskey and A. Janotti, "Defects in semiconductors," *Journal of Applied Physics*, vol. 127, no. 19, p. 190401, 2020. [Online]. Available: https://doi.org/10.1063/5.0012677

[31] G. Papadimitriou, D. Gizopoulos, A. Chatzidimitriou, T. Kolan, A. Koyfman, R. Morad, and V. Sokhin, "Unveiling difficult bugs in address translation caching arrays for effective post-silicon validation," in *2016 IEEE 34th International Conference on Computer Design (ICCD)*, 2016, pp. 544–551.

[32] S. Mukherjee, *Architecture Design for Soft Errors*. San Francisco, CA, USA: Morgan Kaufmann Publishers Inc., 2008.

[33] G. Papadimitriou and D. Gizopoulos, "Avgi: Microarchitecture-driven, fast and accurate vulnerability assessment," in *2023 IEEE International Symposium on High-Performance Computer Architecture (HPCA)*, 2023, pp. 935–948. [Online]. Available: https://doi.org/10.1109/HPCA56546.2023.10071105

[34] P. R. Bodmann, G. Papadimitriou, R. L. R. Junior, D. Gizopoulos, and P. Rech, "Soft error effects on arm microprocessors: Early estimations versus chip measurements," *IEEE Transactions on Computers*, vol. 71, no. 10, pp. 2358–2369, 2022.

[35] I. Tsiokanos, G. Papadimitriou, D. Gizopoulos, and G. Karakonstantis, "Boosting microprocessor efficiency: Circuit- and workload-aware assessment of timing errors," in *2021 IEEE International Symposium on Workload Characterization (IISWC)*, 2021, pp. 125–137.

[36] G. Papadimitriou and D. Gizopoulos, "Demystifying the system vulnerability stack: Transient fault effects across the layers," in *2021 ACM/IEEE 48th Annual International Symposium on Computer Architecture (ISCA)*, 2021, pp. 902–915.

[37] Y. Sazeides, A. Gerber, R. Gabor, A. Bramnik, G. Papadimitriou, D. Gizopoulos, C. Nicopoulos, G. Dimitrakopoulos, and K. Patsidis, "Idld: Instantaneous detection of leakage and duplication of identifiers used for register renaming," in *2022 55th IEEE/ACM International Symposium on Microarchitecture (MICRO)*, 2022, pp. 799–814.

[38] R. Leveugle, A. Calvez, P. Maistri, and P. Vanhauwaert, "Statistical fault injection: Quantified error and confidence," in *2009 Design, Automation and Test in Europe Conference and Exhibition*, 2009, pp. 502–506.

[39] S. Mukherjee, C. Weaver, J. Emer, S. Reinhardt, and T. Austin, "A systematic methodology to compute the architectural vulnerability factors for a high-performance microprocessor," in *Proceedings. 36th Annual IEEE/ACM International Symposium on Microarchitecture, 2003. MICRO-36.*, 2003, pp. 29–40.

[40] S. Mitra, N. Seifert, M. Zhang, Q. Shi, and K. Kim, "Robust system design with built-in soft-error resilience," *Computer*, vol. 38, no. 2, pp.

43–52, 2005.

[41] N. Binkert, B. Beckmann, G. Black, S. K. Reinhardt, A. Saidi, A. Basu, J. Hestness, D. R. Hower, T. Krishna, S. Sardashti, R. Sen, K. Sewell, M. Shoaib, N. Vaish, M. D. Hill, and D. A. Wood, "The gem5 simulator," *SIGARCH Comput. Archit. News*, vol. 39, no. 2, p. 1–7, aug 2011. [Online]. Available: https://doi.org/10.1145/2024716.2024718

[42] A. Chatzidimitriou and D. Gizopoulos, "Anatomy of microarchitecture-level reliability assessment: Throughput and accuracy," in *2016 IEEE International Symposium on Performance Analysis of Systems and Software (ISPASS)*, 2016, pp. 69–78.

[43] A. Chatzidimitriou, P. Bodmann, G. Papadimitriou, D. Gizopoulos, and P. Rech, "Demystifying soft error assessment strategies on arm cpus: Microarchitectural fault injection vs. neutron beam experiments," in *2019 49th Annual IEEE/IFIP International Conference on Dependable Systems and Networks (DSN)*, 2019, pp. 26–38.

979-8-3503-4631-2/23 $31.00 © 2023 IEEE

A Novel LBIST Signature Computation Method for Automotive Microcontrollers using a Digital Twin

Daniel Tille*

Leon Klimasch*

Sebastian Huhn[†§]

*Infineon Technologies AG
85579 Neubiberg, Germany
{Daniel.Tille,Leon.Klimasch}@infineon.com

[†]University of Bremen, Germany
huhn@uni-bremen.de

[§]Cyber-Physical Systems
DFKI GmbH
28359 Bremen, Germany

Abstract—**LBIST has been proven to be an effective measure for reaching functional safety goals for automotive microcontrollers. Due to a large variety of recent innovative features, every customer can adjust LBIST settings in a way that fits their use case. The downside of these user-defined configurations is the handling of their golden signatures: Traditionally, they can be computed only with access to the gate-level netlist. This is typically not possible for MCU customers because a netlist contains protected IP, which cannot be disclosed to third parties.**

This paper proposes a digital twin of the LBIST functionality that can overcome this drawback. It is an executable model that can be delivered together with the product. As a result, for the first time, a customer can compute a golden signature without knowledge of the netlist or other support of the supplier. We prove the efficacy of the digital twin in an industrial environment on an automotive microcontroller.

I. Introduction

In the last two decades, the digitization has heavily impacted the automotive industry. Nowadays, the majority of the functionality is implemented using *Microcontrollers* (MCUs) – even safety-critical functions such as airbag control, power steering, and braking systems. As a consequence, the correct functionality of these chips is more important than ever since malfunctions could lead to large collateral damage. It is not sufficient anymore to just test circuits for defects after manufacturing. Instead, functional safety standards, such as the ISO26262 [1], demand so-called in-field tests that regularly check an MCU for defects (e.g. during each power-up).

Logic Built-In Self-Test (LBIST) is a state-of-the-art structural in-field test method, which is widely used for automotive MCUs [2], [3]. This approach applies a deterministic sequence of pseudo-random test patterns to the circuit logic. The test responses are continuously compacted into a *signature*, which can be evaluated after the end of the test. Since the circuit behavior is deterministic during LBIST, there is an unambiguous *golden* signature for a correct chip for each specific test sequence. This golden signature can be computed by *Gate-Level Simulation* (GLS). Usually, there is a default LBIST configuration whose golden signature is determined and provided in the MCU's datasheet.

At the same time, state-of-the-art automotive MCUs employ LBIST controllers which are highly configurable. The user can

program various parameters such as the seed of the pseudo-random sequence, the number of test cycles, or advanced power reduction measures [4]. Each *Electronic Control Unit* (ECU) family, i.e., the system that integrates MCUs, can use a distinct LBIST configuration in order to satisfy its specific system requirements (w.r.t. runtime, power, ...). Since each configuration yields a different golden signature, this leads to some practical challenges.

The golden signature for a specific LBIST configuration is required in order to evaluate the correctness of the test result. However, MCU customers that program LBIST applications do not have the capabilities to compute the golden signature of a user-defined configuration. They do not have access to the MCU's gate-level netlist, which is required for executing GLS, because MCU suppliers cannot disclose it. The protection of (own and third-party) *Intellectual Property* (IP) as well as the need for confidentiality of hardware security implementations prevent a disclosure.

There are currently three practical possibilities to solve this dilemma. Firstly, the LBIST execution can be limited to a small set of default configurations for which the MCU designer provides the respective golden signatures in the datasheet. This option was common practice in the past when the LBIST application was straightforward and affected only a few use cases. Nowadays, however, there are too many advanced LBIST functions and too many different application scenarios. Therefore, MCU customers no longer accept such a limitation.

Secondly, the MCU designer computes and provides golden signatures for user-defined configurations on demand. While this was common practice in the past, the growth of the automotive market with many new customers and the significant expansion of LBIST applications render this option impractical.

Thirdly, the golden signature can be experimentally determined on a *known good die*. This is a practical solution for the prototyping phase of a product. However, in an environment that demands high quality, some MCU customers are not satisfied with this experimental approach. Also, if the product needs to be certified according to functional safety standards, some auditors might require the golden signature to be computed independently.

In this paper, we overcome this challenge by introducing a **digital twin** of the LBIST functionality. It is an executable

---

979-8-3503-4631-2/23 $31.00 © 2023 IEEE

model that already integrates all relevant circuit information. That means, the novelty over existing simulation approaches is that no separate netlist is required because it is already an integral part of the program itself. The digital twin can be provided together with the product. Consequently, for the first time, MCU customers can compute golden signatures of their user-defined LBIST configurations without further information or support from the MCU supplier. This significantly improves turn-around time of ECU development and gives the customer a higher degree of freedom when choosing an LBIST configuration.

The contribution of this paper includes

- a novel concept of a digital twin that emulates the LBIST functionality without possession of the gate-level netlist,
- a fully-automated framework for extracting and compiling a digital twin from an arbitrary circuit,
- a prototype implementation of the extraction method, and
- an experimental evaluation on an automotive MCU benchmark.

The remainder of the paper is structured as follows. The next section presents the concept of the digital twin. Section III shows how the digital twin is automatically generated. Afterwards, Section IV gives an overview of our prototype implementation. Experimental results are reported in Section V. Finally, conclusions are drawn in Section VI.

II. Novel Digital Twin

In this section, we propose the new digital twin concept. The main idea of our approach is to provide an executable model that "emulates" the LBIST functionality. (For more details of LBIST, we refer to [5].) In particular, a C++ program computes all relevant LBIST steps like they are executed on a chip. This includes the pseudo-random pattern generation using the *Linear Feedback Shift Register* (LFSR), shift and capture operations, and the determination of the signature in the *Multiple-Input Shift Register* (MISR). The input of the program is an arbitrary LBIST configuration which then controls and influences its concrete execution by providing the content of *Special Function Registers* (SFRs), such as the number of shift cycles. The output is the golden signature for this specific configuration.

The key factor of such a model is the correct program architecture. This is because a C++ program executes all instructions sequentially, whereas in hardware, all operations are performed simultaneously. In the following, we describe how such a model can be obtained.

A. Combinational Logic

Representing the combinational part of a circuit in C++ instructions is quite straightforward. Each combinational sub-circuit with n inputs and m outputs computes a Boolean function[1]

$$f : \mathbb{B}^n \rightarrow \mathbb{B}^m. \qquad (1)$$

[1]For the sake of simplicity, we only consider Boolean logic in the following. An enhancement to a multi-valued logic (e.g. including X and Z values) increases the complexity but is possible through a Boolean encoding [6].

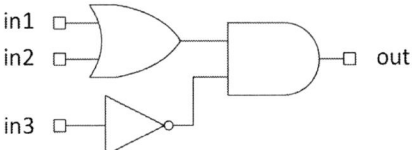

Figure 1: Example circuit

TABLE I: Overview of the variables stored for each FF

Variable	Description
FF_Q	current value
FF_D	next value from D-input
FF_TI	next value from TI-input
FF_TE	indicates active scan-enable signal
FF_CLK	indicates clock pulse
FF_R	indicates reset pulse

Since a combinational circuit has on gate-level a zero-delay timing and does not contain (combinational) feedback loops, it is relatively easy to find a C++ statement that computes its Boolean function. Figure 1 gives an example of a circuit with three inputs and one output. The associated C++ statement is as follows

$$\texttt{out = (in1 || in2) \&\& !in3;} \qquad (2)$$

where `in1`, `in2`, `in3`, and `out` are Boolean variables. (Variables will be explained in the next section). That means the output value is determined by Boolean operations – the same operations as given in the gate-level netlist – applied to the input values. In the case of a combinational sub-circuit with m outputs, there will be m such statements.

B. Sequential Elements

In hardware, *Flip Flops*[2] (FFs) store values from one clock cycle to the next one. We use Boolean variables to implement this behavior in our C++ program. Table I gives an overview of all information that is stored for each FF. The *current value* of a FF is stored in the variable `FF_Q`. The *next value*, which is propagated through the combinational logic to the D-input, is stored in `FF_D`. For *Scan FFs* (SFFs), there are two further inputs, TI and TE, which represent the *scan-in* and *scan-enable* input, respectively. The variables that implement the clocking scheme are explained in the next section.

Figure 2 shows the netlist of the previous example where the inputs and output of the combinational circuit are modeled with FFs. The output of *FF3* is fed back to one of the combinational circuit's inputs. The corresponding C++ statement is as follows:

$$\texttt{FF3\_D = (FF3\_Q || FF1\_Q) \&\& !FF2\_Q;} \qquad (3)$$

That means, the value of the output-FF in the next clock cycle, i.e., its D-input-value, is computed through a series of Boolean operations applied to the current values of the input-FFs, i.e., their Q-output-values.

[2]In this work, we consider only flip flops as sequential elements due to the advantages of full-scan designs for LBIST. However, modeling other elements, such as latches and memories, is possible with this approach.

979-8-3503-4631-2/23 $31.00 © 2023 IEEE

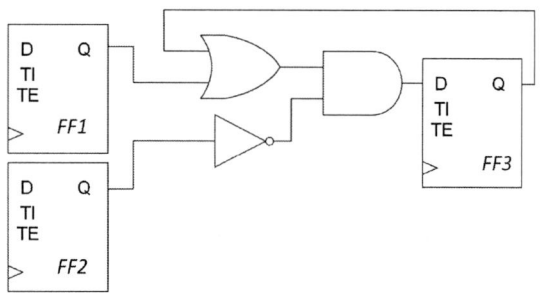

Figure 2: Example circuit with sequential elements

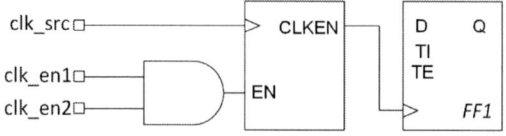

Figure 3: Example for a clock signal controlled by a clock gate

The sequence of the statements for all FFs in a circuit is irrelevant since all values are stable and a D-value depends on the logic in its fan-in cone only, as given within one clock cycles in this abstraction level. Furthermore, the scan path is modeled according to the same principles: The FF_TI values are determined based on the FF_Q values of the preceding SFFs.

Primary Inputs (PIs) to the chip are also modeled by variables. Since inputs are constrained to a constant value in an LBIST context (otherwise, there would be no stable signature), the corresponding variables are also assigned a constant value.

Finally, also the LFSR and the MISR are modeled using Boolean variables.

C. Clocking Scheme

In hardware, a FF takes over the value at its D-input with a rising clock edge. However, our program does not explicitly model the "continuous" clock signal. Instead, we abstract and just encode the following essential information.

Firstly, within one clock cycle, all values are stable and the actual timing information is irrelevant to the LBIST context. Therefore, we always take into account a complete clock cycle as one discrete entity. In the program context, this is accomplished as one loop iteration where the C++ statements explained above are all executed. Secondly, we model the clock-enabling conditions for each FF using the FF_CLK variable. This is done during each clock cycle, i.e., at the beginning of each loop. That means FF_CLK is true if, and only if, the FF receives a clock edge in this specific cycle.

Figure 3 shows an example. We see a FF whose clock input is connected to a clock source that provides a scan clock. As it is common practice, the clock input is not directly connected to the source, but there is a *clock gate* that can block the clock signal. The enable pin is connected to some arbitrary enabling

logic. The *clk_en1* and *clk_en2* inputs that we see in this example are an abstraction. This enabling logic can usually be complex or even depend on one specific parameter of the user-defined LBIST configuration.

In our approach, we only model the enabling logic when computing the FF_CLK, which means for this example:

$$FF1\_CLK = clk\_en1\ \&\&\ clk\_en2; \qquad (4)$$

The *clk_src* input is used to identify the clock domain. The signal itself can be disregarded. In the case of a hierarchical instantiation of clock gates, the complete enabling logic cone is taken into account for the computation of the FF_CLK.

The scan-reset network is generated similarly. Due to page limitations, we will not discuss details here.

D. Program Architecture

This section gives a brief and high-level overview of the architecture of the C++ program. We abstract from most implementation details in order to maintain readability.

Algorithm 1 presents an outline of the main function of the C++ program. During init(), all parameters are read and variables that reflect SFRs (such as the number of LBIST cycles) as well as PIs are set. The outer for-loop (line 3) realizes an LBIST cycle of shift and capture. The number of iterations can be specified through the program parameters. Each iteration of the inner for-loops (lines 4 and 12) corresponds to one clock cycle where one shift or one capture cycle, respectively, is being emulated. At the end of the program, the current state of the MISR contains the golden signature and is printed.

Algorithm 1: Main function of the C++ program

```
1  int main(){
2    init();
3    for (i=0;i<SFR.lbist_cycles;i++){
4      for (j=0;j<chain_length;j++){
5        update_lfsr();
6        shift();
7        update_misr();
8      }
9      if (SFR.scan_reset){
10       reset();
11     } else {
12       for (k=0;k<SFR.capture_cycles;k++){
13         capture();
14       }
15     }
16   }
17   print_misr();
18 }
```

In Algorithm 2, we depict how one capture cycle is modeled in C++. Firstly, the D-values for all FFs are determined (see Eq. 3). Secondly, all CLK-values are determined (see Eq. 4). At the end of the capture cycle, the Q-value is updated for each FF in the circuit, if it receives a clock-edge.

While the skeleton of the program is independent of the actual circuit, the functions which compute the D- and CLK-values (along with the scan-chains in the shift function, which

Algorithm 2: Capture function of C++ program

```
1  void capture(){
2    compute_D_values();
3    compute_CLK_values();
4    for (i=0;i<FF.size();i++){
5      if (FF[i].CLK){
6        FF[i].Q = FF[i].D;
7      }
8    }
9  }
```

TABLE II: Examples of gate library functions

Gate	Function
and	`bool AND2(bool a,bool b){return a&&b;}`
or	`bool OR3(bool a,bool b,bool c){return a||b||c;}`
not	`bool NOT(bool a){return !a;}`
mux	`bool MUX(bool a,bool b,bool s){return (s)?b:a;}`

are not reported here) implement actual circuit behavior. This is how the program incorporates netlist knowledge into the digital twin. Therefore, such a program determines the signature for one specific MCU product. For a new design (even if it is just a minor change in the netlist functionality), a new C++ program is required.

E. Advanced LBIST Features

Due to page limitations, we can only briefly mention that our model is compliant with all major advanced LBIST features. Power reduction measures, both for shift and capture, are present in the circuit structure and are therefore taken into account during the generation of the model. The same is true for *Q-gating* [7] and *observe scan technology* [8].

Re-seeding [9] or *bit-flipping* techniques [10] can be implemented by enhancing the `update_lfsr()` function. More sophisticated methods such as *full-scan LBIST* [11], that stores a set of LBIST seeds in an MCU's memory, can be realized as well: an external file containing this set of seeds is processed during `init()` and provides this data for re-seeding.

The application of at-speed LBIST makes the clocking scheme more complex, especially if there are several different clock domains. However, it is generally possible to enhance the model accordingly.

X-tolerant LBIST [12] and X-canceling MISRs [13] are not implemented by the current model. However, with an enhancement to a multi-valued logic containing X-values and the fact that all masking logic is part of the circuit and, hence, part of the program, the integration of such techniques is generally feasible.

III. AUTOMATIC MODEL EXTRACTION

In this section, we show how the C++ model can be automatically generated for an arbitrary netlist. The fundamental idea of our extraction algorithm is, first, to have a library of primitive Boolean functions that represent the functions of all possible gate types in C++. Secondly, a structural circuit traversal method recursively generates C++ statements based on these primitive functions.

A. Gate Library Construction

In practice, a gate-level netlist is synthesized based on the gates contained in a *standard cell library*. The first step of our

extraction method is to generate a Boolean function description for each of the individual gate types in such a library.[3]

Table II presents a few illustrative examples of functions in the library. The function `AND2` receives the two inputs of an AND gate as a parameter and returns their conjunction. The next two functions show that it is also possible to implement functions with more respectively fewer than two inputs. The final example presents the Boolean function of a multiplexer which returns the value of `b` if the `s` input is true, and `a` otherwise.

Some standard cell libraries contain special gates with more than just one output. Their Boolean function is represented by multiple library functions – one for each output. *Full adder* gates are an example: There are two corresponding library functions, one computing the sum-output and one computing the carry-out-output.

B. Netlist Conversion

The generation of the actual C++ model is explained in the following. The conversion algorithm works on the gate-level netlist and has as output C++ statements. In order to accomplish this, we follow the basic principles explained in Section II-B.

As a first step, all FFs and PIs are assigned Boolean variables (see Table I). In order to generate a C++ statement that computes the FF_D value of a FF, a recursive circuit traversal procedure is called for its predecessor gate. An abstract pseudo code[4] is presented in Algorithm 3. When the recursion reaches a terminal gate, i.e., a FF or a PI, it prints the associated variable. Otherwise, it prints the primitive Boolean function introduced above and calls itself recursively for each predecessor gate in order to determine the function's parameters. This way, the C++ statement is created that computes the *next value* of a FF depending on the *current values* of the FFs in its input cone.

Let us revisit Figure 2 for an illustrative example. In order to generate the statement that computes FF3_D, the trace procedure is called for the AND gate. This prints the primitive function AND2 and calls the procedure recursively for the OR gate. Again, the corresponding primitive function is printed and the procedure is recursively called for FF3. This time we meet the termination criterion and the variable FF3_Q is printed. The program continues with the next calls until all terminal nodes are finally reached. The final statement looks as follows:

`FF3_D=AND2(OR2(FF3_Q,FF1_Q),NOT(FF2_Q));`

---

[3]There are usually multiple standard cells for each gate type, e.g., representing different driver strengths. We only consider one representative Boolean function for the complete equivalence class of gates of the same type.

[4]We explain here only basic functionality and do not consider all necessary measures to produce syntactically correct code, e.g., correct parentheses.

Algorithm 3: Netlist traversal procedure (Tcl script to generate C++ program)

```
1  proc trace (gate){
2    if (gate.type == TERMINAL){
3      print gate.var_Q
4    } else {
5      print gate.primitive_function
6      foreach pred_gate in gate.inputs
7        trace (pred_gate)
8    }
9  }
```

TABLE III: Benchmarks

Benchmark	gates	chains	max chain length
LEON3	55,416	20	85
TriCore	1,016,308	650	101

TABLE IV: Generation of Digital Twin

Benchmark	runtime [m:s]		file size [kB]	
	generation	compile	source	binary
LEON3	3:48	6:45	6,042	4,429
TriCore	45:00	78:32	26,569	2,482

Due to its recursive nature, this implementation of the C++ statement looks different than the one presented in Eq. 3. However, when resolving the functions, it is easy to see that they are functionally equivalent.

The tracing algorithm is applied to all FFs in the circuit. This creates the `compute_D_values()` function from Algorithm 2. Similar techniques are applied to the input cones of all other input-pins (see Table I) in order to generate statements that compute the complete circuit functionality, including scan chains, clock control, and the reset network.

C. Conversion Improvement

The recursive algorithm explained above scales with the number of paths in the circuit. This can be squared in the number of gates in the worst case. In order to decrease this complexity, we propose an optimization. Each time a *Fanout-Free Region* (FFR) is traversed for the first time, the resulting sub-statement is stored. This is easily possible due to the recursive nature of the traversal algorithm. When the FFR is re-visited, the sub-statement can be returned immediately. This reduces the complexity such that the algorithms scales linearly with the number of gates.

IV. PROTOTYPE IMPLEMENTATION

The proposed method has been implemented in Infineon's environment. This work's focus has been on providing a proof-of-concept for a realistic scenario. That means, in the current stage, reaching an optimal runtime was not our priority.

The construction of the gate library (see Section III-A) is straightforward and accomplished using a Python script. This script traverses the relevant standard cell library files and translates the description of the Boolean functions into C++ functions.

The generation of the C++ model (see Section III-B) follows a two-step approach: firstly, since the skeleton of the program is independent of the actual netlist, it can be obtained from a pre-generated template. Secondly, the actual netlist-dependent functions, such as `compute_D_values()`, are generated through a netlist traversal as explained in Algorithm 3. For the sake of simplicity, we decided to implement these algorithms on top of a commercial simulation tool. This is only to prove general feasibility – a final product does not require such a third-party tool.

A. Proof-of-Concept

We executed the prototype on two benchmarks: the publicly available LEON3 processor [14] and a TriCore CPU core which is part of Infineon's AURIX automotive MCU [15] (details will be reported in the next section). Both benchmarks were synthesized using state-of-the-art automotive CMOS technology. The LBIST scheme consists of one clock domain with one synchronous scan clock. The designs are designed to be X-free, i.e., they can be modeled using Boolean logic only.

We generated the digital twin for both benchmarks and determined their golden signatures for four different shift cycle configurations. Afterwards, we compared them to the simulation results and validated the signature match.

B. Safety and Security Discussion

From a functional safety point of view, there are no relevant safety violations to consider. If the digital twin provides a wrong golden signature, the mismatch will be apparent immediately. The likelihood that an incorrect golden signature will mask a defect in an MCU is negligible: for the implemented 32-bit MISR, it is equal to 2^{-32}. This is the same probability that is usually accepted, e.g., for aliasing [16].

Security considerations and protection of intellectual property are of serious concern – after all, this is the reason why the netlist cannot be provided to the customer in the first place. Our approach, however, does not pose any relevant threat: firstly, the C++ model is a very coarse abstraction of the MCU. It covers only the LBIST functionality. The *mission mode*, especially its timing information, is not extracted. Secondly, for critical circuit parts, such as hardware implementations of encryption logic, there is the possibility of applying code obfuscation [17] during a post-process. This can effectively prevent reverse engineering [18].

V. EXPERIMENTAL EVALUATION

Table III reports details of the used benchmarks. We conducted our experiments on a Linux machine with 2.6 GHz and 512GB RAM. We used the gcc compiler in version 4.8.5.

For our experiments, we used a state-of-the-art automotive CMOS standard cell library. It contains 159 different gate types in total; this is also the number of functions in our gate library. The generation of this library requires a runtime of about one second and is a one-time effort.

TABLE V: Execution of Digital Twin

Benchmark	runtime [s]				
	shift cycle	capture cycle	LBIST cycles		
			100	1,000	10,000
LEON3	∅ 6.25e-5	∅ 0.063	6.4	94.0	1,080.0
TriCore	∅ 2.8e-3	∅ 0.025	1.7	21.1	210.3

Table IV presents information about the generation of the C++ program. Columns *generation* and *compile* give the runtimes required for the complete generation of the C++ file and for compiling it, respectively. Columns *source* and *binary* report the file size of the C++ source file and the resulting binary, respectively. We enabled full compiler optimization. If code obfuscation is employed, the compile time and resulting file size will increase.

The fact that we use a commercial tool as the back-end for our generation framework has a significant negative impact on the runtime required for generating the model. A native implementation with data structures that are optimized for this specific use case would yield significantly better results. At the same time, although a low runtime for the C++ program generation is obviously desirable, it is not crucial. Firstly, this generation has to be executed only once for a product. Secondly, this step is not on the critical path of the project: it can be started directly after tape-out, but has to be finished only when customer samples are shipped (which is usually in the order of months after tape-out).

Table V reports the experimental results of executing the digital twin. It gives the average runtimes for one single shift and capture cycle, i.e., the time required for executing the `shift()` and `capture()` functions, respectively. Afterwards, it presents the runtime required for the complete program run for different numbers of LBIST cycles.

Since the program is executed on the customer side, a low runtime matters. This can be achieved through further optimization: The current implementation employs a straightforward translation of the gate-level logic into C++ statements. The number of operations, and with this the required runtime, can be reduced by optimizing the circuit logic. There are multiple optimization approaches available that potentially achieve such an optimization of the circuit logic, e.g., [19], [20].

VI. CONCLUSIONS

In this paper, we presented a novel digital twin that allows for emulating an LBIST run. This enables MCU customers, for the first time to independently compute the golden signature for their user-defined LBIST configurations without access to the gate-level netlist or requiring additional support from the supplier. Furthermore, we provided a proof-on-concept of our method demonstrating its feasibility for state-of-the-art automotive MCUs.

Several conceivable directions exist in which the digital twin model could be enhanced. Currently, customers do not have any knowledge about the fault coverage of their user-defined configuration. However, this information is invaluable during the preparation of the *safety case*. A new fault simulation technique using the digital twin could provide an estimation. Another application could be the LBIST diagnosis. Our model can be enhanced with techniques such as [21]–[23]. The resulting approach could significantly speed up the search for a failure in an ECU when an LBIST run fails in a car. MCU customers that employ such a diagnostics method would have a significant advantage over their competition.

REFERENCES

[1] "ISO26262: Road vehicles functional safety-part 5," 2018.

[2] F. Reimann, M. Glaß, J. Teich, A. Cook, L. Rodríguez Gómez, D. Ull, H.-J. Wunderlich, U. Abelein, and P. Engelke, "Advanced diagnosis: SBST and BIST integration in automotive E/E architectures," in *Design Automation Conference*, 2014.

[3] T. McLaurin, "Periodic online LBIST considerations for a multicore processor," in *IEEE International Test Conference in Asia*, 2018, pp. 37–42.

[4] D. Czysz, M. Kassab, X. Lin, G. Mrugalski, J. Rajski, and J. Tyszer, "Low power scan shift and capture in the EDT environment," in *IEEE International Test Conference*, 2008, pp. 1–10.

[5] V. D. Agrawal, C. R. Kime, and K. K. Saluja, "A tutorial on built-in self-test. I. principles," *IEEE Design & Test of Computers*, vol. 10, no. 1, pp. 73–82, 1993.

[6] R. K. Brayton and S. P. Khatri, "Multi-valued logic synthesis," in *International Conference on VLSI Design*, 1999, pp. 196–205.

[7] X. Lin and J. Rajski, "Test power reduction by blocking scan cell outputs," in *IEEE Asian Test Symp.*, 2008, pp. 329–336.

[8] N. Mukherjee, D. Tille, M. Sapati, Y. Liu, J. Mayer, S. Milewski, E. Moghaddam, J. Rajski, J. Solecki, and J. Tyzer, "Time and area optimized testing of automotive ICs," *IEEE Transaction on VLSI Systems*, vol. 29, no. 1, pp. 76–88, 2021.

[9] S. Venkataraman, J. Rajski, S. Hellebrand, and S. Tarnick, "An efficient BIST scheme based on reseeding of multiple polynomial linear feedback shift registers," in *International Conference on CAD*, 1993, pp. 572 – 577.

[10] H.-J. Wunderlich and G. Kiefer, "Bit-flipping BIST," in *International Conference on CAD*, 1996, pp. 337 – 343.

[11] D. Czysz, M. Kassab, X. Lin, G. Mrugalski, J. Rajski, and J. Tyszer, "Full-scan LBIST with capture-per-cycle hybrid test points," in *IEEE International Test Conference*, 2018, pp. 1 – 9.

[12] P. Wohl, J. A. Waicukauski, G. A. Maston, and J. E. Colburn, "XLBIST: X-tolerant logic BIST," in *IEEE International Test Conference*, 2018, pp. 1–9.

[13] J.-S. Yang and N. A. Touba., "X-canceling MISR architectures for output response compaction with unknown values," *IEEE Transaction on CAD of Integrated Circuits and Systems*, vol. 31, no. 9, pp. 1417–1427, 2012.

[14] "LEON3 processor," https://www.gaisler.com/index.php/products/processors/leon3.

[15] "AURIX™ TriCore™," https://www.infineon.com/aurix.

[16] T. W. Williams, W. Daehn, M. Gruetzner, and C. W. Starke, "Aliasing errors in signature analysis registers," *IEEE Design & Test of Computers*, vol. 4, no. 2, pp. 48–57, 1987.

[17] C. S. Collberg, C. D. Thomborson, and D. W. K. Low, "Obfuscation techniques for enhancing software security," US patent no. 6668325B1, 1998.

[18] S. Schrittwieser and S. Katzenbeisser, "Code obfuscation against static and dynamic reverse engineering," in *Information Hiding*, ser. Lecture Notes in Computer Science, vol. 6958, 2011, pp. 270–284.

[19] N. Eén, A. Mishchenko, and N. Sörensson, "Applying logic synthesis for speeding up SAT," in *International Conference on Theory and Applications of Satisfiability Testing*, 2007, pp. 272–286.

[20] D. Tille, S. Eggersgluß, R. Krenz-Bååth, J. Schloeffel, and R. Drechsler, "Improving CNF representations in SAT-based ATPG for industrial circuits using BDDs," in *IEEE European Test Symp.*, 2010, pp. 176–181.

[21] I. Pomeranz and S. M. Reddy, "On dictionary-based fault location in digital logic circuits," *IEEE Transaction on Comp.*, vol. 46, no. 1, pp. 48–59, 1997.

[22] A. Cook, M. Elm, H.-J. Wunderlich, and U. Abelein, "Structural in-field diagnosis for random logic circuits," in *IEEE European Test Symp.*, 2011, pp. 111–116.

[23] D. Tille, B. Gottinger, U. Pfannkuchen, H. Graeb, and U. Schlichtmann, "On enabling diagnosis for 1-pin test fails in an industrial flow," in *ASP Design Automation Conference*, 2018, pp. 233–238.

V_{min} Prediction Using Nondestructive Stress Test

Chun Chen, Jeng-Yu Liao, James Chien-Mo Li
Graduate Institute of Electronics Engineering
National Taiwan University
Taipei 106, Taiwan
{r09943107, r11k41002, cmli}@ntu.edu.tw

Harry H. Chen, Eric Jia-Wei Fang
MediaTek Inc.
Hsinchu 300, Taiwan
harry-h.chen@mediatek.com,
eric.fang@mediatek.com

Abstract—We propose a novel minimum operating voltage (V_{min}) prediction method using nondestructive stress test to avoid area overhead and reduce test time. We process stress-test fail-logs and generate summation values by three strategies to predict V_{min}. In addition, we select important test patterns by Spearman correlation and simulated annealing to reduce test time. Two regression models are adopted in our experiment. Experimental results on advanced 7nm and 3nm chip designs show that the average RMSE of our predict V_{min} can be as low as 4.48 mV to 8.66mV by one strategy with linear regression model. Our method gives smaller RMSE than process monitor prediction methods with no area overhead, and is 50 to 62.5 times faster compared to conventional testing.

Index Terms—Chip performance prediction, Nondestructive stress test

I. INTRODUCTION

As technologies advance, variations of chip performance keep increasing [1]. We have to monitor performance of chips to deal with the performance variation issue. For mobile device chips, *minimum operating voltage(V_{min})* is one of the important metrics to evaluate chip performance. Traditionally, we perform functional tests on ATE to obtain V_{min}, but it is very time-consuming [2]. To reduce test time, many machine learning(ML)-based techniques are proposed to predict chip performance [3] [4]. These ML techniques usually predict chip performance using process monitors, and thus we called them *process monitor methods*. However, since they have to place process monitors on chips, they have area overhead. If we want to avoid area overhead and to further reduce test time, we have to develop methods from other points of view.

Nondestructive stress test (hereinafter referred to as *stress test*) can help us observe chip performance from other aspects. In this paper, we use *Stressed Onchip-clock test MismAtch Count(SOMAC)* to test chips at lower-than-normal power supply voltage and at higher-than-normal operating frequency [5]. This is similar to other stress tests, such as *very-low-voltage(VLV)* testing [6]. Fig. 1 illustrates how it can help us solve the problem. Suppose we have two chips, one has normal performance and the other has poor performance. We apply delay test patterns to two chips, and show their output waveforms in Fig. 1. In the nominal condition, both chips are able to pass the test while the poor performance one has a tighter time margin. In the stress condition, we make the capture time earlier if we test at higher-than-normal frequency,

or the unstable state later if we test at lower-than-normal power supply voltage. Thus, we can force the poor performance chip to fail while the other works as normal. Therefore, we can use stress test to monitor chip performance.

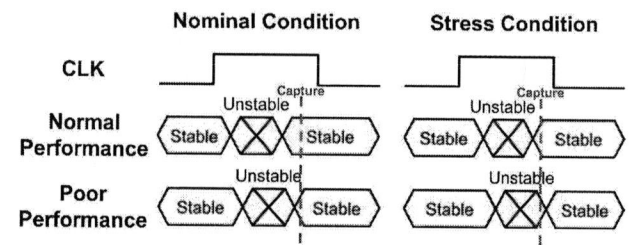

Fig. 1. Stress Test and Chip Performance

In this paper, we propose a method to predict V_{min} by stress test. We process stress-test fail-logs and generate summation values by three strategies to predict V_{min}. In addition, we select important test patterns that correlated to V_{min} to reduce stress test time. We evaluate the relationship using *Spearman correlation* to reduce the impact of outlier values and focus on monotonic relationship [7]. Unlike process monitor methods, our method does not have any area overhead. Moreover, our method can save functional V_{min} test time. We experiment on advanced 7nm and 3nm chip designs to validate our method with two regression models. The results show that the best average RMSE of our predict V_{min} can be as low as 4.48 mV to 8.66mV. Also, the test application time of our method is 50 to 62.5 times shorter than the time of functional V_{min} measurement. Please note that to measure functional V_{min} requires functional test patterns, which are very long.

The rest of the paper is organized as the following. Section II introduces previous researches about predicting V_{min} with ML techniques and researches about stress test. Section III presents the proposed technique, and section IV shows the experimental results. Finally, Section V draws a conclusion of this paper.

II. BACKGROUND

There were several previous works about predicting the performance of chips. Kuo proposed an accumulative learning flow to predict V_{min} with the speed of ring oscillators, on-state current of transistors and the flatness of wafers [3]. Their flow helps to reduce predicted errors which come from lot-to-lot variations. Cantaro used speed monitors to predict maximum operating speed [4]. Their flow is able to screen out under-performing devices. However, these previous works have area overhead since they have to place process monitors on chips.

There were also some previous works that utilize stress tests. Chang used VLV testing to detect flaws in CMOS ICs [6]. Their work showed that VLV testing was a low-cost option to detect marginal timing problems. Chen proposed a new data feature from stress test, which is named SOMAC [5]. They used SOMAC to predict if a chip has system-level test failures. The correlation between V_{min} and SOMAC was also studied [8]. In summary, current chip performance prediction techniques have area overhead. To deal with this problem, we can use stress tests as an effective option to predict chip performance.

III. PROPOSED TECHNIQUE

We propose a method to predict functional V_{min} with stress test. Fig. 2 shows the overall flow, which consists of training phase and testing phase. In the training phase, we perform stress test with all *test patterns* then use fail-logs to generate *MisMatch Count(MMC)*, which would be described in the next section. Next, we select important features, which is named *MMC-Indices*, and important test patterns from MMC. We then regenerate a smaller MMC using stress-test fail-logs from these test patterns. The selection process can not only reduce the data dimension but also help us reduce the test time. Finally, we use the regenerated MMC as input to build a V_{min} prediction model. In the test phase, we only have to perform stress test with important test patterns instead of with all of them. Then, we generate MMC with only important MMC-Indices. Finally, we apply our proposed V_{min} prediction model to estimate functional V_{min}.

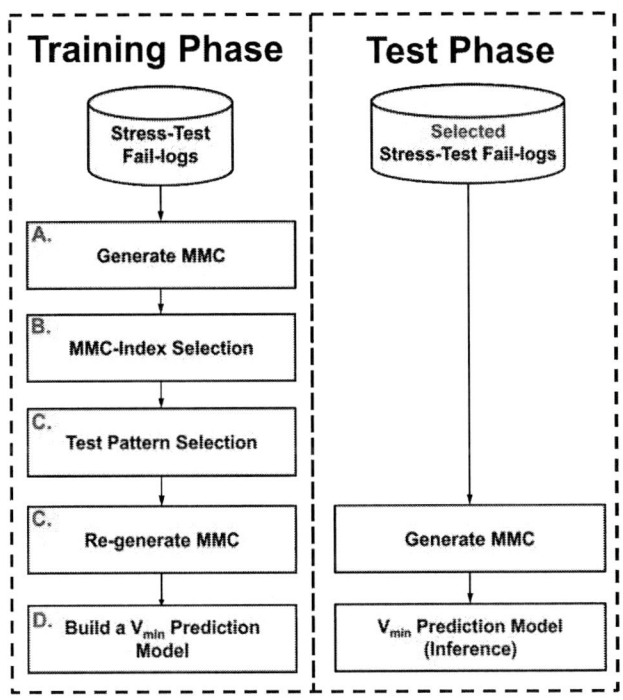

Fig. 2. Overall Flow of the Proposed Method

A. Stress-Test Fail-logs and MMC

This section describes how we collect stress-test fail-logs and use them to generate *MisMatch Count(MMC)*. We apply transition delay fault test patterns using on-chip-clock design-for-testability to the DUT on the automated test equipment(ATE) under stress conditions. In this paper, we applied multiple *Pattern Sets(PS)*, each of which contains many *test patterns*. *Stress conditions* are decided by Shmoo plots. We choose several stress conditions which are on the failing edge of Shmoo plots, so chips with higher V_{min} are more likely to fail than the other chips. We perform continue-on-fail test on each DUT, and collect all the responses in *stress-test fail-logs*. Each response in stress-test fail-logs indicates whether a *scan output (SO)* pin of the DUT failed at a *CYcle(CY)* of a test pattern or not, which can be viewed as binary values. If it is a fail, the value would be 1, otherwise 0 if it is a pass.

However, stress-test fail-logs is hard to analyze directly since it is a high-dimensional binary data. The dimension of stress-test fail-logs is as follows.

$$Dimension = Number\ of\ CY \times Number\ of\ SO \\ \times Number\ of\ Test\ patterns \tag{1}$$

There are tens of thousands of test patterns be used in the stress test. Each test pattern cost about hundreds of cycles. The total scan output pin counts of the DUT can vary from 20 to 50 in our experimental chip designs. Thus, the dimension of stress-test fail-logs is up to millions. For such high-dimensional binary data, it is hard for us to extract useful information directly from it.

Therefore, we transform stress-test fail-logs to a lower-dimensional data. We categorize all the fail responses based on certain strategy, like same CY or same SO. Next, we sum all the failures in the same category. By this, we can reduce the dimension, and also transform fail-logs to numeric values. Each summation value represents the total responses that DUT failed in the stress test based on certain strategy. We call this kind of summation data as *MisMatch Count(MMC)*, and its index is called *MMC-Index*.

In this paper, we tried three different MMC strategies: *CYSO*, *PSCY*, and *PSSO*. *CYcle-ScanOut(CYSO)* strategy is one of the earliest MMC strategies [5]. CYSO strategy categorizes stress-test fail-logs by same CY and same SO, which means each responses in the same category would have the same test cylces and the same scan output pin. By using this strategy, the dimension can be reduced to:

$$Dimension_{CYSO} = Number\ of\ CY \times Number\ of\ SO \tag{2}$$

The basic concept of CYSO is that the fail responses with same CY and same SO potentially come from the same region of the chip. If V_{min} variations are caused by certain tight time-margin regions, we can use CYSO strategy to predict functional V_{min}.

Fig. 3 illustrates the CYSO strategy. To get a better understanding for this strategy, we show the responses of each pattern as a grid. The x-axis of the grid is CY and the y-axis of the grid is SO. A red cell indicates a fail and a green cell indicates a pass. For CYSO strategy, given a cycle i and a SO pin j, its *CYSO-MMC* value $CYSO_{i,j}$ can be calculated as follows.

$$CYSO_{i,j} = \sum_{k=1}^{|P|} Response\ of\ (i,j)\ at\ Test\ Pattern\ k \quad (3)$$

where $|P|$ denotes the total number of test patterns. $CYSO_{i,j}$ represents the total failures in the cycle i and the SO pin j. i,j is its *CYSO-Index*. For example, in Fig. 3, the DUT has 4 SO pins, and there are total 3 test patterns be tested while each pattern take 5 cycles. After executing CYSO strategy, we find that $CYSO_{5,1} = 3$ since the DUT fails on all 3 patterns at cycle 5 on SO pin 1. We also find that $CYSO_{1,2} = 0$ since there are no failures at cycle 1 on SO pin 2.

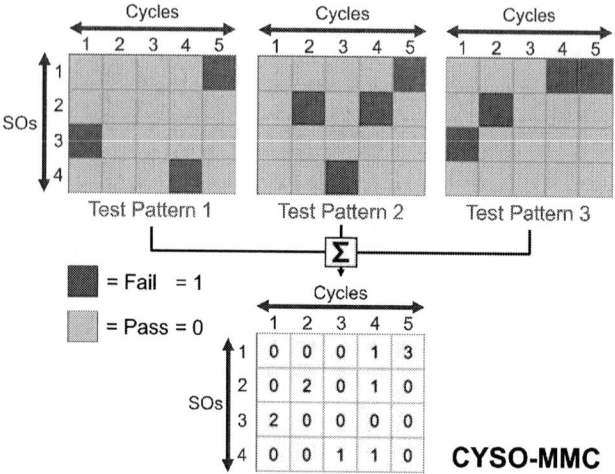

Fig. 3. Example of CYSO-MMC Data

Besides CYSO strategy, we also try the *PatternSet-CYcle(PSCY)* and *PatternSet-ScanOut(PSSO)* strategies. The PSCY strategy categorizes stress-test fail-logs by same PS and same CY, while PSSO strategy categorizes by same PS and same SO [5]. The ideas of these two strategies are that test patterns in the same pattern set are likely to test similar functions. If V_{min} variation is correlated with certain functional failures, PSCY and PSSO strategies can also work well on predicting functional V_{min}.

Fig. 4 shows the PSCY strategy and Fig. 5 shows the PSSO strategy. The x-axis of the grid is test pattern, which is grouped by PS. The y-axis of the grid is CY for the PSCY strategy, and SO for the PSSO strategy. For PSCY strategy, given a PS i and a cycle j, its *PSCY-MMC* value $PSCY_{i,j}$ can be calculated as follows.

$$PSCY_{i,j} = \sum_{k=1}^{|SO|} Response\ of\ (i,j)\ on\ SO\ pin\ k \quad (4)$$

where $|SO|$ denotes the total number of SO pins; for PSSO strategy, given a PS i and a SO pin j, its *PSSO-MMC* value $PSSO_{i,j}$ can be calculated as follows.

$$PSSO_{i,j} = \sum_{k=1}^{|CY|} Response\ of\ (i,j)\ at\ Cycle\ k \quad (5)$$

where $|CY|$ denotes the total cycles for a test pattern.

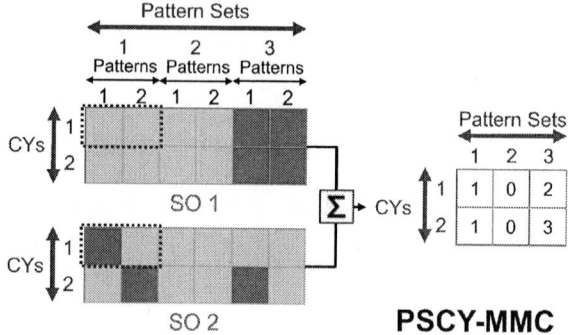

Fig. 4. Example of PSCY-MMC Data

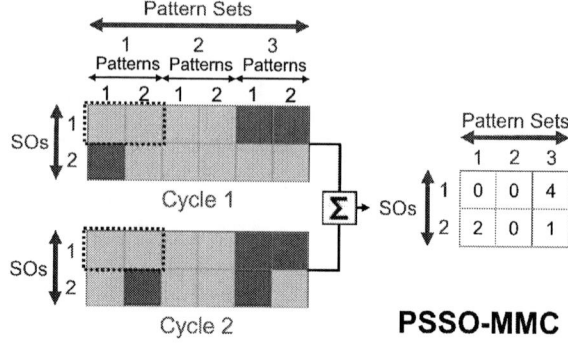

Fig. 5. Example of PSSO-MMC Data

B. MMC-Index Selection

We need to select important MMC-Indices to further reduce the dimension for MMC and to reduce test time. This is because that MMC consists of thousands of MMC-Indices and have to use all test patterns. Therefore, it would take up to minutes to collect the entire MMC. If we can find important MMC-Indices and test patterns related to functional V_{min}, we can collect them only to save test time.

To evaluate which MMC-Indices are important, we use *Spearman correlation* as metrics [7]. Spearman correlation is a nonparametric measure of rank correlation. In comparison with *Pearson correlation* [9], Spearman correlation can show how well the monotonic relationship is between two variables, and not only limited to linear relationship. Also, it is not sensitive to the outlier values. We take CYSO-MMC as an example. Fig. 6 shows the relationship between functional V_{min} and one of the CYSO-MMC value. Each dot in Fig. 6 represents a DUT. The x-axis is the normalized functional V_{min} of the DUT and the y-axis is the CYSO-MMC value of the DUT. From Fig. 6, when functional V_{min} is higher, the CYSO-MMC value also becomes higher. We can also see that there is an extreme high value, which is labeled as red. If we use Spearman correlation to evaluate the relationship, we can focus on the values in the normal range. In this paper, we correlate MMC values with functional V_{min}. We calculate Spearman correlation on all stress conditions of MMC, so that we can compare the importance of MMC-Indices across multiple stress conditions.

We use a greedy algorithm to select important MMC-Indices for multiple stress conditions. Fig. 7 shows an example of this algorithm. In Fig. 7, there are three stress conditions: c_1, c_2 and

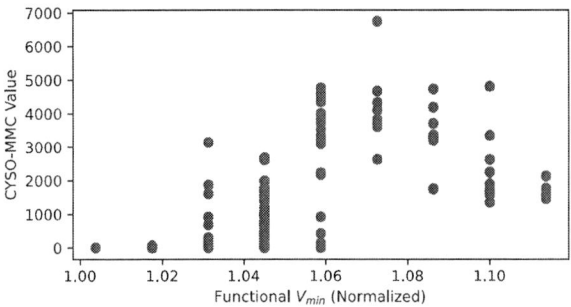

Fig. 6. Relationship between Functional V_{min} and CYSO-MMC Values

c_3. First, we find out top K_c high correlation MMC-Indices for each stress conditions to form MMC data subsets: X_{c_1}, X_{c_2} and X_{c_3}. K_c is a user-defined parameter that decides how many MMC-Indices would be initially chosen. We use $K_c = 6$ in the example shown in Fig. 7. Then, we calculate the *subset count* for each MMC-Indices in subsets. The subset count represents how many subsets a MMC-Index belongs. After the calculation, we sort MMC-Indices by the subset count in descending order. If a MMC-Index belongs to more subsets, it means that the MMC-Index is more important than others in these subsets. For MMC-Indices that have the same subset counts, we sort these MMC-Indices by the average correlation in descending order to determine the importance. Finally, we select top K_f highest importance MMC-Indices. K_f is a user-defined parameter that decides the size of important MMC-Indices subset. We use $K_f = 4$ in the example shown in Fig. 7. Since MMC-Indices a and b have the highest subset count, we select these two MMC-Indices first. Then, we select MMC-Indices c and d since they have top two highest average correlation in MMC-Indices with the subset count equals to two. By this strategy, we are able to select MMC-Indices that are important across stress conditions with high priority.

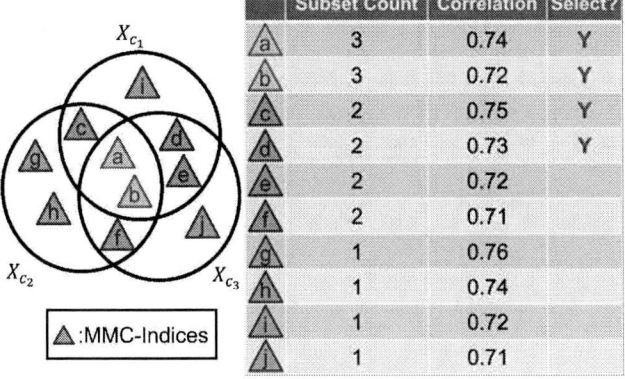

Fig. 7. Find Important MMC-Indices

C. Test Pattern Selection and Re-generate MMC

After we select important MMC-Indices, we have to find which test patterns related the most to the actual V_{min} to further reduce test time. Although we can remove test patterns that with no failures for each MMC-Index, the total counts of test patterns can still be up to thousands. Therefore, it is critical to find out important test patterns for each MMC-Index.

$x_{(i,j)}$	DUT-1	DUT-2	DUT-3	DUT-4	DUT-5	DUT-6	Counts	Rank
Pattern 1							2	4
Pattern 2							2	4
Pattern 3							3	3
Pattern 4							1	6
Pattern 5							4	2
Pattern 6							5	1

Fig. 8. Initial Solution Selection Example

To select important test patterns, we initially determine a solution by a heuristic method. We decide the initial solution by the *failing devices count* of each test pattern. The failing devices count represents how many devices fail in a test pattern. If a test pattern has a high failing devices count, it is possible that it has a tight time margin. Thus, it can be a critical test pattern. Fig. 8 shows an example of the process. First, we calculate the failing devices count for each test pattern. Then, we select test patterns with top K_p high failing devices counts to form the initial solution. K_p denotes a user-defined size limit of the solution. In Fig. 8, assuming $K_p = 2$, then we would select pattern 5 and 6 as the initial solution since they have the top two highest failing devices counts.

After deciding the initial solution, we use *Simulated Annealing(SA)* to improve it. SA is a heuristic algorithm that uses randomization to explore the solution space. It can help up find a solution from a large search space in an acceptable time. The original problem can be viewed as finding the best test patterns subset that are related to functional V_{min}. Since test patterns are a large space, SA is suitable to solve this problem in a short time.

In order to apply SA on our problem, we have to define 4 ingredients: *solution space, neighborhood structure, cost function* and *annealing schedule*.

1) Solution Space: All test patterns subsets.
2) Neighborhood Structure: We randomly perform one of the following operations.
   a) Add a test pattern that not in the current subset to the current subset.
   b) Remove a test pattern from the current subset.
   c) Replace a test pattern in the current subset with another test pattern that not in the subset.
3) Cost Function: We calculate the cost by following.

$$Cost = \begin{cases} 1 - \rho & \text{if } |S| \leq K_p, \\ 1 - \rho + \alpha \cdot (\frac{|S|}{K_p}) & \text{otherwise.} \end{cases} \quad (6)$$

ρ denotes the Spearman correlation of the current subset. $|S|$ denotes the size of the current subset. K_p denotes a user-defined size limit, and α denotes a user-defined penalty parameter. We adjust K_p and α to control the size of the subset.

4) Annealing Schedule: We describe the schedule in Algorithm 1. We use initial temperature T, reduce rate r and frozen temperature T_f to control the schedule.

After applying SA to get a subset S_{best} on every important MMC-Indices, we use S_{best} to regenerate MMC by following the steps described in section III-A. The regenerated MMC

Algorithm 1: Simulated Annealing Schedule

Result: Final solution subset S_{best}

1 Set an initial temperature $T > 0$;
2 Set a reduce rate $r < 1$;
3 Set a frozen temperature $T_f < T$;
4 $S, S_{best} \leftarrow$ Initial solution S_0;
5 **while** $T > T_f$ **do**
6      $S' \leftarrow$ Random neighbor of S;
7      $\Delta \leftarrow \text{Cost}(S') - \text{Cost}(S)$;
8   **if** $\Delta \leq 0$ **then** $S \leftarrow S'$;
9   **if** $\Delta > 0$ **then** $S \leftarrow S'$ with probability $e^{-\frac{\Delta}{T}}$;
10   **if** $\text{Cost}(S) < \text{Cost}(S_{best})$ **then** $S_{best} \leftarrow S$;
11     $T \leftarrow rT$;
12 **end**

would be the final input of our V_{min} prediction model in training phase. In test phase, we can perform stress test with test pattern subset S_{best} only, and use them to generate MMC for our V_{min} prediction model.

D. Build V_{min} Prediction Models

After regenerating MMC, we use these data as input to build a V_{min} prediction model. We adopt two regression models as our V_{min} prediction models. The first one is *Linear regression model*. Based on our past research, linear regression model has great accuracy and low cost compared to other ML models for process monitor data [10]. We would like to observe that if it can work well when using stress test data as input.

In addition to linear regression model, another powerful regression model is used for comparison. *XGBoost regression model* is one of the powerful ML models on regression [11], which based on gradient boosting. XGBoost regression model is able to evaluate non-linear relationship, so it can still work even when linear regression can not fit the model well.

IV. EXPERIMENTAL RESULTS

A. Experimental Setup

Two industrial chip designs, design A and B, are used in our experiments. The details about two designs are shown in Table I. Each PS contains about 1,000 test patterns. Therefore, there are about 34,000 test patterns used on design A, and about 91,000 test patterns used on design B. We implement linear regression with the *Scikit-learn* package [12], and XGBoost regression with the *XGBoost* package [11].

TABLE I
DETAILS ABOUT TWO DESIGNS

Design	Design A	Design B
Technology node	7nm	3nm
Available chips	234	190
Total stress condition tested	8	18
SO pin counts	24	46
PS counts	34	91
CY counts	275	441

The following are our experiment parameters. We set the initial selection parameter $K_c = 50$. For size limit parameters K_f and K_p, we experiment with multiple combinations of the two parameters. The results are shown in the next section. For SA algorithm parameters, we set $T = 1$, $r = 0.999$ and $T_f = 0.0005$. The penalty parameter α is set as 0.01 for CYSO-MMC and PSCY-MMC in design B. For the others, α is set as 0.02.

We perform 10-Fold cross-validation to evaluate our method. We split the original MMC dataset into 10 subsets. Then, we choose one of the subsets as the validation set and the others as the training set. We keep this process until all subsets have run our method. For comparison with process monitor methods, we also use *ring oscillator(RO)* values to build models. To evaluate the performance, we use *root mean squared error(RMSE)* as the metric. We calculate the average RMSE of ten subsets to generate a final score. All of the unit of RMSE in the result are *mV*.

B. Determine K_f and K_p

To figure out the optimal parameters, we perform experiments on one of the stress condition in design A with multiple combinations of K_f and K_p. To determine K_f, we fix K_p at 32 and adjust K_f from 2 to 10. Results are shown in Table II. When K_f increases, the RMSE in the training set becomes smaller while the RMSE in the validation set only has very small difference. This indicates that K_f selection is insensitive to RMSE. To strike a balance between the training set and the validation set, we choose the median $K_f = 6$ for later experiments.

To determine K_p, we fix K_f at 6 and adjust K_p from 8 to 64. Table III shows results. We observe that all of the RMSE are very similar when K_p increases, which means K_p selection is also insensitive to RMSE. We finally choose $K_p = 16$, which is a number close to the median, for later experiments.

TABLE II
PREDICTION RMSE ON DESIGN A WITH DIFFERENT K_f WHEN $K_p = 32$

Set	K_f	CYSO		PSCY		PSSO	
		Linear	XGB	Linear	XGB	Linear	XGB
Training	2	4.10	3.19	3.76	3.00	3.92	2.95
	4	3.94	1.96	3.69	1.64	3.82	1.65
	6	3.87	1.47	3.63	1.11	3.70	1.05
	8	3.82	1.13	3.56	0.85	3.62	0.86
	10	3.79	0.93	3.54	0.73	3.60	0.74
Validation	2	4.72	4.90	4.40	4.75	4.56	5.16
	4	4.62	5.17	4.40	5.03	4.51	5.21
	6	4.56	4.94	4.39	4.93	4.42	4.89
	8	4.50	4.98	4.32	4.86	4.56	4.78
	10	4.55	5.22	4.38	4.93	4.56	4.72

TABLE III
PREDICTION RMSE ON DESIGN A WITH DIFFERENT K_p WHEN $K_f = 6$

Set	K_p	CYSO		PSCY		PSSO	
		Linear	XGB	Linear	XGB	Linear	XGB
Training	8	3.77	1.84	3.71	1.58	3.66	1.45
	16	3.73	1.60	3.62	1.24	3.59	1.21
	32	3.87	1.47	3.63	1.11	3.70	1.05
	64	4.05	1.28	3.70	1.02	3.84	1.04
Validation	8	4.51	4.99	4.63	5.18	4.39	5.09
	16	4.51	5.04	4.47	5.07	4.37	4.89
	32	4.56	4.94	4.39	4.93	4.42	4.89
	64	4.66	4.94	4.49	4.94	4.61	4.87

979-8-3503-4631-2/23 $31.00 © 2023 IEEE

TABLE IV
RMSE OF CYSO-MMC ON ALL STRESS CONDITIONS

(a) Design A

Condition	F_{A1}	F_{A2}
V_{A1}		5.09
V_{A2}		4.68
V_{A3}		**4.67**
V_{A4}	5.31	**4.48**
V_{A5}	4.76	
V_{A6}	**4.67**	
V_{A7}	**4.51**	

(b) Design B

Condition	F_{B1}	F_{B2}	F_{B3}
V_{B1}	12.29		
V_{B2}	12.71		9.81
V_{B3}	11.81		**9.22**
V_{B4}	9.50	9.92	**8.96**
V_{B5}	10.33	**8.66**	9.91
V_{B6}	**8.95**	9.37	12.62
V_{B7}		11.28	11.54
V_{B8}		13.27	
V_{B9}		10.47	

C. Prediction RMSE

We apply our method on all of the stress conditions of design A and design B with $K_f = 6$ and $K_p = 16$. This means that the total test pattern used for each design is $K_f \times K_p = 96$. Table IV shows the overall RMSE of CYSO-MMC for all stress conditions in linear regression. We denote operating frequencies as F_{XY} and power supply voltages as V_{XY}. X represents the chip design, and Y represents the stress level. Bigger Y means larger stress, *e.g.* higher operating frequency or lower power supply voltage. We select stress conditions with the four lowest RMSE for two designs, which are marked as bold. Then, we show the detailed results using three kinds of MMC in Table V and VI. The linear regression model is denoted as *Lin.*, and the XGB regression model is denoted as *XGB*. For each stress condition, the lowest RMSE is marked as bold. The RMSE of process monitor methods is also shown in two tables, which are labeled as *RO*. Please note that RO values are from one nominal condition, so there is only one RMSE value for all stress conditions in RO.

We observe that linear regression models have smaller RMSE than XGB regression models in validation set. This implies us that XGB regression models have overfitting issues. In addition, the cost of linear regression models are also smaller than the one of XGB. Therefore, we find that linear regression models are suitable for MMC.

Finally, we need to decide which is the best MMC strategy for prediction. From the result of design A, three strategies all give low prediction errors. However, from the result of design B, we find that only CYSO strategy predicts V_{min} better than process monitor methods. Therefore, we consider CYSO strategy is the best MMC strategy to predict V_{min}.

TABLE V
PREDICTION RMSE OF DESIGN A

Condition	CYSO		PSCY		PSSO		RO	
	Lin.	XGB	Lin.	XGB	Lin.	XGB	Lin.	XGB
$V_{A6}F_{A1}$	4.67	5.28	4.54	4.90	**4.53**	5.24	4.88	5.14
$V_{A7}F_{A1}$	4.51	5.04	4.47	5.07	**4.37**	4.89		
$V_{A3}F_{A2}$	4.67	5.40	4.58	5.57	**4.42**	5.15		
$V_{A4}F_{A2}$	4.48	5.23	4.59	5.09	**4.43**	5.01		

D. Test Application Time

Table VII shows the estimated test application time for three methods: functional V_{min} measurement, process monitor methods, and our method. The test application time is shown in column *Time*, and its normalized value is shown in column

TABLE VI
PREDICTION RMSE OF DESIGN B

Condition	CYSO		PSCY		PSSO		RO	
	Lin.	XGB	Lin.	XGB	Lin.	XGB	Lin.	XGB
$V_{B6}F_{B1}$	**8.95**	9.52	10.26	10.41	10.97	10.58	9.80	10.94
$V_{B5}F_{B2}$	**8.66**	9.98	11.30	9.99	10.94	11.09		
$V_{B3}F_{B3}$	**9.22**	10.42	10.82	10.34	10.34	10.15		
$V_{B4}F_{B3}$	**8.96**	9.87	12.43	11.23	11.10	10.99		

Norm.. We focus on the test phase for process monitor methods and our method since we only run the training phase once. Functional V_{min} measurement is the test time of measuring V_{min} by functional test patterns. The time of process monitor methods is the summation of process monitor measuring time and V_{min} prediction time. Finally, the time of our method is calculated as follows.

$$T = T_{setup} + \frac{(2 \times \#patterns + 1) \times \#cycle}{F_{shift}} + T_{pred} \quad (7)$$

T_{setup} is the ATE setup time, which is about 2.5ms with our designs. The second term is test patterns test time. We use $K_f = 6$ and $K_p = 16$ as the experiment parameters. F_{shift} denotes the ATE shift frequency, and is 100 MHz in our estimation. The last term, T_{pred} is V_{min} prediction time. We notice that our method has 50 to 62.5 times speed up in comparison to functional V_{min} measurement. Also, the time of our method is 7.5 to 8 times shorter than process monitor methods. Thus, our method can save functional V_{min} test time.

TABLE VII
ESTIMATED TEST APPLICATION TIME

Method	Design A		Design B	
	Time	Norm.	Time	Norm.
Functional V_{min} measurement	500ms	62.5x	500ms	50x
Process monitor methods	60ms	7.5x	80ms	8x
Our method	8ms	1x	10ms	1x

V. CONCLUSION

We propose a novel method using stress test to predict chip performance without area overhead. Our method processes stress-test fail-logs and generate MMC to predict functional V_{min}. To reduce test time, we perform MMC-Index selection and test pattern selection. In MMC-Index selection, we select 6 MMC-Indices from thousands of MMC-Indices by Spearman correlation. In test pattern selection, we use simulate annealing to select 96 test patterns from $34,000$ test patterns in 7nm design A and from $91,000$ test patterns in 3nm design B. We conduct experiments using three kinds of MMC strategies and two kinds of regression models. Our experiments show that CYSO strategy with linear regression model is the best method to predict functional V_{min}. Our method gives smaller RMSE than process monitor methods. Also, we can have 50 to 62.5 times speed up compared with traditional measurement.

ACKNOWLEDGMENT

This research is supported by MediaTek Inc. The authors would like to thank MediaTek Inc. for providing experimental chips and valuable experience in the cutting-edge industry.

REFERENCES

[1] S. Borkar, "Designing reliable systems from unreliable components: the challenges of transistor variability and degradation," *Ieee Micro*, vol. 25, no. 6, pp. 10–16, 2005.

[2] W.-C. Lin, C. Chen, C.-H. Hsieh, C.-M. Li, E. J.-W. Fang, and S. S.-Y. Hsueh, "Ml-assisted vmin binning with multiple guard bands for low power consumption," in *2022 IEEE International Test Conference (ITC)*. IEEE, 2022.

[3] Y.-T. Kuo, W.-C. Lin, C. Chen, C.-H. Hsieh, C.-M. Li, E. J.-W. Fang, and S. S.-Y. Hsueh, "Minimum operating voltage prediction in production test using accumulative learning," in *2021 IEEE International Test Conference (ITC)*. IEEE, 2021.

[4] R. Cantoro, M. Huch, T. Kilian, R. Martone, U. Schlichtmann, and G. Squillero, "Machine learning based performance prediction of microcontrollers using speed monitors," in *2020 IEEE International Test Conference (ITC)*. IEEE, 2020, pp. 1–5.

[5] H. H. Chen, S.-H. Kuo, J. Tung, and M. C.-T. Chao, "Statistical techniques for predicting system-level failure using stress-test data," in *2015 IEEE 33rd VLSI Test Symposium (VTS)*. IEEE, 2015, pp. 1–6.

[6] J. T.-Y. Chang and E. J. McCluskey, "Detecting delay flaws by very-low-voltage testing," in *Proceedings International Test Conference 1996. Test and Design Validity*. IEEE Computer Society, 1996, pp. 367–367.

[7] P. Sprent, *Applied nonparametric statistical methods*. Springer Science & Business Media, 2012.

[8] D. Appello, H. Chen, M. Sauer, I. Polian, P. Bernardi, and M. S. Reorda, "System-level test: State of the art and challenges," in *2021 IEEE 27th International Symposium on On-Line Testing and Robust System Design (IOLTS)*. IEEE, 2021, pp. 1–7.

[9] D. Freedman, R. Pisani, and R. Purves, "Statistics (international student edition)," *Pisani, R. Purves, 4th edn. WW Norton & Company, New York*, 2007.

[10] M.-Y. Su, W.-C. Lin, Y.-T. Kuo, C.-M. Li, E. J.-W. Fang, and S. S.-Y. Hsueh, "Chip performance prediction using machine learning techniques," in *2021 International Symposium on VLSI Design, Automation and Test (VLSI-DAT)*. IEEE, 2021, pp. 1–4.

[11] T. Chen and C. Guestrin, "Xgboost: A scalable tree boosting system," in *Proceedings of the 22nd acm sigkdd international conference on knowledge discovery and data mining*, 2016, pp. 785–794.

[12] F. Pedregosa, G. Varoquaux, A. Gramfort, V. Michel, B. Thirion, O. Grisel, M. Blondel, P. Prettenhofer, R. Weiss, V. Dubourg, J. Vanderplas, A. Passos, D. Cournapeau, M. Brucher, M. Perrot, and E. Duchesnay, "Scikit-learn: Machine learning in Python," *Journal of Machine Learning Research*, vol. 12, pp. 2825–2830, 2011.

979-8-3503-4631-2/23 $31.00 © 2023 IEEE

(Industry Short Paper)

Predicting the Silent Data Error Prone Devices Using Machine Learning

Mohammad Ershad Shaik*
Intel Corporation
mohammadershad.shaik@intel.com

Abhishek Kumar Mishra**
Intel Corporation
abhishekkumar.mishra@intel.com

Yonghyun Kim
Intel Corporation
yonghyun.kim@intel.com

Abstract—**Silent Data Errors (SDEs) are a subset of Defective Parts per Million (DPPM) test escapes that cause unnoticed data corruption. Even at very low levels of DPPM, these are visible at cloud service provider data-center scales. In high-volume manufacturing, some defects manifest as SDEs that are screened at system level test (SLT) which is expensive. Due to subtleness of such defects, semiconductor devices prone to SDEs don't exhibit evident patterns or anomalies in the test data distributions. So, screening such faulty devices with ATE using statistical kill limits is challenging. To accelerate identification of those faulty devices, ahead of system testing, we propose to use Supervised Machine Learning (ML) approach to learn intrinsic patterns in an industrial test dataset. The experimental results illustrate that the embraced supervised learning framework via an ensemble of feature selection methodologies shows a noticeable performance improvement over traditional supervised and unsupervised methods.**

Index Terms—**silent data errors, machine learning**

I. INTRODUCTION

Silent Data Error (SDE) [1] [2] is the insidious and harmful error type, which can impact negatively on large-scale infrastructure services in terms of data loss without an indication. Because of being stealthy in nature, they are not easily detected through detection hardware within a CPU, so it is essential to use special test suite to screen such devices at System Level Tests (SLT) [3].

Several SDE detecting test suits like Data center Diagnostic Project (DCDIAG) [4], [5] are deployed at SLT to detect and screen the devices with defects manifesting as SDE's. However, there is a huge test cost associated with such tests due to long time-to-fail associated with those tests [6]. On the other hand, architectural advancements to implement Lock-step [7] and INFIELD testing [8] techniques can be used to reduce SDE vulnerability but even with such techniques, we still need extended testing to sensitize the failures. All of those SDE failing devices at SLT are already tested on the ATE and are labelled as known good dies. This is due to either lack of effective tests or kill limits to identify such devices.

Therefore, in this work, we strive to predict those devices which are potential candidates to be SDE fails at SLT by using ML modelling on test data from ATE. This prediction can help to screen out such devices before they reach SLT. Related work

---
* Alternate correspondence address : 0000-0002-3102-5249
** Present correspondence address : 0000-0001-6098-015X

in the area of applying ML on SLT fails was conducted on a smaller data set of customer returns [9] but not on SDE fails [10] [11]. In this paper, we discuss our Machine Learning methodology, Labelling Data, pre-processing the Data set, Feature engineering and experimental results from an large industrial dataset.

II. PROPOSED METHODOLOGY

In this preliminary work, we aim to predict devices that manifest as SDEs in the system or platform. Firstly, we discard defective devices screened by traditional ATE structural tests (like SCAN, BIST etc). Then, we use *Supervised learning* approach to learn intrinsic patterns of test data as discussed in the subsection below.

A. Proposed Method Flow

Our proposed methodology is shown in the Figure 1 & 2.

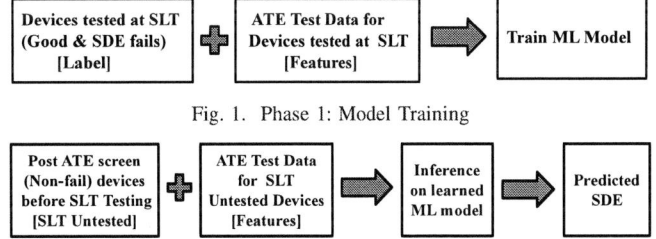

Fig. 1. Phase 1: Model Training

Fig. 2. Phase 2: Inference

B. Labeling Data

In our work, we are employing results of multivariate tests (features) to learn a predictive function based on their ground truth label. Figure 3 demonstrates the format of our data set. Assume we have n SLT tests $T_1, T_2, ..., T_n$ with M good (passing) devices $G_1, ..., G_M$ and W SDE failing devices $D_1, ..., D_W$. For each device, G_j & D_j, ATE tests result data (both Wafer level and package level) is stored as $\vec{f_j} = [f_{j1}, ..., f_{jn}]$ and $\vec{k_j} = [k_{j1}, ..., k_{jn}]$ respectively. Also, We label each passing device with 1 and each failing device due to SDE with 0. We discard the devices that are neither SDE fails nor Good devices. Our objective is to use the information from the above dataset to generate a robust binary classifier H where $H(\vec{T} = T_1, T_2, ..., T_n) \to \{0, 1\}$, having generalizability in nature. The classifier accepts input as a test result vector $\vec{T}$ and gives the output a prediction label (either 0 or 1).

979-8-3503-4631-2/23 $31.00 © 2023 IEEE

$$\begin{bmatrix} G_1 \\ \vdots \\ G_M \end{bmatrix} \begin{array}{ccc} T_1 & \cdots & T_n \\ \begin{bmatrix} f_{11} & \cdots & f_{1n} \\ \vdots & \ddots & \vdots \\ f_{M1} & \cdots & f_{Mn} \end{bmatrix} \end{array} \begin{array}{c} Label \\ \begin{bmatrix} 1 \\ \vdots \\ 1 \end{bmatrix} \end{array}$$

$$\begin{bmatrix} D_1 \\ \vdots \\ D_W \end{bmatrix} \begin{bmatrix} k_{11} & \cdots & k_{1n} \\ \vdots & \ddots & \vdots \\ k_{W1} & \cdots & k_{Wn} \end{bmatrix} \begin{bmatrix} 0 \\ \vdots \\ 0 \end{bmatrix}$$

Fig. 3. Data Mapping

C. Class Balancing

Our Training dataset is highly skewed in which SDE failing devices have very low number of samples compared to passing devices ($W <<< M$). In order to balance both classes of devices, we have opted different kind oversampling techniques like KNearestNeighbor(KNN) synthetic minority over-sampling technique (SMOTE) [12], K-Means SMOTE, and Adaptive Synthetic (ADASYN) [13] but none of them gave good training performance. So, we made a first order assumption to choose wafers only from LOT's that have devices that exhibit SDE fails as these LOTs will have higher probability of SDE fails. Then, we have used KNN SMOTE oversampling to balance both classes of devices.

D. Preprocessing the data

To deal with missing values in test data we have imputed with mean value and also discarded those features which had more than 12% missing values [14]. We have removed duplicate features. We also have removed low variance features, that is those features which may not contribute to the model and are dropped with threshold of $p = 0.95$ following Equation 1.

$$Var[T_1, \ldots, T_n] = \left[\frac{\sum_{v=1}^{M+W} (T_u^v - T_u^{Mean})^2}{M + W}; u \in [1, n] \right] \quad (1)$$

E. Feature Engineering

Feature engineering is a critical step that maximizes relevance & minimize redundancy in input features. Applying Principal Component Analysis (PCA) [15] for feature reduction to 300 principal components by keeping variance of 95% using classifiers like tree based, MLP etc [16] produced only 54% area under ROC (Receiver Operating Characteristic) via CatBoost model. To improve model's performance, we have used ensemble of feature selection methods: L1-norm [17] & Gini impurity [18] which are described below.

1) L1-Norm: With the goal to reduce the dimensionality of the data and find the relevant features, we have penalized classifiers with L1-Norm to estimate the sparse weights ($\vec{W} = [w_1, w_2, \ldots, w_n]$) of tests ($\vec{T}$). Then after, we segregate those weights which have non-zero values and their corresponding tests are separated to consider as relevant features. In order to find relevant features, we have expressed the L1-Norm to the cost function of binary classifier like Equation 2.

$$\min_{W} \sum_{j=1}^{M+W} (-y_j * log(\hat{p}) - (1 - y_j) * log(1 - \hat{p})) + ||\vec{W}||_1 \quad (2)$$

where $\hat{p}$ is the hypothesis function, y_j is the ground truth, and $||\vec{W}||_1$ is the L1-Norm.

2) Gini-Impurity: This is a measurement through which we can compute impurity-based feature importances, which in turn can be used to discard irrelevant features using tree-based classifiers. From Equation 3, we can measure the impurity of the split, and the feature with the lowest impurity would determine the best feature.

$$Gini(X) = 1 - \sum_{j=0}^{|C|-1} p_j^2 \quad (3)$$

This is how, we are selecting top-n tests (features) based on impurity score. Here, p_j is the probability of samples belonging to class j and $|C|$ is the cardinality of classes.

III. EXPERIMENTAL RESULTS

To evaluate the efficacy of our proposed methodology, we have conducted experiments targeting two failing signature buckets from SLT : "SDE fails" and "Non-SDE fails". SDE fails are the silent data errors that are screened at SLT using special DCDIAG test suits, whereas Non-SDE fails are detected deterministically at SLT with signatures from self-checking internal fault detection hardware. We have used sklearn [19] & Keras framework [20] to build the ML models. We describe those experiments in the following subsections.

A. Experiment I

In this experiment, we have taken a total volume of 21K devices that are screened at SLT. Based on the SLT test results, we have labeled the devices as "Good" & "Non-SDE fails". We then extracted features from ATE test data for all these devices. These features comprise V_{min} (minimum operating voltage) from memory BIST tests, logic at speed testing, all structural tests, analog test parameters, power & thermal readings, process monitor data etc. We have split the dataset into training ($\sim$13K), validation ($\sim$4K), & test ($\sim$4K).

1) Training: By following feature selection discussed in II-E, we have reduced our features to top 50 significant features. We then trained multiple models like RandomForest, LightGBM, CatBoost, XGBoost, Variational autoencoder (unsupervised approach based on mean construction loss), and neural network. But, of them, *an XGBoost (number of trees=100, booster learning rate = 0.2, booster = gbtree) using the all test features collected from feature selection, outperforms the other models.*

2) Testing: We tested the above trained XGBoost Model on unseen test dataset of 4K devices. To evaluate the model's performance, we have opted ROC. This metric shows graphically how good our model is to distinguish between classes in terms of true positive & false positive. Fig. 4 shows area under ROC curve is **0.98**, which implies that the model can differentiate between classes with **98%** confidence. The features that are collected from Gini impurity & L1-Norm can be model agnostic but concatenating those features with other features and feed into XGBoost helped the model to achieve higher area under ROC. Our trained model is able to predict

"Non-SDE fails" labelled devices with True Negative rate of 96% & "Good" devices at a True positive rate of 100% .

B. Experiment II

In this experiment, we have considered higher volume of devices than experiment I, consisting of 1.1 Million devices that are screened at SLT. Based on the SLT test results, we have labeled the devices as "Good" and "SDE fails". We then extracted even more features from ATE test data for all these devices. These features comprise Vmin (minimum operating voltage) from memory BIST tests, logic at speed testing, all structural tests, analog test parameters, power & thermal readings, process frequency monitor data etc. We have split the dataset in to training (~80K) and validation (~20K) (for hyper-tuning the parameters). We have used 1 Million devices as a test set. The reason to pick a large volume of test set is to include data drift (due to process) [21], [22] happening over a longer duration. This helps to test the generalizability and efficacy of the model on unseen data.

1) Training: For making the training faster, we reduced 100K features to ~15K by following feature selection method discussed in II-E. We then trained the dataset with multiple models like RandomForest, LightGBM, CatBoost, XGBoost and neural networks. Out of them, *Multi Layer Perceptron (MLP) model with 6 layers outperformed the other models.* It has dropout layer at each hidden layer (having different number of nodes) with different probabilities to prevent over-fitting. Gaussian Error Linear Unit activation function is used at the output of each hidden layer. The model is trained with entropy loss with a learning rate of $1e-4$ and it's convergence is shown in Figure 5, with respect to epochs. We can see that the training data loss closely matches with validation data loss at 9th epoch, which serves as a sweet spot to stop the training.

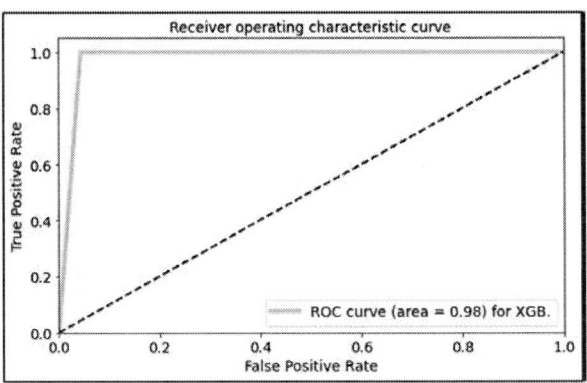

Fig. 4. ROC of XGBoost model

2) Testing: We tested the above trained MLP Model on unseen test dataset of 1 Million devices. Figure 6 shows area under ROC curve is **0.61**, which implies that the model can differentiate between classes with **61%** confidence. *We have observed an increment of ~7% following feature selection discussed in II-E approach compared to other tree-based supervised and neural network-based unsupervised methods.* The features that are collected from Gini impurity and L1-Norm can be model agnostic but concatenating those features with other features and feed into MLP helped the model to

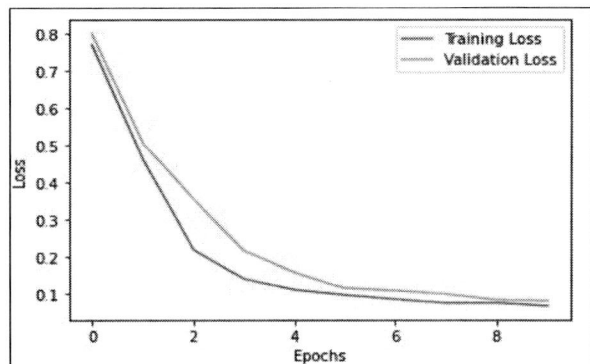

Fig. 5. Convergence of the MLP model

achieve higher area under ROC. Our trained model is able to predict "SDE fail" labelled devices with True Negative rate of 26.61% and "Good" labelled devices at a True positive rate of 95.01% . We have hidden the complete confusion matrix for confidentiality to not give away actual count of failing devices.

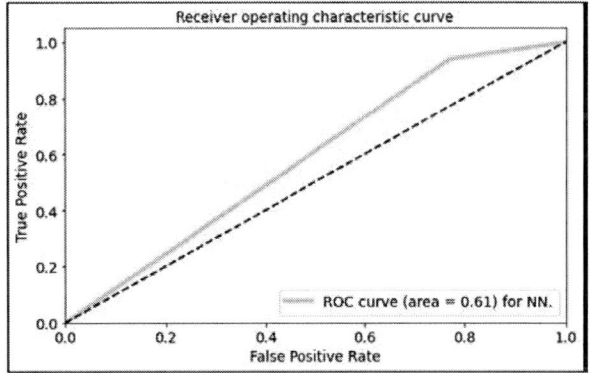

Fig. 6. ROC of MLP model

IV. CONCLUSION

In this paper, we attempted to predict SDE failing devices at SLT by training and inference on large scale industry ATE test data set using ML. To the best of our knowledge, this is the first such an attempt. With our proposed approach, when we build a model to predict between Good and Non-SDE (functional detectable fail) devices, we obtain a good performance score, but when we use the same methodology to predict SDE failing devices, 26.61% of the SDE failing devices are predicted correctly with available test data. In future, we hope to extend the framework using larger dataset by including additional features from pre-silicon design data, architectural Error detection hardware capabilities, Burn-In parameters, advanced structural fault models [23] [24] etc. to improve the ML model performance. Training and inference of ML models will happen offline (meaning after the ATE test while in devices are in transit to SLT test) so that there is no additional test time cost associated with this approach. With improved performance scores, we can use this methodology to predict early a subset of devices that are prone to SDE fails and apply extensive SDE testing to that subset. This helps to lower the risk of SDE DPPM in that subset. In addition to it, we can lower the test application time on "Good" devices which leads to lower effective test time cost at SLT.

979-8-3503-4631-2/23 $31.00 © 2023 IEEE

REFERENCES

[1] Peter H. Hochschild, Paul Turner, Jeffrey C. Mogul, Rama Govindaraju, Parthasarathy Ranganathan, David E. Culler, and Amin Vahdat. 2021. Cores that don't count. In Proceedings of the Workshop on Hot Topics in Operating Systems (HotOS '21). Association for Computing Machinery, New York, NY, USA, 9–16.

[2] Dixit, H. D., Pendharkar, S., Beadon, M., Mason, C., Chakravarthy, T., Muthiah, B. and Sankar, S. (2021). "Silent data corruptions at scale". arXiv preprint arXiv:2102.11245.

[3] D. Appello, H. H. Chen, M. Sauer, I. Polian, P. Bernardi and M. S. Reorda, "System-Level Test: State of the Art and Challenges," 2021 IEEE 27th International Symposium on On-Line Testing and Robust System Design (IOLTS), 2021, pp. 1-7, doi: 10.1109/IOLTS52814.2021.9486708.

[4] https://www.intel.com/content/www/us/en/support/articles/000058107/processors/intel-xeon-processors.html

[5] https://github.com/opendcdiag/opendcdiag

[6] Lerner, David P., Benson Inkley, Shubhada H. Sahasrabudhe, Ethan Hansen, Luis D. Rojas Munoz, and Arjan van de Ven. "Optimization of Tests for Managing Silicon Defects in Data Centers." In 2022 IEEE International Test Conference (ITC), pp. 578-582. IEEE, 2022.

[7] A. Sinha, "Innovative Practices Track: Silent Data Errors," 2022 IEEE 40th VLSI Test Symposium (VTS), 2022, pp. 1-1.

[8] R. T. Veetil, R. Sharma and S. Gundeboyina, "Comprehensive In-field Memory Self-Test and ECC Self-Checker -Minimal Hardware Solution for FuSa," 2021 IEEE International Test Conference India (ITC India), 2021, pp. 1-6.

[9] H. -C. Hsu, C. -C. Lu, S. -W. Wang, K. Jones, K. -C. Wu and M. C. . -T. Chao, "Rule Generation for Classifying SLT Failed Parts," 2022 IEEE 40th VLSI Test Symposium (VTS), 2022, pp. 1-7, doi: 10.1109/VTS52500.2021.9794184.

[10] H. Hu, N. Nguyen, C. He and P. Li, "Advanced Outlier Detection Using Unsupervised Learning for Screening Potential Customer Returns," 2020 IEEE International Test Conference (ITC), 2020, pp. 1-10, doi: 10.1109/ITC44778.2020.9325225.

[11] Lv, S.; Kim, H.; Zheng, B.; Jin, H. A Review of Data Mining with Big Data towards Its Applications in the Electronics Industry. Appl. Sci. 2018, 8, 582. https://doi.org/10.3390/app8040582

[12] Nitesh V. Chawla, Kevin W. Bowyer, Lawrence O. Hall, and W. Philip Kegelmeyer. 2002. SMOTE: synthetic minority over-sampling technique. J. Artif. Int. Res. 16, 1 (January 2002), 321–357.

[13] Haibo He, Yang Bai, E. A. Garcia and Shutao Li, "ADASYN: Adaptive synthetic sampling approach for imbalanced learning," 2008 IEEE International Joint Conference on Neural Networks (IEEE World Congress on Computational Intelligence), 2008, pp. 1322-1328, doi: 10.1109/IJCNN.2008.4633969.

[14] Roderick J A Little and Donald B Rubin (1986). "Statistical Analysis with Missing Data". John Wiley & Sons, Inc., New York, NY, USA.

[15] Mac kiewicz, Andrzej and Ratajczak, Waldemar, Principal components analysis (PCA), Computers & Geosciences journal, volume=19,number=3, pages=303-342, year=1993.

[16] V. Borisov, T. Leemann, K. Seßler, J. Haug, M. Pawelczyk, and G. Kasneci, "Deep neural networks and tabular data: A survey," arXiv preprint arXiv:2110.01889, 2021

[17] Richard G. Baraniuk "Compressive Sensing", IEEE Signal Processing Magazine [120] July 2007

[18] Breiman, L., J. H. Friedman, R. A. Olshen, and C. J. Stone. Classification and Regression Trees. Boca Raton, FL: Chapman & Hall, 1984.

[19] Pedregosa, F et al., "Scikit-learn: Machine Learning in Python" in Journal of Machine Learning Research, 2011, pp. 2825-2830.

[20] Gulli, A., & Pal, S. (2017). Deep learning with Keras. Packt Publishing Ltd.

[21] Kim, Y., Lee, H., & Kim, C. O. (2021). A variational autoencoder for a semiconductor fault detection model robust to process drift due to incomplete maintenance. Journal of Intelligent Manufacturing, 1-12.

[22] Mallick, A., Hsieh, K., Arzani, B. and Joshi, G., 2022. Matchmaker: Data Drift Mitigation in Machine Learning for Large-Scale Systems. Proceedings of Machine Learning and Systems, 4, pp.77-94.

[23] F. Hapke et al., Defect-Oriented Test: Effectiveness in High Volume Manufacturing, in IEEE Transactions on Computer-Aided Design of Integrated Circuits and Systems, vol. 40, no. 3, pp. 584-597, March 2021, doi: 10.1109/TCAD.2020.3001259.

[24] P. G. Ryan, I. Aziz, W. B. Howell, T. K. Janczak and D. J. Lu, Process defect trends and strategic test gaps, 2014 International Test Conference, 2014, pp. 1-8.

Functional Test Generation for AI Accelerators using Bayesian Optimization*

Arjun Chaudhuri, Ching-Yuan Chen, Jonti Talukdar, and Krishnendu Chakrabarty

Department of Electrical and Computer Engineering, Duke University, Durham, NC

Abstract—We propose a black-box optimization method to generate functional test patterns for AI inferencing accelerators. Functional testing is faster than structural testing as scan chains are not used for shifting in patterns and shifting out test responses. Moreover, functional testing reduces "over-testing" by targeting the detection of functionally critical faults for a given application workload. We use Bayesian Optimization for targeted test-image generation for stuck-at faults in a systolic array-based accelerator. Our framework supports test-pattern compaction and leverages various types of error regularization for enforcing functional-likeness of the generated test images. We achieve high fault coverage using a small set of test images for pin-level faults in 16-bit and 32-bit floating-point processing elements of the systolic array achieves high fault coverage with a small set of test images.

I. Introduction

The rise in the use of deep neural networks (DNNs) has led to rapid advances in artificial intelligence (AI) accelerators. These accelerators, typically based on the systolic array and its variants, are being used in commercial products such as Google's Tensor Processing Unit (TPU). We propose a black-box optimization driven methodology for functional test generation for stuck-at faults in systolic array-based inferencing accelerators. Functional testing is faster than structural testing as scan chains are not used for shifting in patterns and shifting out test responses. Moreover, functional testing reduces "over-testing" by targeting the detection of functionally critical faults for a given application workload.

Heavy array reuse in inferencing accelerators during model inferencing results from the mapping of all DNN layers to the same underlying array hardware. The output of one DNN layer, after array processing, becomes the input to the same array. As a result, a single hardware fault maps to multiple faulty neurons in the mapped DNN model. Complex error propagation due to non-linearity in the DNN and dataflow through multiple faulty neurons make it difficult to evaluate array behavior in the presence of faults without carrying out explicit fault simulation. Additionally, the inherent fault tolerance of DNN classification-type applications implies that the impact of a fault depends on dataflow, model parameters, and the fault location in the array.

The objective of the proposed test-generation framework is to determine a test image that can produce functional misbehavior in the presence of a stuck-at fault in the systolic array. Here, *misbehavior* is defined as the misclassification of an input test image. For functional test generation, compu-

tationally expensive fault simulation needs to be carried out to understand the faulty behavior. Therefore, it is necessary to generate the test image in a minimum possible number of iterations. The pattern space consists of m^2 variables for an $m \times m$ input grayscale image to the accelerator. Typically, the pixel value is stored in a byte format; an 8-bit integer is used to represent the intensity of a given pixel. Consequently, each pixel variable in the grayscale image can assume an integral value between 0 and 255 (both included). Therefore, the size of the pattern space that needs to be explored during pattern generation is 256^{m^2}. The multi-dimensional nature of the pattern space makes brute-force search infeasible.

In order to advance intelligent exploration and iterative pruning of the pattern space, we present a method for guided test-pattern search using black-box optimization. Here, the black box refers to the faulty AI inferencing system (systolic array) and the optimization engine is Bayesian Optimization (BO). The key contributions of this paper are as follows:

- Formulation of targeted functional test generation as a black-box optimization problem that can be solved iteratively using BO;
- Functional test generation (with fault dropping) for stuck-at faults in a DNN-mapped inferencing accelerator;
- Integration of a non-negative matrix factorization decoder model to enable image updates in a low-dimensional latent space for accelerating BO convergence for test-image generation;
- Support for test-pattern compaction to reduce the required number of test images for covering a target fault list;
- Evaluation of distributional similarity (referred to as *functional-likeness*) between the generated test image and representative images in the application workload;

The remainder of this paper is organized as follows. Section II discusses functional criticality analysis and prior work on functional testing. Section III presents the BO-driven functional test-generation framework and test-pattern compaction. Section IV describes the incorporation of a pre-trained discriminator model for checking whether the test images lie in the distribution of workload images. Section V presents experimental results and Section VI concludes the paper.

II. Background and Prior Work

A. AI Accelerators and Fault Injection Framework

Systolic-array accelerators offer high throughput, high compute density, and improved performance per Watt for MAC operations. We designed a 128×128 systolic array, supporting

---

*This work was supported in part by the Semiconductor Research Corporation under GRC Task 3106.001.

16-bit and 32-bit IEEE FP data formats, in Verilog HDL. The array was synthesized using Nangate 45 nm cell library. This design is based on state-of-the-art accelerator architectures such as Google's TPU [1]. We used the LeNet-5 DNN trained on the MNIST dataset as our target application [2]. We developed a structural fault-injection framework for injecting stuck-at (s-a-0, s-a-1) faults, delay faults and bridging faults in the gate-level netlist of the PE. While the framework is general, the evaluation presented in this work is aimed at the impact of single stuck-at faults.

B. Efficient Criticality Assessment of Structural Faults

The inherent fault-tolerance of DNNs can be attributed to the robustness of training on large datasets, non-linearity of activation units, and the use of regularization features such as dropout [3]. Furthermore, the application of DNNs to different use-cases provides flexibility in terms of acceptable performance thresholds required to ensure functional correctness. This flexibility can be leveraged to develop metrics to grade structural faults in terms of their functional criticality.

The *functional criticality* of a fault is determined by the *severity* of its impact. For inferencing, this impact is quantified in terms of the loss in inferencing accuracy for the given use-case. Functional fault-criticality assessment can be used to guide test effort and quality assessment of AI accelerators throughout the product life-cycle. As functional fault criticality depends on the type of workload, criticality threshold of the application, and model-to-hardware mapping, fault criticality can be cataloged for different domain-specific use-cases.

Fault criticality can be assessed by performing functional simulations for structural faults. However, brute-force functional simulation is prohibitively expensive due to high CPU runtime. In [4], a deep learning-enabled framework is used to classify faults as functionally critical or benign with respect to a representative application workload (comprising training data images of the mapped DNN model). Faults that are functionally benign can become critical when variations or perturbations are present in the input image. Hence, targeted functional test generation can uncover faults that are potentially critical under certain variants ("corner cases") of the application workload; the resulting test images can augment the representative workload used to train the criticality classifier. If a test image cannot be generated for a fault after significant test-generation effort, the fault is likely to remain benign for workload images encountered by the inferencing accelerator in the field. The generated test images can be stored in on-chip memory and used for cyclic functional tests performed at regular intervals for monitoring the health of the chip [5].

C. Related Prior Work

In [6], the authors carried out structural testing for stuck-at faults in an AI accelerator by mapping the deterministic ATPG patterns to the input weight and activation matrices for the accelerator array. As the compute units, also referred to as processing elements (PEs), of the accelerator are combinational logic units, the input weight and activation matrices comprising the ATPG patterns are passed through the accelerator and the combinational cloud's outputs are recorded as test responses for fault detection. However, structural testing can be an over-kill, especially for the functionally benign faults and in-field testing. In [4], a large majority of stuck-at faults are found to be functionally benign and the functional criticality is shown to be dependent on application workload, DNN-to-accelerator mapping, DNN model parameters, fault location, and accelerator architecture. The functional dataflow through the fault site determines fault criticality and functional test generation aims to uncover faults that are likely to cause functional misbehavior for a given mapped model.

In [7], functional testing is carried out by selecting test images from the application workload that are prone to misclassification under faulty conditions. Model-agnostic test-image selection picks images lying at the class boundaries of the dataset as they are more susceptible to misclassification. Model-aware test-set selection identifies the images that are classified correctly by the mapped model albeit with a low confidence. However, both these approaches do not consider the target fault location and the underlying hardware, especially when array reuse during inferencing plays a significant role in fault activation and error propagation.

In [8], conditional generative adversarial networks (cGANs) are used to generate a fake distribution of the original workload and the cGAN-generated images are used as test images to carry out fault dropping. However, there is no regularization or condition imposed on cGAN to generate images that can distinguish between fault-free and faulty accelerators. It is possible that an image generated by cGAN can lead to the same classification output for fault-free and faulty accelerators for a given fault. Therefore, targeted test generation is needed to detect faults that are critical with respect to the mapped model, underlying hardware, and the application workload.

III. TARGETED TEST GENERATION USING BLACK-BOX OPTIMIZERS

A. Formulation of Bayesian Optimization

In this work, we propose a test generation framework that uses a Bayesian optimization (BO) algorithm as the engine for black-box optimization; Fig. 1 describes this framework. BO has been shown to be efficient in optimizing computationally expensive black-box objective functions [9]. BO algorithms are also commonly used for hyperparameter tuning in the training of deep neural networks [10]. BO methods focus on solving the problem $\max_{x \in A} f(x)$, where A is the feasible set of optimization and f is the objective function to be optimized, i.e., maximized or minimized. A is typically defined as a d-dimensional hyper-rectangle $A = \{x \in \mathbb{R}^d | a_i \le x_i \le b_i, i = 1, 2, ..., d\}$, where a_i and b_i is the lower and upper bound of x in i-th dimension. For each optimization iteration, the algorithm samples a point $x^* \in A$ to maximize $f(x^*)$.

In the proposed framework (Fig. 1), we use a BO-based black-box optimization engine to generate test images for a DNN-mapped systolic array (SA). The generated test images

979-8-3503-4631-2/23 $31.00 © 2023 IEEE

attempt to differentiate between fault-free and faulty SA accelerators. In other words, the classification output, Y, of the fault-free accelerator should be different from the classification output, Y_f, of the faulty one. In order to generate a test image that can distinguish between fault-free and faulty accelerators, we use the cross-entropy error, CSE, between Y and Y_f as the loss value to be maximized as part of the objective function, i.e., $f = CSE(Y, Y_f)$. During each iteration of the optimization process, the BO algorithm samples an image I from the input image space (A) and obtains Y and Y_f through inferencing on fault-free and faulty accelerators, respectively. In this work, we use LeNet-5 DNN trained on (32×32)-pixel gray-scaled MNIST images (pixel values range from 0 to 255) and the input space (feasible set) $A = \{I \in \mathbb{R}^{32 \times 32} | 0 \leq I_{ij} \leq 255\}$. Note that the fault simulation procedure used to obtain Y_f is time-consuming. It is therefore appropriate to use BO for intelligently guiding the optimization flow to solve $\max_{I \in A} CSE(Y, Y_f)$ in minimum possible number of iterations. However, the high-dimensional nature of the input image space ($d = 32 \times 32 = 1024$ is the dimensionality of a 32×32 image) can degrade the convergence performance of the iterative BO algorithm and lead to high runtimes for test generation. To address the challenge posed by the high-dimensional input space, we propose in Sec. III-B a framework to sample images from a low-dimensional latent space and deploy a decoder function inside the test-generation loop for restoring the sampled low-dimensional image to a full 32×32 image prior to inferencing and computation of CSE.

The goal of functional test generation is to generate test images for detecting faults that are critical and cause misclassification of images in the application workload. Therefore, it is desired to generate functional-like test images that exhibit reasonable overlap with the distribution of functional workload images. These functional-like test images can be viewed as "adversarial" or "corner-case" images that are likely to trigger misbehavior of the accelerator in the presence of a fault. In the scenario where the BO-driven test-generation flow fails to generate functional-like test image for a target fault F, there is a strong possibility of F being functionally benign and not causing misclassification of an input image that belongs to the general distribution of the application workload.

To obtain functional-like test images, the user can specify an image I^0 from the application workload as the initial point of BO optimization. In addition, the user can assign a bounded feasible set A, with upper and lower bounds defined for the domains of individual image pixels, to ensure functional-likeness of generated images throughout the iterative test-generation flow. In particular, the user can specify $A = \{I \in \mathbb{R}^{32 \times 32} | \max(0, I_{ij}^0 - \Delta) \leq I_{ij} \leq \min(I_{ij}^0 + \Delta, 255)\}$ as the feasible set of the optimization problem, where Δ is the largest possible deviation from the initial image I^0 for each pixel and the value of Δ can be determined by users. In Fig. 2, we showcase the test images generated to detect single stuck-at-1 (s-a-1) faults in 32 pins of the partial-sum input bus in a PE. The top image is generated by targeting s-a-1 fault in the fourth significant bit (MSB-4) of the exponent

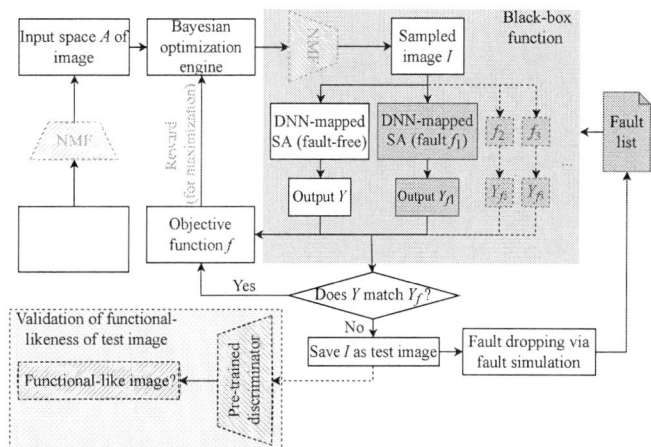

Fig. 1: Proposed BO-based test generation framework, including optional steps (those with dashed outlines) that users can carry out to improve performance of the optimization system.

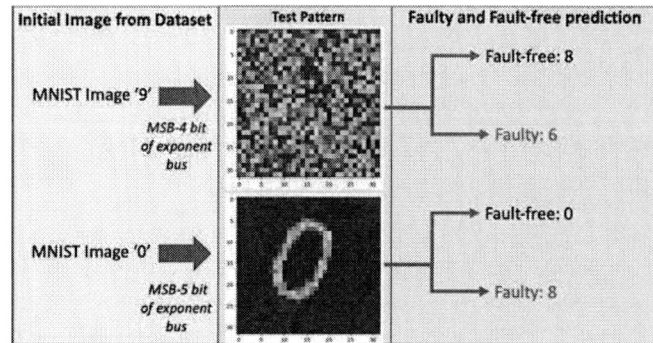

Fig. 2: Illustration of generated test images using A with and without bounds to enforce functional-likeness.

bus with $A = \{I \in \mathbb{R}^{32 \times 32} | 0 \leq I_{ij} \leq 255\}$. In contrast, the bottom image is generated using A with tight bounds to enforce functional-likeness. As shown in Fig. 2, the test image generated using tightly bounded A is more functional-like.

B. Black-Box Optimization in Latent Space

As described in Sec. III-A, it is desirable to reduce the dimensionality of the search space of the BO engine to ensure rapid convergence. As a rule of thumb described in [11], the BO algorithm works better with feasible sets with $d \leq 20$ (compared to the original feasible set A with $d = 1024$). We therefore need to introduce an encoder/decoder into the proposed framework to achieve this extent of dimensionality reduction. In this work, we use non-negative matrix factorization (NMF) [12] as the method to encode/decode BO-sampled images into/from the low-dimensional latent space. Fig. 1 shows the proposed framework with decoder and encoder. With the decoder and encoder, the BO engine can optimize the objective function by identifying the optimal combination of function variables in the low-dimensional latent space that now constitutes A. In each iteration, the engine samples an image in the latent space and decodes it into the full 1024-dimensional space before carrying out fault simulation of the inferencing accelerator.

979-8-3503-4631-2/23 $31.00 © 2023 IEEE

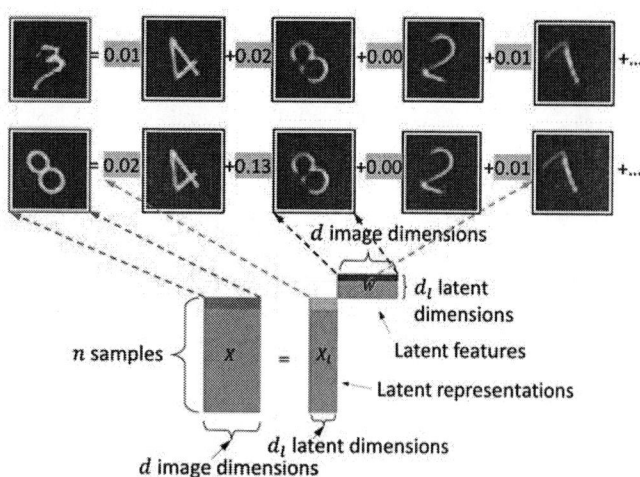

Fig. 3: Illustration of NMF-based encoding and the factorization procedure.

NMF is similar to principal component analysis (PCA) in that it can be used for dimensionality reduction [13, Ch. 3]. As in PCA, the NMF algorithm attempts to encode each data point (or sample) into a weighted sum of extracted components, as illustrated in Fig. 3. To be more specific, an image dataset matrix $\mathbf{X}$ of size $N \times d$ can be factorized using the following formulation; see Fig. 3: $\mathbf{X} \approx \mathbf{X}_l \times \mathbf{W}$, where $\mathbf{X}_l$ is the latent representation of $\mathbf{X}$ and $\mathbf{W}$ is composed of extracted basis components. Each row of $\mathbf{X}$ corresponds to an image I in the original 1024-dimensional space and the number of rows N indicates the number of image samples used to train the NMF-based encoder. Each row of $\mathbf{X}_l$ corresponds to the d_l-dimensional latent representation I_l ($d_l < d$) and each row of $\mathbf{W}$ corresponds to an extracted d-dimensional component that represents a basis image. The number of rows in $\mathbf{W}$ equals d_l which is the number of basis images required to reconstruct the original image in $\mathbf{X}$ from the latent-space image in $\mathbf{X}_l$. The scaling coefficients required for reconstruction act as pixel values of the d_l-dimensional latent-space image (I_l) which forms a row in $\mathbf{X}_l$.

NMF is different from PCA as NMF requires both latent image representation (scaling coefficients) and basis image components to be non-negative. This way, NMF can have human-interpretable basis components as shown in Fig. 3; here, basis images are extracted as strokes of hand-written digits. On the contrary, basis images obtained using PCA usually have negative elements that do not have any physical interpretation in our use case where images have non-negative pixel values. To recover the original image I from the latent representation I_l, the NMF-based decoder can simply carry out vector-matrix multiplication to obtain I, i.e., $I = \mathbf{W}^T \cdot I_l$.

C. Test Pattern Compaction

The proposed test-generation framework is equipped with the ability for test compaction when multiple single faults are present in the fault list. Each fault f_i is represented by the corresponding fault-injected version of the inferencing accelerator. Note that $1 \leq i \leq N_f$, where N_f is the number

of faults to be detected from the fault list. Therefore, the test-generation framework operates on a batch of N_f fault-injected accelerators. The test-generation process begins with an input image (I^0) randomly chosen from the set of images in the application workload. The black-box optimizer (BO-based) updates the input image in each iteration (or timestep) with the objective function f of maximizing the average CSE calculated across all N_f faults in the fault list; the average CSE (referred to as CSE_{avg}) is computed as:

$$f = CSE_{avg} = \frac{\sum_{j=1}^{N_f} CSE(Y, Y_{f,j})}{N_f}$$

Here Y denotes the fault-free accelerator's classification output and $Y_{f,j}$ denotes the classification output of the accelerator injected with fault f_j. Fig. 1 illustrates the concurrent fault simulation carried out inside the black-box function for obtaining the average cross-entropy error across all N_f faults.

After k timesteps (k is a user-defined parameter), we pause the test-generation flow and carry out fault-dropping with the updated image I_k. Typically, k is set as the number of iterations of the BO-based test generation after which CSE_{avg} starts to saturate. The detected faults are removed from the fault list following which we proceed with test generation for the remaining faults; the input image is reset to I^0 which is randomly picked from the set of workload images. The test generation completes when the fault list is empty or $N_{ex} \cdot k$ iterations have completed; N_{ex} is a user-defined parameter.

IV. FUNCTIONAL-LIKE TEST IMAGE VALIDATION USING DISCRIMINATOR MODEL

The objective of generative adversarial networks (GANs) is to generate "fake" samples that exhibit close resemblance to the distribution of a target workload whose copy distribution is being produced [14]. For example, in the case of adversarial image generation, the input to the GAN model is usually white noise or a randomly initialized image which is then converted into an image that lies near or inside the distribution of a set of real images from an application workload. The training of GAN involves co-optimization of two coupled multi-layer perceptron (MLP) models - the generator and the discriminator. The generator model, represented by a differentiable function $G(\cdot)$, generates a sample image z from a noisy input image. The discriminator model, represented by a separate differentiable function $D(\cdot)$, tries to discriminate between a real image x and the generated sample z. The discriminator's outputs $D(z)$ and $D(x)$ indicate the probabilities that z and x belong to the generated and real distributions, respectively. Therefore, by maximizing the probability $D(x)$ and by minimizing the probability $D(z)$ via backpropagation, the discriminator is trained to distinguish between real and fake samples. On the other hand, by minimizing the binary cross-entropy loss $\log(1 - D(z))$ via backpropagation, the generator model $G(\cdot)$ learns to generate z that causes $D(z)$ to approach 1 which, in turn, implies that the discriminator is beginning to interpret z as part of the real image set.

Through iterative two-player minmax optimization of the GAN model, the non-linear function $G(\cdot)$ is learnt such that image z generated by $G(\cdot)$ from noise is successful in confusing the discriminator and $D(z)$ approaches 0.5. In other words, the discriminator is unsure if the input sample belongs to the set of real or generated images. The trained generator model can be used for generating images (z) belonging to the target distribution of x. On the other hand, the trained discriminator model can be used to identify images that are outliers with respect to the distribution of x.

The ability of a pre-trained discriminator model to quantify the likelihood of an image belonging to a given workload distribution is leveraged to evaluate functional-likeness of images generated by the BO engine. First, we train a GAN model to learn the distribution of images. Next, we use the pre-trained discriminator model ($D(\cdot)$) inside GAN to estimate the probability of a BO-generated image I to lie in the distribution of MNIST images. If I has a high (low) distributional similarity with MNIST images, $D(I)$ is expected to approach 1 (0). Therefore, the pre-trained model $D(\cdot)$ can be used to validate the functional-likeness of a test image generated by the BO-based framework; see Fig. 1. If $D(I)$ approaches 1 (0), the corresponding test image I is considered to be functional (non-functional).

In addition to placing upper and lower bounds ($\pm\Delta$) on the initial image's pixels for enforcing functional-likeness of BO-generated test images, we can add a regularization term ($\log D(I)$) to the BO engine's error (or loss) function which is to be maximized. Consequently, the objective function for BO now consists of two weighted components — cross-entropy error between faulty and fault-free accelerator outputs and the above-mentioned regularization term: $f = \alpha \cdot CSE(Y, Y_{flt}) + (1-\alpha) \cdot \log D(I^*)$, where $\alpha \in [0,1]$. Evaluation of such regularization for test generation is left for future research.

V. EXPERIMENTAL RESULTS

A. Experimental Setup

We use a LeNet-5 DNN trained using the MNIST dataset for evaluation of the proposed test generation framework. The LeNet-5 model is mapped to a 128×128 systolic array that supports both 16-bit and 32-bit IEEE floating-point computations. The proposed framework attempts to generate test images to detect both pin-level and gate-level structural faults in PEs of the systolic array. In this work, we demonstrate the proposed BO-based test generation for the pin-level faults, similar to the test-generation flows presented in [7], [8]. We randomly select 10 PEs to detect single stuck-at faults in their partial-sum and multiplier-output buses.

For each 32-bit bus, we primarily target faults on the sign and exponent bits ($1 + 8 = 9$ bits in total) as single stuck-at faults in the mantissa pins of all four buses (weight, activation, multiplier output, and partial sum) have been shown to be benign across multiple 16-bit and 32-bit PEs for a representative MNIST workload [4]. The randomly selected PE locations are $(20,0)$, $(21,70)$, $(25,16)$, $(45,8)$, $(77,7)$, $(14,0)$, $(79,1)$,

TABLE I: Test-generation results obtained for sign and exponent faults in individual 32-bit PEs (36 faults per PE).

PE location	Number of detected faults								
	MAG	MAW	cGAN	BO (∞)	COMP (∞)	BO (63)	COMP (63)	BO (127)	COMP (127)
$(20,0)$	15	14	12	35	35	36	36	36	36
$(21,70)$	7	7	7	11	11	8	13	7	8
$(25,16)$	7	7	7	35	35	36	35	36	36
$(45,8)$	16	16	12	35	35	35	35	35	35
$(77,7)$	16	16	15	30	30	36	36	35	36
$(14,0)$	14	13	13	36	36	36	36	36	36
$(79,1)$	16	16	16	34	34	33	32	36	36
$(34,2)$	16	16	16	26	27	28	28	30	30
$(7,3)$	14	15	13	33	36	36	36	36	36
$(28,4)$	16	16	14	29	30	36	36	36	36
Total	137	136	125	304	306	320	323	323	325
FC (%)	38.1	37.8	34.7	84.4	85.8	88.9	89.7	89.7	90.3
Pattern count	18	20	10	7	7	7	11	11	10

The numbers within parentheses ($\infty, 63, 127$) indicate the value of Δ used during test generation for enforcing functional-likeness.

$(34,2)$, $(7,3)$, and $(28,4)$; there are $10 \times 2 \times 9 \times 2 = 360$ pin-level faults in the fault list (36 faults per 32-bit PE). In case of 16-bit PEs, the fault list consists of single stuck-at faults in the sign and exponent bits ($1 + 5 = 6$ bits in total per bus) of the partial-sum and multiplier output buses. Therefore, the fault list for each 16-bit PE contains $2 \times 6 \times 2 = 24$ faults, totalling to 240 faults for the 10 PEs.

We use the open-source software package `bayes_opt` [15] as the BO engine in our implementation of the proposed test generation framework. Given the target fault(s), the BO engine runs at most 100 iterations ($N_{exit} \cdot k \leq 100$) to generate a test image. If the targeted fault(s) are detected during these iterations, the BO engine terminates the optimization session prematurely and returns the sampled image as the test image for fault dropping. The NMF-based encoder and decoder are used to reduce the dimensionality of the BO search space. We set the dimensionality of the latent space as $d_l = 20$ as it sufficiently captures the distributional variance of the original workload images. To expedite convergence, we apply a domain-reduction scheme [16] to dynamically resize the feasible set (search space) of BO optimization during iterations. Experiments were run on a 40-core Intel Xeon E5 2.4 GHz CPU.

B. Evaluation of BO-based Test Generation

We generated test images using two embodiments of the proposed framework: 1) BO-based test generation targeting one fault in each optimization session, and 2) pattern compaction-aware (COMP) BO-based test generation targeting sign and exponent faults concurrently on a PE bus (six and nine faults for 16-bit and 32-bit PE buses, respectively) in each session. Following each generated test image (optimization session), we carry out fault simulation to drop detected faults from the fault list. We also implemented three test generation methods — (i) model-agnostic (MAG) and (ii) model-aware (MAW) proposed in [7]; (iii) conditional GAN-based image generation (cGAN) proposed in [8] — as the baselines for comparison. Table I shows the results obtained for faults in 32-bit PEs using different methods. The BO-based frameworks achieve more than twice the fault coverage (FC) compared to the baseline method. Furthermore, the proposed frameworks

TABLE II: Test-generation results obtained for sign and exponent faults in 16-bit PEs and mantissa faults in 32-bit PEs using different methods.

Test generation method	Fault list	Fault count	Pattern count	FC (%)
MAG			18	35.8
MAW			20	37.8
cGAN	16-bit, sign, exp	240	10	36.7
BO ($\Delta = 63$)			**7**	**87.5**
COMP ($\Delta = 63$)			**9**	**86.7**
MAG			18	0
MAW			50	0
cGAN	32-bit, mantissa	160	10	8.1
BO ($\Delta = \infty$)			**1**	**16.9**
COMP ($\Delta = \infty$)			**1**	**30**

achieve this high degree of coverage using fewer test images.

Table II shows results obtained for pin-level faults in the four higher-order mantissa bits in 32-bit PEs and the sign and exponent bits in 16-bit PEs. The proposed BO-based test-generation flow consistently achieves higher FC with lower (or comparable) pattern count compared to the baseline methods. Faults in mantissa bits are benign with respect to the MNIST workload and are therefore harder to detect using functional patterns which are variations of the original workload.

C. Evaluation of Functional-likeness of Test Images

To evaluate functional-likeness of the generated test images, we first use the t-distributed stochastic neighbor embedding (t-SNE) algorithm to map test sets generated using different BO-based methods to a two-dimensional (2-D) space; Fig 4 shows these results. The t-SNE algorithm is commonly used to embed high-dimensional data points in a low-dimensional space (2-D in our use case) for visualization. Each data point in Fig 4 is either the 2-D representation of an image from the original MNIST dataset or the 2-D representation of a test image generated for 32-bit PEs. As shown in the figure, more functional-like test images have better approximation to the MNIST data points in the t-SNE-mapped 2-D space. The figure shows that tests generated using BO with a smaller value of Δ are more functional-like. We also report the average L2-norm distances between the MNIST workload distribution and the distribution of BO-generated test images. Having the bounds $\pm\Delta$ on the BO optimization variables (in latent space) result in a lower L2 distance compared to the BO flow without any input bounds ($\Delta = \infty$). Moreover, a smaller value of $\Delta = 63$ leads to a lower L2 distance of 7.2 compared to the L2 distance of 17.5 for $\Delta = 127$. This confirms that smaller Δ leads to the generation of more functional-like test images.

Next, we evaluate functional-likeness of the generated test images using a pre-trained GAN-based discriminator. The discriminator used for evaluation $D(\cdot)$ is trained with its generator counterpart $G(\cdot)$ using MNIST dataset. To evaluate the performance of $D(\cdot)$, we test $D(\cdot)$ against MNIST images and randomly generated noise. We observe that the pre-trained $D(\cdot)$ can correctly classify 99.20% of the MNIST images as being functional and 98.45% of the noise images as being non-functional. Finally, the pre-trained $D(z)$ classifies 96.90% of all test images (generated by the BO frameworks for the targeted faults in 16-bit and 32-bit PEs) as being functional.

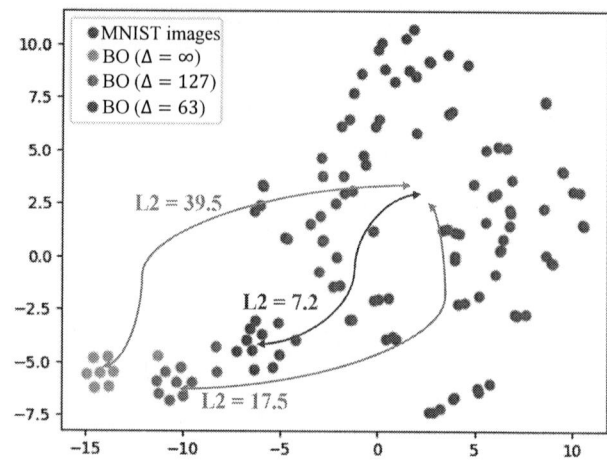

Fig. 4: t-SNE visualization of test images generated using different embodiments of the proposed method.

VI. CONCLUSION

We have presented a Bayesian Optimization driven framework for targeted functional test generation for stuck-at faults in AI inferencing accelerators. Our framework supports test-pattern compaction and leverages various approaches for enforcing functional-likeness of the generated test images.

REFERENCES

[1] N. P. Jouppi et al., "In-datacenter performance analysis of a tensor processing unit," in *ISCA*, 2017, pp. 1–12.

[2] Y. LeCun et al., "Gradient-based learning applied to document recognition," *Proc. IEEE*, vol. 86, no. 11, pp. 2278–2324, 1998.

[3] G. Li et al., "Understanding error propagation in deep learning neural network (DNN) accelerators and applications," in *ACM SC*, 2017.

[4] A. Chaudhuri et al., "Functional criticality analysis of structural faults in AI accelerators," *IEEE TCAD*, 2022.

[5] A. Chunduri et al., "An effective verification strategy for testing distributed automotive embedded software functions: A case study," in *Int. Conf. Product-Focused Software Process Improvement*, 2016.

[6] Y. He, T. Uezono, and Y. Li, "Efficient functional in-field self-test for deep learning accelerators," in *2021 IEEE International Test Conference (ITC)*, 2021, pp. 93–102.

[7] S. Kundu et al., "Toward functional safety of systolic array-based deep learning hardware accelerators," *IEEE Trans. VLSI*, vol. 29, pp. 485–498, 2021.

[8] S. Kundu and K. Basu, "Detecting functional safety violations in online ai accelerators," in *Proc. IOLTS*, 2022, pp. 1–4.

[9] J. Snoek et al., "Practical Bayesian optimization of machine learning algorithms," in *NIPS*, 2012.

[10] F. et al., "BOHB: Robust and efficient hyperparameter optimization at scale," in *Proc. ICML*, 2018.

[11] P. I. Frazier, "A tutorial on Bayesian optimization," 2018. [Online]. Available: https://arxiv.org/abs/1807.02811

[12] A. Cichocki and A.-H. Phan, "Fast local algorithms for large scale nonnegative matrix and tensor factorizations," *Trans. Fund. Electronics, Communications and Computer Sciences*, vol. E92.A, pp. 708–721, 2009.

[13] A. C. Müller and S. Guido, Eds., *Introduction to Machine Learning with Python*, 2016.

[14] I. Goodfellow et al., "Generative adversarial nets," in *NIPS*, 2014.

[15] F. Nogueira, "Bayesian Optimization: Open source constrained global optimization tool for Python," 2014–. [Online]. Available: https://github.com/fmfn/BayesianOptimization

[16] N. Stander and K. J. Craig, "On the robustness of a simple domain reduction scheme for simulation-based optimization," *Engineering Computations*, vol. 19, pp. 431–450, 2009.

979-8-3503-4631-2/23 $31.00 © 2023 IEEE

Expanding a Pool of Functional Test Sequences to Support Test Compaction

Irith Pomeranz
School of Electrical and Computer Engineering
Purdue University
West Lafayette, IN 47907, U.S.A.
E-mail: pomeranz@ecn.purdue.edu

Abstract—**When a pool of functional test sequences is created for simulation-based verification of a design, the same sequences can be used as manufacturing tests to complement scan-based tests. Without otherwise changing the pool, earlier compaction procedures selected sequences or subsequences as manufacturing tests from a given pool. This article is based on the new observation that the pool can be compacted further if it is first expanded by adding new sequences. New sequences are obtained without performing test generation from pairs of sequences in the pool that reach common states. The new sequences preserve functional properties of sequences in the pool. They also combine fault detection capabilities of sequences from the pool, allowing faults to be detected at earlier clock cycles. The ability of the new sequences to contribute to test compaction is evaluated based on their effect on the clock cycles where faults are detected. The article describes a procedure that applies these concepts to a pool iteratively. Experimental results demonstrate significant reductions in the size of an already-compacted pool for benchmark circuits with pools that contain common states.**

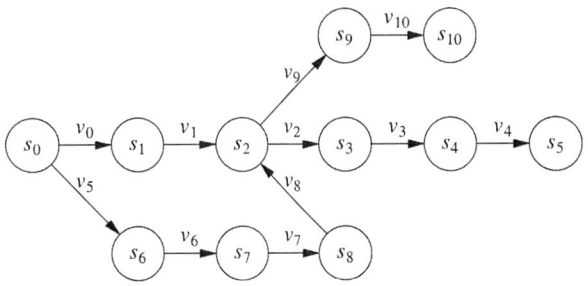

Fig. 1. Functional test sequences.

I. INTRODUCTION

When a pool of functional test sequences is created for simulation-based verification of a design [1], the same sequences can be used as manufacturing tests to complement scan-based tests. Functional test sequences have two advantages. (1) They can detect defects that are manifested only after several clock cycles of at-speed operation [2]. (2) They avoid the non-functional operation conditions because of which scan-based tests may cause a fault-free circuit to exhibit faulty behavior (overtesting) [3]-[5].

Without otherwise changing the pool, the compaction procedures described in [6]-[11] select sequences or subsequences as manufacturing tests from a given pool. When a subsequence is selected, it is obtained by truncating a sequence from the pool, and using only its first clock cycles. This maintains the functional properties of the sequences in the pool. Other procedures achieve test compaction by concatenating and modifying sequences in the pool [12]-[13]. In this article the functional properties of the pool are preserved during test compaction by only removing or truncating sequences.

To enhance the ability to compact a pool of functional test sequences under these conditions, the article makes the new observation that the pool can be compacted further if it is first expanded by adding new sequences that have the following properties. (1) New sequences are obtained without

performing test generation from pairs of sequences in the pool that reach common states. The use of common states ensures that the new sequences preserve the functional properties of the sequences in the pool. (2) The new sequences combine fault detection capabilities of sequences in the pool, allowing faults to be detected at earlier clock cycles. This allows shorter subsequences to be selected. (3) The ability of the new sequences to contribute to test compaction is evaluated based on their effect on the clock cycles where faults are detected.

Specifically, new sequences are obtained from the pool based on the following observation, illustrated by Figure 1. Let P_i and P_j be two sequences in a pool P. In Figure 1, $P_i = v_0 v_1 v_2 v_3 v_4$ takes the circuit through the sequence of states $s_0 s_1 s_2 s_3 s_4 s_5$, and $P_j = v_5 v_6 v_7 v_8 v_9 v_{10}$ takes the circuit through the sequence of states $s_0 s_6 s_7 s_8 s_2 s_9 s_{10}$. Here, s_0 is the initial state of the circuit for functional operation.

Suppose that at clock cycle a under P_i, and at clock cycle b under P_j, the circuit is in the same state. For a sequence P_i, the state reached under P_i at clock cycle a is denoted by $s_{i,a}$. We have that $s_{i,a} = s_{j,b} = s_c$, where s_c is referred to as a common state. In the case of P_i and P_j from Figure 1, P_i takes the circuit to state s_2 at clock cycle $a = 2$, and P_j takes the circuit to state s_2 at clock cycle $b = 4$. Thus, $s_{i,2} = s_{j,4} = s_2$.

The fact that P_i and P_j meet at a common state $s_{i,a} = s_{j,b} = s_c$ implies that the subsequences of P_i and P_j ending at clock cycles a and b, respectively, are interchangeable in a functional test sequence. This is justified by the fact that a state captures all the information needed about the earlier clock cycles, and the two subsequences contain the

979-8-3503-4631-2/23 $31.00 © 2023 IEEE

same information, captured by s_c. In a similar manner, the subsequences of P_i and P_j starting at clock cycles a and b, respectively, are interchangeable. This can be used for defining two new functional test sequences based on P_i and P_j, without performing test generation. The new sequences combine the subsequence of P_i (P_j) that ends at clock cycle a (b) with the subsequence of P_j (P_i) that starts at clock cycle b (a). The new sequences are denoted by $P_{i,a,j,b}$ and $P_{j,b,i,a}$, respectively.

Considering Figure 1 as a partial state diagram, and the functional test sequences as paths in the partial state diagram, the sequences $P_{i,a,j,b}$ and $P_{j,b,i,a}$ are defined by paths in the partial state diagram similar to P_i and P_j. In Figure 1, the new sequences are $P_{i,2,j,4} = v_0 v_1 v_9 v_{10}$ and $P_{j,4,i,2} = v_5 v_6 v_7 v_8 v_2 v_3 v_4$.

This article describes a procedure that computes new sequences based on common states of sequences in the pool. Generation of new sequences requires only logic simulation to find common states. The use of common states ensures that the new sequences preserve functional properties of the sequences in the pool. The new sequences obtained from a pair, P_i and P_j, combine some of their fault detection capabilities. As a result, fault detections may occur earlier under the new sequences, $P_{i,a,j,b}$ and $P_{j,b,i,a}$, than under P_i and P_j. The procedure evaluates each new sequence to decide whether or not it should be added to the pool. After expanding the pool, the procedure compacts the pool by selecting sequences and subsequences, using some of the new sequences and their subsequences. The procedure applies this process iteratively. Whereas each new sequence uses a single common state, after several iterations, it is possible to obtain sequences that use several common states.

The concept of expanding a pool to help compact it was not considered earlier. The procedure described in this article is not limited to a specific source for the pool of functional test sequences. Functional test sequences can also be obtained from test programs under software-based self-test [14]-[23]. In this context, the data registers may assume arbitrary values and prevent common states from being obtained. However, other flip-flops are likely to go through common states. Thus, the results presented in the paper are relevant to a processor in two ways. (1) By considering two states to be equal if they differ only in the values stored in data registers. (2) When considering functional units that do not contain data registers.

Experimental results for benchmark circuits are presented using a pool that is not optimized by sequential test generation as manufacturing tests. This is the case with pools that target simulation-based design verification. The results demonstrate the ability to find common states for different sequences, and significant reductions in the size of an already-compacted pool when new sequences are added based on common states. For completeness, experimental results are also included for a pool produced by sequential test generation.

The article is organized as follows. Section II describes a procedure for expanding a pool of functional test sequences with new sequences based on common states. Section III describes a procedure that compacts a pool by selecting

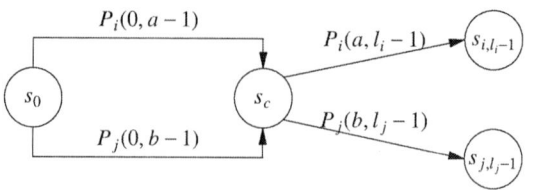

Fig. 2. Common state and new sequences.

sequences and subsequences. Section IV presents experimental results for benchmark circuits. Section V concludes the article.

II. EXPANDING THE POOL

This section describes a procedure that accepts a pool $P = \{P_0, P_1, ..., P_{N-1}\}$ of functional test sequences. The procedure finds common states and adds new sequences to the pool to support further test compaction.

A. New Sequences

The following notation is used for describing the derivation of new test sequences. The length of the sequence $P_i \in P$ is denoted by l_i. The primary input vector at clock cycle a of P_i is denoted by $v_{i,a}$. The subsequence of P_i that starts at clock cycle a and ends at clock cycle b is denoted by $P_i(a, b) = v_{i,a} v_{i,a+1} ... v_{i,b}$.

Logic simulation is applied to the sequences in P starting each sequence from the initial state of the circuit for functional operation. Logic simulation yields the states that the circuit traverses. For a sequence $P_i \in P$, the state at clock cycle a is denoted by $s_{i,a}$.

Considering all the pairs of sequences in the pool, a common state is defined based on two sequence, P_i and P_j, and two clock cycles, a and b. It is possible to use $i = j$. In this case it is required that $a \neq b$. A common state s_c is obtained based on P_i, a, P_j and b, such that $i \neq j$ or $a \neq b$, if $s_{i,a} = s_{j,b} = s_c$.

With $s_{i,a} = s_{j,b} = s_c$, the following subsequences of P_i and P_j are important. The subsequences are shown in Figure 2. The subsequences $P_i(0, a - 1)$ and $P_j(0, b - 1)$ take the circuit from its initial state to state s_c. In a functional test sequence, these two subsequences are interchangeable because the information provided by both is captured by s_c.

The subsequences $P_i(a, l_i - 1)$ and $P_j(b, l_j - 1)$ are applied starting from the same state, s_c. In a functional test sequence, these two subsequences are interchangeable because they are applicable starting from the same initial state.

Based on this discussion, we have that $P_i = P_i(0, a - 1) P_i(a, l_i - 1)$ and $P_j = P_j(0, b - 1) P_j(b, l_j - 1)$. It is also possible to define two new sequences where the pairs of subsequences above are interchanged. The new sequence $P_{i,a,j,b} = P_i(0, a - 1) P_j(b, l_j - 1)$ consists of the subsequence $P_i(0, a - 1)$ of P_i that takes the circuit from its initial state to state s_c, followed by the subsequence $P_j(b, l_j - 1)$ of P_j that takes the circuit from state s_c to the final state of P_j.

The new sequence $P_{j,b,i,a} = P_j(0, b - 1) P_i(a, l_i - 1)$ consists of the subsequence $P_j(0, b - 1)$ of P_j that takes

979-8-3503-4631-2/23 $31.00 © 2023 IEEE   98

the circuit from its initial state to state s_c, followed by the subsequence $P_i(a, l_i - 1)$ of P_i that takes the circuit from state s_c to the final state of P_i.

The procedure described in this article represents the new sequences $P_{i,a,j,b}$ and $P_{j,b,i,a}$ by the quadruples (i, a, j, b) and (j, b, i, a), respectively. All the quadruples obtained based on P are entered into a set denoted by Q.

The number of quadruples in Q can be large. The procedure selects a subset of the quadruples, and adds the corresponding new sequences to P. The selection procedure considers the quadruples from Q in a random order to ensure that different selections are made in different iterations. For parameters $N_{Q,0}$ and $N_{Q,1}$, the procedure evaluates up to $N_{Q,0}$ quadruples from Q. The evaluation criterion indicates whether or not a new sequence should be added to P based on every quadruple. The procedure adds at most $N_{Q,1}$ new sequences to P before the selection stops.

The parameter $N_{Q,0}$ limits the computational effort of evaluating new sequences. The parameter $N_{Q,1}$ ensures that the number of sequences in the expanded pool is not excessive, and the compaction procedure can utilize them effectively.

B. Evaluating a New Sequence

The criterion for the addition of new sequences to P is described next.

The pool P detects a set of target faults denoted by F. Using fault simulation with fault dropping, a fault $f \in F$ is detected by a sequence whose index is denoted by $s(f)$. The clock cycle where $P_{s(f)}$ detects f is denoted by $u(f)$.

To compact the pool, the procedure described in Section III adjusts the variables $s(f)$ and $u(f)$ such that each fault would be detected at the earliest possible clock cycle. For this purpose, the procedure simulates the sequences from the pool again. A sequence P_i is considered with every fault $f \in F$ for which $s(f) \neq i$. The procedure simulates f under P_i until clock cycle $u(f)$. If the fault is detected at a clock cycle $a < u(f)$, the procedure assigns $s(f) = i$ and $u(f) = a$. Using the updated values of $s(f)$ and $u(f)$, the procedure selects sequences and subsequences from the pool.

To evaluate a quadruple $(i, a, j, b) \in Q$, the procedure adds the new sequence $P_{i,a,j,b} = P_i(0, a-1)P_j(b, l_j-1)$ to P as P_n temporarily. It then checks whether P_n reduces the detection clock cycle of any fault $f \in F$ by simulating every $f \in F$ under P_n until clock cycle $u(f)$. If f is detected at a clock cycle $a < u(f)$, the procedure assigns $s(f) = n$ and $u(f) = a$.

After considering all the faults from F, if $s(f) = n$ for any fault $f \in F$, P_n detects the fault f earlier than any sequence in P. In this case, the procedure keeps P_n in P, and increments n. Otherwise, it removes P_n from P.

The procedure evaluates every quadruple $(i, a, j, b) \in Q$ relative to the expanded pool, ensuring that every new sequence added to P is potentially effective for test compaction.

C. Example

For illustration, benchmark circuit $s27$ is shown in Figure 3 and considered next. The pool P consists of six functional test

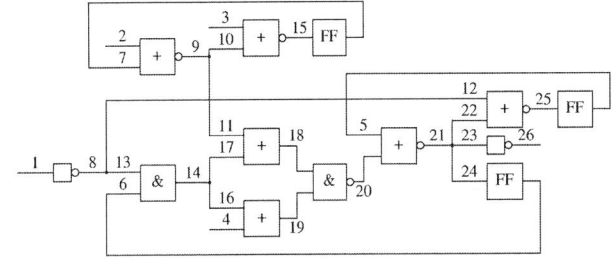

Fig. 3. Benchmark circuit $s27$.

TABLE I
INITIAL DETECTION SEQUENCES AND CLOCK CYCLES

$s(f)$	$u(f)$	faults
0	0	f_{15} f_{22} f_{24} f_{30}
0	1	f_3 f_7 f_{28}
1	0	f_2 f_8 f_{11} f_{23} f_{31}
1	1	f_{13}
2	0	f_0 f_9 f_{20}
2	2	f_{21} f_{27}
3	3	f_1 f_{12}
4	0	f_4
4	3	f_5 f_{16}
5	1	f_{10} f_{17} f_{25} f_{29}
5	2	f_6 f_{14} f_{18} f_{19} f_{26}

TABLE II
ADDING NEW SEQUENCES

n	quadruple	f	$s(f)$	$u(f)$
6	(4,0,4,1)	f_5	4→6	3→2
		f_{16}	4→6	3→2
7	(2,1,3,2)	f_1	3→7	3→2
		f_{12}	3→7	3→2
8	(0,0,2,2)	f_{21}	2→8	2→0
9	(5,1,5,2)	f_6	5→9	2→1
		f_{14}	5→9	2→1
		f_{18}	5→9	2→1
		f_{19}	5→9	2→1
		f_{26}	5→9	2→1
10	(4,1,4,0)	f_{27}	2→10	2→1
11	(4,1,1,0)	f_1	7→11	2→1
		f_{12}	7→11	2→1

sequences of lengths $l_0 = 2$, $l_1 = 2$, $l_2 = 3$, $l_3 = 4$, $l_4 = 4$ and $l_5 = 3$. The pool detects all the 32 single stuck-at faults of the circuit. Table I shows the variables $s(f)$ and $u(f)$ for every fault $f \in F$. For example, the faults f_{15}, f_{22}, f_{24} and f_{30} are detected by P_0 at clock cycle 0.

The set Q consists of 60 quadruples, which are evaluated in a random order. The quadruples that result in sequences, which reduce detection clock cycles of target faults, are shown in Table II. These sequences are added to P. The first such quadruple is (4,0,4,1) that results in the new sequence $P_6 = P_{4,0,4,1} = P_4(0,-1)P_4(1,3)$. The first subsequence is empty in this case, yielding $l_6 = 3$. This sequence reduces the detection clock cycles of two faults.

The second quadruple for which a new sequence is added to P is (2,1,3,2). The corresponding sequence is $P_7 = P_{2,1,3,2} = P_2(0,0)P_3(2,3)$ of length $l_7 = 3$. This sequence reduces the detection clock cycles of two additional faults.

Additional new sequences are added to P as shown in Table II to obtain a pool that contains 12 sequences.

D. Iterative Procedure

The iterative procedure for expanding and compacting a given pool P is summarized as Procedure 1.

Procedure 1: Expanding and compacting a pool P

1) Assign $Q = \emptyset$.
2) For every $0 \le i < n$, $0 \le a < l_i - 1$, $0 \le j < n$, and $0 \le b < l_j - 1$, if $a > 0$ or $b > 0$, $i \ne j$ or $a \ne b$, and $s_{i,a} = s_{j,b}$, add to Q the quadruples (i, a, j, b) and (j, b, i, a).
3) Assign $n_{Q,0} = 0$ and $n_{Q,1} = 0$. As long as $n_{Q,0} < N_{Q,0}$ and $n_{Q,1} < N_{Q,1}$:
   a) Select a quadruple $(i, a, j, b) \in Q$ randomly, remove it from Q, and assign $n_{Q,0} = n_{Q,0} + 1$.
   b) Add the sequence $P_n = P_{i,a,j,b}$ to P.
   c) Use P_n to update detection sequences and clock cycles for the faults in F by calling Procedure 2 with $i = n$.
   d) If $s(f) = n$ for any fault $f \in F$, assign $n = n + 1$ and $n_{Q,1} = n_{Q,1} + 1$; else remove P_n from P.
4) Compact the pool P.
5) If the total length of the sequences in P decreased in the last N_{ITER} iterations, go to Step 1.
6) Stop

Procedure 2: Adjusting detection clock cycles based on P_i

1) For every fault $f \in F$ such that $s(f) \ne i$:
   a) Simulate f under P_i until clock cycle $u(f)$.
   b) If f is detected at a clock cycle $a < u(f)$, assign $s(f) = i$ and $u(f) = a$.

Procedure 1 constructs the set Q in Steps 1 and 2. It evaluates quadruples from Q one by one in Step 3, and adds new sequences to P. In Step 4 it compacts the expanded pool P with the new sequences. The compaction procedure is described in Section III. For a constant N_{ITER}, the procedure terminates after N_{ITER} iterations that do not reduce the total length of the sequences in P.

Applying Procedure 1 iteratively allows the procedure to find sequences that are based on more than one common state. Considering Figures 1 and 2 as partial state diagrams of the circuit, a functional test sequence traces a path through the partial state diagram. A common state, such as s_2 in Figure 1 or s_c in Figure 2, has more than one incoming and outgoing subpaths in the partial state diagram. This creates new paths that can be traced to define new sequences. An arbitrary path through the partial state diagram may use more than one common state. Such a path is obtained when Procedure 1 uses one common state in one iteration, and a second common state in a later iteration.

III. Compacting the Pool

This section describes how a pool P is compacted. The procedure described in this section is applied to the initial pool, before expanding it with new sequences. It is also applied after the pool is expanded in every iteration of Procedure 1.

Fault simulation with fault dropping of F under P yields a detecting sequence $s(f)$ and a detection clock cycle $u(f)$

TABLE III
Adjusting Sequence Lengths

f	$s(f)$	$u(f)$	$l_{s(f)}$	detected
f_5	6	2	$l_6 = 3$	9
f_1	11	1	$l_{11} = 2$	20
f_3	0	1	$l_0 = 2$	23
f_{10}	5	1	$l_5 = 2$	25
f_{13}	1	1	$l_1 = 2$	26
f_{14}	9	1	$l_9 = 2$	30
f_{27}	10	1	$l_{10} = 2$	31
f_{21}	8	0	$l_8 = 1$	32

for every fault $f \in F$. The detecting sequences and clock cycles are adjusted by calling Procedure 2 with every sequence $P_i \in P$. This results in a minimal detection clock cycle for every fault. The minimal fault detection clock cycles will result in minimal lengths for the sequences in the compacted pool. Initially, the procedure assigns $l_i = 0$ for every $P_i \in P$.

The procedure considers the faults in F from high to low value of $u(f)$. When a fault $f \in F$ is considered, the procedure uses $s(f)$ and $u(f)$ to assign $l_{s(f)} = u(f) + 1$. This sets the length of $P_{s(f)}$ to $l_{s(f)}$, and ensures that $P_{s(f)}$ will detect f. After updating $l_{s(f)}$, the procedure simulates F under $P_{s(f)}$ with fault dropping. This removes from consideration faults that are detected by $P_{s(f)}$ with its updated length.

By considering the faults from high to low value of $u(f)$, the procedure ensures that the length of every sequence is updated at most once. By considering all the faults that remain in F, the procedure ensures that all the faults will be detected. After obtaining $F = \emptyset$, the procedure removes from P every sequence P_i for which $l_i = 0$. In this case, P_i is not used for detecting any fault.

In the example of $s27$, the procedure considers the faults in the order shown in the leftmost column of Table III. For each fault f, the values of $s(f)$ and $u(f)$ are shown. These values determine the length of $P_{s(f)}$ as shown under column $l_{s(f)}$. Column *detected* shows the number of detected faults after the length of $P_{s(f)}$ is updated. Of the 12 sequences in P, the procedure keeps eight with lengths shown under column $l_{s(f)}$ of Table III. The total length of the sequences in P is reduced from 18 to 16.

In general, it is possible that the total length of the sequences in the pool P will not be decreased after expanding the pool and applying the compaction procedure. In this case, the procedure restores the previous lengths of the sequences, before it attempted to adjust them, and removes the new sequences that were added for expanding the pool.

The compaction procedure uses the smallest detection clock cycle $u(f)$ for every fault $f \in F$ to determine the lengths of the sequences in the pool. This yields the following property.
Lemma: After compacting the pool, a fault $f \in F$ is detected either by the shortest subsequence from P that detects it, or by the shortest subsequence from P that detects another fault.

IV. Experimental Results

Procedure 1 was applied to single stuck-at faults and transition faults in benchmark circuits as described in this section.

979-8-3503-4631-2/23 $31.00 © 2023 IEEE

A. Setup

The initial state of the circuit for functional operation is assumed to be the all-0 state. Every sequence in the pool is applied starting from the all-0 state.

The benchmark circuits are such that the single stuck-at fault coverage achievable for them is at least 50%. This target is not achieved for benchmark circuits with high levels of redundancy, where more than 50% of the faults are sequentially-undetectable. Such circuits are not considered.

The initial pool P consists of 64 sequences of length 512 generated by the procedure from [24]. For many of the benchmark circuits considered, this pool achieves the maximum or close to the maximum achievable single stuck-at fault coverage. The sequences are not optimized as manufacturing tests. This is also the case when functional test sequences that are available from another application, such as simulation-based design verification, are used as manufacturing tests. This is the scenario addressed in this article.

The initial pool is compacted by applying the procedure described in Section III. This ensures that the initial pool to which Procedure 1 is applied is already compacted. Based on the analysis in Section III, the compacted pool consists of minimal subsequences. It is thus suitable for representing the extent to which the pool can be compacted using earlier procedures that do not expand the pool to compact it further. Application of Procedure 1 demonstrates the ability to compact the pool further by first expanding it.

Procedure 1 was applied to the compacted initial pool with the following parameter values. The procedure considers at most $N_{Q,0} = 1024$ quadruples in every iteration. It adds at most $N_{Q,1} = |P|$ new sequences to the pool in every iteration. Thus, the number of sequences in the expanded pool at most doubles in every iteration. Compaction reduces the number of sequences such that it remains approximately the same from one iteration to the next. The bound given by $N_{Q,1} = |P|$ was found experimentally to be effective in ensuring that the procedure is able to utilize the new sequences well for compacting the pool. The procedure terminates after $N_{ITER} = 8$ iterations where the total length of the sequences in the pool does not decrease.

B. Results

The results for stuck-at faults are shown in Table IV. The results for transition faults are shown in Table V. In Table IV, the first row for every circuit describes the initial compacted pool. Up to two additional rows describe the first and last iteration of Procedure 1 where the procedure reduces the total length of the sequences in the pool. In Table V, only the final pool produced by Procedure 1 is reported.

In every case, after the circuit name, column sv shows the number of state variables. Column pi shows the number of primary inputs. Column $iter$ shows the iteration of Procedure 1. Column $quad$ shows the number of quadruples in Q. For the initial pool, columns exp and $comp$ show the number of sequences in the initial compacted pool. For a pool compacted by Procedure 1, column exp shows the number of sequences in the expanded pool, and column $comp$ shows the number of sequences in P after the expanded pool is compacted.

Column max shows the maximum length of a sequence in the pool. Column tot shows the total length of all the sequences in the pool. Column $ratio$ shows the total length as a fraction of the total length for the initial compacted pool. The circuits are arranged from high to low value of this ratio

TABLE IV
EXPERIMENTAL RESULTS (STUCK-AT FAULTS)

circuit	sv	pi	iter	quad	exp	comp	max	tot	ratio	sa	ntime
s5378	179	35	0	0	21	21	496	2783	1.000	74.777	1.00
s35932	1728	35	0	0	15	15	43	310	1.000	89.809	1.00
s38584	1452	12	0	0	57	57	509	18681	1.000	53.588	1.00
aes_core	530	258	0	0	44	44	55	1671	1.000	99.175	1.00
des_area	128	239	0	0	53	53	15	333	1.000	100.000	1.00
systemcaes	670	258	0	0	22	22	506	4625	1.000	94.945	1.00
systemcdes	190	130	0	0	13	13	43	326	1.000	99.985	1.00
wb_dma	523	215	0	0	31	31	505	11028	1.000	70.430	1.00
sasc	117	15	0	0	9	9	482	3065	1.000	97.272	1.00
sasc	117	15	1	46	18	9	481	3059	0.998	97.272	7.36
sasc	117	15	5	4	10	9	479	3040	0.992	97.272	32.56
b05	34	2	0	0	3	3	221	329	1.000	59.306	1.00
b05	34	2	1	1506	6	5	137	324	0.985	59.306	5.37
b05	34	2	3	1436	10	5	137	321	0.976	59.306	16.02
spi	229	45	0	0	27	27	481	8724	1.000	95.322	1.00
spi	229	45	1	1534	54	27	479	8677	0.995	95.322	7.42
spi	229	45	21	206	50	27	454	8099	0.928	95.322	147.99
usb_phy	98	14	0	0	11	11	369	1607	1.000	67.769	1.00
usb_phy	98	14	1	16	13	11	369	1480	0.921	67.769	3.95
b04	66	12	0	0	20	20	46	314	1.000	86.999	1.00
b04	66	12	1	1188	29	23	46	275	0.876	86.999	59.58
b04	66	12	2	696	25	23	46	273	0.869	86.999	97.85
simple_spi	131	15	0	0	14	14	508	3803	1.000	76.916	1.00
simple_spi	131	15	1	1850	28	14	506	3762	0.989	76.916	5.81
simple_spi	131	15	37	106	28	14	432	3199	0.841	76.916	243.60
b07	51	2	0	0	5	5	85	297	1.000	70.921	1.00
b07	51	2	2	1802	10	5	85	289	0.973	70.921	11.75
b07	51	2	22	944	8	4	85	238	0.801	70.921	99.92
s1423	74	17	0	0	23	23	375	3840	1.000	91.485	1.00
s1423	74	17	1	120	43	23	371	2782	0.724	91.485	7.74
s1423	74	17	2	488	26	23	369	2780	0.724	91.485	27.08
b03	30	5	0	0	10	10	47	208	1.000	74.336	1.00
b03	30	5	1	716	20	15	29	185	0.889	74.336	7.00
b03	30	5	6	458	21	15	19	145	0.697	74.336	114.91
i2c	128	17	0	0	18	18	404	2357	1.000	82.371	1.00
i2c	128	17	1	1288	36	20	290	2076	0.881	82.371	4.42
i2c	128	17	8	726	46	23	175	1531	0.650	82.371	43.05
b22	709	33	0	0	58	58	511	21399	1.000	79.950	1.00
b22	709	33	1	400554	116	76	502	19334	0.904	79.953	3.72
b22	709	33	22	72556	128	122	302	11498	0.537	79.961	37.39
b15	447	36	0	0	52	52	512	20008	1.000	70.864	1.00
b15	447	36	1	1407186	104	72	501	19553	0.977	70.864	4.11
b15	447	36	45	159668	128	128	222	9780	0.489	70.864	69.50
b11	30	8	0	0	7	7	225	842	1.000	86.134	1.00
b11	30	8	1	1514	14	8	154	770	0.914	86.134	4.48
b11	30	8	6	420	28	19	57	405	0.481	86.134	38.64
b21	494	33	0	0	53	53	505	18144	1.000	78.820	1.00
b21	494	33	1	254800	106	66	505	15815	0.872	78.820	3.82
b21	494	33	27	9622	128	117	270	8369	0.461	78.820	49.65
b20	494	33	0	0	55	55	504	19045	1.000	82.949	1.00
b20	494	33	1	278460	110	73	495	16308	0.856	82.949	3.93
b20	494	33	59	8396	128	122	285	8369	0.439	82.949	93.00
b09	28	2	0	0	11	11	271	1372	1.000	81.905	1.00
b09	28	2	1	10276	22	12	154	999	0.728	81.905	6.31
b09	28	2	5	748	15	14	105	559	0.407	81.905	71.05
b14	247	33	0	0	34	34	498	9766	1.000	80.102	1.00
b14	247	33	1	97788	68	41	477	7341	0.752	80.102	3.26
b14	247	33	9	3554	83	74	175	3652	0.374	80.102	24.38
tv80	359	13	0	0	52	52	512	17715	1.000	56.734	1.00
tv80	359	13	1	748788	104	64	510	17274	0.975	56.734	3.86
tv80	359	13	342	38086	127	109	115	3819	0.216	56.964	379.71
s953	29	16	0	0	13	13	402	2174	1.000	98.703	1.00
s953	29	16	1	146520	26	14	279	1566	0.720	98.703	7.39
s953	29	16	13	8690	44	41	22	460	0.212	98.703	281.04

for the last iteration reported for every circuit.

In Table IV, column *sa* shows the single stuck-at fault coverage. In Table V, column *trans* shows the transition fault coverage. Column *ntime* shows the cumulative runtime, divided by the runtime for fault simulation with fault dropping of the initial compacted pool. This is referred to as the normalized runtime.

The following points can be seen from Tables IV and V. Different benchmark circuits benefit from the application of Procedure 1 to different extents. For the circuits at the beginning of Tables IV and V, the total length of the sequences in the pool is not reduced. For the circuits that appear later in Tables IV and V, the reduction is significant.

Based on column *quad*, large numbers of common states are obtained even for the circuits with the largest numbers of state variables. Moreover, there is no reduction in the number of common states when the number of state variables is larger.

The results for transition faults are similar to those obtained for stuck-at faults. The differences result from the fact that transition faults have more detection conditions than stuck-at faults. As a result, the transition fault coverage is lower than the stuck-at fault coverage. The pool for transition faults may be larger than the pool for stuck-at faults because of the additional detection conditions, or smaller because of the lower fault coverage.

The normalized runtime for the first iteration does not increase with the size of the circuit. Thus, the procedure scales similar to a sequential fault simulation procedure.

C. Additional Results

Although the scenario targeted in this article is that where a pool of functional test sequences is not produced by sequential test generation, Procedure 1 is applicable in such a case. To demonstrate this point, Procedure 1 is applied to a pool that was generated by sequential test generation, and compacted by the procedure from [25]. The procedure from [25] omits unnecessary test vectors as well as entire sequences that are not necessary.

The sequential test generation procedure, as well as the procedure from [25], use the all-unspecified state as the initial state of the circuit. For consistency, Procedure 1 is applied with the same initial state. Quadruples (i, a, j, b) and (j, b, i, a) are defined when $s_{i,a}$ and $s_{j,b}$ are compatible.

The results of Procedure 1 when applied to the pool from [25] are shown in Table VI. Only the final pool produced by Procedure 1 is shown in Table VI for circuits that are available from [25], and for which Procedure 1 reduces the total length of the sequences in the pool.

The pool from [25] is significantly more compact than the pool generated by the procedure from [24]. When Procedure 1 is applied to the initial pool from [25] without expanding it, it is typically unable to reduce the total length of the sequences it contains. Nevertheless, by expanding the pool, Procedure 1 is able to compact the pool further in several cases. The level of additional compaction in Table VI is not as high as in Table IV because of the difference in the initial pool. Table IV is

TABLE V
EXPERIMENTAL RESULTS (TRANSITION FAULTS)

circuit	sv	pi	iter	quad	exp	comp	max	tot	ratio	trans	ntime
s5378	179	35	0	0	33	33	497	6800	1.000	68.244	1.00
s38584	1452	12	0	0	63	63	509	26760	1.000	41.305	1.00
aes_core	530	258	0	0	47	47	207	2698	1.000	99.195	1.00
wb_dma	523	215	0	0	35	35	505	12990	1.000	58.988	1.00
sasc	117	15	5	16	30	29	479	6307	0.992	80.875	21.61
des_area	128	239	2	238	116	58	24	651	0.991	100.000	26.77
systemcdes	190	130	1	2858	54	27	73	716	0.986	99.660	83.11
s35932	1728	35	1	1214	44	22	66	632	0.983	87.211	5.80
systemcaes	670	258	3	1828	83	42	506	6262	0.951	80.641	31.85
usb_phy	98	14	2	590	31	29	294	3910	0.944	49.173	12.36
b07	51	2	3	5278	16	8	89	530	0.935	46.798	15.06
b22	709	33	4	667358	128	119	509	25069	0.929	66.132	10.16
spi	229	45	41	244	76	38	478	13096	0.906	82.679	171.60
b20	494	33	12	517156	128	128	503	22442	0.890	70.110	28.58
b21	494	33	17	464740	128	128	503	22370	0.878	64.845	37.26
b15	447	36	16	1480900	128	128	492	19917	0.845	43.145	34.68
b05	34	2	3	5850	20	15	134	586	0.823	38.206	36.35
simple_spi	131	15	36	266	29	21	407	4733	0.784	55.864	152.45
s1423	74	17	6	1212	41	38	360	4714	0.684	78.285	71.38
b03	30	5	3	948	34	18	23	188	0.608	56.771	13.17
b14	247	33	99	55566	128	128	255	8168	0.414	69.744	126.05
i2c	128	17	19	3302	82	43	173	2627	0.407	67.203	82.31
b11	30	8	11	1432	41	30	60	593	0.301	71.475	52.11
tv80	359	13	44	37626	128	88	241	5028	0.288	25.480	45.97
b04	66	12	7	2422	46	40	59	516	0.267	79.904	47.01
b09	28	2	10	2468	26	20	97	784	0.279	70.501	76.76
s953	29	16	11	17534	100	60	23	773	0.177	93.442	82.20

TABLE VI
POOL PRODUCED BY SEQUENTIAL TEST GENERATION

circuit	sv	pi	iter	quad	exp	comp	max	tot	ratio	sa	ntime
b03	30	5	5	6138	22	18	28	166	0.954	73.894	270.25
s1423	74	17	1	28730	32	15	311	828	0.952	93.333	9.22
b11	30	8	14	9758	14	11	354	442	0.923	92.195	2134.19
s35932	1728	35	2	7556	24	10	48	249	0.915	89.784	13.68
usb_phy	98	14	7	51916	26	13	334	1394	0.841	71.231	45.17

more representative of the scenario for which Procedure 1 is designed.

V. CONCLUDING REMARKS

To compact a pool of functional test sequences, earlier procedures selected sequences or subsequences from the pool without otherwise changing the pool. This article made the new observation that the pool can be compacted further if it is first expanded by adding new sequences that have the following properties. New sequences are obtained from sequences in the pool that reach common states without performing test generation. The new sequences exist in a partial state diagram of the circuit, and preserve the functional properties of the sequences in the pool. The new sequences also combine fault detection capabilities of sequences in the pool, allowing earlier fault detection. Moreover, the new sequences can be evaluated based on their ability to reduce detection clock cycles for target faults, and thus, potentially contribute to test compaction. The procedure described in the article based on these concepts expands the pool with new sequences, and then compacts the pool by selecting sequences and subsequences. Experimental results demonstrated significant reductions in the size of an already-compacted pool for benchmark circuits that have pools with common states.

REFERENCES

[1] W. K. Lam, *Hardware Design Verification : Simulation and Formal Method − Based Approaches*, Prentice Hall, 2008.

[2] P. C. Maxwell, R. C. Aitken, K. R. Kollitz and A. C. Brown, "IDDQ and AC Scan: The War Against Unmodelled Defects", in Proc. Intl. Test Conf., 1996, pp. 250-258.

[3] J. Rearick, "Too Much Delay Fault Coverage is a Bad Thing", in Proc. Intl. Test Conf., 2001, pp. 624-633.

[4] J. Saxena, K. M. Butler, V. B. Jayaram, S. Kundu, N. V. Arvind, P. Sreeprakash and M. Hachinger, "A Case Study of IR-Drop in Structured At-Speed Testing", in Proc. Intl. Test Conf., 2003, pp. 1098-1104.

[5] S. Sde-Paz and E. Salomon, "Frequency and Power Correlation between At-Speed Scan and Functional Tests", in Proc. Intl. Test Conf., 2008, Paper 13.3, pp. 1-9.

[6] F. Corno, P. Prinetto, M. Rebaudengo and M. Sonza Reorda, "New Static Compaction Techniques of Test Sequences for Sequential Circuits", in Proc. European Design and Test Conf., 1997, pp. 37-43.

[7] M. Dimopoulos and P. Linardis, "Accelerating the Compaction of Test Sequences in Sequential Circuits Through Problem Size Reduction", in IEEE Trans. on Computer-Aided Design, Oct. 2003, pp. 1443-1449.

[8] M. Dimopoulos and P. Linardis, "Efficient Static compaction of Test Sequence Sets through the Application of Set Covering Techniques", in Proc. Design, Autom. and Test in Europe Conf., 2004, pp. 194-199.

[9] S. Park, L. Chen, P. Parvathala, S. Patil and I. Pomeranz, "A Functional Coverage Metric for Estimating the Gate-Level Fault Coverage of Functional Tests", in Proc. Intl. Test Conf., Paper 27.1, 2006, pp. 1-10.

[10] I. Pomeranz, P. K. Parvathala and S. Patil, "Estimating the Fault Coverage of Functional Test Sequences Without Fault Simulation", in Proc. Asian Test Symp., 2007, pp. 25-32.

[11] H. Fang, K. Chakrabarty, A. Jas, S. Patil and C. Tirumurti, "RT-Level Deviation-Based Grading of Functional Test Sequences", in Proc. VLSI Test Symp., 2009, pp. 264-269.

[12] R. K. Roy, T. M. Niermann, J. H. Patel, J. A. Abraham and R. A. Saleh, "Compaction of ATPG-Generated Test Sequences for Sequential Circuits", in Proc. Intl. Conf. on Computer-Aided Design, 1988, pp. 382-385.

[13] I. Pomeranz, "Restoration-Based Merging of Functional Test Sequences", in IEEE Trans. on Computer-Aided Design, Oct. 2017, pp. 1739-1749.

[14] L. Chen and S. Dey, "Software-Based Self-Testing Methodology for Processor Cores", IEEE Trans. on Computer-Aided Design, Mar. 2001, pp. 369-380.

[15] P. Parvathala, K. Maneparambil and W. Lindsay, "FRITS - A Microprocessor Functional BIST Method" in Proc. Intl. Test Conf., 2002, pp. 590-598.

[16] N. Kranitis, A. Paschalis, D. Gizopoulos and G. Xenoulis, "Software-Based Self-Testing of Embedded Processors" IEEE Trans. on Computers, Apr. 2005, pp. 461-475.

[17] M. Nakazato, S. Ohtake, M. Inoue and H. Fujiwara, "Design for Testability of Software-Based Self-Test for Processors", in Proc. Asian Test Symp., 2006, pp. 375-380.

[18] A. Apostolakis, D. Gizopoulos, M. Psarakis and A. Paschalis, "Software-Based Self-Testing of Symmetric Shared-Memory Multiprocessors", in IEEE Trans. on Computers, Dec. 2009, vol. 58, no. 12, pp. 1682-1694.

[19] S. Di Carlo, G. Gambardella, M. Indaco, I. Martella, P. Prinetto, D. Rolfo and P. Trotta, "A Software-based Self Test of CUDA Fermi GPUs", in Proc. European Test Symposium, 2013, pp. 1-6.

[20] G. Theodorou, N. Kranitis, A. Paschalis and D. Gizopoulos, "Power-aware Optimization of Software-based Self-test for L1 Caches in Microprocessors", in Proc. Intl. On-Line Testing Symp., 2014, pp. 154-159.

[21] P. Bernardi, R. Cantoro, S. De Luca, E. Sanchez and A. Sansonetti, "Development Flow for On-Line Core Self-Test of Automotive Microcontrollers", in IEEE Trans. on Computers, March 2016, vol. 65, no. 3, pp. 744-754.

[22] A. Riefert, R. Cantoro, M. Sauer, M. Sonza Reorda and B. Becker, "Effective Generation and Evaluation of Diagnostic SBST Programs", in Proc. VLSI Test Symp., 2016, pp. 1-6.

[23] A. Riefert, R. Cantoro, M. Sauer, M. Sonza Reorda and B. Becker, "A Flexible Framework for the Automatic Generation of SBST Programs", in IEEE Trans. on VLSI Systems, Oct. 2016, Vol. 24, No. 10, pp. 3055-3066.

[24] I. Pomeranz and S. M. Reddy, "Primary Input Vectors to Avoid in Random Test Sequences for Synchronous Sequential Circuits", in IEEE Trans. on Computer-Aided Design, Jan. 2008, Vol. 27, No. 1, pp. 193-197.

[25] I. Pomeranz, "Modeling a Set of Functional Test Sequences as a Single Sequence for Test Compaction", IEEE Trans. on VLSI Systems, Nov. 2015, pp. 2629-2638.

A guided debugger-based fault injection methodology for assessing functional test programs

Francesco Angione, Paolo Bernardi,
Nicola di Gruttola Giardino
Dip. di Automatica e Informatica, Politecnico di Torino, Turin, Italy
name.surname at polito.it

Davide Appello, Claudia Bertani,
Vincenzo Tancorre
STMicroelectronics, Italy
name.surname at st.com

Abstract—**Functional test programs are increasingly used as flexible approaches to verify device functionality, both online (e.g., Software-Based Self Tests encapsulated in Software Test Library) and during the manufacturing test flow (e.g., System-Level Test). However, a traditional integrated development environment seldom analyzes functional test program weaknesses regarding fault-masking and propagation in CPU registers. Therefore, understanding the presence of errors in the programs due to logic faults at the end of the test program is only assessed by a signature computation, e.g., xor operation between all registers. The presence of program weaknesses in terms of data or control flow is not assessed, and it may lead to fault escapes and/or masking. In particular, those are perfect conditions for proliferating Silent data corruption and unrecoverable errors at the system level.**

This work aims to introduce a guided debugger-based fault injector framework to verify the fault propagation and masking capabilities of functional test programs by eventually catching program errors without relying on time-consuming simulation-based approaches. The fault-free instruction trace is dumped and analyzed to provide information about possible target registers of specific instructions for injecting faults, e.g., in jump instructions, where errors into registers may change the execution flow or not, to verify the presence of silent data corruptions and errors.

The experimental results are carried out on an automotive device from the SPC58 family manufactured by STMicroelectronics, and faults are injected through a script running on Power Debug E40 from Lauterbach.

Index Terms—**Debugger-injector, automotive SoC, fault-injection, functional test programs evaluations, silent-data corruption, errors evaluation.**

I. INTRODUCTION

In the last decades, functional safety requirements dictated by standards such as ISO26262 [1] have improved and become more stringent. As a matter of fact, new approaches have been developed to achieve these requirements safely. DfT-based approaches lack in testing system component interactions and communication peripherals [2] as well as online testing capabilities. Therefore, they have been introduced side by side with functional testing for both online testing and during the manufacturing test flow to overcome such limitations. In addition, due to the shrink of transistors, automotive devices are more susceptible to transient faults [3]. They could manifest at the system level as silent data corruption and/or generation of silent, unrecoverable and non, errors.

Fault simulations grade functional test programs by providing the test metrics representing the strength of such functional test programs to capture faulty behavior [4]. Although fault simulations provide a very precise and effective way of grading functional test programs, they have limitations. A fault simulation based on functional test patterns is achieved by injecting one fault at a time from the fault list and then by applying the functional test patterns; the coverage is computed by observing the faulty behavior on the outputs. Therefore, a crucial and critical limitation of fault simulations is the computing time, which is dramatically rising due to the growing complexity of devices and functional test programs.

Despite such limitations on the grading, functional test programs have become a common approach, especially for online testing [5], and the need for system-level functional test programs is also rising [2]. However, their development and grading are only supported by a fault simulator, which provides the test engineers with only the test coverage without any other information. Understanding the presence of errors in the program due to logic faults at the end of the test program is only assessed by a signature computation, e.g., xor operation between all registers, and if the test program can reach its end. Moreover, it may become cumbersome, especially after functional test programs written in high-level languages are processed by compilers which could introduce additional instructions or change them and their sequence of execution for some optimization reasons. Therefore, the presence of program weaknesses in terms of data or control flow, i.e., situations in which a code snippet does not contribute to the final signature, is not assessed, and it may lead to escapes and masking of faults during a time-consuming fault simulation.

This work aims to provide an alternative to expensive, time-consuming fault simulations during the early development stages of functional test programs by providing a methodology to assess functional test programs directly on the manufactured device.

It is a guided, from an assembly instruction perspective, debugger-based tool to inject transient faults, analyze and help the development of functional test programs in order to overcome limitations of Integrated Development Environment (IDE) on the analysis of program weaknesses, such as identifying data or control flow corruptions, their masking and propagation capabilities.

In such a manner, the number of fault simulations for grading a functional test program can be reduced since an early analysis phase before fault simulation is introduced due to the presence of a low-cost, in terms of time, alternative.

979-8-3503-4631-2/23 $31.00 © 2023 IEEE

However, the flaw of a low-cost alternative is the reduction in controllable sites for the injections (e.g., pipelines control registers are uncontrollable in this work) compared to the capabilities of a fault simulation of injecting faults in every portion of the device.

The proposed fault injection methodology is based on executing a guided fault injector script on a debugger. The script analyzes and injects transient faults into registers of weak instructions; the weak instructions list is generated by an instruction *Connectivity analysis tool* [6] in order to guide the fault injection, and it contains the list of instructions that do not propagate information toward the end of the test program. The *Connectivity tool* analyzes only the propagation of data and control flow at the assembly level of the fault-free execution of a functional test program.

Afterward, that information is used in order to perform faults injections to registers in instructions identified as weak instructions. As the last step, the tool generates an information file that contains each weak instruction if the injected fault leads to the different behavior of the functional test programs in terms of data or control flow, i.e., if the fault injection has generated silent data corruption or an unrecoverable error. That information can be used as feedback to the test engineers developing the functional test programs.

The experimental results are carried out on an automotive device from the SPC58 family manufactured by STMicroelectronics, and faults are injected through a script running on Power Debug E40 from Lauterbach for accessing Software Test Libraries (STL) and System-Level Test (SLT) applications.

The paper provides in Section 2 some background about the addressed issue. Section 3 explains the proposed approach, and Section 4 shows the results obtained from the case study. Section 5 draws some conclusions.

II. BACKGROUND

The following subsections give a general idea about functional test programs and the connectivity metric.

A. Functional Test programs

They are in charge of testing the functional behavior of CPU-based devices [7] either during the mission mode or during the manufacturing test steps. Different approaches have been developed in past years, ranging from Software-Based Self Test [5], [8]–[10], which are collected in Software Test Library (STL) to System-Level Test [2], advanced holistic functional test programs mainly based on operating system bootstrap. The general approach of a functional test program is to exploit CPU-based instruction to compute operations within a set of test data values and assess the correctness of the operation at the end of the test program by means of a signature computation. A signature is an arithmetic accumulation in a single register of a custom subset of the overall registers. In the literature, there exist different approaches for computing the signature:

- CRC-based, they involve CRC algorithms for computing the signature between memory words or registers.

- XOR-based, they are the simplest; the approach consists of exoring the memory words or registers involved.

As Figure 1 depicts, the test program structure for online or manufacturing testing is generally the same. It mainly consists of a data section for storing the initial test data of the program and the final golden signature.

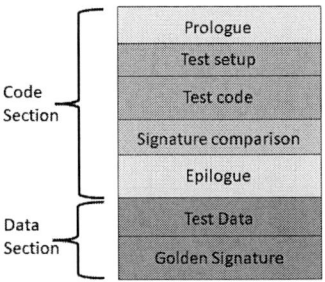

Fig. 1: Generic structure of a Functional Test Program.

The code section is further subdivided into a setup phase to prepare the execution of the test code, and it ends with the signature computation. This part is surrounded by a prologue and epilogue code in order to save the current state of the registers before entering the functional test program.

B. Connectivity metric

The connectivity metric presented in [6] is a strong approximation of lower-level metrics, such as the one representing faults and toggle coverage. However, it is faster to compute, and it can effectively guide the development of test programs, especially during the early stages. Nevertheless, fault simulation is still needed for grading functional or non-functional test patterns. Indeed, low connectivity indicates that some flaws may affect the program, i.e., previously computed results are overwritten, or memory locations are not included in the signatures. This behavior is not beneficial for the fault coverage and grows stout the code footprint.

The connectivity metric is computed on the assembly trace obtained from the execution of a test program. It measures the percentage of assembly instructions that carry their computed values towards the end of the test programs,i.e., in a proper signature. In more detail, a Control and Data flow graph (CDFG) is generated starting from the instruction trace with a vertex for every instruction. Afterward, the temporal dependency of writing and reading operations is analyzed to calculate the percentage of propagated values to the end.

The connectivity analysis tool computes the *Connectivity metrics* that represents how many instructions propagate their computed values towards the program end (green instruction), as it can be seen in Figure 2, in the signature, eventually over the entire executed instructions.

Instructions that do not propagate information toward the end of the test program are marked as black or undecided (for branches without alternatives in the golden instruction trace). For the sake of this work, black or undecided instructions are referred as weak instructions.

Therefore, a program with very high connectivity can be considered promising from the perspective of its potential fault

979-8-3503-4631-2/23 $31.00 © 2023 IEEE

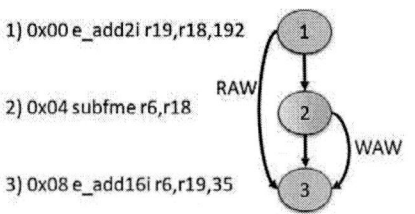

Fig. 2: Control data flow graph generated from connectivity analysis

coverage. Conversely, if the program shows a lot of blocked instructions and a low connectivity, it needs to be revised to avoid useless and time-expensive fault simulation stages.

III. PROPOSED APPROACH

Fault simulations provide a very precise and effective way of grading functional test programs [4], but they have limitations. A crucial and critical limitation of fault simulations is the computing time expensiveness, which is dramatically rising due to the growing complexity of devices and functional test programs.

Therefore, new low-cost alternatives capable of providing preliminary metrics and identifying weaknesses of a functional test program are essential in order to reduce the number of fault simulations during the development of test programs. The injection of transient faults is used in this work to verify the fault resilience capabilities of a functional test program.

This work aims at providing a guided from an assembly instruction perspective, a debugger-based tool to inject transient faults, analyze and help the development of functional test programs in order to overcome the limitations of Integrated Development Environment (IDE) on the analysis of program weaknesses, such as identifying data or control flow corruptions, as Figure 3 summarizes.

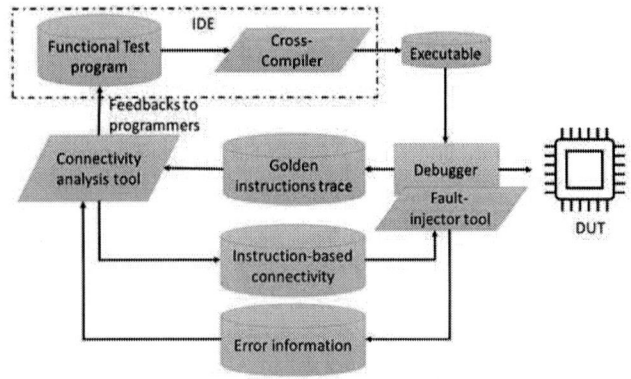

Fig. 3: Proposed development workflow.

The proposed approach is based on injecting transient faults by means of a script running on the debugger on previously identified program weaknesses, in terms of data and control flow, by a *connectivity analysis tool* proposed in [6].

By generating an instruction-based file representing if a given instruction is capable of propagating or not its computed value, fault injections can be performed. Therefore, based on the information related to the propagation of data and control flow for each instruction in a functional test program proposed in [6], the proposed *Injection guided algorithm* performs faults injections to registers in weak instructions in order to classify eventual errors based on the different behavior of the functional test programs.

The temporal flow of the proposed approach in Figure 3 is based, firstly, on a preliminary analysis of the golden instruction trace, from the fault-free execution of the program [6]. Afterward, the identification of weak instructions, such as black or undecided instructions, by the *connectivity analysis tool*, the proposed fault-injector methodology is guided to perform injections on the identified weak instructions. Ultimately, the fault-injector tool generates a report file with errors for each instruction.

Eventually, generated errors or data corruption within registers of weak instructions are saved. That information can be used as feedback to the test engineers developing the functional test programs in the Integrated Development Environment (IDE).

A. Functional Test programs Error Classifications

From a system perspective, errors are discrepancies between the expected output of a program and its actual output. A fault in a register is not always observed or propagated to the program output to capture an erroneous program behavior. As suggested in [11], the error classification used in this work is enhanced and represented in Figure 4.

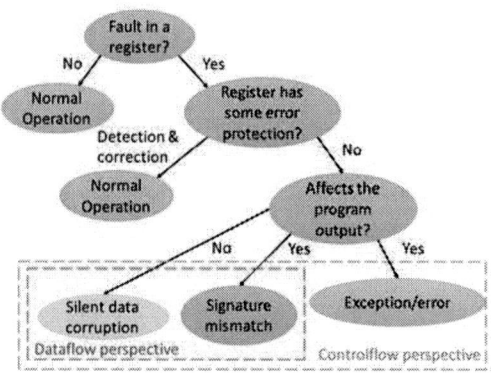

Fig. 4: Errors classification from a system perspective.

Therefore, even by injecting faults into registers, it is not guaranteed that a system failure will be observed.

The most important aspect in Figure 4 is when the program output is affected since there is no error protection at the register level from faults. The expected output of a program can be classified from different perspectives, from a dataflow to a controlflow perspective.

More in detail, errors can be classified according to the focus of the program output and further subdivided:

- On the computed data (from a dataflow perspective), they focus on the correctness of the values within the registers, and they can be classified as:

979-8-3503-4631-2/23 $31.00 © 2023 IEEE

- Silent Data Corruption (SDC), corruption of data invisible from the given CPU perspective (they might be data used by another CPU or peripheral).
- Signature Mismatch (SM), corruption of data into registers captured by signature computation.

- On the flow of instructions (from a controlflow perspective), they focus on the correct flow of executed instruction, depending mainly on branches, and they can be classified as:
  - Silent Data Corruption (SDC), corruption of data invisible from CPU perspective (they might be data used by another CPU or peripheral).
  - Exception/Error (detected unrecoverable error, DUE), triggering of a generic exception handler or system error handler. In case redundant-registers information is stored in a different memory of the device, it is considered a Detected recoverable error.

The instruction type affects the nature of the error, e.g., a branch falls into a controlflow perspective. Meanwhile, the color, from the *Connectivity analysis* of arithmetic instruction, determines the error outcome, e.g., silent data corruption (black instruction) or signature mismatch (green instruction), by the construction of the CDFG graph.

B. Guided injection methodology

The core idea of the guided injection methodology is based on injecting transient faults in weak instructions identified by the connectivity analysis tool. In this manner, the on-chip test program behavior is observed and registered. Figure 5 represents the methodology used hereinafter for the fault injection into the CPU registers.

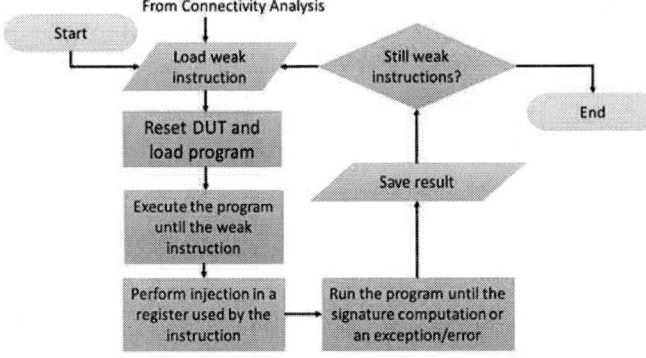

Fig. 5: Fault injector methodology.

The methodology is based on executing the functional program from the start every time an injection is performed. It stops until a given address, corresponding to a weak instruction, is reached. Afterward, it injects a transient random fault into one of the source registers used by the targeted instruction. Consequently, it restores the normal execution of the functional test program. At this point, the signature check is reached and returns a mismatch with the golden signature, or the test program falls into an error handler or an exception. Then, it saves the type of error. The methodology continues until all the weak instructions have been injected with transient faults.

C. Example on injection

The scope of this section is to illustrate possible scenarios during the guided fault injections and how they are treated from the system error classification perspective. The starting point is invariably the CDFG generated from the *connectivity analysis tool* from the golden instruction trace as represented in Figure 6.

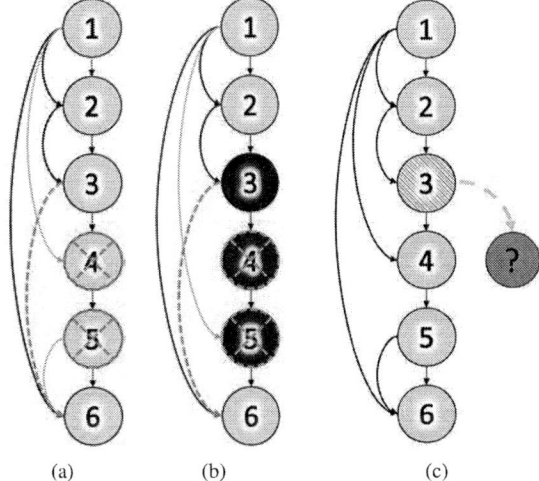

Fig. 6: Analyzed instruction execution traces and possible scenarios.

By supposing to inject a fault into a register used in conditional instruction as instruction in node 3 in all CDFGs in Figure 6 in order to branch to a different address already present in the golden execution trace (Figures 6(a) and 6(b) representing if statement where the instruction within the if statement could contribute to the final signature or not) or not(Figure6(c) representing if-then-else statement, where in the golden instruction trace only a given branch of the if-then-else is taken and dumped in the trace). It generates different situations:

- SM as in Figure 6(a), when the instructions (4,5) in the if-statement contribute to the end signature, and they belong to the golden instruction trace.
- SDC as in Figure 6(b), when the instructions (4,5) in the if-statement do not contribute to the end signature, and they belong to the golden instruction trace.
- DUE as in Figure 6(c), when the next instruction causes a system exception or error and does not belong to the golden instruction trace. On the other hand, when a system exception or error is not generated and the faulty instruction trace, at some point, reconverges to the golden execution trace, depending on the previously executed instruction, it can cause an SM if they affected the end signature in some way, or an SDC if they did not.

Moreover, it is also important to highlight that an injection into a green instruction is going to cause a Signature mismatch by construction of the CDFG. On the other hand, an injection into a black instruction is going to generate Silent Data Corruption.

979-8-3503-4631-2/23 $31.00 © 2023 IEEE

TABLE I: Fault injection campaign on Functional Test program.

Test Program	Execution Time [ms]	Connectivity [%]	Connectivity Grading Time[s]	Fault Coverage [%]	Fault Simulation time [h]	Impr. Connectivity [%]	# Injections (weak instructions)	Tot. Fault Injection time [s]
Adder	0.018	91.39	40	92.56	24.39	93.80	53	0.424
Multiplier	0.038	68.96	48.45	92.3	46.09	81.55	336	0.287
Floating Point	0.435	96.4	680.58	90.78	1407.11	97.79	532	0.409
Shifter	0.040	92.38	80.23	86.19	39.86	95.06	106	0.823
Count-zeros	0.021	94.49	42.89	86.81	12.19	95.19	47	0.368
Bit-wise Logical	0.002	97.63	5.63	95.00	5.65	98.31	2	0.015
Load-Store	0.023	53.28	40.54	91.85	11.20	62.41	474	0.370
Branch Target Buffer	0.173	68.45	133.54	71.16	66.83	78.11	886	0.703
RTOS v1	1.171	75.61	1108.43	NA	7,154.92 (est.)	76.63	5,151	7,688.13
RTOS v1 + canary	2.599	76.91	1349.97	NA	10,655.26 (est.)	77.60	5,684	7,900.76
RTOS v2	1.634	88.42	1430.82	NA	9,982.67 (est.)	88.46	6,605	9,643.3
RTOS v2 +canary	2.754	89.33	1546.50	NA	11,293.39 (est.)	89.37	5,786	9,084.02
RTOS v2 +canary + Loop opt.	2.985	88.41	1550.67	NA	12,239.33 (est.)	88.43	6,632	2,573.21
RTOS v2 +canary + L/S opt.	2.977	88.41	1546.54	NA	12,206.75 (est.)	88.43	6,607	2,444.59
RTOS v2 +canary + Loop unrolling	2.264	86.46	1175.93	NA	9,281.56 (est.)	86.48	6,195	1,920.45
RTOS v2 +canary + Reduced register live range	2.970	88.56	1542.64	NA	12,176.00 (est.)	88.58	6,494	2,467.72
RTOS v2 +canary + Min distance variable L/S	2.977	88.42	1546.50	NA	12,206.45 (est.)	88.44	6,605	2,509.90
RTOS v2 +canary + all opt.	2.439	85.99	1266.96	NA	10,000.07 (est.)	86.01	6,474	2,524.86

IV. EXPERIMENTAL RESULTS

The approach proposed in this work is applied to an industrial automotive case study and resorts to the Power Debug E40 debugger from Lauterbach for fault injections.

A. Experimental setup

The Device Under Test (DUT), in Figure 7 is an automotive SoC belonging to the SPC58 family from STMicroelectronics, ISO26262 ASIL-D compliant. The SoC has a multicore architecture with three 32-bit cores using the PowerPC Variable-Length Encoding (VLE) instruction set. It has 6Mbyte of Flash memory and 128Kbyte of general-purpose SRAM.

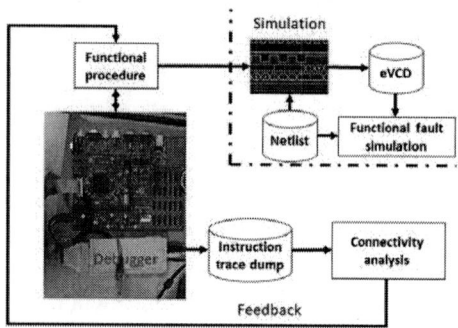

Fig. 7: Experimental setup.

The experimental setup described in Figure 7 is divided into a simulation part and a development part on the manufactured device.

Fault-injection campaigns are performed in previously developed Software Test Library [12] [8], written in assembly code, for online testing of CPU modules. System-level Test applications are based on booting a Real-Time Operating System (RTOS), written in high-level language and assembly, *Micrium C OS-III* [13] in a different version, with and without the canary on the task stack (i.e., a signature computation between CPU registers) and different compiler optimizations (e.g., loop optimizations, register live range, load/store optimizations). Regarding the fault simulation, only STL are fault simulated for the stuck-at models on the targeted CPU modules, while RTOS-based programs are unfeasible to fault simulate.

B. Fault injection campaign

Previously developed functional test programs in STL, and different flavors of the RTOS are analyzed by the *Connectivity analysis tool* to identify weak instructions for guiding the fault injection campaign accordingly. In addition, information after the faults injections campaign is further used from the *Connectivity analysis tool* to remove undecided branch instructions in the CDFG and improve the connectivity metric.

Table I reports the execution time for every functional test program, the *connectivity metric*, and its grading time, which include the generation processes of the instruction trace and the analysis time. Moreover, for STL, it is presented their Fault Coverage on a given CPU unit. Meanwhile, for the SLT application, it is presented an estimated fault simulation time over the whole CPU.

In addition, Table I is presented the improved Connectivity post optimizations on test programs, the number of injections represents the overall weak instructions identified by the connectivity analysis tool, and they are used as targeted instructions in which fault is injected. On the other hand, the total injection time represents the amount of time for injecting transient faults in all the weak instructions by the debugger (i.e., reaching the weak instruction for injecting the fault and continuing the execution of the test program). It depends on the number of executed instructions for a given test program and if it generates an error (if not, the program has to be executed, increasing the injection time).

The fault injection campaign permits eliminating, where it is possible in the assembly programs of STL, weak instructions that do not contribute to the signature. Since those weak instructions are within the scope of the test program, they can be safely removed. Consequently, it directly impacts the Connectivity metric, the code size(average code size reduction for STL around 6%), and the fault simulation time while maintaining the same fault coverage. Moreover, all the arithmetic instructions identified by the connectivity analysis tools are not entirely removed since some synchronize the CPU pipelines, memory accesses, and status registers. It is crucial to mention that improvements in STL in terms of code size and the number of executed instructions have been possible due to the presence of a low-cost and less detailed alternative to fault simulation. More in detail, the fault-injection performed in seconds, combined with the *connectivity analysis tool*,

979-8-3503-4631-2/23 $31.00 © 2023 IEEE

TABLE II: System-Error rate for System-Level Test program.

SLT program	Improved Connectivity [%]	# Executed instructions	# Weak instructions	DUE rate [%]	SM rate [%]	SDC rate [%]
RTOS v1	75.65	28,736	5,151	0.04	0.00	99.96
RTOS v1 + canary	76.93	37,998	5,684	45.57	30.95	23.49
RTOS v2	88.46	37,094	6,605	9.81	0.00	90.19
RTOS v2 + canary	89.37	40,093	5,786	12.69	81.56	5.76
RTOS v2 + canary + Loop opt	88.43	40,201	6,632	42.60	45.28	12.12
RTOS v2 + canary + L/S opt	88.43	40,094	6,607	42.60	45.28	10.38
RTOS v2 + canary + Loop unrolling	86.48	30,486	6,195	47.34	44.71	7.94
RTOS v2 + canary + Reduced register live range	88.58	39,993	6,494	43.96	45.49	10.55
RTOS v2 + canary + Min distance variable L/S	88.44	40,093	6,605	44.53	45.07	10.40
RTOS v2 + canary + all opt	86.01	32,846	6,474	44.79	44.73	10.47

allowed to spot instructions masking or not propagating faults to the final signature computation and directly impacting the final fault coverage. On the other hand, this could have been possible from a fault simulation perspective but in a widened time frame (hours of fault simulation vs. seconds of fault injection). In the development flow, introducing a new assessment methodology before the fault simulation allows to reduce the number of fault simulations.

Regarding SLT applications (RTOS-based applications), a fault injection campaign leads to a slight increase in connectivity metrics due to the presence of undecided branches without alternatives in the execution trace since RTOS is built in such a way to have error control flow during its execution. On the other hand, since it is written mainly in high-level code, there is less space of maneuver for removing weak arithmetic instructions that, in this case, potentially do not contribute to the signature computation but could affect other cores behaviors by uploading computed values into shared memory. Therefore, a possible solution is to play with the compiler optimization goals to change the underlying assembly code.

At a glance, regarding the STL, the proposed approach allows to reduce the number of executed instructions (impacting at the same time code size and execution time) for already developed test programs (STL) and helps the development of new functional test programs; for example, by adding signature computation (RTOS-based test programs for SLT) or changing compiler optimizations affecting the number of weak instructions. Regarding single-thread fault simulations for STL, the main important aspect downstream of the optimizations is the reduced execution time (on average 20% less). Meanwhile, the fault coverage is preserved and, in some cases, improved. For SLT applications, as Table I shows, it is unfeasible to execute a fault simulation in the early development stages. However, in order to have feedback on SLT test applications, the proposed approach can guide their development and provide a different assessment methodology before time-consuming fault simulations.

C. System-Error rate

Hereinafter, errors and exception rates from a system point of view are analyzed in order to understand how the functional test programs behave due to fault injection. System-error rates are calculated only for SLT applications because they could be meaningful from a system assessment perspective. On the

other hand, test routines in STL are only focused on dataflow propagation within their scope.

Moreover, system-error rates are calculated by injecting into weak instructions, seeing that an injection into a green instruction is going to generate a system error thanks to how the CDFG is built.

A given rate is calculated with the following formula:

$$Rate = \frac{Faulty\_Behavior(SM/DUE/SDC)}{TotInjected_{WEAK\ INSTRUCTION}} \quad (1)$$

Table II reports the calculated rates (SM, DUE, SDS) for each SLT application. As it can be seen, in RTOS v1 and v2 without the canary, the SDC rate is extremely high. It is important to highlight that introducing a canary mechanism significantly impacts the SDC rate reduction, increasing the DUE and SM rates. On the other hand, compiler optimizations play a crucial role in the number of weak instruction and their position in the execution trace. Consequently, they distribute quite evenly error rates between DUE and SM compared to the RTOS version 2 with only the canary enhancement.

An important aspect of Table II is that it shows how the faulty behavior rates (SM/DUE/SDC) are directly dependent on the *Connectivity metric*. More in detail, a high connectivity metric leads to a low SDC rate while having high DUE and SM rates, for example, in the RTOS v2 with the canary enhancement.

Overall, an SLT application with high DUE and SM rates can capture faulty behavior during manufacturing or online testing. Therefore, an SLT application can be considered good as its SM, and DUE rates are as high as possible while it has to minimize the SDC rate.

V. CONCLUSIONS

Functional test programs are based on data and control flow across registers to propagate computed values into a signature computation, which is afterward compared with an offline computed signature. The fault coverage drop in fault simulations can be correlated with code snippets that mask faults or do not propagate them. However, IDEs do not provide an assessing methodology before a time-consuming fault simulation. Therefore, a guided framework for analyzing and injecting faults in code snippets is presented to reduce the number of useless time-consuming fault simulations, improve the quality of functional test programs and provide a low-cost alternative to unfeasible fault-simulation of SLT applications.

979-8-3503-4631-2/23 $31.00 © 2023 IEEE

REFERENCES

[1] "Iso 26262-[1-10], road vehicles – functional safety," 2011.

[2] I. Polian *et al.*, "Exploring the mysteries of system-level test," in *2020 IEEE 29th Asian Test Symposium (ATS)*, 2020, pp. 1–6.

[3] D. Rossi *et al.*, "Multiple transient faults in logic: an issue for next generation ics?" in *DFT*, Oct 2005, pp. 352–360.

[4] P. Bernardi *et al.*, "Fault grading of software-based self-test procedures for dependable automotive applications," in *DATE*, March 2011, pp. 1–2.

[5] A. Paschalis *et al.*, "Effective software-based self-test strategies for on-line periodic testing of embedded processors," *IEEE Transactions on Computer-Aided Design of Integrated Circuits and Systems*, 2005.

[6] F. Angione *et al.*, "An innovative strategy to quickly grade functional test program," in *ITC*, 2022.

[7] Thatte and Abraham, "Test generation for microprocessors," *IEEE Transactions on Computers*, vol. C-29, no. 6, pp. 429–441, June 1980.

[8] P. Bernardi *et al.*, "Development flow for on-line core self-test of automotive microcontrollers," *IEEE Transactions on Computers*, 2016.

[9] P. Bernardi and et al., "On the functional test of the register forwarding and pipeline interlocking unit in pipelined processors," in *Proceedings of the 2013 14th International Workshop on Microprocessor Test and Verification*, ser. MTV '13. USA: IEEE Computer Society, 2013, p. 52–57. [Online]. Available: https://doi.org/10.1109/MTV.2013.10

[10] K. Christou *et al.*, "A novel sbst generation technique for path-delay faults in microprocessors exploiting gate- and rt-level descriptions," in *26th IEEE VLSI Test Symposium (vts 2008)*, 2008, pp. 389–394.

[11] A. Biswas *et al.*, "Computing architectural vulnerability factors for address-based structures," in *ISCA*, 2005.

[12] D. Piumatti *et al.*, "An efficient strategy for the development of software test libraries for an automotive microcontroller family," *Microelectronics Reliability vol. 115*, vol. 115, dec 2020.

[13] J. J. Labrosse, *UC/OS-III, The Real-Time Kernel, or a High Performance, Scalable, ROMable, Preemptive, Multitasking Kernel for Microprocessors, Microcontrollers amp; DSPs*. Weston, FL, USA: Micrium Press, 2009.

979-8-3503-4631-2/23 $31.00 © 2023 IEEE

Refreshing the JTAG Family

Michele Portolan
Univ Grenoble Alpes, CNRS
Grenoble INP, TIMA
Grenoble, France

Martin Keim
Siemens
Digital Industries Software.
Wilsonville, OR

Jeff Rearick
Advanced Micro Devices
Fort Collins, CO

Heiko Ehrenberg
GOEPEL Electronics
Austin, TX

Abstract — Following IEEE Standard Association (IEEE SA) guidelines, actively used standards should be revised every ten years. IEEE 1149.1 was published in 2013, while IEEE 1687 was published in 2014. Hence, both standards were revisited, and for both, it was determined that a refresh would be beneficial to the community. Consequently, working groups were established and are ongoing. Here, we outline their agenda and progress. Since 2014, other standards that based on 1687 were initiated. In this paper IEEE P1687.1, IEEE P1687.2, and P2654 report on their progress. An in-depth update of IEEE P1687.1 was given during the International Test Conference (ITC) in 2022.

Keywords— JTAG, IJTAG, IEEE 1687, IEEE 1687.1, IEEE 1687.2, IEEE 1149.1, IEEE 2654

I. INTRODUCTION

IEEE 1149.1-1990 [1] (JTAG), twice amended and finally revised in 2001 [2] is one of the foundational standards in the Design-for-Test (DFT) community. Despite its success over more than a decade, in the early 2000s, it became clear that the network architecture the standard describes will not scale well with the ever-increasing number of embedded IP. This gave rise to a race between an IEEE 1149.1 [3] refresh and a new standard, specifically designed to access and control embedded instruments (IP), IEEE 1687 [4] (IJTAG). 1149.1 won the race by being published in 2013, while 1687 was published a year later in 2014.

Both standards introduced new languages to describe the network architecture hardware as well as the patterns that come with each IP. Although the two standard working groups tried to align, the outcome was unfortunately not fully consistent, with small variations in the common pattern description language.

As Figure 1 shows, IJTAG draws from both, IEEE 1149.1-2001 and IEEE 1500-2005 [5]. IJTAG was designed to integrate both of those standards in the sense that any design that is compliant, would be describable with IJTAG. Without doubt, this enabled the very fast adoption of IJTAG in industry since it allowed users a no-risk adoption path of IJTAG. In 2022, the revision [6] of IEEE 1500 made this co-existence with IJTAG even smoother.

Realizing the industrial adoption, the opportunities of IJTAG, but also its limitations, two subsequent standards were proposed each addressing one of the limitations of IJTAG. IEEE P1687.1 [7] initiated in 2016 provides means to describe device interfaces that are not the common IEEE 1149.1 TAP controller. IEEE P1687.2 [8] initiated in 2018, extends IJTAG to describe analog IP and analog networks.

IEEE P2654 [9] carries the IJTAG idea forward to the system level, building especially on top of P1687.1. Both working groups put huge efforts into their alignment, so that a device with an IEEE P1687.1 described interface seamlessly integrates into a system, described by IEEE P2654.

IEEE 1838-2019 [10], a standard that describes the test access architecture for 3D devices, chose not to publish their own hardware description language or to mandate one by references. Nonetheless, the working group architected the (serial) access network to be compliant (enough) with IJTAG, so that it is describable [11], but without demanding it.

Without any doubt both the IEEE P1687 refresh initiated in 2022 and the IEEE 1149.1 refresh also initiated in 2022 have their challenges. On the one side, it would be good for the JTAG community to realign both standards, overcoming the incompatibility created in 2013/2014. On the other side, IJTAG must look out for coherence across its refresh and its children, P1687.1 and P1687.2, each having their own list of change requests to the base standard. This will be discussed in Section II.

P1687.1, P1687.2, the analog variant of IJTAG, and P2654 are discussed in Section III. The recently started IEEE P1149.1 refresh is discussed in Section IV. For an in-depth discussion of P1687.1, see for example [12], which describes the initial idea, while [13] presents the latest update as of end-2022.

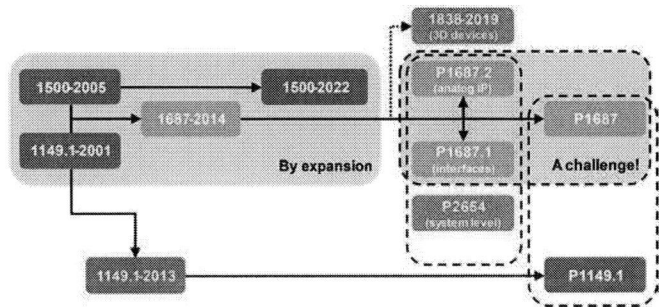

Figure 1: IEEE 1149.1 / IEEE 1687 Family and Friends

II. THE REVISION OF IEEE 1687-2014

IEEE 1687, also known as IJTAG or internal JTAG, was born from the need to operate an ever-increasing number of embedded test IP. Finally released by the end of 2014, it revolutionized the embedded test world, gaining industry acceptance very fast, competing with IEEE 1149.1-2013 and connecting with IEEE 1500-2005. As a natural consequence of the broad acceptance of IJTAG, there is today a great need for updates and upgrades of the standard.

A. IJTAG MEETS WORLD

As the IJTAG standard was finally published by the end of 2014, it had already been subject of academic and industrial research. The best example is the publication of a special edition of the IEEE Design & Test magazine in 2013 [14] featuring five articles centered around IJTAG, one year before the release of the standard. This magazine also showed that the EDA industry already had tools in place prior to the official start of IJTAG. Similarly, academia and industry explored the depth of the standard in many directions. Without hope for completeness and only to name some published work, see e.g. on-chip fault monitoring [15], diagnosis of the IJTAG network [16], access time optimized networks and test scheduling [17]-[20], and network security [21]-[24]. Also, a standard set of benchmark test cases was collected and released [25].

Industrial adoption of IJTAG followed, see e.g. [26]. At the same time, the boundaries of IJTAG were pushed and bent by the desire of the user to apply the IJTAG principles to designs for which it was not meant, see e.g. [27] and [28].

B. REFRESHINBG THE STANDARD

In the early 2020s it became clear that IJTAG became a very successful standard, supported by all EDA vendors, and widely used in industrial applications, especially DFT. It also became clear that it had gaps and must be refreshed to keep-up with the types of designs, design architecture, and usage. Consequently, a PAR for IEEE P1687 was proposed and approved in 2022, commencing the working group.

Two guiding principles were defined right from the start: Firstly, the new IJTAG standard shall retain backward compatibility to IEEE 1687-2014. Only in rare cases and with clear need, shall this be broken. Secondly, the new IJTAG standard shall retain being a descriptive standard. This means for example, it will not describe any specific hardware or software implementing IJTAG. Any given IP example is only for illustration purposes, informatively provided in the standard. This principle of a descriptive standard allowed academia and industry to research and optimize implementation aspects for the benefit of the standard and its users, continuously pushing the standard forward.

C. CATEGORIES OF WORK

The work ahead can largely be divided into three groups:

1. Minor error correction and clarifications

2. Expansion within the existing IJTAG concept and major corrections

3. Expansion that would add new principles and methodologies.

An example of the first group are simple errors like a missing whitespace in the grammar, an incorrect label at a figure, or a request for another example or further explanation of a stated clause. Each one of these are easy to work into the revised standard. It is the number of more than 80 of such reported issues that make these corrections and clarifications time consuming.

The working group has spent most of its time so far on reviewing items of the second group. Besides the topics collected since the initial release in 2014, EDA vendors as well as power users have reported on what they had to do to make IJTAG work for modern, complex designs. Topics in this group include pipelining and clock stretching for faster shift operation and support of large networks, sophisticated reset protocols (the IJTAG one is rather simplistic), the concept of bitfields, ICL to RTL mapping, revisiting call-back and call-back registers, revisiting broadcasting, overshifting scan registers, and non-modeled inputs to the IJTAG network (e.g. from a security IP), just to name a few of the topics.

One particular omission of the initial standard was given to a tiger team: The addition of bidirectional pins and ports. More about this in the next section.

The third group includes a set of topics that ask for tool interaction and introspections. This includes more user control over the result of the IJTAG retargeting engine to influence which IJTAG network path to avoid or select. An example for the latter is defining a preferred input port of a data or scan mux. A further example of user-tool interaction is introspection. This may allow dynamic PDL generation, tailored e.g. to match the actual ICL instance parameters. Another set of topics expands this even further, asking for an API, morphing PDL into a programming language, or adding flow control to PDL (working title "PDL level 0.5"). Interestingly, IEEE P1687.2 already added parts of the latter.

D. SUPPORTING REQUESTS FROM OTHER STANDARDS

The last point illustrates an important problem. A (valid) argument can be made that "PDL 0.5" should be part of the revised base standard and not locked in P1687.2. This way, it could serve all standards that derive from or build onto IJTAG. However, IEEE P1687.1 and P1687.2 are much further ahead in their respective timeline, while P1687 just started. It would be unreasonable for these two standard developments to wait until P1687 is ready and released. The best the working groups can do is to cooperate.

Similarly, both P1687.1 and P1687.2 require the support of bidirectional pins and ports. Also, the former IEEE 1838 working group submitted a set of topics to be added to the IJTAG refresh. Bidirectional pins and ports are also on their list. Again, this impasse can only be overcome by close cooperation across all working groups. In the case of bidirectionals, all three 1687 standards will add wording that is mutually compatible. P1687 / P1687.1 will describe only the digital part of such pins and ports but worded in a way so that P1687.2 can expand from there adding the analog aspect in a consistent manner.

979-8-3503-4631-2/23 $31.00 © 2023 IEEE

E. SHORTENING THE TIME TO RELEASE OF THE 1687 REFRESH

Both examples described in the previous section illustrate the fundamental problem of non-aligned timelines between the three IJTAG standards. Following the common modus operandi of standard development, there are many years between the PAR and the publication of the standards. P1687.1 and P1687.2 are expected to be published years before P1687. Because of interdependencies and concurrent development of the same final solution, this common way of working would be detrimental to the IJTAG family and any other standard that would like to utilize IJTAG. Therefore, the P1687 working group, together with TTSC is exploring ways of accelerating the release of P1687, defining a path for all standards that have a high urgency of a quick release, and continued need for frequent updates. We hope to report on this in a future update.

F. OUTLOOK

Many important upgrades are ahead for P1687. These ensure the continued growth and adoption of the standard for designs to come. Working on this refresh in coordination with related standards also sets the stage for a more universal and coordinated JTAG world going forward. Everyone is invited to participate!

III. COHERENCE ACROSS THE 1687 FAMILY

The previous three subsections point out a key issue: maintaining a coherent approach across multiple standards which each have a unique feature set and language syntax and are being developed in parallel but on different time schedules. This is indeed a challenge, and to understand how it is being addressed it will be instructive to take a quick tour through three of the related standards that appear in Figure 1: P1687.1, P1687.2, and P2654.

A. P1687.1: INTERFACES AND CONTROLLERS

The original 1687 standard anticipated that the IJTAG network would often be connected to an 1149.1 Test Access Port (TAP), and the AccessLink statement explicitly teaches how to do that. In addition, 1687 anticipated that there could be other device interfaces to the IJTAG network, but only left a "generic" placeholder for those, fully expecting that another standard would be needed to address the multiplicity of such interfaces. That is exactly the scope of P1687.1, which is defining the mechanism for converting a stream of retargeted actions at the IJTAG network edge into a corresponding stream of actions which obey the protocol of the selected device interface (e.g. I2C, SPI, MDIO, etc.).

In the spirit of the original 1687, P1687.1 does not prescribe a specific hardware circuit to perform this conversion; in fact, doing so would severely restrict the number of interfaces that could be supported and could exclude future interfaces. Rather, P1687.1 resolves this problem by requiring the IP provider who builds that circuitry (known as the "Transformation Engine" or TE for short) to describe the behavior of the TE in a set of software "transformation procedures" which convert the primitive operations performed by an IJTAG network (capture, shift, update, parallel load, reset, idle, etc.) into the corresponding operations at the device interface (e.g. i2c_write, i2c_read, etc.). These procedures may be invoked as part of a static translation flow or interactively by software agents running in an embedded system (more on that in section C below).

The compatibility of P1687.1 with the rest of the 1687 family is addressed in four important ways. First, the grammar for the primitive operations at the IJTAG network edge is defined to match the actions allowed in 1687 (in Figure 53 in IEEE Std 1687-2014 [4]). Second, the language in which the body of each transformation procedure is written is left to the provider, but the API to call the procedure is standardized in such a way that it can be called from many other languages, including Tcl (which is syntactically compatible with IEEE 1687 PDL). Third, the invocation of these procedures is placed in a request-response messaging framework that is a subset of P2654. And fourth, the "AccessLink Generic" feature of 1687 is leveraged to point to the mechanisms described in 1687.1.

B. P1687.2: ANALOG TEST ACCESS

The original 1687 is strictly limited to digital circuits, but the real world is full of analog and mixed-signal circuitry that would greatly benefit from the concepts of structured test access and retargetable hierarchical patterns that 1687 codifies. Providing those analog extensions is the purpose of P1687.2, which extends PDL with new commands to control (iForceVoltage, iForceCurrent) and observe (iMeasureVoltage, iMeasureCurrent, iMeasureTime) analog values in much the same way that the existing commands (iWrite, iRead) are used for digital values. Similarly, ICL is extended to enable the description of analog instruments (ADCs, DACs, PMUs, etc.) and analog test access mechanisms (e.g. analog test buses).

The trick used in P1687.2 to bridge between analog tests (in PDL) and analog instruments (in ICL) is the introduction of a set of 80 or so analog properties which serve as "requirements" in PDL (e.g. a sine wave with a voltage amplitude and a particular frequency and phase) and "specifications" in ICL (e.g. a signal generator whose voltage and frequency ranges are listed). One of the tasks of the analog retargeting tool is to select instruments whose specs are capable of delivering or capturing the required values in the test. The other tasks of the retargeter are remarkably similar to those in the original 1687: staging the operations in an iApply group, flagging resource constraints, and producing a stream of actions at the edge of the IJTAG network.

Most of the compatibility issues between P1687.2 and the rest of the 1687 family are trivial due to the orthogonality of the original digital and the new analog commands – P1687.2 is largely an augmentation of the existing feature set rather than an alteration of it. However, there is overlap between some changes being considered for the revision of 1687 and features that are planned for P1687.2 (bidirectional ports being the primary example, as discussed above).

979-8-3503-4631-2/23 $31.00 © 2023 IEEE

C. P2654: System Test Access and Manageemnt

The title of the original 1687 reflects is scale: it is aimed at instruments embedded within a semiconductor device. However, as Moore's Law has found a new path to continue its exponential growth through the heterogeneous integration of chiplets into multi-die packages, it is increasingly difficult to draw a crisp boundary between a device and a system. Last decade's printed circuit board is this year's chip, and instruments that used to be accessible from a device interface are now buried deep inside the hierarchy of another device. This, among other reasons, necessitates a system-level view of test access, and that is the driving force behind P2654.

P2654 can be abstractly thought of as a mechanism to send test control and data in the form of serialized messages from one interface to another. At one end is a test controller, and at the other end is the target of a particular test, and there may by an arbitrary number of hops through interfaces in between. At each hop, the payload of the test must be translated into the protocol associated with each interface. A library of transformation procedures exists for each interface, and once the management software determines the route through the tree of nodes to connect the controller to the target, the appropriate procedures are invoked to perform the translations.

If some of this sounds familiar, it is because it is a superset of the approach described above for P1687.1. Specifically, an IJTAG network is a target endpoint of a P2654 node tree, and the device interface in P1687.1 is the "last hop" for P2654.

The approach to ensure compatibility here was fairly obvious: use the P2654 framework to guide the implementation of P1687.1 since the latter is a subset of the former. Interestingly, this added an extra layer (the messaging layer which processes the requests and responses) to P1687.1 that wasn't originally envisioned (since we thought we could manage with only a "content layer" to translate payloads). However, deeper analysis revealed corner cases where multiple hops and multiple interfaces may exist even within a single device, so the more abstract approach for compatibility was essential. The other aspects of compatibility are being addressed primarily through language-neutral approaches for the transformation library.

D. Proceeeding Coherently

The final paragraph in each of the previous three sections described the tactical details of maintaining compatibility across the family of standards. These considerations emerged during working group meetings and were shared across the different parallel teams, which fortunately include several common members. This highlights the main strategy for ensuring coherence across the family of standards: coordination.

There are three aspects to how the teams execute in a coordinated manner. The first is the presence of a high degree of communication, which is greatly facilitated by those common members, but also by the expectation that any issues raised are given a high priority as new business in each of the affected working groups. That leads to the second pillar: a collaborative mindset for each of the teams. Rather than each working group just focusing on the delivery of one standard in isolation, they each take into account the collective offering as a whole. This approach is encouraged and nurtured by our sponsor organization, the Test Technology Standards Committee. That, in turn, leads to the third item: issue resolution. Since most of the common members shared between working groups are also longstanding members of the TTSC, we have developed relationships and have a regular forum to meet, escalate, and resolve cross-team issues. It really does take a village to raise a standard, and we're lucky to have created a community of dedicated professionals committed to building an entire family of standards.

Further evidence of this broad approach can be found in the developments for the revision of the foundational standard in the TTSC portfolio: IEEE 1149.1.

IV. Revision of IEEE 1149.1-2013 - What to Expect?

The last revision of IEEE Std 1149.1 was approved and published in 2013. Therefore, according to the IEEE SA guidelines, the standard was due for a review. Subsequently, it was determined that the standard requires a refresh, and a working group was established in 2022.

Those of us familiar with this JTAG / boundary-scan standard, as IEEE Std 1149.1 is often referred to, will recall that it gained a significant amount of new, albeit optional, capabilities at 2013. What can we expect this time around? Are there significant changes in the works? Is this standard still relevant after 30 years after its original release? Where does IEEE Std 1149.1 fit into the big picture of test-related standards utilized in the industry? This paper attempts to answer these and other questions and solicit feedback from the various stakeholders of IEEE Std 1149.1.

A. Introduction

Every new generation of Integrated Circuits (ICs) features higher complexity and more stringent demands for power consumption than the previous one. Moreover, modern ICs and System-on-Chip (SoC) devices often collaborate and incorporate third-party suppliers' Intellectual Property (IP). As a result, reigning over that complexity requires a well-developed testing and monitoring apparatus.

In modern ICs and SoC devices, this requirement is met through a wide selection of on-chip instrumentation, primarily targeted for chip-level design validation and manufacturing test but also suitable for in-field testing and diagnostics. The standardization of embedded test resources, defined through IEEE Std 1149.1, enables board designers to reduce the amount of physical probe access (test points) required for a comprehensive connectivity test by relying on standard-compliant ICs.

Introduced more than thirty years ago, the primary mission of IEEE Std 1149.1 is still to detect and diagnose manufacturing defects on Printed Circuit Board Assemblies (PCBA). Over

979-8-3503-4631-2/23 $31.00 © 2023 IEEE

these years, IEEE Std 1149.1 received multiple updates. The latest set of significant updates in 2013 focused on device initialization, boundary-scan register (BSR) segmentation, power domain control, and access to chip-embedded instrumentation. While there is substantial overlap with the related IEEE 1687 standard, separate inclusion of access to that chip-embedded instrumentation in the IEEE 1149.1 specification allows designers who already implement the latter in their ICs to facilitate board-level connectivity tests to also provide standardized access to other chip-embedded instrumentation without the need to implement the full scope of IEEE Std 1687 as well.

As stated in the project authorization request (PAR) for this latest IEEE 1149.1 revision, this standard defines a serial digital access methodology and architecture that allows for the ability to control and/or observe data inside an integrated circuit for a variety of applications, including testing, programming, configuring, and debugging IP blocks and logic contained within that integrated circuit.

B. IEEE 1149.1 IS IN REVISION – WHY?

IEEE SA guidelines require active standards to be reviewed at least once every ten years to determine if the standard should remain active and be revised or be classified inactive. The last revision of IEEE Std 1149.1 was balloted and approved in 2013. In 2022 the Test Technology Standards Committee (TTSC) formed a study group to determine if IEEE Std 1149.1 should remain active (and be revised) or become inactive. (Allowing it to become inactive would mean that the standard is still valid and available for purchase and implementation, but it would not be actively maintained for the time being.)

Initially developed in the late 1980s, the first edition of IEEE Std 1149.1 was approved and published in 1990. Clarifications, corrections, and enhancements were incorporated and approved in 1993. Shortly after that, in 1994, the Boundary Scan Description Language (BSDL) was introduced and added to the specification. The next revision was approved and published in 2001, featuring minor changes and clarifications. Lastly, the 2013 revision provided a major expansion of the standard, adding a vast set of new, optional capabilities and defining associated extensions for BSDL. That revision also introduced Procedural Description Language (PDL) to describe component-specific test procedures in the IEEE Std 1149.1 domain.

Seasoned as it is, IEEE Std 1149.1 is still widely implemented and used in most digital and mixed-signal integrated circuits. Board test engineering relies on IEEE Std 1149.1 to detect manufacturing defects on complex, compact board designs. Design engineers benefit from the easy access to debug, configuration, and test resources during design validation and prototyping at chip-level, board-level, and system-level (here, "system-level" refers to an assembly of modules and boards, making it a functionally complete entity). Field service engineers can utilize IEEE 1149.1 access for troubleshooting and re-configuration if the board or system is designed accordingly.

A few related IEEE standards rely on the test access port (TAP) interface and the TAP controller defined in IEEE Std 1149.1. These standards include IEEE 1149.4 [29], IEEE 1149.6 [30], IEEE 1149.8.1 [31], IEEE 1532 [32], IEEE 1687 [4], and IEEE 1838 [10]. Other IEEE standards have been designed to be used in conjunction with IEEE 1149.1, such as IEEE 1500 [6] and IEEE 1581 [33], while IEEE 1149.10 [34] makes intensive use of BSDL and PDL, and IEEE 1149.7 [35] can be considered a superset of IEEE 1149.1. Furthermore, new standard developments are ongoing that likely will at least partially rely on IEEE 1149.1 conformant device capabilities, including IEEE P2654 [9] and IEEE P2929 [36].

With all this in mind, the TTSC study group determined that IEEE 1149.1 should remain active. After soliciting and collecting feedback from the industry, the group submitted a project authorization request (PAR), which was granted in mid-2022. A working group was formed, tasked with revising the document based on feedback received from stakeholders. Most of the feedback calls for clarifications or corrections in some examples in the standard document. The working group expects to propose some improvements to select features, but no major changes or additions are planned for this revision of IEEE 1149.1.

One of the main issues being considered in this revision is that both IEEE 1149.1 and IEEE 1687 define procedural description language (PDL) in their specifications. Still, some PDL keywords have slightly different connotations/implications in the two standards.

Another concern brought to the working group is that chipsets and IPs with embedded TAPs present complications related to the BSDL description when incorporated into an SoC. The BSDL must present only a single bit at the SoC level in the bypass register. But if chiplets and IP with embedded TAPs are daisy-chained inside the SoC, each has its bypass register bit. Tools typically cope with this by representing the various "components" inside the SoC as daisy-chained virtual ICs, each with their associated BSDL file, and mapping those to package pins on the SoC. The question was posed whether BSDL should permit the definition of bypass registers with more than one bit and possibly multiple ID Codes.

A third example of a concern being discussed in the working group relates to the device ID register, specifically the manufacturer ID value specified in the standard. IEEE 1149.1 refers to a manufacturer ID defined in JEDEC standard JEP106. In addition to a seven-bit manufacturer ID value, the 11-bit manufacturer identity code in IEEE 1149.1 has a 4-bit field to indicate the "bank" of manufacturers defined in JEP106. This 4-bit field allows the encoding of 16 banks; manufacturer ID values defined in banks 17 and up (bank M) in JEP106 would pose a problem for the current device ID specification in IEEE 1149.1 in the sense that one would not be able to differentiate them from a manufacturer with the same ID in an earlier bank $(M - 16)$. At the time of this writing, JEP106 is assigning manufacturer IDs in bank 15, which means the above-mentioned problem is expected to become real within the next

979-8-3503-4631-2/23 $31.00 © 2023 IEEE

few years. How big of a problem this potential ambiguity would present is debatable.

C. OUTLOOK

To summarize: a TTSC study group met in early 2022 to determine if a revision is in order or if the standard should become inactive. The study group decided the standard should be revised and submitted a respective PAR. The IEEE P1149.1 working group has been meeting weekly since August 2022. The plan is to address all issues brought forward to the study group or the working group by the end of December 2023 and submit a revised draft for a ballot in 2024.

If interested, join the working group, and participate in the revision process. Please submit any issues with the current version of the standard you are aware of, even if you don't feel like you have the resources to participate in the working group.

IEEE Std 1149.1 is here to stay for the foreseeable future. Expect to see collaboration between different working groups to ensure the interoperability of standardized resources that rely on each other. As technology evolves, IEEE 1149.1 may also need to be revised in the future to be able to satisfy new requirements and remain relevant for state-of-the-art designs.

Those interested in participating in the IEEE 1149.1 working group (WG) are encouraged to contact the WG chair through IEEE https://sagroups.ieee.org/1149-1/.

REFERENCES

[1] IEEE Std 1149.1-1990 - IEEE Standard Test Access Port and Boundary-Scan Architecture, IEEE, USA, 1990, https://standards.ieee.org/ieee/1149.1/1727/

[2] IEEE Std 1149.1-2001 - IEEE Standard Test Access Port and Boundary-Scan Architecture, IEEE, USA, 2001, https://standards.ieee.org/ieee/1149.1/1728/

[3] IEEE Std 1149.1-2013 - IEEE Standard Test Access Port and Boundary-Scan Architecture, IEEE, USA, 2013, https://standards.ieee.org/ieee/1149.1/4484/

[4] IEEE Std 1687-2014 - IEEE Standard for Access and Control of Instrumentation Embedded within a Semiconductor Device", IEEE, USA, 2014, https://standards.ieee.org/ieee/1687/3931/

[5] IEEE 1500-2005 - IEEE Standard Testability Method for Embedded Core-based Integrated Circuits, IEEE, USA, 2005, https://standards.ieee.org/ieee/1500/2238/

[6] IEEE 1500-2022 - IEEE Standard Testability Method for Embedded Core-based Integrated Circuits, IEEE, USA, 2022, https://standards.ieee.org/ieee/1500/7704/

[7] IEEE P1687.1 – Standard for the Application of Interfaces and Controllers to Access 1687 IJTAG Networks Embedded Within Semiconductor Devices, https://standards.ieee.org/project/1687_1.html

[8] IEEE P1687.2 - Standard for Describing Analog Test Access and Control, https://standards.ieee.org/ieee/1687.2/7232/

[9] IEEE P2654 - Standard for System Test Access Management (STAM) to Enable Use of Sub-System Test Capabilities at Higher Architectural Levels, https://standards.ieee.org/ieee/2654/7426/

[10] IEEE 1838 - IEEE Standard for Test Access Architecture for Three-Dimensional Stacked Integrated Circuits, IEEE, USA, 2019, https://standards.ieee.org/ieee/1838/5073/

[11] J. -F. Côté et al., "Affordable and Comprehensive Testing of 3-D Stacked Die Devices," in IEEE Design & Test, vol. 39, no. 5, pp. 17-25, Oct. 2022, doi: 10.1109/MDAT.2022.3191016.

[12] A. Crouch, M. Laisne, M. Keim, "Generalizing Access to Instrumentation Embedded in a Semiconductor Device", Computer Magazine, July 2017.

[13] M. Laisne et al., "IEEE P1687.1: Extending the Network Boundaries for Test," 2022 IEEE International Test Conference (ITC), Anaheim, CA, USA, 2022, pp. 382-390, doi: 10.1109/ITC50671.2022.00084.

[14] IEEE Design & Test, vol. 30, no. 5, pp. 36-43, Oct. 2013

[15] F. G. Zadegan, D. Nikolov and E. Larsson, "On-Chip Fault Monitoring Using Self-Reconfiguring IEEE 1687 Networks," in IEEE Transactions on Computers, vol. 67, no. 2, pp. 237-251, 1 Feb. 2018, doi: 10.1109/TC.2017.2731338.

[16] R. Cantoro, M. Montazeri, M. S. Reorda, F. G. Zadegan and E. Larsson, "On the diagnostic analysis of IEEE 1687 networks," 2016 21th IEEE European Test Symposium (ETS), Amsterdam, Netherlands, 2016, pp. 1-2, doi: 10.1109/ETS.2016.7519294.

[17] F. G. Zadegan, E. Larsson, A. Jutman, S. Devadze and R. Krenz-Baath, "Design, Verification, and Application of IEEE 1687," 2014 IEEE 23rd Asian Test Symposium, Hangzhou, China, 2014, pp. 93-100, doi: 10.1109/ATS.2014.28.

[18] R. Krenz-Baath, F. G. Zadegan and E. Larsson, "Access time minimization in IEEE 1687 networks," 2015 IEEE International Test Conference (ITC), Anaheim, CA, USA, 2015, pp. 1-10, doi: 10.1109/TEST.2015.7342408.

[19] M. A. Ansari, J. Jung, D. Kim and S. Park, "Time-Multiplexed 1687-Network for Test Cost Reduction," in IEEE Transactions on Computer-Aided Design of Integrated Circuits and Systems, vol. 37, no. 8, pp. 1681-1691, Aug. 2018, doi: 10.1109/TCAD.2017.2766146.

[20] P. Habiby, S. Huhn and R. Drechsler, "Optimization-based Test Scheduling for IEEE 1687 Multi-Power Domain Networks Using Boolean Satisfiability," 2021 16th International Conference on Design & Technology of Integrated Systems in Nanoscale Era (DTIS), Montpellier, France, 2021, pp. 1-4, doi: 10.1109/DTIS53253.2021.9505098.

[21] J. Dworak, A. Crouch, J. Potter, A. Zygmontowicz, M. Thornton, "Don't forget to lock your SIB: hiding instruments using P1687", IEEE International Conference (ITC), November 2013.

[22] H. Liu and V. D. Agrawal, "Securing IEEE 1687-2014 Standard Instrumentation Access by LFSR Key," 2015 IEEE 24th Asian Test Symposium (ATS), Mumbai, India, 2015, pp. 91-96, doi: 10.1109/ATS.2015.23.

[23] S. K. K., N. Satheesh, A. Mahapatra, S. Sahoo and K. K. Mahapatra, "Securing IEEE 1687 Standard On-chip Instrumentation Access Using PUF," 2016 IEEE International Symposium on Nanoelectronic and Information Systems (iNIS), Gwalior, India, 2016, pp. 56-61, doi: 10.1109/iNIS.2016.024.

[24] M. Portolan, V. Reynaud, P. Maistri, R. Leveugle and G. Di Natale, "Security EDA Extension through P1687.1 and 1687 Callbacks," 2021 IEEE International Test Conference (ITC), Anaheim, CA, USA, 2021, pp. 344-353, doi: 10.1109/ITC50571.2021.00050.

[25] A. Tšertov et al., "A suite of IEEE 1687 benchmark networks," 2016 IEEE International Test Conference (ITC), Fort Worth, TX, USA, 2016, pp. 1-10, doi: 10.1109/TEST.2016.7805840.

[26] H. Ma et al., "Fast Bring-Up of an AI SoC through IEEE 1687 Integrating Embedded TAPs and IEEE 1500 Interfaces," 2020 IEEE International Test Conference (ITC), Washington, DC, USA, 2020, pp. 1-5, doi: 10.1109/ITC44778.2020.9325251.

[27] M. Baby et al., "IJTAG Through a Two-Pin Chip Interface," 2020 IEEE International Test Conference (ITC), Washington, DC, USA, 2020, pp. 1-5, doi: 10.1109/ITC44778.2020.9325232.

[28] H. M. v. Staudt, M. A. Benhebibi, J. Rearick and M. Laisne, "Industrial Application of IJTAG Standards to the Test of Big-A/little-d devices," 2020 IEEE International Test Conference (ITC), Washington, DC, USA, 2020, pp. 1-10, doi: 10.1109/ITC44778.2020.9325267.

[29] IEEE Std. 1149.4-2010 - IEEE Standard for a Mixed-Signal Test Bus, IEEE, USA, 2010, https://standards.ieee.org/ieee/1149.4/4022/

[30] IEEE Std. 1149.6-2015 - IEEE Standard for Boundary-Scan Testing of Advanced Digital Networks, IEEE, USA, 2015, https://standards.ieee.org/ieee/1149.6/4706/

[31] IEEE Std. 1149.8.1-2012 - IEEE Standard for Boundary-Scan-Based Stimulus of Interconnections to Passive and/or Active Components, IEEE, USA, 2012, https://standards.ieee.org/ieee/1149.8.1/4567/

[32] IEEE Std. 1532-2002 - IEEE Standard for In-System Configuration of Programmable Devices, IEEE, USA, 2002, https://standards.ieee.org/ieee/1532/3366/

[33] IEEE Std. 1581-2011 - IEEE Standard for Static Component Interconnection Test Protocol and Architecture, IEEE, USA, 2011, https://standards.ieee.org/ieee/1581/4212/

[34] IEEE Std. 1149.10-2017 - IEEE Standard for High-Speed Test Access Port and On-Chip Distribution Architecture, IEEE, USA, 2017, https://standards.ieee.org/ieee/1149.10/5786/

[35] IEEE Std 1149.7-2022 - IEEE Standard for Reduced-Pin and Enhanced-Functionality Test Access Port and Boundary-Scan Architecture, IEEE, USA, 2022, https://standards.ieee.org/ieee/1149.7/7703/

[36] IEEE P2929 - Standard for System-level State Extraction for Functional Validation and Debug, https://standards.ieee.org/ieee/2929/10385/

Innovation Practices Track: Silicon Lifecycle Management Challenges and Opportunities

Fei Su	Xiankun (Robert) Jin	Nilanjan Mukherjee	Yervant Zorian
Intel Corporation	NXP Semiconductor	Siemens	Synopsys
U.S.	U.S.	U.S.	U.S.
fei.su@intel.com	robert.jin@nxp.com	n.mukherjee@siemens.com	Yervant.Zorian@synopsys.com

I. Introduction (Fei Su)

We need to address the multifaceted challenges about silicon and system quality throughout the life cycle of silicon-based systems, spanning from design, production to in-field deployment. Innovations spanning across various parts of the silicon ecosystem are needed. In this session we invited industry experts to discuss technical trends and challenges in semiconductor industry driving a pressing need of more innovations in the emerging field of silicon lifecycle management (SLM). They will present the state-of-the-art SLM methodologies and share their perspectives of this emerging field and thoughts of future R&D directions.

II. Silicon Health Monitoring for Automotive Predictive Maintenance (Robert Jin)

Predictive maintenance is gaining traction recently in automotive to deliver enhanced functional safety beyond ASIL-D (ISO 26262) – system availability and fail-operational. The ability to detect and address imminent failures of the large number of complicated electronic devices found in today and future electronic systems (e.g., Robo-taxi) is needed to provide an additional and critical layer of assurance to consumers. The silicon health monitoring and data analytics are key parts of this solution to enable this capability, while providing many additional benefits, e.g., quality improvement through outlier detection, enhanced reliability with unprecedented post-silicon visibility, and design methodology improvement. There is an ongoing effort within the Functional Safety standardization work group to report the state of the art which will become a technical report (TR 9839) with the potential to be included in the 3rd edition of ISO 26262. This talk will review the motivation and challenges from both business and technical perspectives, development of the ISO26262 standard toward predictive maintenance, demonstrate the novel health monitors, show how the solution is being tailored on a system-on-a-chip, and discuss the open points e.g., data ownership and analytics.

III. Hardware/Software Infrastructure to enable Silicon Monitoring throughout the IC Lifecycle (Nilanjan Mukherjee)

Silicon lifecycle solutions provide design augmentation and linked applications that detect, mitigate, and eliminate risks throughout the IC lifecycle. This enables customers to address their debug, test, yield, safety, security, and optimization requirements for today's most complex SoCs. By creating an infrastructure that makes designs more testable, silicon lifecycle management solutions achieve high-quality test, identify defects and hidden yield limiters, and move beyond test into system debug and validation. This ecosystem of tools effectively consumes and analyzes data to provide critical system insights. This data can then be used for in-life monitoring.

IV. The Need for End-to-end SLM Highway (Yervant Zorian)

This talk presents why end-to-end Silicon Lifecycle Management (SLM) "highway" (i.e., infrastructure) is needed. There are a variety of SLM application challenges driving this need, including AVFS (Adaptive Voltage & Frequency Scaling), software defined vehicle architectures, functional safety (FuSa) subsystem for post-silicon test/monitoring, and multi-die system with UCIe and HBM, etc.

979-8-3503-4631-2/23 $31.00 © 2023 IEEE

Special Session: Approximation and Fault Resiliency of DNN Accelerators

Mohammad Hasan Ahmadilivani[1], Mario Barbareschi[2], Salvatore Barone[2], Alberto Bosio[3],
Masoud Daneshtalab[4,1], Salvatore Della Torca[2], Gabriele Gavarini[5], Maksim Jenihhin[1],
Jaan Raik[1], Annachiara Ruospo[5], Ernesto Sanchez[5], and Mahdi Taheri[1*]

[1]Tallinn University of Technology, Tallinn, Estonia
[2]University of Naples Federico II, Naples, Italy
[3]Ecole Centrale de Lyon, Lyon, France
[4]Mälardalen University, Västerås, Sweden
[5]Politecnico di Torino, Torino, Italy

Abstract—Deep Learning, and in particular, Deep Neural Network (DNN) is nowadays widely used in many scenarios, including safety-critical applications such as autonomous driving. In this context, besides energy efficiency and performance, reliability plays a crucial role since a system failure can jeopardize human life. As with any other device, the reliability of hardware architectures running DNNs has to be evaluated, usually through costly fault injection campaigns. This paper explores approximation and fault resiliency of DNN accelerators. We propose to use approximate (AxC) arithmetic circuits to agilely emulate errors in hardware without performing fault injection on the DNN. To allow fast evaluation of AxC DNN, we developed an efficient GPU-based simulation framework. Further, we propose a fine-grain analysis of fault resiliency by examining fault propagation and masking in networks.

Index Terms—deep neural networks, approximate computing, fault emulation, reliability, resiliency assessment

I. INTRODUCTION

Deep Neural Networks (DNNs) have evolved to be increasingly applied to assist different aspects of human life, e.g., healthcare, transportation, security, IoT and edge applications [1]. In this context, energy efficiency and performance are the key constraints to be taken into account in designing DNN accelerators. Approximate Computing (AxC) is an emerging paradigm applied for improving their efficiency that produces acceptable results despite inaccuracies in the computations [2], [3].

Employing DNN accelerators in safety-critical applications has raised hardware reliability concerns. In compliance with ISO 26262 functional safety standard for road vehicles, the FIT (Failures In Time) rate of particular hardware components has to be 10 failures in 1 billion hours of operation at maximum to meet the target safety integrity level, which necessitates very circumspect design [4], [5]. The reliability of DNN accelerators is boosted by their ability to function correctly even in the presence of environment-related faults (soft errors, electromagnetic effects, temperature variations) or faults in the underlying hardware (manufacturing defects, process variations, nanoelectronics aging effects) [6]. DNNs are known

to be resilient to faults due to their numerous interconnected layers and the ability to mask faults [7]. However, several studies in recent years have shown that the accuracy of DNNs may still drop significantly in the presence of faults [6], [8]–[11]. These observations demonstrate that the reliability of DNN accelerators must be considered alongside efficiency. Some research works studied the reliability of approximated DNNs to show the trade-off between reliability and efficiency [12], [13].

The key challenge for DNN efficiency and reliability is the exploration of the huge design space. As mentioned, employing AxC units in DNN accelerators is one of the eminent approaches to gaining efficiency. However, the design space for approximated DNNs is too large [14], and implementing different AxC units to find an optimum efficiency is impracticable for FPGA accelerators. Notably, Graphic Processing Units (GPUs) that are widely applied for accelerating the DNN training can be utilized to assist this process as well. To tackle the task of exploiting AxC in DNNs, we present a GPU-accelerated framework for DNN approximation exploration.

Addressing accelerators' reliability issues starts with architecture-level fault-resiliency evaluation. Fault Injection (FI) is a conventional method for this purpose that has been vastly applied for DNNs as well [15], [16]. The main approaches for FI experiments are fault simulation in software and fault emulation in hardware, both implying a huge fault space. Fast fault emulation in accelerators (especially in FPGAs, which are widely used for DNNs [17]) is still a challenge because of its iterative procedure, including numerous extra memory accesses as well as huge fault injection campaigns. To tackle this issue, we leverage AxC units in DNNs as a non-conventional use of both FI and AxC, to emulate errors in the accelerator hardware. In this method, AxC units and their variants are a substitution for FI targeting the fault resilience analysis of DNN architectures.

Moreover, reducing fault space can also be done at the software level. We have carried out an empirical study on the inherent resilience to faults and errors of DNNs, with the aim of investigating how they can mask a large portion of faults. In line with this, we propose the adoption of three different metrics to compute in advance (right after the injection of the

---

*The authors are sorted in alphabetic order.

979-8-3503-4631-2/23 $31.00 © 2023 IEEE

fault) the effect the fault will have on the output vector score. In this way, it might be possible to both reduce the fault space and lower the FI time.

The paper is organized as follows: Section II introduces the GPU-accelerated framework for DNNs approximation exploration, Section III presents a method for harnessing approximation for agile analysis of fault resiliency in DNN accelerators, Section IV provides a fine-grain DNNs fault resiliency study by examining fault propagation and masking in networks, and Section V concludes the paper.

II. GPU ACCELERATED FRAMEWORK FOR CNN APPROXIMATION

A. Motivations and Related Works

As stated in the introduction, the Approximate Computing paradigm is widely used to improve the energy efficiency of hardware accelerators for DNNs. In particular, one promising solution is to use approximate arithmetic circuits [18]–[20]. However, quantifying the error introduced by these circuits requires expensive hardware prototyping, and, as a result, a software emulator of the DNN accelerator is often executed on a CPU or General Purpose - Graphic Processing Unit (GP-GPU) instead. Nevertheless, this emulation is typically much slower than a software DNN implementation running on a CPU or GP-GPU that uses the standard floating-point arithmetic instructions and common DNN libraries because CPUs and GP-GPUs lack hardware support for approximate arithmetic operations; therefore, the latter operations must be emulated, that is costly.

To address this issue, we propose Inspect-NN (I-NN), that provides efficient emulation for approximate circuits to be deployed in DNNs accelerator: approximate circuits are implemented as look-up tables and accessed through the memory mechanism of CUDA-capable GP-GPUs, reducing the inference time of the emulated DNN accelerator by approximately 200 times compared to an optimized CPU version on complex DNNs.

In the following, we present the I-NN framework in Section II-B, while Section II-C discusses case studies concerning the use of the mentioned framework to assess the accuracy loss due to approximate multipliers in Artificial Neural Networks (ANNs).

B. Proposed method

The main purpose of the I-NN framework is to investigate the impact of erroneous components on Artificial Intelligence (AI) applications. In particular, it allows investigating how the accuracy of DNNs-based applications is affected by imprecise components, i.e., those that do not meet their nominal behavioral specifications either because of faults, or because they have been specifically designed to differ in a controlled way from that behavior, while pursuing performance advantages. Examples are arithmetic components designed while exploiting the Approximate Computing (AxC) design paradigm [21]. The behavior of imprecise components are modeled at the behavioral level by exploiting lookup tables, in which input operands select the corresponding output of the component. I-NN exploits parallelism allowed by GP-GPUs: the inference phase is split in blocks, each assigned to a thread block on the GP-GPU and

TABLE I: Error characterization and hardware requirements for approximate circuits taken from the EvoApproxLib-Lite library, as reported in [22]

Circuit name	MAE (%)	AWCE (%)	MRE (%)	Power (nW)	MAE (μm^2)
mul8s_1KV6	0.00	0.00	0.00	0.425	729.8
mul8s_1KV8	0.0018	0.0076	0.28	0.422	711.0
mul8s_1KV9	0.0064	0.026	0.90	0.410	685.2
mul8s_1KVA	0.019	0.075	2.53	0.391	641.1
mul8s_1KVM	0.049	0.20	2.40	0.369	652.8
mul8s_1KVP	0.051	0.21	2.73	0.363	635.0
mul8s_1KVQ	0.056	0.25	3.64	0.351	599.8
mul8s_1KX5	0.15	0.69	8.93	0.289	543.0
mul8s_1KXF	0.34	1.37	15.72	0.237	482.4
mul8s_1L2J	0.081	0.39	4.41	0.301	558.9
mul8s_1L2L	0.23	1.16	12.26	0.200	411.6
mul8s_1L2N	0.52	2.66	27.44	0.126	284.9
mul8s_1L12	3.08	12.30	135.77	0.052	172.2

executed independently and parallelly from the others. I-NN does the latter computation through a kernel, i.e., a CUDA function called by the CPU and executed on the GP-GPU: operations within each layer are parallelized so that each thread block execute a part of the overall operation; then, if needed, the output is normalized to be represented using n bits, with n being configurable. Data exchange between the CPU and the GP-GPU are minimized: data is copied from the GP-GPU memory to the CPU ones when strictly required; hence, if two consecutive layers are working on the GP-GPU, the first one feeds the GP-GPU memory address of the computed data to the next layer, rather than coping them back and forth from/to the CPU.

C. Experimental Results

Case studies discussed in this Section concern the evaluation of the accuracy loss due to the use of multipliers taken from the EvoApproxLib-Lite [22] library of approximate circuits while targeting several pre-trained DNNs. In particular, through I-NN (i) we import the DNN to be analyzed directly from the most common machine learning frameworks, such as TensorFlow, TensorFlow LITE, and (ii) we define which specific approximate components have to be used, and (iii) we specify whether the analysis has to be performed at either coarse or fine grain. In coarse grain analysis, a single approximate component is deployed in the whole network. Conversely, in fine-grain analysis, each layer of the target DNN can use a different imprecise component.

We deploy multipliers from [22] – whose error characterization and hardware overhead are reported in Table I, for the reader convenience – to LeNet5 Convolutional Neural Network (CNN) [23], to MinNet, and to ResNet-8 [24], that, although trained using floating-point arithmetic, are all quantized to use 8-bit integer. The first CNN, i.e., LeNet5, has been trained to classify images from the Modified National Institute of Standards and Technology (MNIST) benchmark [25], on which it exhibits 99.07% accuracy. The MinNet CNN is a custom-made CNN inspired by the LeNet5 architecture: as for the latter, it consists of two Convolutional Layers (CLs), a Fully-Connected Layers (FCLs) and one Pooling Layers (PLs) between each CL, and it consists of approximately 160 thousand parameters. Despite its small size w.r.t. state-of-the-art networks, it exhibits 80.07% accuracy on the CIFAR-10 dataset [26]. Last,

979-8-3503-4631-2/23 $31.00 © 2023 IEEE

TABLE II: Accuracy loss and computational time for approximate circuits taken from the EvoApproxLib-Lite library [22].

Circuit Name	LeNet5			MinNet		ResNet8	
	Acc. Loss (%)	GPU Time	CPU Time	Acc. Loss (%)	GPU Time	Acc. Loss (%)	GPU Time
mul8s_1KV6	0	13.23s	≈10h	0	13.0s	0	31.07s
mul8s_1KV8	0.07	13.19s	≈10h	-0.3	13.6s	-0.19	31.1s
mul8s_1KV9	0.15	13.27s	≈10h	0.3	13.6s	-0.42	31.3s
mul8s_1KVA	0.51	13.22s	≈10h	2.5	13.5s	-0.08	31.3s
mul8s_1KVM	0.16	13.23s	≈10h	-0.4	13.5s	0.12	31.5s
mul8s_1KVP	0.27	13.17s	≈10h	-0.8	13.7s	-0.18	31.4s
mul8s_1KVQ	0.61	13.18s	≈10h	0.5	13.5s	0.09	31.4s
mul8s_1KX5	1.77	13.18s	≈10h	5.5	13.5s	5.48	31.3s
mul8s_1KXF	1.57	13.18s	≈10h	-1.2	13.6s	8.45	31.2s
mul8s_1L2J	0.79	13.2s	≈10h	46.6	14.2s	74.61	31.5s
mul8s_1L2L	3.81	13.14s	≈10h	61.5	14.2s	73.73	31.8s
mul8s_1L2N	15.92	13.11s	≈10h	65.9	14.0 s	74.52	32.06s
mul8s_1L12	75.66	13.15s	≈10h	66.4	14.4s	74.49	33.6s

the ResNet-8 CNN, instead, has been trained while targeting images taken from the CIFAR-10 dataset [26], which consists of 60 thousand RGB images, each belonging to one among ten classes. The network, that consists of more than 300 thousand learned parameters, and it exhibits 84.31% accuracy on the mentioned dataset. During the inference phase, these three architectures require performing 400 thousand, 4 million and 40 million multiplications each, respectively; hence, they represent a good test case for the evaluation of execution time.

To estimate the error introduced by the approximation, we execute the approximate CNN to obtain its classification accuracy on the whole test data set, reporting the accuracy-loss and computational time required for the inference phase in Table II. The latter table also reports the error and hardware parameters for each of the considered approximate multipliers. We performed the inference phase on an NVIDIA RTX A5000 GP-GPU, that is built on the NVIDIA Ampere architecture and combines 256 Tensor Cores and 8192 CUDA cores with 24 GB of graphics memory. Furthermore, for comparison purpose, the computational time of the inference phase while resorting to a CPU-only implementation is reported in Table II. In this case, we leverage two 3.20 GHz Intel Xeon Silver 4210 CPUs, providing 20 cores / 40 threads computing power. We reported CPU time only for the LeNet5 case. For the MinNet and ResNet8, the CPU execution time was higher than 10 hours and we were not able to complete the experiments.

As it is easy to foresee, the speed-up provided by the GP-GPU is crucial: we can state that by exploiting the GP-GPU through our look-up table implementation of approximate multiplier allows for tremendous performance improvements, even though we compared the execution time. Furthermore, it can be noticed that the execution time increases as the number of multiplications performed during the inference phase increases, and it is independent of the particular approximate multiplier being deployed, as it can be observed in Table II.

III. HARNESSING APPROXIMATION FOR FAULT INJECTION IN DNN ACCELERATORS

A. Motivations and Related Works

A major consequence of single or multiple accumulated soft-error-caused bitflips affecting the weights of a given layer is their propagation as errors at the layer outputs (also known as layer Output Feature Map) and further throughout the subsequent layers, leading to incorrect DNN predictions. *Fault*

resilience is the ability to tolerate the impact of faults on the output accuracy, and, in practice, it is one of the contributors to the final DNN accelerators' reliability. A relevant mitigation strategy at the architecture level can be a hardening of the DNN, e.g., by layer redesign or selective hardening of neurons, such as hardened Processing Elements (PEs) or Triple Modular Redundancy (TMR) variants [8]. These imply the assessment of layers' fault resiliency or identification of critical neurons in a neural network that are the most vulnerable to faults [27]–[29]. Fig. 1 presents a taxonomy for DNN reliability assessment methods. Along with analytical and hybrid methods [29], Fault Injection (FI) is a commonly used method for evaluating the fault resilience of DNNs [11], [30], [31]. The industry often employs fault injection by emulation in hardware, particularly in FPGAs, as it allows for evaluating real-scale DNN accelerator designs in significantly shorter run times than software-based simulations [9].

Fiji-FIN [32] is a representative framework implemented on the embedded Processing System for evaluating the resiliency of DNNs by emulating FI on FPGA. It measures accuracy degradation as a metric to study the impact of soft errors on network parameters. Designing fault injection campaigns for such frameworks requires significant effort, as each injection halts inference execution to manipulate DNN parameters. This interrupts classification time for a batch of inputs.

The state-of-the-art approaches for FI by emulation in FPGA using the embedded Processing System often require iterative procedures for each injected fault. In particular, such an iterative approach breaks the pipeline execution of the accelerator, requires a complex FI controller, and needs an extra FI control interconnection to handle the injection [32]–[34]. These procedures also involve multiple additional memory accesses, resulting in time-consuming processes and complex implementation.

Unlike the works mentioned above, our proposed method can be classified as fault injection by emulation in Programmable Logic. It leverages the functional approximation as a substitute for the errors generated by FI to improve processing and design time as well as the control complexity in the DNN fault resiliency analysis process. This approach allows the inference pipeline to be executed on a batch of inputs without interruption. This agile method enables a fast and efficient exploration of different options for network architecture, training, dataset selection, and more, to study the fault resilience of DNNs. Specifically, the introduced errors mimic single or multiple accumulated faults in weights. The method allows for efficient analysis of how subsequent layers in the network tolerate errors in the Output Feature Map of an assumed compromised layer are affected by faults in the weights of a compromised layer.

To the best of our knowledge, this is the first time that AxC units are utilized to enhance the efficiency and reduce the complexity of resilience analysis for DNNs.

B. Proposed method

AxC is commonly used to approximate hardware components to improve compute efficiency while maintaining functional accuracy. However, in practice, the errors induced by approximation can be used to mimic the errors caused by faults in logic circuits. These errors affect the outputs of the corresponding

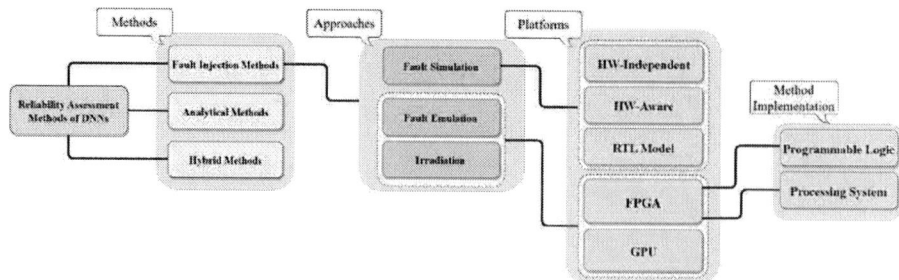

Fig. 1: A taxonomy of DNN reliability assessment methods

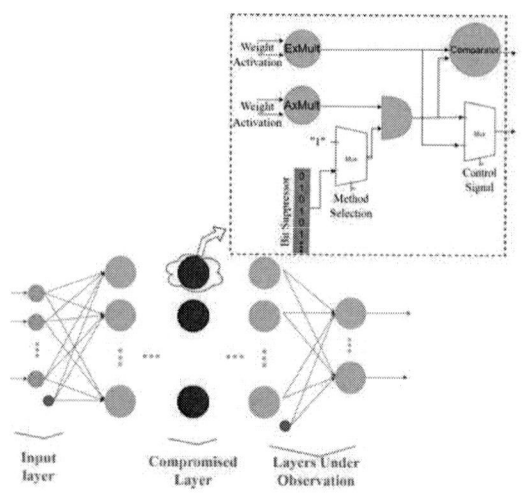

Fig. 2: Proposed method evaluation

units and propagate to subsequent layers, impacting their activations (Fig. 2). The proposed approach for evaluating DNN's fault resiliency using approximate computing (AxC) units is presented in Fig. 2. To implement our proposed method, an AxMult, or an AxMult + a bit suppression unit (AxMult+) is implemented along with the exact implementation of the multipliers (ExMult) in the network, depending on whether the network is being run in functional or fault resilience assessment mode. The golden inference for the validation dataset is run only once, and the layer outputs are stored and compared with a Comparator unit. The Bit Suppressor unit is meant to increase the probability of more significant bits of the neuron being impacted by faults. The less significant bits of the layer Output Feature Map are already affected by the AxMult with proper randomness depending on the data distribution in the network and layers.

The overall flow of the proposed method is illustrated in Fig. 3. In Step 1, the user initializes the method by selecting the compromised layer in the DNN structure, the validation dataset (i.e., DNN inputs), and the application-specific target fault rate assumed for the analysis. In Step 2, suitable AxC units are selected for Approximate Processing Elements (AxPEs), such as the AxC multipliers from a relevant library, e.g., the EvoApproxLib [35], or their variants with bit suppression. In Step 3, the selected AxMults started executing the compromised layer by enabling corresponding AxPEs along with the Exact Processing Elements (ExPE) in the DNN architecture. The DNN inference is run while keeping the network pipeline intact, and the resulting DNN output accuracy drop is recorded as the

primary metric for analyzing DNN fault resilience. A more significant drop in accuracy with induced errors implies a less fault-resilient DNN implementation. At the same time, the outputs of the AxMults are compared with the ExMults outputs to calculate the actual error at each neuron. The rest of the inference is executed by ExMults for both erroneous and exact outputs, and the comparison is performed for all the subsequent neurons of the network.

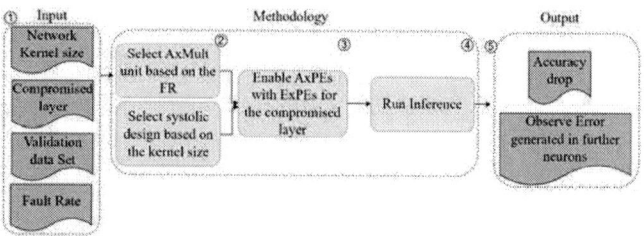

Fig. 3: Methodology flow

The characteristics of the approximation-induced errors can be evaluated using different metrics such as normalized error, number of flipped bits, and impact on the neural network classification accuracy drop. In this study, we rely on a simple set of metrics that includes:

- Normalized error: the average error on the output of each layer is calculated by subtracting the neurons' outputs of that layer from the golden output and dividing all the error values by the maximum value.
- Network accuracy: calculated by executing the network under different circumstances (faulty, AxMult, AxMult + bit suppressor and bit suppressor) over the test set.
- Bitflips in subsequent layers: calculated by comparing all bits in the next layers' outputs with the golden model and counting the bits that do not match as flipped bits.

1) Accelerator Model: Fig. 4 illustrates the accelerator model to perform resilience analysis on FPGA. It consists of two different systolic architecture designs based on the network under test. The $N \times N$ systolic architecture is used based on the convolution layers' kernel size to perform the most optimum dot matrix. At the same time, all designs have ExPE and AxPE to perform the resilience analysis and benefits of a dual register to store the results of both approximate systolic and exact systolic for further comparisons. An Error Detector (ED) module is also provided to compute the error generated at each neuron's output compared to the exact output and can be used for the neuron's vulnerability evaluation.

979-8-3503-4631-2/23 $31.00 © 2023 IEEE

This implementation provides us following features:

(a) Understanding the vulnerability of neurons by computing the error generated through the hardware and further layers by comparing the exact and approximate systolic design outputs;

(b) Increasing the controllability for enabling errors in each layer individually and keeping the other layers correct;

(c) Eliminating the need for designing and deploying an extra complex controller for the fault injection procedure. A simple approximate unit enabling circuitry is employed instead;

(d) The inference pipeline process executes a batch of inputs with no need to break this process;

(e) The resilience assessment process is performed without an extra interconnect for weight sampling;

(f) The proposed approach is not iterative for each potential fault location (unlike the traditional fault injection). Thus, the analysis complexity is vastly reduced.

Note that the features (c)-(f) are specific for FI emulation in Programmable Logic and generally not available in Processing Logic based methods such as Fiji-FIN.

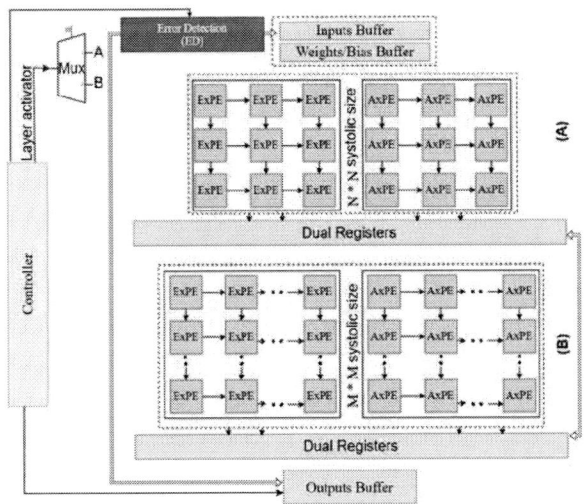

Fig. 4: Proposed systolic architecture for our Resiliency assessment DNN accelerator framework

C. Experimental Results

1) Evaluation methodology: To assess the feasibility of the proposed method, we implemented the same flow as shown in Fig. 2 with fault injection (FI). Using Table I, we narrowed down the list of candidate approximate multipliers from the EvoApproxLib library [35] based on several relevant metrics, with a primary focus on two established features, namely, the Variance of Error Distance (Var-ED) and Root Mean Square (RMS-ED) presented in [36]. These metrics are crucial in determining the approximation-induced errors that affect the performance of an AxC unit in DNNs. We selected mul8s_1L2N for the experiment based on these metrics and results achieved from the high-level experiments on the network through the proposed GPU accelerated framework for CNN approximation in Section II.

For the reference part, we repeated the fault resiliency evaluation on the original network, which was instrumented with a state-of-the-art FI method [32]. In this study, we considered the injection of multiple bitflips at a random location in all OFM' bits of the compromised layer for every input in the DNN validation test set. In this case, we assumed that 10% of the weights' bits were faulty.

To achieve a high FI confidence level using the statistical fault injection approach [37], we repeated the experiment for each fault model with 1000 random faults per image. The average accuracy of all repetitions was then reported.

We evaluated the impact of AxMult, AxMult + Bit Suppression (AXMult+), Bit Suppression alone, and fault injection, along with normalized error and the number of flipped bits, on the DNN accuracy. The results show a drop in DNN accuracy due to these factors. We compared the normalized error and the number of flipped bits for each scenario.

2) Experimental Setup: To evaluate the feasibility of the proposed method, a case-study Convolutional Neural Network (CNN) with two convolutional layers, two max-pooling, and one Fully-Connected (FC) layer was implemented and trained. The simulations were performed on an Intel® Core™ i7-6800K CPU @ 3.40GHz × 12, and the proposed method was implemented with Python 3. The hardware synthesis and implementation results are produced by the Xilinx Vivado HLS tool on a Xilinx Versal VCK190 FPGA (xcvc1902-vsva2197-2MP-e-S) at 166 MHz operational frequency.

The CNN under study is trained on a dataset of 2000 images of animals (cats and dogs) and humans for binary classification. The accuracy of the network over the test set (including 450 images of animals and humans) is 93.34%. Bit truncation quantization is applied in network parameters during training, and data precision is reduced to 8-bit.

3) Evaluation Results: We analyzed the similarity of the fault resiliency analysis results obtained by fault emulation and our proposed method using the metrics identified in Section III-B.

Fig. 6 shows the distribution of *normalized error* in the output of the second convolutional layer (Conv2) in the presence of 10% random faults in the first convolution layer (grey), errors induced by AxMult (blue), and errors induced by AxMult + bit suppressor (orange) enabled in the first convolution layer, respectively. Fig. 5 reports the impact of applying FI and our proposed method on the same convolutional layer and its effect on the second pooling layer of the network. These results demonstrate the similarity in error propagation trends between the proposed and reference methods.

In practice, by analyzing these charts, users can set a criticality threshold on the output error of the neurons based on their application and determine the number and indices of neurons to be used for any protection techniques. Generally, if we set the threshold at some error value, all methods suggest some neuron indices for mitigation techniques. As it can be concluded, both AxMult + bit suppression and FI show very similar behaviors. However, relying solely on the AxMult or bit suppression techniques is quite inaccurate for high fault ratios like this case study here.

For example, by setting the error threshold to 0.7, FI will recommend the user to protect 50 out of 1024 neurons of the

979-8-3503-4631-2/23 $31.00 © 2023 IEEE

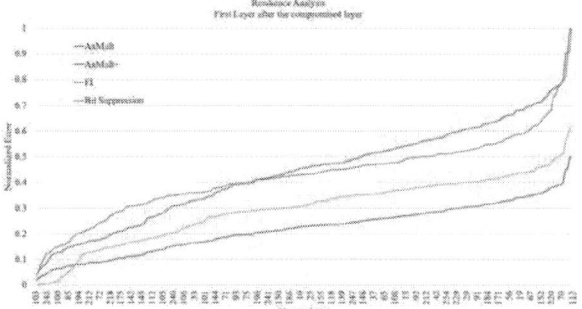

Fig. 5: Normalized output error of Pool2: Applying Ax-Mult, AxMult+ , Bit Suppression and FI on the Conv1

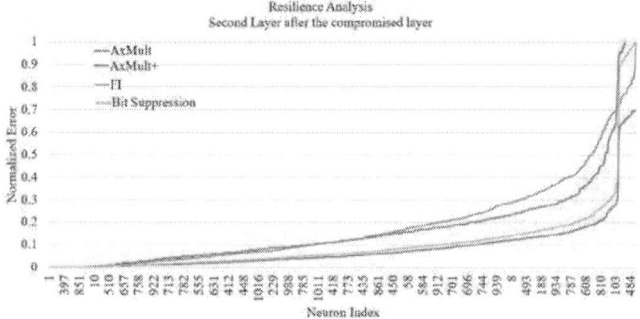

Fig. 6: Normalized output error of Conv2: Applying AxMult, AxMult+ , Bit Suppression and FI on the Conv1

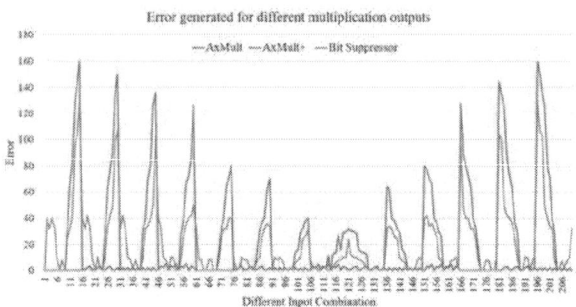

Fig. 7: Multiplication output error generated by AxMult, AxMult+ and Bit Suppression

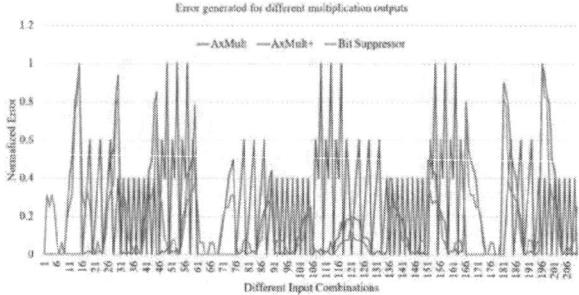

Fig. 8: Normalized Multiplication output error generated by AxMult, AxMult+, and Bit Suppression

Conv2 network's second CONV's neurons, while AxMult + bit suppression will recommend 53 out of 1024 neurons, including all the critical neurons recognized by FI. Fig. 7 and Fig. 8 show the error distribution of the three different methods, i.e., AxMult, AxMult + bit suppressor, and bit suppressor on the output of a multiplication operation with all the combinations of two 8-bit inputs. From Fig. 7, it is evident that the error values generated by AxMult + bit suppressor can almost cover a vast range of different values, and Fig. 8 shows that the error is evenly distributed on all different input combinations.

Table. III is reporting the number of bitflips and accuracy drop in subsequent layers caused by the compromised first convolution layer. These results also demonstrate the strong similarity of the trends in error propagation by the AxMult and its variants with the reference method. In case of accuracy drop, AxMult + bit suppression shows a strong correlation with the FI method and surpasses the other two methods.

Table IV reports details of the hardware accelerator implementation. Based on the results, the proposed implementation can be executed on the FPGA at 166 MHz clock frequency, and only by using ~16% of the available LUTs on the board all three mentioned systolic-array size architectures can be implemented to improve the efficiency of the accelerator. The timing comparison of the proposed method and the state-of-the-art fault injection method are presented in Table. V. As it can be concluded, by keeping an acceptable accuracy of FI in identifying the critical neurons, we get thousands of times speed-up in the resilience assessment of the DNNs. (Specifically, it is 5417 times in this example). At the same time, the proposed method does not need extra interconnects to

TABLE III: Bitflips and Accuracy drop induced by our proposed method vs. the reference fault injection method by fault rate 10% in OFM of the first convolution layer

Measured Layer	Bitflips in subsequent layers			
	FI (reference) [%]	AxMult [%]	AxMult+ [%]	Bit suppressor [%]
Conv1	10	10.30	10	10.20
Pool1	9.07	9.20	9.06	9.15
Conv2	16.76	16.80	16.77	16.83
Pool2	16.51	16.66	16.53	16.62
Accuracy drop [%]				
	16.73	9.33	18.33	24.73

TABLE IV: Hardware implementation of the proposed hardware accelerator

Conv2D systolic size	Resource Utilization (%)			Data Path Delay	CLK Frequency
	LUT	FF	BRAM		
3*3	0.03	0.00	0.83	Logic: ~20% Route: ~80%	166 MHz
5*5	0.09	0.00	0.83		
32*32	15.30	0.91	0.85		

manage the assessment process, and the original controller of the accelerator can take care of the fault resiliency assessment process.

TABLE V: Timing overheads of the proposed method vs. the reference fault injection method (Conv1 layer)

Network	Analysis Control Circuitry	Interconnects	DNN execution time in FPGA
Base CNN	N/A	Data Exchange Interconnect	~120ms
Fault Resilience Assessment			
CNN instrumented with FI	Complex FI Controller	(Data Exchange + FI) Interconnect	~650,000ms
CNN instrumented with AxMult+	Accelerator Controller	Data Exchange Interconnect	~120ms

979-8-3503-4631-2/23 $31.00 © 2023 IEEE

IV. FAULT RESILIENCY IN DNNs

A. Motivations and Related Works

In the last few years, researchers have investigated the theory behind brain-inspired computational models to build artificial structures capable of addressing highly complex computational problems. Today, DNNs are considered attractive solutions in several fields due to their outstanding computational capabilities as well as their human-level performance. The human brain is a complex and fascinating system able to bear synapses or neuron faults and still keep working properly, thanks to its plastic ability to remodel, repair, and reorganize its neural functions. Similarly, artificial neural networks possess in their structure a certain degree of redundancy that leads to intrinsic robustness and resilience against the occurrence of faults. This is caused by two aspects: the first is related to their distributed and parallel structure; the second to the redundancy resulting from the over-provisioning [38]. Indeed, neural networks are furnished with a quantity of artificial neurons higher than the minimal number required to perform a computation. It means that they can bear a bounded number of errors thanks to the excessive neuron budget: once this number is exceeded, the precision degrades gracefully as the number of errors increases [39].

This structural feature allows them to have an attractive property known as *masking ability*, which corresponds to the ability of DNNs to stop the propagation of some faults by masking their effects. As an example, it has been shown that the presence in DNNs of the Rectified Linear Unit (ReLU) activation function halves the percentage of critical faults by stopping the propagation of faults on negative weights [40]. Understanding how faults propagate through the neural network is very important, as it may influence: the reliability assessment procedure; efficient fault detection and mitigation strategies.

The analysis of fault propagation in DNNs has been conducted in the literature by different perspectives. A preliminary theory-driven analysis is proposed in [41], where the authors explore inherent characteristics of fault propagations in DNNs from the theoretical aspect. They propose a formula to compute the perturbation caused by the i-th bit flip on a weight represented in a 32-bit floating-point format. The authors in [42] characterize the propagation of soft errors from the hardware to the application software of DNN systems. Based on this, they devise cost-effective solutions to mitigate Silent Data Corruption (SDC) in software and hardware. Further studies on faults propagations in DNNs are described in [43] and [44].

Nevertheless, it is important to underline that the major effort in the above-mentioned research works consists in understanding how critical faults (i.e., those that lead to application failures) propagate through the hardware-software system.

The intent of this section is twofold. On the one hand, it aims to show how a critical fault spreads through a network. On the other hand, this section tackles the problem from a different angle, showing how a *masked fault* is propagated within the system, analysing the role DNNs have in this process. The investigation of this latter category of faults is important for the following reasons:

- In a fault injection process, the identification of sets of faults that are masked may reduce the fault space and, as a consequence, lower the costs of the reliability assessment;
- In the design of DNN models, the knowledge of architectural elements that favour the masking ability of DNNs can lead to the design of more robust models.

This section presents an analysis on masked faults with the goal of identifying at what point in the computation their propagation is stopped and if it is possible to know in advance their effects on the output of the DNN.

B. Proposed Method

CNNs are a subset of DNNs composed of a set of convolutional layers. The output of each layer is a multidimensional tensor, often referred to as the *Output Feature Map* (OFM). In the field of Image Classification, the output of the network is represented by a vector called *logit*. A fault affecting a CNN can be classified as:

- **Critical**, if it causes a change in the network prediction;
- **Non-Critical**, if it impacts the logit without changing the prediction;
- **Masked**, if it does not modify the logit.

When a fault affects the parameters of a layer (i.e., weights), it may change its OFM, as well as the one of all the following layers. If the fault is masked, the difference between the *golden Output Feature Map* (gOFM) and the *faulty Output Feature Map* (fOFM) of the impacted layer should be small or zero. Contrarily, it is logical to assume that a critical fault also produces a fOFM that is radically different from the gOFM.

As a consequence of these two observations, it is possible to predict the impact of a fault without needing to carry out a complete inference. In fact, this section aims at showing that:

1) Masked faults, once triggered, rarely propagate for more than one layer. Thus, the only different OFM is the one of the layer directly affected by the fault;
2) Critical faults, can be immediately identified by performing some early measures, using some metrics that can be computed by comparing the fOFM and the gOFM of the affected layer.

The OFM of a layer l can be interpreted as a collection of n filtered images, where n is the number of filters applied in layer l. Furthermore, the fOFM resulting from a fault in the network parameters can be interpreted as the gOFM plus a Gaussian noise. Therefore, it is possible to apply well-known objective image quality metrics, such as the Peak signal-to-noise Ratio (PSNR) and the Structural Similarity Index Metric (SSIM) [45].

This section proposes to use three different metrics to predict the criticality of a fault, starting from the OFM of the affected layer.

1) Max Difference: This first metric computes the maximum distance between the gOFM and the fOFM. This metric is presented as a baseline since, to the best of the authors' knowledge, there are no metrics that correlate the criticality of a fault with the changes in the OFM.

2) PSNR: This metric is directly proportional to the ratio between the peak signal (i.e., the maximum element of the gOFM) and the power of the corrupting noise, represented by the mean square error between the gOFM and the fOFM. The value can be computed as follows:

$$PSNR = 10 \cdot \log_{10} \frac{\max(gOFM)^2}{MSE(gOFM, fOFM)} \quad (1)$$

979-8-3503-4631-2/23 $31.00 © 2023 IEEE

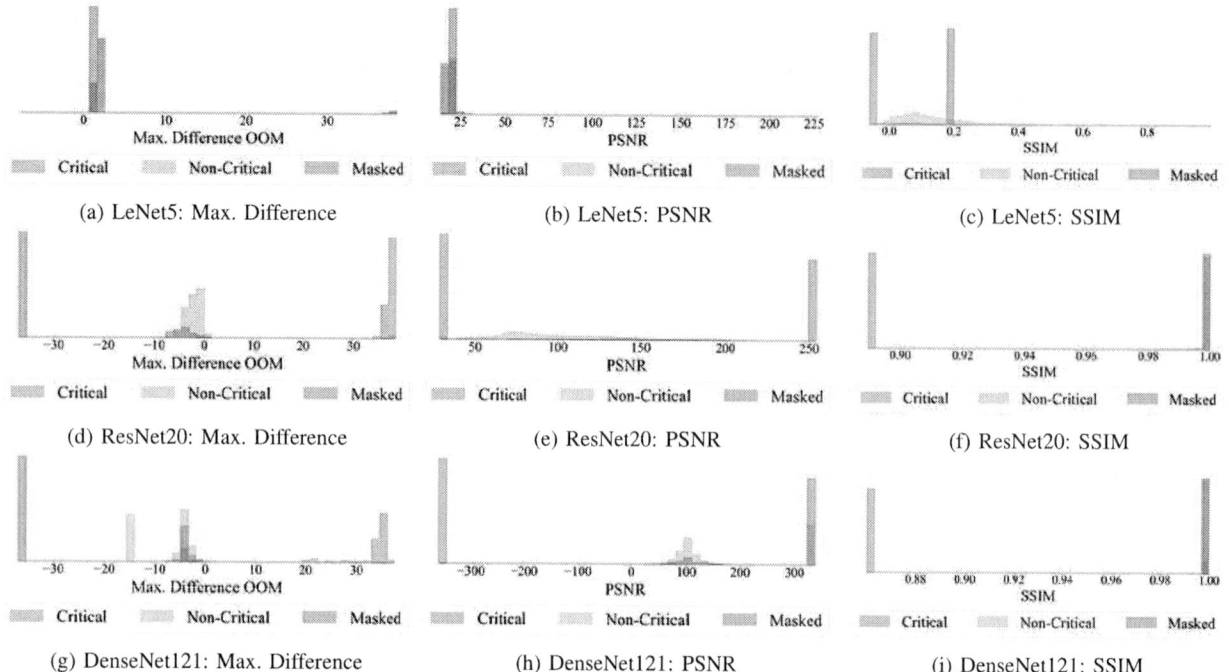

(a) LeNet5: Max. Difference

(b) LeNet5: PSNR

(c) LeNet5: SSIM

(d) ResNet20: Max. Difference

(e) ResNet20: PSNR

(f) ResNet20: SSIM

(g) DenseNet121: Max. Difference

(h) DenseNet121: PSNR

(i) DenseNet121: SSIM

Fig. 9: Metric probability distributions for the CNN under exam, observed in the layer where the fault is injected. Figures (a)-(c) refer to LeNet-5, Figures (d)-(f) refer to ResNet20 and Figures (g)-(i) refer to DenseNet121.

Where $\max(gOFM)$ is the maximum value of the gOFM and MSE is the Mean Square Error between the gOFM and the fOFM.

3) SSIM: this metric improves the PSRN, by including the concept of *structural information*, represented by the relationship of a neuron with its neighbours. The formula is composed by the product of three terms, the *luminance*, the *contrast* and the *structural* term. In the context of the study of the OFM, the simplified formula can be expressed as:

$$SSIM = \frac{(2\mu_f\mu_g + C_1)(2\sigma_{fg} + C_2)}{(\mu_f^2 + \mu_g^2 + C_1)(\sigma_f^2 + \sigma_g^2 + C_2)} \quad (2)$$

Where μ_g, μ_f are the mean of the gOFM and of the fOFM, σ_g, σ_f their standard deviation, σ_{fg} their cross-covariance. C_1 and C_2 are two regularization parameters.

C. Experimental Results

This section analyses three different CNNs used for Image Classification to study how a fault can propagate. The networks under analysis are: LeNet-5 with the MNIST dataset, ResNet20 with CIFAR-10 and Densenet-121 with ImageNet. For each network, we performed a statistical FI as described in [46]. The tool used to carry out the FI campaign is SCI-FI [47], that allows to speed up the FI process using the Fault Dropping and the Delayed Start techniques. The faults injected are single bit-flips in the network parameters, represented as 32-bit floating points. Further details on the networks under exam and on the FI campaigns are reported in Table VI.

Firstly, to demonstrate that Masked faults only modify the OFM of the layer affected by the fault, we report the percentage of Masked faults that affect more than one layer. In particular, for LeNet5, all the Masked faults do not modify any OFM

TABLE VI: The networks under analysis

Network	Dataset	Dataset Size	Acc. [%]	Weights	Injected Faults
LeNet5	MNIST	10,000	98.85	61,706	2,212
ResNet20	CIFAR-10	10,000	91.72	269,722	15,675
DenseNet121	ImageNet	50,000	74.43	7,978,856	16,685

besides the one of the impacted layer. For ResNet20, 87.99% of Masked faults show no effect in the OFM of the layer immediately after the impacted one, while for DenseNet this number rises to 99.17%.

To show that Critical faults have a strong impact early on, we compute the metrics introduced in Section IV-B on the OFM of the layer affected by the fault. Figure 9 reports the metrics distributions for the Max Difference, the PSNR and the SSIM. Each image shows, for each network, the distribution of a metric computed for all the FI campaigns. In particular, the distribution is further subdivided according to the impact of the fault affecting the network when they were measured. This means that the distribution labelled 'Critical' reports only the value measured when a Critical fault is affecting the network. For a metric, the more separable the three distributions are, the better the metric is at predicting the effect of a fault.

In particular, we can observe a stark contrast between the metrics computed for LeNet5 and the other networks. This can be imputed to the lack of batch-normalization layers, that normalize the value of the weights (and of the OFM) between $[-1, 1]$. Consequently, even a bit-flip in the mantissa bits of the weight can have a large impact. Nonetheless, SSIM performs sufficiently well, as it correctly separates Critical and Non-Critical faults.

For the other two CNNs, we can notice that both the Max. Difference and the PSNR separate Masked faults from Critical and Non-Critical faults. However, for ResNet20, SSIM outperforms the other metrics, as it completely splits ups Critical and Non-Critical faults while providing a good degree of separation between Masked and Non-Critical faults. Contrarily, for DenseNet-121, SSIM does not completely separate Masked from Critical. For this latter network, the best solution is offered by the PSNR.

Therefore, we observe how different metrics can correctly predict Masked and Critical faults without the need for a complete inference, by simply analysing the fOFM of the layer affected by the fault.

As a final note, we want to highlight that the cost of the computation of the metric is quite small, requiring only a portion of the time required for the computation of a whole layer. On average, the per-layer overhead added by the computation of one of the metrics is 76.51% for LeNet5 74.28% for ResNet20 and 73.54% for DenseNet121.

V. CONCLUSIONS

This paper explored approximation and fault resiliency of DNN accelerators. To allow fast evaluation of AxC DNN, an efficient GPU-based simulation framework was developed. The paper proposed a method for employing approximate (AxC) arithmetic circuits to agilely emulate errors in hardware without performing fault injection on the DNN. Finally, it presented a fine-grain analysis of fault resiliency by examining fault propagation and masking in networks.

ACKNOWLEDGMENTS

This work was supported in part by the European Union through European Social Fund in the frames of the "Information and Communication Technologies (ICT) programme" ("ITA-IoIT" topic), by the Estonian Research Council grant PUT PRG1467 "CRASHLES" and by Estonian-French PARROT project "EnTrustED".

REFERENCES

[1] V. Sze, Y.-H. Chen, T.-J. Yang, and J. S. Emer, "Efficient processing of deep neural networks: A tutorial and survey," *Proceedings of the IEEE*, vol. 105, no. 12, pp. 2295–2329, 2017.

[2] G. Armeniakos, G. Zervakis, D. Soudris, and J. Henkel, "Hardware approximate techniques for deep neural network accelerators: A survey," *ACM Computing Surveys*, vol. 55, no. 4, pp. 1–36, 2022.

[3] A. Bosio, D. Ménard, and O. Sentieys, Eds., *Approximate Computing Techniques*. Springer International Publishing, 2022. [Online]. Available: https://doi.org/10.1007/978-3-030-94705-7

[4] A. Nardi and A. Armato, "Functional safety methodologies for automotive applications," in *2017 IEEE/ACM International Conference on Computer-Aided Design (ICCAD)*. IEEE, 2017, pp. 970–975.

[5] M. Jenihhin, M. S. Reorda, A. Balakrishnan, and D. Alexandrescu, "Challenges of reliability assessment and enhancement in autonomous systems," in *2019 IEEE International Symposium on Defect and Fault Tolerance in VLSI and Nanotechnology Systems (DFT)*, 2019, pp. 1–6.

[6] M. Shafique, M. Naseer, T. Theocharides, C. Kyrkou, O. Mutlu, L. Orosa, and J. Choi, "Robust machine learning systems: Challenges, current trends, perspectives, and the road ahead," *IEEE Design & Test*, vol. 37, no. 2, pp. 30–57, 2020.

[7] A. Bosio, P. Bernardi, A. Ruospo, and E. Sanchez, "A reliability analysis of a deep neural network," in *2019 IEEE Latin American Test Symposium (LATS)*. IEEE, 2019, pp. 1–6.

[8] S. Mittal, "A survey on modeling and improving reliability of dnn algorithms and accelerators," *Journal of Systems Architecture*, vol. 104, p. 101689, 2020.

[9] Y. Ibrahim, H. Wang, J. Liu, J. Wei, L. Chen, P. Rech, K. Adam, and G. Guo, "Soft errors in dnn accelerators: A comprehensive review," *Microelectronics Reliability*, vol. 115, p. 113969, 2020.

[10] C. Torres-Huitzil and B. Girau, "Fault and error tolerance in neural networks: A review," *IEEE Access*, vol. 5, pp. 17322–17341, 2017.

[11] F. Su, C. Liu, and H.-G. Stratigopoulos, "Testability and dependability of ai hardware: Survey, trends, challenges, and perspectives," *IEEE Design & Test*, 2023.

[12] L. M. Luza, D. Söderström, G. Tsiligiannis, H. Puchner, C. Cazzaniga, E. Sanchez, A. Bosio, and L. Dilillo, "Investigating the impact of radiation-induced soft errors on the reliability of approximate computing systems," in *2020 IEEE International Symposium on Defect and Fault Tolerance in VLSI and Nanotechnology Systems (DFT)*. IEEE, 2020, pp. 1–6.

[13] M. Taheri, M. Riazati, M. H. Ahmadilivani, M. Jenihhin, M. Daneshtalab, J. Raik, M. Sjödin, and B. Lisper, "Deepaxe: A framework for exploration of approximation and reliability trade-offs in dnn accelerators," in *24th International Symposium on Quality Electronic Design*. https://doi.org/10.48550/arXiv.2303.08226, 2023.

[14] M. Pinos, V. Mrazek, F. Vaverka, Z. Vasicek, and L. Sekanina, "Acceleration techniques for automated design of approximate convolutional neural networks," *IEEE Journal on Emerging and Selected Topics in Circuits and Systems*, pp. 1–1, 2023.

[15] A. Ruospo, L. M. Luza, A. Bosio, M. Traiola, L. Dilillo, and E. Sanchez, "Pros and cons of fault injection approaches for the reliability assessment of deep neural networks," in *2021 IEEE 22nd Latin American Test Symposium (LATS)*. IEEE, 2021, pp. 1–5.

[16] A. Bosio, I. O'Connor, M. Traiola, J. Echavarria, J. Teich, M. A. Hanif, M. Shafique, S. Hamdioui, B. Deveautour, P. Girard *et al.*, "Emerging computing devices: Challenges and opportunities for test and reliability," in *2021 IEEE European Test Symposium (ETS)*. IEEE, 2021, pp. 1–10.

[17] M. A. Talib, S. Majzoub, Q. Nasir, and D. Jamal, "A systematic literature review on hardware implementation of artificial intelligence algorithms," *The Journal of Supercomputing*, vol. 77, pp. 1897–1938, 2021.

[18] M. Barbareschi, S. Barone, and N. Mazzocca, "Advancing synthesis of decision tree-based multiple classifier systems: an approximate computing case study," *Knowledge and Information Systems*, pp. 1–20, Apr. 2021, company: Springer Distributor: Springer Institution: Springer Label: Springer Publisher: Springer London. [Online]. Available: https://link.springer.com/article/10.1007/s10115-021-01565-5

[19] M. Barbareschi, S. Barone, A. Bosio, J. Han, and M. Traiola, "A Genetic-algorithm-based Approach to the Design of DCT Hardware Accelerators," *ACM Journal on Emerging Technologies in Computing Systems*, vol. 18, no. 3, pp. 1–25, Jul. 2022. [Online]. Available: https://dl.acm.org/doi/10.1145/3501772

[20] M. Barbareschi, S. Barone, N. Mazzocca, and A. Moriconi, "A Catalog-based AIG-Rewriting Approach to the Design of Approximate Components," *IEEE Transactions on Emerging Topics in Computing*, 2022.

[21] A. Bosio, D. Ménard, and O. Sentieys, Eds., *Approximate Computing Techniques: From Component- to Application-Level*. Cham: Springer International Publishing, 2022. [Online]. Available: https://link.springer.com/10.1007/978-3-030-94705-7

[22] V. Mrazek, Z. Vasicek, L. Sekanina, H. Jiang, and J. Han, "Scalable Construction of Approximate Multipliers With Formally Guaranteed Worst Case Error," *IEEE Transactions on Very Large Scale Integration (VLSI) Systems*, vol. 26, no. 11, pp. 2572–2576, Nov. 2018, conference Name: IEEE Transactions on Very Large Scale Integration (VLSI) Systems.

[23] Y. Lecun, L. Bottou, Y. Bengio, and P. Haffner, "Gradient-based learning applied to document recognition," *Proceedings of the IEEE*, vol. 86, no. 11, pp. 2278–2324, Nov. 1998, conference Name: Proceedings of the IEEE.

[24] K. He, X. Zhang, S. Ren, and J. Sun, "Deep Residual Learning for Image Recognition," in *2016 IEEE Conference on Computer Vision and Pattern Recognition (CVPR)*. Las Vegas, NV, USA: IEEE, Jun. 2016, pp. 770–778. [Online]. Available: http://ieeexplore.ieee.org/document/7780459/

[25] Y. LeCun, C. Cortes, and C. Burges, "MNIST Handwritten digit database," 1998. [Online]. Available: http://yann.lecun.com/exdb/mnist/

[26] A. Krizhevsky, V. Nair, and G. Hinton, "CIFAR-10 (canadian institute for advanced research)," 2010. [Online]. Available: https://www.cs.toronto.edu/ kriz/cifar.html

[27] C. Schorn *et al.*, "Accurate neuron resilience prediction for a flexible reliability management in neural network accelerators," in *2018 DATE*. IEEE, 2018, pp. 979–984.

[28] A. Ruospo and E. Sanchez, "On the reliability assessment of artificial neural networks running on ai-oriented mpsocs," *Applied Sciences*, vol. 11, no. 14, p. 6455, 2021.

[29] M. H. Ahmadilivani, M. Taheri, J. Raik, M. Daneshtalab, and M. Jenihhin, "Deepvigor: Vulnerability value ranges and factors for dnns' reliability assessment," *arXiv preprint arXiv:2303.06931*, 2023.

[30] A. Ruospo, E. Sanchez, L. M. Luza, L. Dilillo, M. Traiola, and A. Bosio, "A survey on deep learning resilience assessment methodologies," *Computer*, vol. 56, no. 2, pp. 57–66, 2023.

[31] M. Taheri, M. H. Ahmadilivani, M. Jenihhin, M. Daneshtalab, and J. Raik, "Appraiser: Dnn fault resilience analysis employing approximation errors," in *26th International Symposium on Design and Diagnostics of Electronic Circuits and Systems*. In press, 2023.

[32] N. Khoshavi, C. Broyles, Y. Bi, and A. Roohi, "Fiji-fin: A fault injection framework on quantized neural network inference accelerator," in *2020 19th IEEE International Conference on Machine Learning and Applications (ICMLA)*. IEEE, 2020, pp. 1139–1144.

[33] M.-C. Hsueh, T. K. Tsai, and R. K. Iyer, "Fault injection techniques and tools," *Computer*, vol. 30, no. 4, pp. 75–82, 1997.

[34] N. Khoshavi, A. Roohi, C. Broyles, S. Sargolzaei, Y. Bi, and D. Z. Pan, "Shieldenn: Online accelerated framework for fault-tolerant deep neural network architectures," in *2020 57th ACM/IEEE Design Automation Conference (DAC)*. IEEE, 2020, pp. 1–6.

[35] V. Mrazek, R. Hrbacek, Z. Vasicek, and L. Sekanina, "Evoapprox8b: Library of approximate adders and multipliers for circuit design and benchmarking of approximation methods," in *Design, Automation Test in Europe Conference Exhibition (DATE), 2017*, March 2017, pp. 258–261.

[36] M. S. Ansari, V. Mrazek, B. F. Cockburn, L. Sekanina, Z. Vasicek, and J. Han, "Improving the accuracy and hardware efficiency of neural networks using approximate multipliers," *IEEE Transactions on Very Large Scale Integration (VLSI) Systems*, vol. 28, no. 2, pp. 317–328, 2019.

[37] R. Leveugle, A. Calvez, P. Maistri, and P. Vanhauwaert, "Statistical fault injection: Quantified error and confidence," in *2009 Design, Automation & Test in Europe Conference & Exhibition*. IEEE, 2009, pp. 502–506.

[38] V. Piuri, "Analysis of fault tolerance in artificial neural networks," *Journal of Parallel and Distributed Computing*, vol. 61, no. 1, pp. 18 – 48, 2001.

[39] E. M. El Mhamdi and R. Guerraoui, "When neurons fail," in *2017 IEEE International Parallel and Distributed Processing Symposium (IPDPS)*, 2017, pp. 1028–1037.

[40] F. Angione *et al.*, "Test, reliability and functional safety trends for automotive system-on-chip," in *2022 IEEE European Test Symposium (ETS)*, 2022, pp. 1–10.

[41] R. Sun, J. Zhan, and W. Jiang, "An insight into fault propagation in deep neural networks: Work-in-progress," in *2020 International Conference on Embedded Software (EMSOFT)*, 2020, pp. 20–21.

[42] G. Li, S. K. S. Hari, M. Sullivan, T. Tsai, K. Pattabiraman, J. Emer, and S. W. Keckler, "Understanding error propagation in deep learning neural network (dnn) accelerators and applications," in *SC17: International Conference for High Performance Computing, Networking, Storage and Analysis*, 2017, pp. 1–12.

[43] J. E. R. Condia, J.-D. Guerrero-Balaguera, F. F. Dos Santos, M. S. Reorda, and P. Rech, "A multi-level approach to evaluate the impact of gpu permanent faults on cnn's reliability," in *2022 IEEE International Test Conference (ITC)*, 2022, pp. 278–287.

[44] F. F. Dos Santos, P. Rech, A. Kritikakou, and O. Sentieys, "Evaluating the impact of mixed-precision on fault propagation for deep neural networks on gpus," in *2022 IEEE Computer Society Annual Symposium on VLSI (ISVLSI)*, 2022, pp. 327–327.

[45] A. Horé and D. Ziou, "Image quality metrics: Psnr vs. ssim," in *2010 20th International Conference on Pattern Recognition*, 2010, pp. 2366–2369.

[46] A. Ruospo, G. Gavarini, C. D. Sio, J. Guerrero, L. Sterpone, M. S. Reorda, E. Sanchez, R. Mariani, J. Aribido, and J. Athavale, "Assessing convolutional neural networks reliability through statistical fault injections," in *2023 Design, Automation & Test in Europe Conference & Exhibition (DATE)*, 2023, [In press].

[47] G. Gavarini, A. Ruospo, and E. Sanchez, "Sci-fi: a smart, accurate and unintrusive fault-injector for deep neural networks," in *2023 European Test Symposium*, 2023, In press.

Fully Deterministic Storage Based Logic Built-In Self-Test

Subashini Gopalsamy
School of Electrical and Computer Engineering
Purdue University
West Lafayette, IN 47907, U.S.A.
sgopalsa@purdue.edu

Irith Pomeranz
School of Electrical and Computer Engineering
Purdue University
West Lafayette, IN 47907, U.S.A.
pomeranz@ecn.purdue.edu

Abstract—**This paper presents a fully deterministic storage based logic built-in self-test (LBIST) approach that stores, on chip, reduced deterministic uncompressed test data sufficient for achieving complete fault coverage. The goal of this approach is to eliminate the need for pseudo-random tests, thereby reducing the test application time by reducing the number of tests required to achieve complete fault coverage. Under this approach, two types of test data are stored on chip. 1) Subsets of scan vectors obtained from a reduced set of deterministic tests, one subset per test and, 2) permutations of scan vectors stored as sets of indices, to indicate how to combine scan vectors to form tests on chip. The same permutations are applied to all the subsets, magnifying the effectiveness of each stored permutation and each subset, allowing fewer subsets as well as fewer permutations to be used. This helps in reducing the storage requirements. Experimental results are presented for single stuck-at faults in benchmark circuits and logic blocks of the OpenSPARC T1 microprocessor to demonstrate the effectiveness of this approach.**

I. INTRODUCTION

With the rapid advancement in the automotive industry and other safety critical applications, the highly integrated complex ICs used in these applications need high quality test solutions to ensure their functional safety and long-term reliability. Besides a high fault coverage, these test solutions should meet requirements such as an ability to perform in field testing and low test application time (TAT). Logic Built In Self-Test (LBIST) is a testing technique where the hardware required to test a chip is built into the chip [1]-[19]. This built-in hardware takes care of test generation, test application and response verification avoiding the need for an external tester, therefore allowing in field testing. In addition to enhancing security by removing the need for transfer of test data to and from the chip [11], LBIST also provides the ability to test at the chip frequency.

Basic LBIST techniques generate pseudo-random patterns which on their own result in low fault coverage because of the presence of random pattern resistant (RPR) faults with low detection probabilities. To increase the fault coverage, LBIST solutions based on pseudo-random patterns typically require test points to be inserted [5], [16]. A large class of methods modify pseudo-random tests to increase the fault coverage

Research supported in part by NSF Grant Number CCF-2041649

they achieve. These methods include bit-flipping [10], [13] or bit-fixing [3] and weighted random pattern generation [2]. Hybrid LBIST, another class of solutions [6], [12], [15], stores the top up deterministic patterns needed to detect RPR faults in a compressed form on a tester, and decompresses them on-chip using existing test data compression hardware [21]-[23]. Another class of solutions stores the compressed deterministic top up patterns on-chip. LFSR reseeding [7], [9] can be considered a solution of this class. With LFSR reseeding, multiple seeds for the LFSR are stored on-chip and used for test application. Stellar BIST [17] stores compressed deterministic parent patterns on-chip and decompresses them using the on-chip test data compression logic. To reduce the storage requirements, transformed derivatives of the parent patterns are obtained in [17] by complementing multiple scan slices at uniform intervals using on-chip test logic.

Another class of LBIST approaches where all the test data required to achieve complete fault coverage are stored on-chip is described in [8], [18] and [19]. This class of approaches is based on partitioning a precomputed deterministic test set into test data entries, for example scan vectors, small enough to be stored on-chip. The on-chip test generation is performed by combining the test data entries (the scan vectors) either randomly or deterministically to achieve complete fault coverage. The strength of this approach stems from the large number of tests that can potentially be formed by combining deterministic test data entries. From this large set it is possible to select a subset for detecting target faults. In [18] and [19] this property is used for achieving complete fault coverage not only for stuck-at faults but also for single-cycle gate-exhaustive faults.

In the approach described in [8], scan vectors obtained by partitioning a precomputed deterministic test set are stored separately for each scan chain. The subsets are reduced for on-chip storage. The Cartesian product of these stored subsets is performed on-chip for test generation. The number of tests generated by the Cartesian product depends on the size of each subset of scan vectors (N) and the number of subsets (n). In [8], the Cartesian product was feasible since N and n were small enough. The more recent approaches in this class do not rely on the Cartesian product to accommodate circuits for which N and n are large.

979-8-3503-4631-2/23 $31.00 © 2023 IEEE

In [18], a special type of pseudo-random tests is formed by pseudo-random combinations of stored deterministic test data entries (scan vectors). They are complemented by tests referred to as deterministic formed by deterministic combinations of stored scan vectors. To generate deterministic tests, additional test data are stored on-chip representing which combinations of scan vectors (indices of scan vectors in the stored set) are needed to detect the target faults. Together the two types of tests achieve complete fault coverage for single stuck-at and single-cycle gate-exhaustive faults. All the scan vectors for all the scan chains are stored in a single set in [18]. In [19], the set of scan vectors is partitioned into subsets, allowing the on-chip test generation process to focus on specific combinations of scan vectors. As a result, only pseudo-random combinations of deterministic scan vectors are used to achieve complete fault coverage in [19]. As with other types of pseudo-random tests, the number of tests needed in [18] and [19] is significantly larger than the number of deterministic tests.

Whereas [18] and [19] rely on the use of pseudo-random tests (pseudo-random combinations of stored scan vectors), the goal of this paper is to eliminate the need for pseudo-random tests. This will reduce the number of tests required to achieve complete fault coverage and allow the LBIST solution to meet the test time constraints that exist during system startup and periodic in-field testing. Thus, this paper proposes a fully deterministic storage based LBIST approach from the class of approaches described in [8], [18] and [19]. The proposed approach stores two types of test data entries on-chip. 1) Subsets of scan vectors, obtained from uncompressed deterministic tests, one subset per test. 2) Permutations of scan vector indices, stored to indicate how to combine scan vectors to form tests. The same permutations are applied to all the subsets, magnifying the effectiveness of each stored permutation and each subset, allowing fewer subsets as well as fewer permutations to be used. The permutation 0, 1, ..., n-1, where n is the number of scan vectors in a subset, results in the original deterministic tests. Other permutations result in different tests referred to as deterministic derivatives. Together the original tests and their deterministic derivatives achieve complete fault coverage, eliminating the need for pseudo-random tests. Since the use of every permutation with every subset of scan vectors reduces the number of subsets as well as the number of permutations, it also reduces the storage requirements associated with input stimuli. The proposed scheme is based on the concepts from other solutions of the same class [8], [18], [19] but avoids the limitations of [8]. It adopts the underlying principles of [8], [18], [19], such as partitioning precomputed deterministic tests into scan vectors, and achieving complete fault coverage without disturbing the values of the stored deterministic test data. The permutation operation in the proposed scheme enables reuse of the stored test data entries from each subset several times in their corresponding derivatives without altering the values of the stored scan vectors.

The target faults in this paper are single stuck-at faults. The approach described in this paper can be extended to other fault models as well. Encoding [4] or test data compression [21]-[23] can be used on the original deterministic scan vectors to be stored on-chip to further reduce the on-chip storage required. Test points can be used for increasing the fault coverage without storing additional subsets. These options are not considered in this paper.

The rest of the paper is organized as follows. Section II describes the test data stored on-chip. The on-chip test generation logic is described in section III. A software procedure for computing the subsets of scan vectors and the permutations is described in Section IV. Section V describes a software procedure to reduce the number of permutations. Section VI presents the experimental results.

II. ON-CHIP STORAGE

This section describes the on-chip storage of test data entries for on-chip test application. A circuit under consideration is assumed to have n scan chains each of length k. The shorter scan chains are padded to bring their length to k. This is done only to simplify the discussion. As in [18], $n \approx k$ is used to make the scan vectors small enough to be stored on-chip.

The proposed scheme stores m subsets of scan vectors, $S_0, S_1, \ldots, S_{m-1}$. Each subset consists of n scan vectors, corresponding to one deterministic test. Thus, $S_i = \{sv_{i,0}, sv_{i,1}, \ldots, sv_{i,n-1}\}$.

The proposed scheme also stores the permutations of scan vector indices used to construct original deterministic tests and their deterministic derivatives on-chip. Let Y denote the collection of p permutations, $Y = \{X_0, X_1, \ldots, X_{p-1}\}$. A permutation X_j is represented as a set of indices of scan vectors $X_j = (r_{j,0}, r_{j,1}, \ldots, r_{j,n-1})$. The permutation X_0 is always the original permutation $(0, 1, \ldots, n$-1$)$ that results in constructing the original deterministic patterns. The remaining permutations, $X_1, \ldots, X_{p-1}$, obtained by permuting the original index set, are used to construct the deterministic derivatives.

For $0 \leq i \leq m-1$ and $0 \leq j < p$, a test t_{ij} is formed by applying permutation X_j to subset S_i. The subsets of scan vectors and the permutations are chosen by a software procedure discussed in Sections IV and V.

Tables I and II illustrate the on-chip storage components and the test set generated on-chip. In this example, the circuit is assumed to have n=3 scan chains each of length k=4. Table I shows the scan vectors in two subsets S_0 and S_1. Each subset contains n=3 scan vectors. Table II shows the set of permutations Y= $\{X_0, X_1, X_2, X_3, X_4\}$ and the tests constructed from S_0 and S_1 using Y. Here t_{00} and t_{10} are the original deterministic tests reconstructed by applying the permutation X_0 on S_0 and S_1, respectively. The other tests $t_{01}, \ldots, t_{04}$ and $t_{11}, \ldots, t_{14}$ are the deterministic derivatives generated from S_0 and S_1, respectively, by applying the permutations X_1, X_2, X_3 and X_4 to both the subsets. Since every permutation is applied to every subset of scan vectors, the number of tests generated on-chip is $p*m$, which is 5*2 = 10 tests for the above example. Both p and m will be

979-8-3503-4631-2/23 $31.00 © 2023 IEEE

TABLE I
SUBSETS OF SCAN VECTORS

i	$sv_{i,0}$	$sv_{i,1}$	$sv_{i,2}$
0	0100	1101	0010
1	1011	0000	0110

TABLE II
TEST SET CONSTRUCTED FROM SCAN VECTORS USING PERMUTATIONS

j	$r_{j,0}$	$r_{j,1}$	$r_{j,2}$	t_{0j}	t_{1j}
0	0	1	2	$t_{00}=\{0100,1101,0010\}$	$t_{10}=\{1011,0000,0110\}$
1	2	0	1	$t_{01}=\{0010,0100,1101\}$	$t_{11}=\{0110,1011,0000\}$
2	1	2	0	$t_{02}=\{1101,0010,0100\}$	$t_{12}=\{0000,0110,1011\}$
3	1	0	2	$t_{03}=\{1101,0100,0010\}$	$t_{13}=\{0000,1011,0110\}$
4	0	2	1	$t_{04}=\{0100,0010,1101\}$	$t_{14}=\{1011,0110,0000\}$

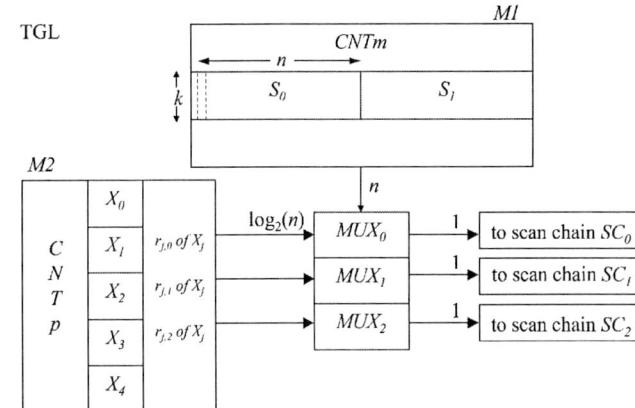

Fig. 1. On-chip Test Generation Logic.

minimized to keep the number of applied tests as small as possible.

III. ON-CHIP TEST GENERATION LOGIC

The on-chip test generation logic (TGL) for the proposed scheme is described in this section. The TGL is adopted from [19]. The TGL for the example in Tables I and II is illustrated in Figure 1. Here $M1$ and $M2$ are two on-chip memories storing subsets of scan vectors and permutations, respectively. These memories are dedicated to LBIST. The memory $M1$ is partitioned into two blocks storing the subsets S_0 and S_1. The subset in each block has $n = 3$ scan vectors each of length $k = 4$. A single scan vector of length k is shown by the dashed box inside the memory block containing S_0. It takes k clock cycles to shift a scan vector out of the memory and into a scan chain bit by bit. The storage requirement of memory $M1$ in bits is $m * n * k$. A counter denoted by $CNTm$ selects a memory block thereby selecting which subset of scan vectors will be used for test application.

The memory $M2$ is partitioned into p blocks storing p permutations $X_0, X_1, \ldots, X_{p-1}$. The storage requirement of memory $M2$ in bits is $p * n * log_2 n$. A counter denoted by $CNTp$ selects which permutation will be applied to the subset chosen by $CNTm$.

When the counter $CNTm$ selects the memory block containing S_i, the n scan vectors of S_i are available in the output lines of the memory $M1$. From these n scan vectors, to select the scan vector to be shifted into the scan chain SC_q, a multiplexer MUX_q is used. MUX_q has n data lines routed from the output lines of memory $M1$ and $log_2 n$ select lines routed from the output lines of memory $M2$. When the counter $CNTp$ selects the permutation X_j, the n indices in X_j are available in the output of the memory $M2$. These indices $r_{j,0}, r_{j,1}, \ldots, r_{j,n-1}$ are routed as select inputs to the n multiplexers MUX_0, $MUX_1, \ldots, MUX_{n-1}$, respectively. The $log_2 n$ select inputs of a multiplexer MUX_q point to the scan vector to be loaded to the scan chain SC_q. The selected scan vector is then shifted out of the selected memory block and into scan chain SC_q bit by bit over k clock cycles.

During on-chip test application, every time $CNTm$ is incremented, $CNTp$ counts from 0 to p-1. This allows p permuta-

tions in $M2$ to be applied to the subset selected by $CNTm$ to produce p tests. The total number of tests applied to the circuit is $p*m$. The software procedure discussed in the next section computes and optimizes the number of subsets of scan vectors as well as the number of permutations to reduce the memory sizes of $M1$ and $M2$ and hence the overall area overhead, as well as the number of tests. The routing overhead from the TGL to the scan chains is similar to the routing overhead incurred with test data compression. In the LBIST approach, the memories $M1$ and $M2$ (with counters $CNTm$ and $CNTp$) and n multiplexers replace the on-chip decompression logic. The stored test data entries can be adjusted if the scan chains are re-ordered.

For the output response, it is assumed that an output compaction logic such as a MISR [1] is used on the output side to reduce the volume of captured results.

IV. SOFTWARE PROCEDURE

This section describes a software procedure for computing the subsets of scan vectors and the permutations that reconstruct the original tests and their corresponding deterministic derivatives. This software procedure targets complete fault coverage for single stuck-at faults. The set of all the detectable stuck-at faults is denoted by F_{sa}. Undetectable faults are eliminated from consideration to simplify the procedure.

A. Overview

An overview of the procedure is shown in Figure 2. The procedure starts by initializing the undetected fault list $F_{usa} = F_{sa}$. In each iteration $i = 0, 1, \ldots,$ the software procedure creates a new subset of scan vectors S_i and a set of permutations $Y_i = \{X_{i,0}, X_{i,1}, \ldots, X_{i,p-1}\}$. To create a new subset of scan vectors S_i in an arbitrary iteration i, the procedure generates a deterministic test set T that detects all the faults in F_{usa}. Each test in the set T is simulated under F_{usa}. From the set T, the procedure chooses the test that detects the greatest number of faults in F_{usa}. The chosen test pattern is then partitioned into n scan vectors each of length k. These n scan vectors are stored in a subset S_i. Next, the procedure computes a set of

979-8-3503-4631-2/23 $31.00 © 2023 IEEE

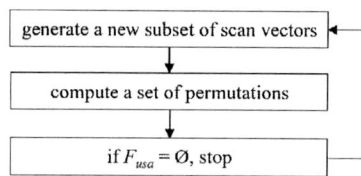

Fig. 2. Overview of software procedure.

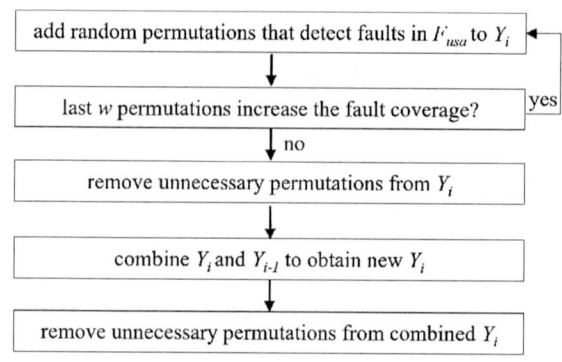

Fig. 3. Steps in permutation computation.

permutations Y_i that can be applied to every existing subset of scan vectors S_0, S_1, ..., S_{i-1}, S_i. The procedure computes the set $Y_i = \{X_{i,0}, X_{i,1}, ..., X_{i,p-1}\}$ in such a way that the deterministic derivatives produced by each permutation in Y_i detect as many faults as possible from F_{usa}. This property of the permutations magnifies their effectiveness thereby allowing fewer subsets as well as fewer permutations to be used.

At the end of the permutation computation step, the procedure generates a test set T_i. The tests in T_i are constructed by applying every permutation in Y_i to every existing subset of scan vectors S_0, S_1, ..., S_{i-1}, S_i. Every test in T_i is then fault simulated with fault dropping under F_{sa}. After fault simulation, the undetected fault list F_{usa} is updated. The procedure terminates when F_{usa} is empty which implies that all the faults from F_{sa} are detected. The steps involved in the permutation computation are described next.

B. Permutation Computation

Figure 3 shows the steps involved in the computation of a set of permutations. In an arbitrary iteration i, the software procedure computes a set of permutations $Y_i = \{X_{i,0}, X_{i,1}, ..., X_{i,p-1}\}$. The first permutation of Y_i is the original permutation $X_{i,0} = (0, 1, ..., n\text{-}1)$. The rest of the permutations $X_{i,1}, ..., X_{i,p-1}$ are selected randomly to avoid the complexity of forming permutations using a deterministic procedure. The procedure produces a random permutation by randomly permuting the original index set (0, 1, ..., n-1). It checks for every random permutation whether the deterministic derivatives obtained from its application on existing subsets of scan vectors detect any faults from F_{usa}. It then adds only permutations that detect new faults to the set Y_i. The procedure stops producing the permutations when the last w random permutations do not increase the fault coverage.

The value of w should be small enough to avoid long run times, and large enough to find as many useful permutations as possible. For the experimental results reported in this paper, w is set to 15000 for all the circuits. This value was selected by experimenting with different values and attempting to balance the run time and the quality of the results.

C. Removing Unnecessary Permutations

The set Y_i may contain permutations that are not necessary for the fault coverage attained by S_0, S_1, ..., S_{i-1}, S_i and Y_i. For example, the random permutation $X_{i,1}$ may not be necessary after adding the rest of the permutations $X_{i,2}, ..., X_{i,p-1}$. The original permutation $X_{i,0}$ is excluded from this analysis since the original tests are typically necessary for achieving complete fault coverage.

The software procedure removes the unnecessary permutations as described next. The concepts are adopted from [20]. The procedure constructs test sets $T_{i,1}, ..., T_{i,p-1}$ for the permutations $X_{i,1}, ..., X_{i,p-1}$, respectively. A test set $T_{i,j}$ contains all the deterministic derivatives constructed from applying the permutation $X_{i,j}$ to every existing subset S_0, S_1, ..., S_{i-1}, S_i. The procedure considers the test sets in different orders to remove unnecessary test sets and their corresponding permutations. Let T_i represent the collection of all the test sets. First the procedure considers the test sets in T_i in the original order followed by a random order and finally the reverse order. While considering the test sets in a particular order, the procedure simulates the ordered test sets in T_i under F_{sa} with fault dropping. It then associates with every test set $T_{i,j}$ the number of faults it detects. This number is denoted by $f(T_{i,j})$. The procedure rearranges the test sets by decreasing order of $f(T_{i,j})$ and simulates them again. If the number of faults detected by a test set becomes zero, the test set and the corresponding permutation are removed from T_i and Y_i, respectively. The procedure repeats the rearranging and fault simulation three times under each initial order. The sets T_i and Y_i are updated every time. The procedure reinitializes F_{sa} to include all the target faults before changing the order.

D. Combining Sets of Permutations

Although Y_i targets the subsets S_0, S_1, ..., S_i directly, it is possible that permutations included in Y_{i-1}, targeting S_0, S_1, ..., S_{i-1}, would be useful after S_i is added. It is important to take advantage of effective permutations from Y_{i-1} since this will reduce the number of iterations, and the number of subsets. For this purpose, the procedure combines Y_i with Y_{i-1} as follows. The procedure computes the average number of faults $f_a(T_i)$ detected by a subset $T_{i,j} \in T_i$. It also computes the average $f_a(T_{i-1})$, where T_{i-1} represents the collection of test sets from the previous iteration. The procedure then combines the permutations from Y_i and Y_{i-1} in the following fashion. Every $X_{i,j}$ with $f(T_{i,j}) \geq f_a(T_i)$ followed by every $X_{i-1,j}$ with $f(T_{i-1,j}) \geq f_a(T_{i-1})$ followed by $X_{i,j}$ with $f(T_{i,j}) < f_a(T_i)$ followed by $X_{i-1,j}$ with $f(T_{i-1,j}) < f_a(T_{i-1})$. This order ensures that the permutations that detect the largest numbers of faults, from both the sets Y_i and Y_{i-1}, are preserved. Unnecessary permutations are removed from the combined set using the same procedure described earlier, thereby allowing fewer permutations at the end of each iteration. The faults that remain undetected at the end of iteration i are assigned to F_{usa}.

V. IMPROVING THE PERMUTATIONS

The improvement procedure described in this section considers the permutations one by one for improvement. The order of the permutations ensures that permutations detecting more faults are considered earlier, making it more likely that permutations at the end of the set Y will become unnecessary. The procedure adds the improved permutations to a set Y_{mod}. If F_{mod}, which is a set of all the faults detected by the deterministic derivatives, is not empty after considering all the permutations in Y for improvement, the procedure repeats the same process for as many passes as needed, until F_{mod} becomes empty. The entire procedure is repeated for several iterations where at the end of each iteration Y_{mod} is assigned to Y. Each iteration tries to improve the set Y obtained from the previous iteration. The procedure stops when M consecutive iterations do not improve any permutation or reduce the number of permutations. For the experimental results reported in this paper, M is set to 4 for all the circuits.

The steps involved in an arbitrary iteration are shown in Figure 4. In an arbitrary iteration, the procedure attempts to improve the permutations X_1, X_2, ..., X_{p-1} from Y. When X_j is considered, the procedure constructs the test set T_j by applying X_j to every subset of scan vectors S_0, S_1, ..., S_{m-1}. It then simulates T_j under F_{mod} to compute the number of faults T_j detects, denoted by $f(T_j)$. It is important to simulate T_j because it is possible that the permutations that were modified before X_j detect the faults that were originally detected by X_j. If $f(T_j) = 0$, no permutation based on X_j is added to Y_{mod}. If $f(T_j) > 0$, the improvement procedure attempts to modify the permutation $X_j = (r_{j,0}, r_{j,1}, \ldots, r_{j,n-1})$ as follows. It considers the indices $r_{j,0}, r_{j,1}, \ldots, r_{j,n-1}$ one by one. It replaces the index under consideration with every other index from the set $\{0, 1, \ldots, n-1\}$. For every index in X_j, there are $n-1$ options available for replacement. The total number of modifications performed on X_j is $n * (n-1)$. The procedure considers these options one at a time. For every replacement option for the index $r_{j,m}$ where $0 \leq m \leq n - 1$, a modified permutation X_j^{mod} and its corresponding test set T_j^{mod} are obtained. The procedure simulates T_j^{mod} under F_{mod} and performs the following check. If the modified permutation X_j^{mod} detects more faults than its previous version, the replacement option is accepted, and the procedure continues to modify X_j^{mod} to improve it further. If not, the index under consideration is restored to its previous value. After exploring all the replacement options, the improved permutation X_j^{mod} is added to Y_{mod} and the faults detected by T_j^{mod} are dropped from F_{mod}.

VI. EXPERIMENTAL RESULTS

The software procedure that computes the subsets of scan vectors S_0, S_1, ..., S_{m-1} and set of permutations Y was

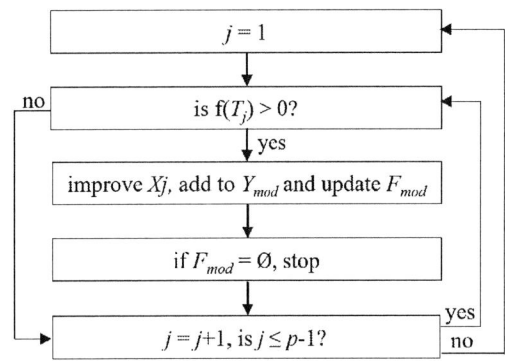

Fig. 4. Steps in an arbitrary iteration for improving the permutations.

applied to single stuck-at faults in ISCAS-89, ITC-99 and IWLS-05 benchmark circuits, and three logic blocks (*ffu*, *spu* and *exu*) of the OpenSPARC T1 microprocessor [24]. Consideration of these logic blocks demonstrates the applicability of the proposed approach to logic blocks of a large processor. A commercial tool was used for test generation, fault simulation and reordering. The experimental results are shown in Table III. Table III also compares the proposed method with [19] and a pseudo-random LBIST method available in the same commercial tool used for the implementation of the software procedure.

In Table III, after the circuit name, column n shows the number of scan chains which is equal to the number of scan vectors in each subset S_i. Column k shows the length of each scan chain. Column *ATPG tests* shows the number of uncompressed deterministic tests that achieves complete fault coverage. Column m shows the number of subsets of scan vectors. Column p shows the number of permutations in the set Y. Column $p*m$ shows the total number of tests generated on-chip to achieve complete fault coverage. Column *bits* shows the number of storage bits required for m subsets of scan vectors (subcolumn S), number of bits required for storing p permutations (subcolumn Y) and total number of bits for storing both the subsets and permutations (subcolumn *tot*). Column *red%* shows the storage reduction percentage, computed as *(ATPG bits-tot bits)/ATPG bits*. Here *ATPG bits* is the number of storage bits required for ATPG tests. The ATPG test set is generated by the same commercial ATPG tool used for implementing the software procedure. Column *FC%* shows the complete fault coverage achieved by the proposed LBIST scheme. Column *rt* shows the run time in minutes. This is the total run time taken to execute the entire software procedure. Along with the commercial tool, the implementation uses python scripts which are slow.

Column [19] and column *random tests* present the results from [19] and a pseudo-random LBIST method, respectively, to compare with the proposed LBIST scheme. Column [19] subcolumn *tests* shows the number of test patterns generated by [19], on chip, to achieve complete stuck-at fault coverage. Subcolumn *bits* shows the number of bits required for storing the scan vectors used in [19]. Subcolumn *red%* shows the

979-8-3503-4631-2/23 $31.00 © 2023 IEEE

TABLE III
EXPERIMENTAL RESULTS

circuit	n	k	ATPG tests	m	p	p*m	bits S	bits Y	bits tot	red%	FC%	rt	[19] tests	[19] bits	[19] red%	random tests
s35932	42	43	59	1	23	23	1806	5796	7602	92.71	100.0	246	-	-	-	128
sasc	12	12	35	4	11	44	576	528	1104	76.46	100.0	13	384	144	95.04	256
systemcdes	18	18	104	3	31	93	972	2790	3762	88.84	99.74	38	768	216	99.15	704
des_area	18	18	195	5	30	150	1620	2700	4320	92.85	100.0	41	1152	720	98.34	768
usb_phy	11	11	39	4	14	56	484	616	1100	75.47	100.0	12	1024	176	95.09	1728
aes_core	28	29	376	2	253	506	1624	35420	37044	87.55	100.0	2580	-	-	-	2688
systemcaes	31	31	208	13	28	364	12493	4340	16833	91.32	100.0	546	4096	992	99.12	13504
b04	9	9	78	5	25	125	405	900	1305	79.34	96.91	18	49152	108	96.85	20032
s1423	10	10	70	7	17	119	700	680	1380	79.25	100.0	35	8192	320	86.48	25856
b05	6	7	90	21	17	357	882	306	1188	66.15	100.0	16	20480	120	94.54	34432
s5378	14	15	129	11	48	528	2310	2688	4998	80.72	99.89	93	16384	1920	91.03	49344
s13207	16	16	60	13	14	182	3328	896	4224	71.61	95.75	78	53248	5616	96.59	213056
wb_conmax	44	44	178	8	68	544	15488	17952	33440	90.12	99.96	3840	-	-	-	310976
s9234	13	14	98	7	36	252	1274	1872	3146	81.34	99.17	144	-	-	-	406144
ffu	38	38	192	6	68	408	8664	15504	24168	91.17	97.73	1860	-	-	-	508224
b07	7	7	71	19	11	209	931	231	1162	66.60	100.0	15	98304	192	93.03	*1M
b14	16	16	284	63	51	3213	16128	3264	19392	72.80	99.91	4422	-	-	-	*1M
b15	21	22	459	73	125	9125	33726	13125	46851	77.62	99.52	12696	-	-	-	*1M
b20	22	22	332	69	58	4002	33396	6380	39776	74.29	99.92	3660	-	-	-	*1M
s38417	39	39	149	19	122	2318	28899	28548	57447	74.19	99.98	8094	-	-	-	*1M
s38584	34	35	127	17	37	629	20230	7548	27778	81.38	99.70	660	-	-	-	*1M
DMA	49	49	351	102	115	11730	244902	33810	278712	66.86	99.64	26849	-	-	-	*1M
i2c	12	13	73	24	15	360	3744	720	4464	58.96	99.91	86	24576	624	90.44	*1M
pci_spoci_ctrl	9	10	174	55	31	1705	4950	1116	6066	59.93	100.0	384	425984	1040	91.42	*1M
simple_spi	12	13	62	11	18	198	1716	864	2580	71.88	100.0	23	28672	364	93.07	*1M
spi	17	17	500	18	72	1296	5202	6120	11322	91.86	99.93	225	32768	1088	99.02	*1M
tv80	19	20	466	76	102	7752	28880	9690	38570	77.93	100.0	3240	-	-	-	*1M
spu	46	46	198	114	28	3192	241224	7728	248952	40.27	99.04	13380	-	-	-	*1M
exu	54	55	421	151	75	11325	448470	24300	472770	61.93	98.37	45372	-	-	-	*1M

storage reduction achieved in [19]. It is to be noted that the circuits are synthesized differently and the test sets are different in [19]. Column *random tests* shows the number of pseudo-random test patterns generated by a pseudo-random LBIST method to achieve complete fault coverage. An asterisk indicates that the pseudo-random test generation is stopped at 1M tests without achieving complete fault coverage. The circuits in Table III are arranged by increasing order of number of random tests.

The following points can be seen from Table III. The proposed fully deterministic LBIST scheme achieves complete fault coverage for all the circuits, thereby eliminating the need for pseudo-random tests. This in turn reduces the number of tests. Compared with [19] and a pseudo-random LBIST method, the proposed scheme uses a significantly smaller number of tests. This is the main purpose of the work. For example, in the case of *spi*, the reduction in number of tests is 32768/1296=25.28.

The software procedure reduces the storage requirements for all the circuits considered. The storage reduction achieved by the proposed scheme is lesser than [19] which is inevitable when no random tests are used.

The improvement software procedure helps in reducing the initial number of permutations by over 52%. This procedure is important for further reducing the storage requirement and thus the hardware overhead.

The circuit *spu* has the lowest storage reduction of 40%. Without the last 1% fault coverage, the storage reduction for *spu* is 92%. Similarly, for *i2c*, the reduction is 82% without the last 2% fault coverage. This points to the possibility that test points will be useful in keeping the storage reduction high.

VII. CONCLUSION

This paper described a fully deterministic storage based logic built-in self-test (LBIST) approach that stores, on chip, reduced deterministic uncompressed test data sufficient for achieving complete fault coverage. This approach eliminated the need for pseudo-random tests, thereby reducing the test application time by reducing the number of tests required to achieve complete fault coverage. Under this approach, two types of test data are stored on chip. 1) Subsets of scan vectors obtained from a reduced set of deterministic tests, one subset per test and, 2) permutations of scan vectors stored as sets of indices, to indicate how to combine scan vectors to form tests on chip. The same permutations are applied to all the subsets, magnifying the effectiveness of each stored permutation and each subset, allowing fewer subsets as well as fewer permutations to be used. This helps in reducing the storage requirements. Experimental results for single stuck-at faults in benchmark circuits and logic blocks of the OpenSPARC T1 microprocessor demonstrated the effectiveness of this approach.

REFERENCES

[1] P. H. Bardell, W. H. McAnney and J. Savir, Built-In Test for VLSI Pseudorandom Techniques, Wiley Interscience, 1987.

[2] R. Kapur, S. Patil, T. J. Snethen and T. W. Williams, "Design of an efficient weighted random pattern generation system," *Proc. ITC*, 1994, pp. 491-500.

[3] N. A. Touba and E. J. McCluskey, "Altering a pseudo-random bit sequence for scan-based BIST," in *Proc. ITC*, 1996, pp. 167-175.

[4] V. Iyengar, K. Chakrabarty and B. T. Murray, "Built-in self testing of sequential circuits using precomputed test sets," in *Proc. VTS*, 1998, pp. 418-423.

[5] M. Nakao, S. Kobayashi, K. Hatayama, K. Iijima and S. Terada, "Low overhead test point insertion for scan-based BIST," *Proc. ITC*, 1999, pp. 348-357.

[6] D. Das and N. A. Touba, "Reducing test data volume using external/LBIST hybrid test patterns," in *Proc. ITC*, 2000, pp. 115-122.

[7] H. G. Liang, S. Hellebrand and H. -J. Wunderlich, "Two-dimensional test data compression for scan-based deterministic BIST," in *Proc. ITC*, 2001, pp. 894-902.

[8] I. Pomeranz and S. M. Reddy, "A partitioning and storage based built-in test pattern generation method for scan circuits," in *Proc. VLSI Design Conf.*, 2002, pp. 677-682.

[9] A. A. Al-Yamani, S. Mitra and E. J. McCluskey, "BIST reseeding with very few seeds," in *Proc. VTS*, 2003, pp. 69-74.

[10] V. Gherman, H. -J. Wunderlich, H. Vranken, F. Hapke, M. Wittke and M. Garbers, "Efficient pattern mapping for deterministic logic BIST," in *Proc. ITC*, 2004, pp. 48-56.

[11] S. Pateras, "Security vs. test quality: fully embedded test approaches are the key to having both," in *Proc. Intl. Test Conf.*, 2004, pp. 1413.

[12] A. Jas, C. V. Krishna and N. A. Touba, "Weighted pseudorandom hybrid BIST," in *IEEE Trans. Very Large Scale Integr. (VLSI) Syst.*, Dec. 2004, Vol. 12, No. 12, pp. 1277-1283.

[13] V. Gherman, H. -J. Wunderlich, J. Schloeffel and M. Garbers, "Deterministic Logic BIST for Transition Fault Testing," in *Proc. ETS*, 2006, pp. 123-130.

[14] R. S. Oliveira, J. Semião, I. C. Teixeira, M. B. Santos and J. P. Teixeira, "On-line BIST for performance failure prediction under aging effects in automotive safety-critical applications," in *Proc.* Latin American Test Workshop, 2011, pp. 1-6.

[15] M. Filipek, G. Mrugalski, N. Mukherjee, B. Nadeau-Dostie, J. Rajski, J. Solecki and J. Tyszer, "Low-Power Programmable PRPG With Test Compression Capabilities", in *IEEE Trans. Very Large Scale Integr. (VLSI) Syst.*, June 2015, Vol. 23, No. 6, pp. 1063-1076.

[16] E. Moghaddam, N. Mukherjee, J. Rajski, J. Tyszer and J. Zawada, "Test point insertion in hybrid test compression/LBIST architectures," *Proc. ITC*, 2016, pp. 1-10.

[17] Y. Liu, N. Mukherjee, J. Rajski, S. M. Reddy and J. Tyszer, "Deterministic Stellar BIST for In-System Automotive Test," in *Proc. ITC*, 2018, pp. 1-9.

[18] I. Pomeranz, "Storage-Based Built-In Self-Test for Gate-Exhaustive Faults," in *IEEE Trans. Comput.-Aided Design Integr. Circuits Syst.*, Oct. 2021, Vol. 40, No. 10, pp. 2189-2193.

[19] I. Pomeranz, "Zoom-In Feature for Storage-Based Logic Built-In Self-Test," in *Proc. Symp. Defect Fault Tolerance*, 2021, pp. 1-6.

[20] X. Lin, J. Rajski, I. Pomeranz and S. M. Reddy, "On static test compaction and test pattern ordering for scan designs," *Proc. ITC*, 2001, pp. 1088-1097.

[21] B. Koenemann, "Care bit density and test cube clusters: multi-level compression opportunities," *Proc.* International Conference on Computer Design, 2003, pp. 320-325.

[22] J. Rajski, J. Tyszer, M. Kassab and N. Mukherjee, "Embedded deterministic test," in *IEEE Trans. Comput.-Aided Design Integr. Circuits Syst.*, May 2004, Vol. 23, No. 5, pp. 776-792.

[23] P. Wohl, J. A. Waicukauski, S. Patel, F. DaSilva, T. W. Williams and R. Kapur, "Efficient compression of deterministic patterns into multiple PRPG seeds," *Proc. ITC*, 2005, pp. 10 pp.-925.

[24] Oracle 2022, *OpenSPARC T1 Specifications*, https://www.oracle.com/servers/technologies/opensparc-overview.html.

Test Generation for Defect-Based Faults of Scan Flip-Flops

Yu-Teng Nien*, Chen-Hong Li*, Pei-Yin Wu*, Yung-Jheng Wang*,
Kai-Chiang Wu[†], and Mango C.-T. Chao*

* Institute of Electronics, National Yang Ming Chiao Tung University, Hsinchu, Taiwan
[†] Department of Computer Science, National Yang Ming Chiao Tung University, Hsinchu, Taiwan

Abstract—**When testing scan flip-flops (SFFs), chain test is first applied to ensure the functionality of scan chains and to detect the majority of stuck-at (SA) and transition delay (TD) faults along scan paths. However, there still exist some defects inside scan cells that cannot be effectively detected by chain test or conventional SA and TD patterns. This paper presents five cell-aware (CA) fault models to explicitly target the defects inside scan flip-flops. The proposed static shift (SS) and dynamic shift (DS) faults identify the defects detectable by chain test. For the defects escaping chain test, static single-capture (SSC) faults target the defects detectable when SFFs are in one-cycle capture mode, while static double-capture (SDC) and dynamic double-capture (DDC) faults target those detectable when SFFs are in two-cycle capture mode. The identified CA faults of SFFs are output in a format compatible with a commercial ATPG tool for pattern generation. Experimental results on large IWLS05 benchmarks demonstrate that our proposed faults cannot be fully covered by conventional SA and TD patterns and hence require dedicated test patterns to detect.**

I. INTRODUCTION

Conventional fault models, such as stuck-at (SA) and transition delay (TD) faults, all define faulty behavior at the I/O ports of standard cells with simple rules of fault activation and propagation. However, some intra-cell defects might remain undetected if the tests based on such conventional fault models are applied. By explicitly SPICE-simulating the faulty behavior, cell-aware (CA) methodology was proposed to further increase the coverage of intra-cell defects [1], [2]. The effectiveness of CA tests was proven in [3]–[5] with silicon data, demonstrating its capability of capturing defective parts missed by tests for conventional SA and TD faults. This CA methodology is supported by the commercial Tessent ATPG [6] with a flexible format, user-defined fault model (UDFM), which enables users to specify complicated conditions for fault-activation and fault-observation. Likewise, CA solutions are also provided by commercial TestMAX [7] and Modus [8] ATPG. Although extensive research has investigated the impact of intra-cell defects on combinational cells, limited effort has been directed toward the defects inside sequential cells, especially scan flip-flops (SFFs).

Formed by connecting scan cells one after another, scan chain serves as the infrastructure that enables scan-based testing and diagnosis [9]. The industry has adopted chain test (aka flush test or shift test) to ensure the functionality of scan chains, during which alternating double zeroes and double ones (0011...) are loaded into and unloaded from all chains. For an SFF instance, during chain test, scan enable (SE) is held at 1 and scan input (SI) is supposed to experience all the four transitions possible over two cycles: 00, 01, 10, and 11. By following this practice, the majority of SA and TD faults along scan paths can be covered. However, [10] reported that chain test or conventional SA and TD tests cannot fully detect the defects inside scan latches and SFFs.

Some previous works have investigated the impact of cell-internal defect within SFFs with SPICE simulation, in an attempt to improve the coverage of such defects. In [11], CA fault models for SFFs were used to improve the diagnostic resolution of static defects within scan chains with conventional SA tests applied. In [12]–[16], various sets of half-speed and slower chain tests with modified toggle sequence were proposed to cover more SFF-internal defects detectable when scan cells are in shift mode. In [17], an efficient CA fault model generation flow was proposed for multi-bit flip-flops, which would incur relatively greater runtime effort to perform exhaustive SPICE simulation due to the large numbers of input pins. In [18], by focusing on the defects affecting the multiplexers of SFFs, CA faults were identified and then test-generated on top of chain test as well as conventional SA and TD tests. While the leading commercial tool of CA fault extraction, CellModelGen [19], also supports sequential cells, no documents or publications from Mentor Graphics explained how CA faults are extracted for SFFs or detailed the settings around its underlying defect-injected SPICE simulation. Moreover, the CA models from CellModelGen are basically not understandable to an end user since the fault conditions are all encrypted in the generated UDFM files.

In this paper, we present five CA fault models to explicitly examine the faulty behavior of defects inside scan flip-flops via SPICE simulation. Based on the results of simulation over three cycles, our proposed fault models define a fault depending on (*i*) the number of capture cycles required for detection and (*ii*) the type (static or dynamic) of fault effect. The proposed static shift (SS) and dynamic shift (DS) faults identify the defects detectable with chain test. The other three, static single-capture (SSC), static double-capture (SDC), and dynamic double-capture (DDC) faults, target the defects that escape chain test and are detectable only when SFFs are in capture mode. The extracted CA faults are output in UDFM format and then test-generated with Tessent ATPG [6]. Lastly, we provide the results of fault extraction for SFFs, coverage of SS and DS faults by chain test, and the results of top-off ATPG on SSC, SDC, and DDC faults. Experimental results on large IWLS05 [20] benchmarks demonstrate that our proposed faults cannot be fully covered by conventional SA and TD tests and hence require dedicated tests to detect.

II. PROPOSED CA FAULTS FOR SCAN FLIP-FLOPS

We propose five CA fault models to better address the defects internal to scan flip-flops (SFFs). For each fault model, the conditions of fault-activation and fault-observation are defined.

Among the defects detectable during chain test (SE=1), static shift (SS) models those causing faulty output voltage while dynamic shift (DS) models those behaving as delay faults. For the defects detectable when scan cells are in capture mode (SE=0) for one cycle, static single-capture (SSC) models those which cause faulty values at output pins. For the defects detectable when scan cells are in capture mode (SE=0) for two cycles, static double-capture (SDC) models the defects causing faulty output voltage, while dynamic double-capture (DDC) models those behaving as delay faults.

For the static fault models (SS, SSC, and SDC), we measure the output voltage under the influence of a defect. As for the dynamic fault models (DS and DDC), the transition time from a clock edge to the time at which the output rising/falling reaches 50% VDD is measured. Based on the measurement made in SPICE simulation, we can evaluate the severity of a defect inside a scan cell, and each detectable defect can be classified into one of the five fault models.

Our study is based on a mature cell library that provides SFF cell types with various driving strengths, trigger types, and support for

979-8-3503-4631-2/23 $31.00 © 2023 IEEE

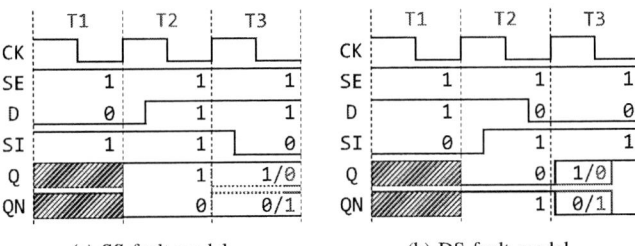

(a) SS fault model   (b) DS fault model

Fig. 1. Waveform example of shift fault models (SE=111).

TABLE I
FAULT CONDITIONS OF SS FAULT MODEL

Fault condition index	Input condition (T1T2T3)			Output fault-free (T2T3)		Output faulty (T3)		# of simulated stimuli
	SE	SI	D	Q	QN	Q	QN	
C1	111	001	000	00	11	1	0	1
C2	111	001	001	00	11	1	0	2
C3	111	001	010	00	11	1	0	4
C4	111	001	011	00	11	1	0	2
C5	111	001	100	00	11	1	0	2
C6	111	001	101	00	11	1	0	4
C7	111	001	110	00	11	1	0	2
C8	111	001	111	00	11	1	0	1
C9	111	011	000	01	10	0	1	1
⋮								
C17	111	100	000	10	01	1	0	1
⋮								
C25	111	110	000	11	00	0	1	1
⋮								

set/reset/enable functions. In this section, the simplest SFF cell from this library, SDFFX1, is used for illustration purposes only. On this SFF cell, there are four input pins, scan enable (SE), scan input (SI), data (D), and clock (CK); output pins include the latched value (Q) and its inversion (QN). Please note that although a relatively simple SFF is chosen, our proposed methodology can extend to SFF cells with set/reset/enable pins, as long as the states of such pins are enumerated in a similar manner as D pin.

A. Static Shift (SS) Fault

During chain test, the toggle sequence 0011... is shifted into and out of each scan chain such that all the SFFs experience each of the two-cycle combination from 00, 01, 10, and 11 at their SI pins with their SE pins kept at 1. This way, some defects internal to SFFs might be detected during the shift-in and shift-out procedure of chain test. To find out the defects potential detected during chain test, we first propose SS and DS fault models, where the arranged stimuli at SI correspond to those applied during chain test.

Fig. 1(a) illustrates the arrangement of stimuli used for SS fault model, where the logic values at input and output pins within three cycles (T1, T2, and T3) are shown. For fault-activation, the values of SE, D, and SI pins at the three cycles are examined. In this case, SE is high across the three cycles, while D is 011 and SI is 110. Upon the two rising edges at CK in T2 and T3, the value on SI is latched into SFF, and values of Q (QN) becomes 11 (00). For fault-observation, the values at Q and QN in T3 are measured. To ensure that fault propagation is possible whichever Q or QN is connected to scan path, both output pins must be faulty for a defect to be considered detectable. In Fig. 1(a), the faulty values are 0 for Q and 1 for QN and thus marked in red. In this work, we define SS faults as those faults that (i) could be activated during chain test and (ii) cause discrepant logic values from the defect-free cases at output pins.

To measure the defective voltage deviation under all input combinations, we apply a set of different 72 stimuli in SPICE simulation, which can be subsumed under 32 fault conditions. For the fault conditions used in SS fault model, the three-cycle input values on SI correspond to the values that an SI pin would undergo during chain test (001, 011, 100, and 110), whereas the sequence on D is enumerated. Table I lists the fault conditions for SS fault model. Column 1 lists the index of fault conditions. Columns 2–4 stand for the logic values applied at the input pins over the three cycles (T1, T2, and T3). The corresponding fault-free output responses in the last two cycles (T2 and T3) are listed in Columns 5 and 6. Columns 7 and 8 list the faulty output values observed in T3. Column 9 lists the number of required stimuli for each condition in simulation. For the sake of brevity, only a selection of fault conditions is shown in this table.

In this work, we also take into consideration the variations in the arrival time of transition on D pin because the timing of transition at input pin could affect the activation/detection of a defect [12]–[16]. A fault condition may thus require multiple input stimuli in simulation as shown in Column 9, to ensure that, whether the transition on D pin occurs early (CK=1) or late (CK=0), a defect would still expose its faulty values onto the output pins and remain detectable. Depending on the number of transition in the condition of D pin, a fault condition requires two stimuli if the number of transition is one (i.e., 001, 011, 100, and 110); once two transitions can be found (i.e., 010 and 101), one could demand up to four stimuli to address the differences in transition times over both T2 and T3. As for SI pin, the incoming transition is assumed to arrive early (CK=1) since the interconnections between two SFF instances in a scan chain tend to be short.

B. Dynamic Shift (DS) Fault

Similar to SS fault model, our DS fault model defines SFF faults that are detectable and would cause faulty values on scan path during chain test, but as opposed to static faults, dynamic (delay) faults are addressed in DS fault model.

Fig. 1(b) illustrates the arrangement of stimuli used for DS fault model, where the logic values at input and output pins within three cycles (T1, T2, and T3) are shown. For fault-activation, SE remains high throughout the three cycles, while 100 is applied to D pin and 011 to SI. Because the value on SI switches from 0 to 1, a rising (falling) transition can be observed on Q (QN) at the beginning of T3. For fault-observation, the transition delays on output pins are examined to identify whether a defect would cause delay faults. In Fig. 1(b), due to the extra delay incurred by an internal defect, the identified DS fault behaves as a slow-to-rise (STR) fault on Q and a slow-to-fall (STF) on QN and is thus marked in red. In this work, we define DS faults as those faults that (i) could be activated during chain test and (ii) exhibit STR/STF behavior at output pins.

For DS fault model, we apply a set of different 36 stimuli to defects under examination, which can be subsumed under 16 fault conditions. Over the three cycles in simulation, the values on D are exhaustive, and values on SI could only be 011 or 100 to launch a transition at each output pin. It can be clearly noted that the fault conditions used for DS fault model are a subset of those used for SS fault model, which are shown in Table I. The primary distinction of DS fault model is that we focus on STR/STF faulty behaviors at output pins rather than static ones (see Columns 7 and 8). Hence, the DS fault conditions are not specifically tabulated here due to page limit.

C. Static Single-Capture (SSC) Fault

We propose SSC fault model to cover the defects that are detectable under the fault conditions involving one capture cycle (SE=0). Due to the number of required capture cycles in fault-activation and fault-propagation, the test patterns needed to target the SSC faults are similar to those based on SA fault model in terms of test application scheme. In this work, we define SSC faults as those faults that (i) are detectable under fault conditions involving one capture cycle and (ii) cause discrepant logic values at output pins.

Fig. 2 illustrates the stimuli arranged for the activation and observation of SSC faults. Fig. 2(a) shows the logic values on the input and output pins within three cycles (T1, T2, and T3) for SSC fault model. SE turns from 1 to 0 in T2 and returns to 1 in the end of T3, bringing the SFF into capture mode during T3. For fault-activation, 011 and 100 are applied to D and SI pins. Output value Q (QN) becomes 11 (00)

979-8-3503-4631-2/23 $31.00 © 2023 IEEE   137

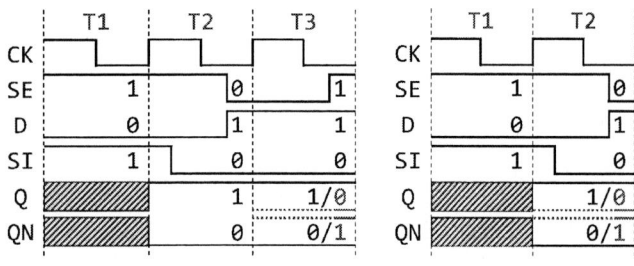

(a) Fault effect observed in T3   (b) Fault effect observed in T2

Fig. 2. Waveform example of SSC fault model.

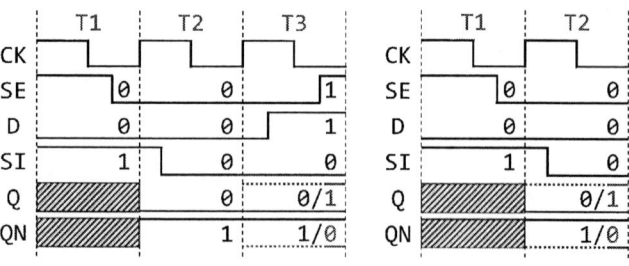

(a) Fault effect observed in T3   (b) Fault effect observed in T2

Fig. 3. Waveform example of SDC fault model.

TABLE II
FAULT CONDITIONS OF SSC FAULT MODEL

Fault condition index	Input condition (T1T2T3)			Output fault-free (T2T3)		Output faulty (T2)		Output faulty (T3)		# of simulated stimuli
	SE	SI	D	Q	QN	Q	QN	Q	QN	
C1	101	000	000	00	11	1	0	1	0	1
C2	101	000	001	00	11	1	0	1	0	2
⋮										
C8	101	000	111	01	10	1	0	0	1	1
C9	101	001	000	00	11	1	0	1	0	1
⋮										
C17	101	010	000	00	11	1	0	1	0	1
⋮										
C57	101	111	000	10	01	0	1	1	0	1
⋮										

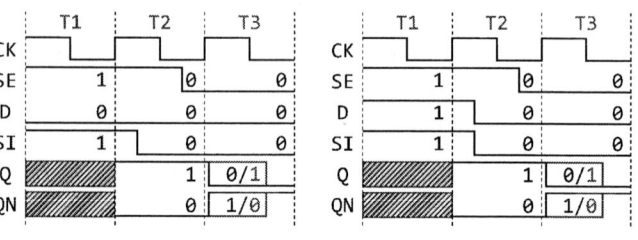

(a) Value on D is 0 in T1   (b) Value on D is 1 in T1

Fig. 4. Waveform example of DDC fault model.

as a result of SI value in T1 and D value in T2. For fault-observation, we inspect the voltage levels on output pins, where the observed faulty values are 0 on Q and 1 on QN, and are thus marked in red. Note that we also inspect the output response in T2 for observing fault effects as shown in Fig. 2(b), where the faulty values are 0 on Q and 1 on QN. For SSC faults observed in T2, fault-activation depends on input values in T1 and T2 instead.

Besides the number of required cycles for fault-activation and fault-observation, the propagation of fault effects required by SSC faults observed in T2 and T3 also differs. For an SSC fault with faulty value observed in T3 (as in Fig. 2(a)), its fault effect is exposed in the capture cycle to scan path and can then be directly shifted out of netlist via scan chain. In contrast, for one with faulty value observed in T2 (as in Fig. 2(b)), the fault effect needs to be first propagated across the combinational part of netlist before it can be captured in the clock cycle following T2 and then be shifted out by scan chain.

For SSC fault model, we apply a set of 144 different stimuli to defects under examination. The applied stimuli can be subsumed under 64 fault conditions where values on D and on SI are both exhaustive and SE is 101. Table II shows a fraction of the SSC fault conditions and the faulty values observed at T2 and T3. Column 1 lists the index of fault conditions. Columns 2–4 list the condition on input pins in the three cycles, and Columns 5 and 6 list the corresponding fault-free output responses in T2 and T3. Columns 7 and 8 list the faulty values observed in T2, while Columns 9 and 10 list those observed in T3. Column 11 lists the numbers of required stimuli for each fault condition.

D. Static Double-Capture (SDC) Fault

We define SDC faults as the faults that (i) are detectable under fault conditions involving two capture cycles and (ii) cause discrepant logic values at output pins. Due to the number of required capture cycles for fault-activation and fault-propagation, the test patterns needed to target the SDC faults are similar to those based on launch-off-capture (LOC) TD fault model in terms of test application scheme.

Fig. 3 illustrates the stimuli arranged for the activation and observation of SDC faults. Fig. 3(a) shows the logic values on the input and output pins within three cycles (T1, T2, and T3). SE pin switches from 1 to 0 in T1 and returns to 1 in the end of T3, bringing the SFF into capture mode during T2 and T3. For fault-activation, 001 and 100 are applied to D and SI pins. Output value Q (QN) becomes 00 (11) in accordance with values on D in T1 and T2. For fault-observation, we inspect the voltage levels on output pins, where the observed faulty values are 1 on Q and 0 on QN, and are thus marked in red. We also inspect the faulty values in T2 as shown in Fig. 3(b), where the faulty values are 1 on Q and 0 on QN. For SDC faults observed in T2, fault-activation depends on input values in T1 and T2 instead. Lastly, like SSC fault model in last subsection, for the SDC faults, the propagation of fault effects observed in T2 and T3 also differs.

For SDC fault model, we apply a set of 144 different stimuli to defects under examination. The applied stimuli can be subsumed under 64 fault conditions where values on D and on SI are both exhaustive while SE is always 001. Clearly, in terms of the arrangement of fault conditions, the only difference between SDC and SSC fault models is in the values assigned to SE pin as well as the corresponding fault-free and faulty values on Q and QN pins. Due to the high similarity to SSC fault conditions, the SDC fault conditions are not specifically tabulated here to minimize redundancy.

E. Dynamic Double-Capture (DDC) Fault

Similar to SDC fault model, our DDC fault model covers the defects that are detectable when SFF instances are switched to and from capture mode of two cycles, but as opposed to static faults, dynamic (delay) faults are addressed in DDC fault model. Due to the number of required capture cycles, the test application scheme of DDC faults and that of LOC TD fault model are identical. In this work, we define DDC faults as the SFF faults that (i) are detectable under the fault conditions involving two capture cycles (SE=0) and (ii) exhibit STR/STF behavior at output pins.

Fig. 4 illustrates the stimuli arranged for the activation and observation of DDC faults. Fig. 4(a) shows the logic values on the input and output pins within three cycles (T1, T2, and T3). SE turns from 1 to 0 at the end of T2 and is held at 0 till the end, bringing the SFF into capture mode during T3. For fault-activation, 000 and 100 are applied to D and SI pins, leading to a falling (rising) transition at Q (QN) pin. For fault-observation, the transition delays on output pins are inspected to identify whether a defect would cause delay faults. In Fig. 4(a), due to the defect-induced extra delay, the identified DDC fault causes STF on Q and STR on QN, and is thus marked in red. The fault effect observed in T3 needs to be propagated through the combinational part of a netlist

TABLE III
FAULT CONDITIONS OF DDC FAULT MODEL

Fault condition index	Input condition (T1T2T3)			Output fault-free (T2T3)		Output faulty (T3)		# of simulated stimuli
	SE	SI	D	Q	QN	Q	QN	
C1	100	000	-10	01	10	0	1	4
C2	100	000	-11	01	10	0	1	2
C3	100	001	-10	01	10	0	1	4
⋮								
C8	100	011	-11	01	10	0	1	2
C9	100	100	-00	10	01	1	0	2
C10	100	100	-01	10	01	1	0	4
C11	100	101	-00	10	01	1	0	2
⋮								
C16	100	111	-01	10	01	1	0	4

```
Cell("SDFFX1") {
  Fault("d1") {
    Test {  // Propagated through Q pin
      DelayFault {"Q": 1;}
      Conditions {"SE": 00; "D": 00; "SI": 00;} }
    Test {  // Propagated through QN pin
      DelayFault {"QN": 0;}
      Conditions {"SE": 00; "D": 00; "SI": 00;} }
  }
  Fault("d2") { DelayEquivalent: "d1"; }
}
```

Fig. 5. UDFM description of a DDC fault detectable with stimuli in Fig. 4.

before it can be captured by scan chain in the next cycle following T3 and then be shifted out.

Since we can only specify the input values in T2 and T3 as well as the direction of output transition in UDFM, the stimulus is accompanied by the one in Fig. 4(b) to ensure that a detectable defect would remain a DDC fault whether the D value in T1 is 0 or 1. Between the two exemplary stimuli, the only difference is the input values on D in T1, whereas the values on SI are both 1 in T1. Note that it is not necessary to make sure that the defect under examination is detectable if SI in T1 is either 0 or 1, because SI in T1 determines the values on output pins in T2 and can hence be implied by the transition direction at Q and QN in UDFM.

For DDC fault model, we apply a set of 48 different stimuli to defects under examination. The applied stimuli can be subsumed under 16 fault conditions where SE is 100, and the values on SI are exhaustive. For the D values within three cycles, the value in T1 is always a don't-care bit (denoted by "-"), the value in T2 is opposite to the SI value in T1 to create transition at output pins, and the value in T3 can be either 0 or 1. Table III lists a fraction of fault conditions for DDC fault model. The table format is almost the same as Table I except for Columns 7 and 8, where "0" and "1" stand for STR and STF, respectively.

Fig. 5 shows an example of UDFM description for our DDC faults. In this case, scan cell "SDFFX1" has two defects that are detectable as DDC faults: $d1$ and $d2$. The UDFM in Fig. 5 first defines the name of the cell. Then each DDC fault (declared with Fault) is defined by specifying all its fault conditions (declared with Test), within which the faulty values at output pins (declared with DelayFault) and the associated activation conditions (declared with Conditions) are defined. Suppose that defect $d1$ exposes faulty values if the stimuli in Fig. 4 are applied and is thus detectable under fault condition C9 listed in Table III. As can be noted from the example, only two bits can be specified in the activation conditions for a delay fault, like DDC, corresponding to the input values in T2 and T3. Please also note that a fault condition may entail multiple Test statements, depending on the number of output pins of the scan cell. Since SDFFX1 has two output pins, this single fault condition contains two Test statements, one for fault propagation through Q and the other for QN. A DDC fault can also be declared equivalent to another fault using DelayEquivalent. In

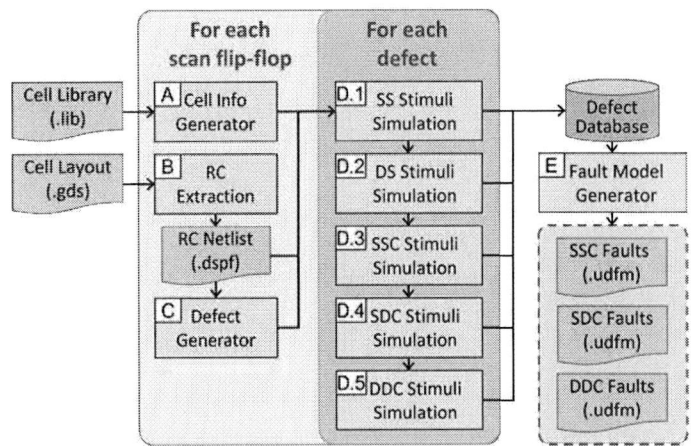

Fig. 6. Flow of extracting CA faults for scan flip-flops.

this case, $d2$ is set equivalent to the $d1$ defined above. Lastly, besides DDC, the SSC and SDC faults can also be represented in UDFM similarly, as exemplified in Fig. 5.

III. PROPOSED FRAMEWORK FOR DETECTING SFF FAULTS

In this section, we will propose a framework for extracting CA faults of SFFs and generating CA tests based on the proposed fault models presented in Section II. This framework incorporates several commercial tools, such as Calibre [21], HSPICE [22], and Tessent [6], and is mainly composed of two parts: CA fault extraction for SFFs (Section III-A) and fault coverage evaluation of chain test (Section III-B).

A. Flow of Extracting CA Faults for Scan Flip-Flops

Fig. 6 depicts the proposed flow of extracting CA faults for scan flip-flops. Step A, Step B, and Step C are performed for each SFF cell, while Step D is performed for each identified defect. Step A (Cell Info Generator) parses a cell library (.lib), fetching input/output pins and boolean functions of SFFs to be later used in SPICE simulation. By running Calibre, Step B (RC Extraction) extracts the parasitic resistance and coupling capacitance from the cell layout (.gds) and creates RC netlist in DSPF (Detailed Standard Parasitic Format). Step C (Defect Generator) identifies the locations of potential defects inside SFFs with four defect types considered: short, open, transistor stuck-on (ton), and transistor stuck-open (toff).

For each defect identified, with the stimuli of the proposed fault models applied in five stages, the defective behaviors are characterized by HSPICE (Step D). Toward this end, a SPICE netlist is created by injecting intra-cell defect into the DSPF of SFF one at a time. HSPICE is then invoked against the defect-injected RC netlist. Next, the defective behaviors derived from the measurements made in simulation are recorded in a database (represented by the blue cylinder). The simulation of defect behaviors is arranged in a stop-on-detection fashion: for a defect to be in a certain simulation stage, it must be undetectable through all the preceding stages. Note that the cycle times specified in the simulation of first three stages (SS, DS, and SSC) is 50ns, which is as slow as a test clock, while for the last two stages (SDC and DDC), simulation is based on fast clock, which is equal to the fastest clock period of the targeted design.

For a defect to be considered detectable, its defective behavior needs to be faulty under at least one of the fault conditions defined in Section II; otherwise, the defect is considered undetectable and will be disregarded. In our framework, for the static fault models proposed (SS, SSC, and SDC), a defect with voltage deviation more than 60% of supply voltage is regarded as faulty. As for the dynamic fault models proposed (DS and DDC), for a defect to be regarded as faulty, the induced extra delay needs to be greater than 50ns in DS, while the extra delay needs to exceed 50% of the fastest clock period of the targeted design in DDC.

Last, Step E (Fault Model Generator), according to the simulation results and the detection thresholds mentioned above, identifies the CA faults of the five proposed fault models, and outputs the

979-8-3503-4631-2/23 $31.00 © 2023 IEEE

TABLE IV
INFORMATION OF IWLS05 BENCHMARKS

Design	Fastest clock period (ns)	Total # scan cells	Total # cell instances
leon2	7	149,507	649,725
leon3mp	5	108,863	406,112
leon3-avnet-3s1500	8	185,108	796,745

TABLE V
RESULT OF CA FAULT EXTRACTION FOR SCAN FLOP-FLOPS

Defect type	# defects	SS	DS	SSC	SDC	DDC
Short (a)	557	428	0	81	0	1
Open (b)	1,472	728	0	171	7	38
Ton (c)	487	308	0	36	0	7
Toff (d)	487	201	0	60	0	26
(a)+(b)+(c)+(d)	3,003	1,665	0	348	7	72
Ratio	100.0%	55.4%	0.0%	11.6%	0.2%	2.4%

TABLE VI
COVERAGE OF SS AND DS FAULTS ACHIEVED BY CHAIN TEST

Design	Shift faults		# detected	FC of chain test
	# SS faults	# DS faults		
leon2	23,605,985	0	23,598,401	99.97%
leon3mp	17,193,063	0	17,188,007	99.97%
leon3-avnet-3s1500	29,025,697	0	29,013,986	99.96%
Average	23,274,915	0	23,266,798	99.97%

identified faults of SSC, SDC, and DDC in UDFM for the purpose of test generation with Tessent. Up to this point, a better methodology of generating CA tests for scan cells can be implemented by following the proposed flow.

B. Flow of Evaluating the Fault Coverage of Chain Test

In last subsection, the extracted SS and DS faults are not exported in UDFM for pattern generation (see `Step E` in Fig. 6). No further ATPG is performed because faults of these two types are all detectable when SE=1 and are likely to be covered by the chain test. During the application of chain test, the toggle sequence 0011... loaded into scan chain ensures that each SFF undergoes all the four possible transitions at SI pin. Conversely, the values on D pins are not guaranteed by chain test but depend directly on the combinational part of the netlist. Therefore, logic simulation is necessitated in order to ascertain, given an SS (or DS) fault, whether any of its fault conditions has been fulfilled.

The flow of evaluating the coverage of SS and DS faults by chain test is mainly composed of two steps. First, we obtain the switching activity (.vcd) by running logic simulation on the target design along with the pattern testbench (.v) for chain test created by Tessent. From the switching activity file, the logic states on the input pins of SFF instances in the shift-in and shift-out procedure during chain test can be determined. Next, given the faults of our shift fault models (SS and DS) identified in `Step E` of Fig. 6, we then compare the fault conditions required to activate the SS and DS faults against the conditions achieved during chain test. An SS (or DS) fault is regarded as "detected" by chain test if at least one of its fault conditions defined in Section II-A (or in Section II-B) can be found in the switching activity file. Last, for SS and DS fault models, their fault coverage by chain test is defined as the ratio of detected faults to the total number of faults in each fault model.

Please note that, if the obtained fault coverage of SS and DS faults turns out to be insufficient, the faults not detected by chain test would not be left out as long as we send the corresponding defects back to the simulation with SSC, SDC, and DDC stimuli (`Step D.3`, `D.4`, and `D.5` in Fig. 6) and create the additionally needed tests.

IV. EXPERIMENTAL RESULTS

The experiments were conducted on a set of three IWLS05 [20] benchmarks synthesized with an industrial cell library from a mature technology node. Please note that in order to obtain netlist profiles in line with [20], only SFFs without set/reset/enable functions are instantiated during synthesis. For each selected design, Table IV lists the fastest clock periods in nanosecond, numbers of SFF instances, and the total numbers of cell instances. Various sets of experimental results will be demonstrated in the following.

A. Result of SFF Fault Extraction

Table V shows the result of CA fault extraction over the 12 SFFs without set/reset/enable functions from the cell library, where 1Ω is used

for injecting short and ton defects, while $100G\Omega$ is used for injecting open and toff defects. With up to 20 HSPICE processes running in parallel, the runtime of the entire flow of CA fault extraction for SFFs, as depicted in Fig. 6, was around 6.4 hours. In Rows 1–5, Column 1 shows the four defect types considered in this work; Column 2 lists the total number of defects for the four defect types; for our shift fault models, Columns 3 and 4 report the number of detectable defects in SS and DS; for our capture fault models, Columns 5–7 report those detectable in SSC, SDC, and DDC. Rows 6 and 7 summarize the statistics of CA fault extraction for scan cells. For each proposed fault model, Row 6 sums the numbers of detectable defects across all defect types, and Row 7 lists the ratio of Row 6 to the total defect number, which is 3,003.

As the table shows, the numbers of total defects are quite unbalanced among the four defect types. Open is the defect type with the most intra-cell defects, accounting for nearly 50% (1,472 out of 3,003) of the intra-cell defects taken into consideration, while short, the one with second most defects, takes up no more than 20%. As for the detection of defects, the majority of defects are covered in SS, followed by SSC, DDC, and SDC. As can be seen, none can be detected by the DS fault model. It is because a detection threshold of relatively large delay (50ns) is adopted during the identification of DS faults to target the faults that exhibit gross delay and hence could cause timing failure during chain test. Also note that among the three capture fault models (SSC, SDC, and DDC), the contribution of SDC to the detection of defects is marginal (0.2%). A main reason is that a large portion of defects detectable in SDC has been already covered in the prior SS and SSC fault models, and the defects escaping all the way to the SDC stage would actually require dynamic stimuli for detection in the later DDC stage.

B. Result of Chain Test Evaluation

In this subsection, we attempt to examine the number of SS and DS faults detected during chain test, by following the flow described in Section III-B. Table VI shows the fault coverage of chain test, where the extracted faults of our shift fault models are targeted. Columns 2 and 3 stand for the number of SS and DS faults. The number of DS faults is constantly zero since not a single detectable defect could be identified as Table V indicates. Column 4 shows the number of detected faults. The last column lists the fault coverage achieved by chain test, which is simply the number of detected faults divided by the sum of SS and DS fault numbers. As demonstrated in Table VI, among the three designs, the fault coverage of chain test is significant without exception, ranging from 99.96% to 99.97%, with an average of 99.97%. This result suggests that although the D values on SFF instances are not guaranteed (unlike the SI values) during chain test, the fault conditions required for detecting our shift faults can still be adequately satisfied.

C. Result of ATPG with SSC Faults

Before running ATPG on SSC faults, we performed fault simulation with the patterns of SA fault model to first drop SSC faults. Next, for those not detected by SA patterns, ATPG was invoked to generate additional test patterns for better test quality. Finally, we fault-simulated SSC faults with the two pattern sets serially: SA tests followed by top-off SSC tests.

Table VII first shows the result of ATPG on SA faults, listing the number of SA faults, the number of SA patterns, test coverage of SA faults, and runtime, respectively. With SA tests as base pattern set, Table VII then shows the result of top-off ATPG on SSC faults, listing the number of SSC faults, the extra number of generated patterns, resulting test coverage of SSC faults, and runtime. On top of the fault

TABLE VII
RESULT OF SA ATPG AND TOP-OFF ATPG ON SSC FAULTS

Design	SA ATPG				Top-off SSC ATPG				SSC FC		(b) / (a)	(d) − (c)
	# faults	# patterns (a)	SA TC	Runtime (hours)	# faults	# patterns (b)	SSC TC	Runtime (hours)	SA pat (c)	+SSC pat (d)		
leon2	5,352,108	15,136	99.93%	1.94	4,780,958	51	99.99%	0.26	99.12%	99.99%	0.00337	0.88%
leon3mp	3,494,570	5,979	99.99%	0.45	3,482,138	42	99.99%	0.12	98.86%	99.99%	0.00702	1.13%
leon3-avnet-3s1500	6,587,792	12,147	97.39%	1.39	5,878,622	65	99.99%	0.26	98.99%	99.99%	0.00535	1.00%
Average	5,144,823	11,087	99.10%	1.26	4,713,906	53	99.99%	0.21	98.99%	99.99%	0.00525	1.00%

TABLE VIII
RESULT OF TD ATPG AND TOP-OFF ATPG ON SDC FAULTS

Design	TD ATPG				Top-off SDC ATPG				SDC FC		(b) / (a)	(d) − (c)
	# faults	# patterns (a)	TD TC	Runtime (hours)	# faults	# patterns (b)	SDC TC	Runtime (hours)	TD pat (c)	+SDC pat (d)		
leon2	5,352,108	75,002	99.82%	8.50	298,806	8	99.99%	1.86	99.97%	99.99%	0.00011	0.03%
leon3mp	3,494,570	39,783	99.32%	3.31	217,618	3	99.99%	0.21	99.97%	99.99%	0.00008	0.02%
leon3-avnet-3s1500	6,587,792	77,017	94.79%	13.48	367,408	7	99.99%	1.38	99.94%	99.99%	0.00009	0.05%
Average	5,144,823	63,934	97.98%	8.43	294,611	6	99.99%	1.15	99.96%	99.99%	0.00009	0.03%

TABLE IX
RESULT OF TD ATPG AND TOP-OFF ATPG ON DDC FAULTS

Design	TD ATPG				Top-off DDC ATPG				DDC FC		(b) / (a)	(d) − (c)
	# faults	# patterns (a)	TD TC	Runtime (hours)	# faults	# patterns (b)	DDC TC	Runtime (hours)	TD pat (c)	+DDC pat (d)		
leon2	5,352,108	75,002	99.82%	8.50	1,195,229	13,758	99.96%	1.02	88.04%	99.96%	0.18344	11.92%
leon3mp	3,494,570	39,783	99.32%	3.31	870,497	8,202	99.93%	0.32	90.17%	99.92%	0.20617	9.75%
leon3-avnet-3s1500	6,587,792	77,017	94.79%	13.48	1,469,642	13,133	95.17%	1.35	86.30%	95.15%	0.17052	8.85%
Average	5,144,823	63,934	97.98%	8.43	1,178,456	11,698	98.35%	0.90	88.17%	98.34%	0.18671	10.17%

coverage achieved with SA pattern set, the fault coverage of SSC faults without and with top-off SSC patterns are reported in the columns labeled by "SSC FC." The last two columns list the ratio of pattern counts and the difference in the achieved fault coverage of SSC faults.

As can be seen from Table VII, SA tests invariably deliver high coverage of our SSC faults on the three designs, ranging from 98.86% to 99.12%, with an average of 98.99%. Despite the high SSC fault coverage attained with SA tests, the coverage of SSC faults can be improved with top-off ATPG runs. For instance, on leon3-avnet-3s1500, an addition of 1.00% SSC faults can be further detected by the top-off pattern set, which is merely 0.0054× (65 / 12,147) in size by comparison with that of SA tests. On average, we can detect 1.00% more SSC faults at the cost of only 0.53% more test patterns. This result demonstrates that, with a small number of top-off ATPG patterns, we can efficiently supplement the coverage of SSC faults.

D. Result of ATPG with SDC and DDC Faults

In this subsection, we generated patterns to test the double-capture faults, where the test patterns of TD fault model based on LOC scheme were applied beforehand to drop SDC and DDC faults. For those not detected by TD patterns, ATPG was invoked to generate additional tests. Finally, we fault-simulated our SDC (or DDC) faults with the two pattern sets serially: TD tests followed by top-off SDC (or DDC) tests.

Table VIII, in a similar format to Table VII, first presents the result of ATPG on TD faults and top-off ATPG on SDC faults, and it then compares the pattern counts and shows the achieved improvement in SDC fault coverage. As Table VIII shows, a considerable portion of SDC faults can be covered by TD tests, ranging from 99.94% to 99.97%, with an average of 99.96%. For SDC faults, the coverage resulting from conventional tests goes even higher than that for SSC faults (see Table VII), which is an average of 98.99% by SA tests. Two main reasons can account for this high coverage: the little number of identified SDC faults, which is 6.25% of SSC faults (294,611 vs. 4,713,906), and relatively large pattern counts from TD ATPG runs, which is 5.77× of SA tests (63,934 / 11,087). As a result, there is limited opportunity for top-off ATPG runs toward better SDC fault coverage. On average, for another 0.03% increase in the coverage of SDC faults, we managed to

further generate six more patterns, which can be considered as negligible test overhead.

Table IX, in a similar format to Table VII, presents the result of ATPG on TD faults and top-off ATPG on DDC faults, and it then compares the pattern counts and shows the achieved improvement in DDC fault coverage. As demonstrated in Table IX, compared with the results of top-off ATPG with SSC or SDC faults, the coverage of DDC faults achieved by conventional tests is much lower (86.30%–90.17%), with an average of 88.17%. Furthermore, the coverage of DDC faults can be improved with additional test patterns on the three designs without exception. Take leon2 as an example, 11.92% more DDC faults can be further covered by using a top-off pattern set of 18.34% (13,758 vs. 75,002) in size. On average, at the cost of only 18.67% of TD tests in size, we can further enhance the DDC fault coverage by 10.17%. The result demonstrates that, with TD tests alone, a substantial portion of DDC faults can hardly be detected and could in turn be regarded as a potential source of DPPM if only conventional tests are applied. More importantly, this leak of DDC fault coverage could be made up by using a dedicated pattern set with affordable size.

V. CONCLUSION

In this paper, we have presented five CA fault models to explicitly target the defects inside SFFs by examining their faulty behavior via SPICE simulation. Unlike conventional fault models, which define faulty behaviors outside standard cells with simple activation rules, our defect-based methodology can cover the SFF-internal defects entailing more complicated conditions for activation. The extraction of CA faults for SFFs is fully automated in the proposed framework, where the identified faults are output in UDFM format for pattern generation with Tessent ATPG [6]. Based on large IWLS05 benchmarks synthesized with an industrial cell library, experimental results demonstrated that our proposed faults could not be effectively covered with tests based on conventional SA and TD fault models. This inadequacy in detecting such faults signifies the need for fault modeling of SFF-internal defects and thus their dedicated test patterns. The coverage of DDC faults, in particular, could be further enhanced by 10.17% with an affordable number of top-off patterns, which were 18.67% of TD tests in size.

979-8-3503-4631-2/23 $31.00 © 2023 IEEE

REFERENCES

[1] F. Hapke *et al.*, "Defect-oriented cell-aware atpg and fault simulation for industrial cell libraries and designs," in *2009 International Test Conference*, Nov 2009, pp. 1–10.

[2] F. Hapke *et al.*, "Cell-aware test," *IEEE Trans. on Computer-Aided Design of Integrated Circuits and Systems*, vol. 33, no. 9, pp. 1396–1409, Sep. 2014.

[3] F. Hapke *et al.*, "Cell-aware analysis for small-delay effects and production test results from different fault models," in *2011 IEEE International Test Conference*, Sept 2011, pp. 1–8.

[4] F. Hapke *et al.*, "Cell-aware production test results from a 32-nm notebook processor," in *2012 IEEE International Test Conference*, Nov 2012, pp. 1–9.

[5] F. Hapke *et al.*, "Cell-aware experiences in a high-quality automotive test suite," in *2014 19th IEEE European Test Symposium (ETS)*, May 2014, pp. 1–6.

[6] *Tessent® Scan and ATPG User's Manual*, v2019.3 ed., Mentor Graphics Corporation, Sep. 2019.

[7] *TestMAX™ ATPG and TestMAX Diagnosis User Guide*, Version P-2019.03, Synopsys Inc.

[8] Z. Gao, S. Malagi, E. J. Marinissen, J. Swenton, J. Huisken, and K. Goossens, "Defect-location identification for cell-aware test," in *Proc. of Latin American Test Symp. (LATS)*, Mar. 2019, pp. 1–6.

[9] Y. Huang, R. Guo, W. Cheng, and J. C. Li, "Survey of Scan Chain Diagnosis," *IEEE Design Test of Computers*, vol. 25, no. 3, pp. 240–248, May 2008.

[10] S. R. Makar and E. J. McCluskey, "ATPG For Scan Chain Latches and Flip-Flops," in *1997 VLSI Test Symposium*, Apr. 1997, pp. 364–369.

[11] R. Guo, L. Lai, H. Yu, and W.-T. Cheng, "Detection and Diagnosis of Static Scan Cell Internal Defect," in *2008 IEEE International Test Conference*, Oct. 2008, pp. 1–10.

[12] F. Yang, S. Chakravarty, N. Devta-Prasanna, S. M. Reddy, and I. Pomeranz, "On the Detectability of Scan Chain Internal Faults An Industrial Case Study," in *26th IEEE VLSI Test Symposium (vts 2008)*, Apr. 2008, pp. 79–84.

[13] F. Yang, S. Chakravarty, N. Devta-Prasanna, S. Reddy, and I. Pomeranz, "Detection of Internal Stuck-open Faults in Scan Chains," in *2008 IEEE International Test Conference*, Oct. 2008, pp. 1–10.

[14] F. Yang, S. Chakravarty, N. Devta-Prasanna, S. M. Reddy, and I. Pomeranz, "Detection of Transistor Stuck-Open Faults in Asynchronous Inputs of Scan Cells," in *2008 IEEE International Symposium on Defect and Fault Tolerance of VLSI Systems*, Oct. 2008, pp. 394–402.

[15] F. Yang, S. Chakravarty, N. Devta-Prasanna, S. Reddy, and I. Pomeranz, "Detectability of internal bridging faults in scan chains," in *2009 Asia and South Pacific Design Automation Conference*, Jan. 2009, pp. 678–683.

[16] F. Yang, S. Chakravarty, N. Devta-Prasanna, S. Reddy, and I. Pomeranz, "Improving the Detectability of Resistive Open Faults in Scan Cells," in *2009 24th IEEE International Symposium on Defect and Fault Tolerance in VLSI Systems*, Oct. 2009, pp. 383–391.

[17] R. Guo, B. Archer, K. Chau, and X. Cai, "Efficient Cell-Aware Defect Characterization for Multi-bit Cells," in *2018 IEEE International Test Conference in Asia (ITC-Asia)*, Aug. 2018, pp. 7–12.

[18] A. Touati, A. Bosio, P. Girard, A. Virazel, P. Bernardi, M. S. Reorda, and E. Auvray, "Scan-Chain Intra-Cell Aware Testing," *IEEE Transactions on Emerging Topics in Computing*, vol. 6, no. 2, pp. 278–287, Apr. 2018.

[19] *Tessent® CellModelGen Tool Reference*, v2019.3 ed., Mentor Graphics Corporation, Sep. 2019.

[20] *IWLS 2005 Benchmarks*. [Online]. Available: http://iwls.org/iwls2005/benchmarks.html

[21] *Calibre® xRC™ User's Manual*, v2012.2 ed., Mentor Graphics Corporation.

[22] *HSPICE® User Guide: Basic Simulation and Analysis*, Version j-2014.09 ed., Synopsys Inc., September 2014.

Design for testability (DFT) for RSFQ circuits

Mingye Li, Yunkun Lin and Sandeep Gupta

Department of Electrical Engineering, University of Southern California, Los Angeles, CA, 90089

mingyel, yunkunli, sandeep@usc.edu

Abstract—Superconducting electronics (SCE), especially Rapid Single Flux Quantum (RSFQ) logic, is being developed due to its high-performance and low power. In [1]–[3], we developed new static and delay fault models and an efficient automatic test pattern generator (ATPG) for testing both delay and static faults in RSFQ logic. However, test pattern application involves moving patterns and responses via long wires from the test equipment at room temperature to the chip under test in liquid helium. Due to the high cost associated with large numbers of such wires, testing is extremely expensive in absence of design for testability.

We present a scan architecture for RSFQ circuits which enables the application of a large number of test patterns. Due to the unique characteristic of RSFQ, this scan architecture includes completely new scan cell design and a new scan control strategy. The on-chip test control logic enables scan chain to shift in test patterns from the test equipment at room temperature via a small number of wires, apply the pattern to the chip under test in parallel and at speed, and shift out the corresponding test response for checking. We demonstrate that our new scan architecture supports testing at low overheads.

Index Terms—RSFQ, Design for testability (DFT)

I. Introduction

Superconducting electronics (SCE) is receiving close attention due to its ultra-low power consumption and ultrahigh performance. Rapid Single Flux Quantum (RSFQ) [4], the most mature SCE technology, can operate at frequencies of 100GHz with switching energy as low as 1aJ [5]. With two organizations supporting the fabrication of RSFQ chips and with the emergence of CAD tools [6]–[8], post-fabrication testing is becoming a practical concern.

Early exploration and prototyping of test and testability of RSFQ circuits has been the subject of some research since [9] and [10]. In our previous research, we have developed models for static and delay faults and corresponding methods for simplifying testing of them. [1]–[3], [11]. Early research [12], [13] explored the scan chain design and test point insertion. However, test controller design, including the shift register to scan in/out input/output patterns is not complete.

Here we present a design-for-testability (DFT) method for RSFQ. We identify the test requirements for RSFQ logic by identifying the barriers to high coverage of static and delay faults, special test application requirements, causes of high test data volume (which impact test time and cost), and the limitations of the external test equipment.

The research is based upon work supported by the Office of the Director of National Intelligence (ODNI), Intelligence Advanced Research Projects Activity (IARPA), via the U.S. Army Research Office grant W911NF-17-1-0120. The views and conclusions contained herein are those of the authors and should not be interpreted as necessarily representing the official policies of the ODNI, IARPA, or the U.S. Government. The U.S. Government is authorized to reproduce and distribute reprints for Governmental purposes notwithstanding any copyright notation.

This paper is organized as follows: In Section II, we identify the unique test requirements for our RSFQ DFT together with the goals of our scan-chain design. Section III shows the architecture and detailed design of our scan design which uses neither the muliplexor-based scan cells (used in most CMOS designs) nor the level-sensitive scan design (LSSD; designed for latch-based designs). Section IV uses extensive circuit-level simulations of example RSFQ circuit blocks with our scan design to show that all our components work correctly in concert with the circuit under test.

II. DFT Requirements for RSFQ

In this section, we analyze all aspects of testing of RSFQ logic circuits and identify the goals of our DFT design by identifying special characteristics of RSFQ testing. Then we identify the key requirements for our DFT for RSFQ logic.

1. Limitations of External Test Equipment: In CMOS testing, the external test equipment's speed and bandwidth are lower than those of the chip under test (CUT). These challenges are significantly accentuated for RSFQ logic. First, the test equipment will be implemented in CMOS, and hence much slower than the RSFQ CUT. Second, much longer interconnects are required between the test equipment at room temperature and the RSFQ CUT at cryogenic temperature.

Hence, DFT architecture must enable at-speed testing of the RSFQ chip under test using slow-speed external test equipment that uses low bit-width low-bandwidth interconnects. Hence, the DFT design must use a small number of pins during testing, use low-speed clocks to scan in each test pattern and scan out each response, yet apply each test pattern at-speed to the CUT. (Note that some of these requirements are similar to those identified by Intel's Quantum Computing group [14].)

2. Special Scan Cell Design for RSFQ: RSFQ logic operation imposes new requirements on how we design scan DFT. First, as every RSFQ logic gate is a pipeline stage, the use of multiplexor-scan cells is extremely challenging as it would alter the sequential timing of the logic circuit *even during the normal mode*. Second, the application of a test pattern produces a response only after D clock cycles, where D is the depth of the fine-grained pipeline.

3. Need Scan to Eliminate Feedback Cycles: In CMOS, high fault coverage for sequential logic is achieved only by eliminating the feedback loops in the test mode via use of scan DFT [15]. Hence, for RSFQ logic, DFT must also support scan to eliminate loops in logic blocks with feedback.

4. No Need to Scan all Flipflops to use ATPG for Combinational Circuits: The fine-grained balance pipelined structure of RSFQ logic enables the use of automatic test pattern generator (ATPG) designed for combinational circuits for feedback-free logic [15]. Hence, in RSFQ logic, it is unnecessary to scan

979-8-3503-4631-2/23 $31.00 © 2023 IEEE

every sequential cell. However, the fault coverage for very deep pipelines can be low, and may require additional scan. We have studied cascaded RSFQ logic blocks and showed that additional levels of cells may need to be scanned to increase the coverage for large logic blocks.

5. *Simplification of DFT Requirements for Delay Testing:* In CMOS, delay testing imposes three major requirements on DFT. First, two-pattern test for delay fault requires a high-overhead two-pattern scan chain design. Second, coverage of path delay faults is typically very low. Third, large number of patterns are required for delay testing and this makes multiple scan chains and test compression necessary. However, as we have reported in [11] [3], RSFQ's pulse-based operation enables the use of one-pattern tests for delay testing. Second, the fine-grained pipeline eliminates the possibility of having delay faults at most circuit lines. This improves delay fault coverage and reduces the test data volume. Hence, in RSFQ logic, we can avoid high-overhead DFT requirements needed for delay testing in CMOS logic.

6. *Simplification of large-fanout nodes and large test controller:* In RSFQ circuits, for one cell to drive multiple fanouts we need a special cell called a *splitter*. Since each splitter drives only two fanouts, a splitter tree is required when a cell has a large fanout. This contributes large area overhead. Further, a splitter is an asynchronous cell and hence increases clock period. Also, process variation can induce significant skew at each fanout node and reflections on the splitter tree can distort the signals. Hence, it is important to avoid use of large splitter trees. While the DFT structure in [13] is area-efficient in terms of device (JJs) count, tight timing requirements and the large splitter trees it uses make it extremely challenging for automatic place-and-route tools to find a solution. Furthermore, the externally applied test signal is also difficult to synchronize with the fast clock.

III. SCAN DFT DESIGN

In this section, we present our RSFQ scan DFT design in a top-down manner by introducing the circuit model, analyzing the overall design and the working mechanism of the whole structure and exposing the details of each component.

A. Circuit model

A general model for a sequential logic block is shown in Figure 1. While we can apply any desired pattern directly at the primary inputs (PIs; represented by PI0, PI1, PI2) of a sequential logic block, we do not have direct control over state inputs (SIs; represented by SI0 and SI1). This typically limits the fault coverage. Further, this increases the memory and average run-time complexity for test generation to such impractically high levels that scan DFT is used universally to avoid the problem [15] and the unique characteristics of RSFQ do not change this fact. Hence, one key application of our DFT design is to support scan for sequential circuits.

In this paper, the description focuses on the above circuit model. However, our method also applies to the other circuit model, namely scan to achieve high fault coverage in circuits

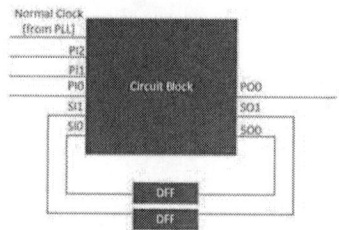

Fig. 1. Generic model of a sequential logic block

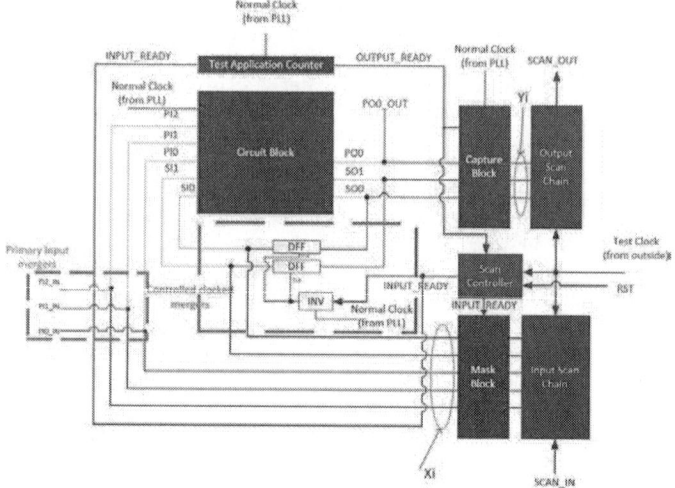

Fig. 2. The circuit under test (grant block) with our scan chain architecture

where multiple feedback-free circuit blocks are cascaded. In that model, we scan level(s) of DFFs in the cascaded blocks.

B. Overall scan design

The top-level design of our scan chain is shown in Figure 2. We present the circuit block under test and our entire scan design in nine parts: the input scan chain, the mask block, the scan controller (a counter), controlled clocked mergers (green block), primary input mergers (orange block), the circuit block (the original circuit under test without the feedback), the test application counter, the capture block, and the output scan chain. The blocks in red are controlled by the test clock, which is provided by the external tester via an additional chip input; blocks in blue are controlled largely by the normal clock, i.e., the clock used during the normal operation of the original circuit. Red lines show the data path when the circuit works in the test mode; blue lines show the data path when the circuit works in the normal mode; and orange lines show the paths that are common to the normal and test modes. The test mode is sub-divided to three sub-modes, reset cycle, scan in/out mode, and test pattern application mode. For connections between circuit blocks, blocking gate [13] is used.

First, consider the inputs to the design during the test mode.

1. The normal clock is used during normal operation as well as during the test pattern application sub-mode of the test mode. (The normal clock can remain active during other sub-modes of the test mode, but is not used.)

2. The primary inputs (exemplified by PI2_IN, PI1_IN, PI0_IN) receive the input values during normal operation.

979-8-3503-4631-2/23 $31.00 © 2023 IEEE

These are either driven by chip input pins or the outputs of other logic blocks on the chip. During the normal mode these receive inputs; during test mode, no SFQ pulses should be applied at these primary input pins.

3. The test clock is used during the test mode and is applied by the external test equipment. The test equipment is implemented in CMOS and hence is significantly slower than the RSFQ logic circuit under test. Further, the long interconnects from the room temperature tester to chip under test at cryogenic temperature limits the speed of interconnects from the tester. Hence, test clock is significantly slower than the normal clock. The two clocks must be appropriately synchronized such that the transitions from scan in/out mode to test pattern application, and the transitions from test pattern application and scan out mode work correctly. The test clock is applied during the entire test mode; during normal mode, SFQ pulses must not applied at the test clock input.

4. SCAN_IN and SCAN_OUT are connected to the external tester. During specific sub-modes of the test mode, external tester applies the bits of test pattern to SCAN_IN and captures the corresponding response at circuit outputs at SCAN_OUT; during normal mode, no SFQ pulses are applied at SCAN_IN.

5. RST input is used whenever the circuit moves from normal mode to test mode. During the first cycle during such mode transition, a single SFQ pulse is applied; no SFQ pulse is applied during other cycles. Here, we still use the DFF and inverter structure rather than the structure in [13]. The reason is that DFF and inverter are synchronous cells and the timing requirements for them are easier to satisfy than the asynchronous structure in [13].

The functionality of each part is as follows.

1. Input scan chain: It is implemented as a FIFO using DFFs and is used to load the test pattern applied bit-by-bit by the external test equipment. Once all the bit values are loaded into this scan chain, these are to be applied to the state and primary inputs of the circuit under test.

2. Mask block: It ensures that the SFQ pulses that appear at the input scan chain during scan in mode are blocked from propagation to primary and state inputs of the circuit block. Once the next test pattern are loaded and the pattern application mode is invoked, the bits of the pattern scanned in are applied to the primary and state inputs of the circuit.

3. Scan controller: A counter that counts the cycles in the scan in/out mode and determines when that mode ends and the test application mode is invoked. Specifically, it sends an SFQ pulse to the mask block and the controlled clocked mergers to enable application of the scanned-in test pattern. Our use of an automatic counter rather than an external signal, eliminates the large splitter tree required in [13].

4. Primary input mergers: Mergers to merge primary inputs and scan chain. During normal mode, no SFQ pulses are applied by the mask block and the primary input values are applied to the circuit block; during test mode, no SFQ pulses are applied at primary inputs and the bits of scanned in test pattern are applied each time the scan in/out mode ends and the test application mode is invoked.

TABLE I
OPERATION OF EACH PART UNDER DIFFERENT WORKING MODE

	Normal mode	Test mode		
		Reset cycle	Scan in/out mode	Pattern application mode
Input Scan Chain	No pulse applied to: RST, Scan_in and Test clock	No pulse applied to Scan_IN	Input pattern is applied to Scan_in per Test clock	No pulse applied to: RST and SCAN_IN
Scan Controller	No pulse is applied to: RST and Test clock	RST pulse is applied to counter	Counter counts the number of test clock cycle	Counter is reset when the output is ready automatically
Mask block	No Input_ready signal is generated, keep masking Input scan chain from circuit	No Input_ready signal is generated, keep masking Input scan chain from circuit	No Input_ready signal is generated, keep masking Input scan chain from circuit	Input_ready is generated when counter counts enough clock cycles, stop masking and apply test pattern to controlled clocked mergers
Controlled clocked merger	No Input_ready signal is applied, controlled clocked merger will only capture state output from circuit block	No Input_ready signal is applied, controlled clocked merger will only capture state output from circuit block	No Input_ready signal is applied, controlled clocked merger will only capture state output from circuit block	When Input_ready is applied, controlled clocked merger captures test pattern from Input scan chain and block state outputs, otherwise, controlled clocked merger will only capture state output from circuit block
Test application counter	No Input_ready signal is applied, Test application counter does not capture any pulse	No Input_ready signal is applied, Test application counter does not capture any pulse	No Input_ready signal is applied, Test application counter does not capture any pulse	Input_ready signal is captured and delayed D normal clock cycles and generates Output_ready signal.
Primary input mergers	Pulse is applied to primary input merger per normal clock	No constraint on PI	No constraint on PI	No input pulse should be applied to PI in the normal clock cycle when INPUT_READY is generated
Capture block	No Output_ready signal is applied; capture block does not capture anything	No Output_ready signal is applied; capture block does not capture anything	No Output_ready signal is applied; capture block does not capture anything	When Output_ready signal is applied, capture block captures corresponding output
Output Scan Chain	No pulse should be applied to test clock and no pulse is generated to SCAN_OUT	No pulse should be applied to test clock.	Pulse is applied to test clock, output pulse is shifted out per test clock	Pulse is applied to test clock, output pulse is shifted out per test clock

5. Controlled clocked mergers: During normal mode, the test controller output does not produce any SFQ pulses and the values at state outputs are applied to the state inputs via the DFFs. In the test mode, when scan in/out mode ends and the test application mode is invoked, the SFQ pulse at the output of the test controller blocks the DFF outputs and instead applies the bits of scanned in patterns to the state inputs.

6. Test application counter: As RSFQ circuit is pipelined at fine-grain, once the test pattern is applied to the primary and state inputs of the circuit block, D cycles (D is the depth of the pipeline in the circuit block) are needed for the block's response to the test to appear at its primary and state outputs. This counter is simply a shift register with D DFFs and its output triggers the capture block when the response is ready.

7. Capture block: This block stops the propagation to the scan out chain of the SFQ pulses appearing at the primary and state outputs of the circuit block during the entire test mode **except** during the cycle when test application mode ends and the response bits to the test pattern becomes available to be loaded onto the scan out chain.

8. Circuit block: The circuit under test without the feedback paths, now controlled and observed via the scan chain.

9. Output scan chain: Along with the input scan chain, this constitutes the core of the classic scan chain. It is implemented as a FIFO using DFFs and used to load the response to the test pattern from the primary and state outputs. Once the response values are loaded into this scan chain, these are scanned out serially to SCAN_OUT and sent to the external test equipment.

Table I summarizes the operation of each block.

During the normal mode, no Test Clock, SCAN_IN, and RST signals must be applied and input values should be applied to the primary inputs and out values will appear at

979-8-3503-4631-2/23 $31.00 © 2023 IEEE

PO0, SO1 and SO0 as shown in fig 2. Under this mode, since no external test signal is applied, scan controller does not count and FIFOs within the scan chains do not store any pulses. Controlled clocked mergers pass state outputs to state inputs.

The test mode starts with the application of a RST signal, which is the reset cycle of the test mode. The scan controller is reset; other parts are not connected to RST and work as in the normal mode. RST is only required when we apply the first pattern; for every subsequent pattern, scan controller is reset automatically. After reset cycle, we enter the scan in/out mode where we apply bits of test pattern to SCAN_IN, one bit per test clock cycle. Hence, if we have a total of five primary and state inputs, we need to apply three arbitrary bits in the first three test cycles. (To significantly reduce the complexity of our DFT design, pattern lengths we apply are a power of 2.) As we shift in input test pattern, we also shift out the response for the previous test pattern captured earlier in the scan out chain. When we finish shifting in the test pattern, scan controller generates INPUT_READY signal to the test application counter and the controlled clocked merger to initiate the test pattern application sub-mode. Under this mode, user is free to apply anything to SCAN_IN and RST, and output scan chain may continue to shift out bits of the response for previous pattern (captured earlier). No input pulse should be applied to the primary input during the first clock cycle of the test application mode. In the subsequent normal clock cycles, there is no constraint on primary inputs. The controlled clocked mergers block the state outputs when INPUT_READY signal is generated. Hence, during the first cycle of the test application mode the state inputs receive the bits of pattern loaded into the scan chain; in the subsequent normal clock cycles, state output values are again applied to the state inputs. Test application counter also captures the INPUT_READY signal. Once the CUT's response to the pattern is generated, test application counter generates OUTPUT_READY signal to allow the capture block pass the response bits to the scan out chain, as well as to reset the Scan controller. The controller is reset when the OUTPUT_READY signal is generated by the test application counter, and the entire BIST circuit returns to scan in/out mode without requiring any external reset pulse.

Synchronization between normal clock and test clock is required, clock domain crossing happens between input scan chain and mask block, between output scan chain and capture cells. The constraint is shown in Equations 1 plus the test clock period must be larger than the normal clock period.

$$T_{TAI} + \Delta_{CQin} + \Delta connection + SU_{mask} < T_{NAI}$$
$$T_{NAO} + \Delta_{CQout} + \Delta connection + SU_{out} < T_{TAO}$$
$$T_{TAI} + \delta_{CQ} + \Delta connection > T_{NAI} + H_{mask} \qquad (1)$$
$$T_{NAO} + \delta_{CQ} + \Delta connection > T_{TAO} + H_{out}$$

where:
T_{TAI}: test clock arrival time for the input scan chain cell
T_{NAI}: normal clock arrival time for the input scan chain cell
T_{NAO}: normal clock arrival time for the output scan chain cell
T_{TAO}: test clock arrival time for the output scan chain cell
$\Delta connection$: Interconnection delay

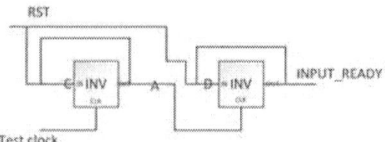

Fig. 3. Scan controller: A low-overhead 2-bit RSFQ counter

Δ_{CQin}: maximum clock to Q delay of input scan chain cell
Δ_{CQout}: maximum clock to Q delay of the capture cell
δ_{CQin}: minimum clock to Q delay of input scan chain cell
δ_{CQout}: minimum clock to Q delay of the capture cell
SU_{mask}: mask cell setup time
SU_{out}: output scan chain cell setup time
H_{mask}: hold time of mask cell
H_{out}: hold time of output scan chain cell

C. Detailed design and validation of modules

1) Notation: We first introduce notation used in the remainder of this paper.
N_{pi}: The number of primary inputs
N_{si}: The number of state inputs
N_{po}: The number of primary outputs
D: The pipeline depth of the circuit under test
T_{test}: The period of the test clock
T_{normal}: The period of the normal clock
N_{inv}: The number of inverters in the test controller
$N_{shiftin}$: The number of test clock cycles required to shift in all bits of a test pattern
$N_{shiftout}$: The number of test clock cycles required to shift out all bits of the response to a test pattern

Next, we present detailed design of every part of our scan design as shown in Figure 2.

2) Input interface:

a) Scan controller:: This controller is a counter which counts the number of bits to be serially scanned in and determines when the bits of a test pattern are ready to be applied. Before the first pattern, its state is reset by RST signal; before every subsequent pattern its state resets automatically. When it completes counting, it sends an SFQ pulse on its INPUT_READY output to initiate the application of the scanned in test pattern to the CUT.

Since the overhead of a general counter is typically very high, we have chosen to design our scan architecture to use a counter design which counts to a power of 2. If our circuit needs to count to m, we design a counter that counts to 2^n, where n is smallest value such that $2^n \geq m$. This increases the test length slightly (by a factor of ≤ 2) and requires the application of $2^n - m$ arbitrary bits at the SCAN_IN pin at the start of each scan in cycle; but saves a lot of circuit area.

An example design for $n = 2$ is shown in Figure 3. (The labels at inputs and outputs correspond to the labels shown in Figure 2.) This design uses n INVs to create the required counter by taking advantage of the fact that every RSFQ logic cell has a built-in pipeline flipflop. We have validated this design via extensive simulations, which show that when RST signal is applied, an output pulse at node INPUT_READY is generated after four test clock cycles.

979-8-3503-4631-2/23 $31.00 © 2023 IEEE

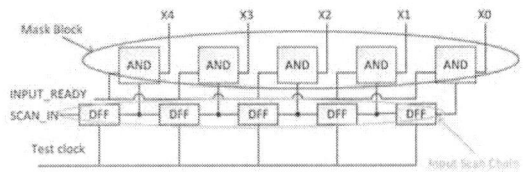

Fig. 4. Input scan chain and mask block

b) Input scan chain and mask block:: The input interface consists of the input scan chain and the mask block shown in Figure 2. The scan chain is a FIFO and is used to serially shift in and store the test pattern and the mask cells are used to block the test pattern from propagating to the circuit block (i.e., the CUT), until all the test pattern are scanned in.

The schematic of a 5-bit input scan chain with the mask block is shown in Figure 4. Taken together, these two blocks have three inputs and five outputs and the labels on the lines correspond to those in Figure 2. The input scan chain is controlled by the test clock and the mask block are controlled by the normal clock. In every test clock, one bit of a test pattern is applied to SCAN_IN. The mask block cells prevent values in the scan chain from propagating to the Circuit Block (the CUT) until all bits of the test pattern are scanned in. The test controller (see Figure 2) counts and outputs a signal at its output IN_READY, which controls the mask block. In the next normal clock cycle, the test pattern stored in the input scan chain is released and applied to the inputs of the Circuit Block. In general, this design contains $N_{pi} + N_{si}$ DFFs and AND cells; the DFFs constitute the input scan chain and the AND cells constitute the mask block.

c) Controlled clocked merger:: When a test pattern is to be applied to the Circuit Block, the controlled clocked merger is used to block the bits generated at the state output (SO_j); this enables the bits of the test pattern at the corresponding outputs of the mask block (X_i) to be propagated to the corresponding state inputs (SI_j). *However, as mentioned earlier, use of a MUX designed using library cells would change the pipeline depth of the circuit even in the normal mode. To avoid this, we design this non-mux based circuit, namely the controlled clocked merger.* The schematic is shown in Figure 5.

In Figure 5, IN_READY is driven by the output of the scan controller mentioned above (see Figure 2). Hence, the line labeled SO_j is connected to a state output of the circuit block, and X_i is connected to the corresponding output of the mask block. During normal mode, no pulse is applied to INPUT_READY, so the inverter provides one clock pulse to the DFF every normal clock cycle. Since mask cells block the data from the input scan chain, no pulses are applied to X_i and the value at SO_j (which come from the corresponding state output of the circuit block) to be applied to corresponding state input, SI_j of the circuit block. In the test mode, in the cycle during which a pulse is applied to IN_READY, the normal clock pulse to the DFF in the clocked merger is blocked and hence the value at SO_j is blocked and the bit of the scanned test pattern at X_i is applied to SI_j, the corresponding state input of the circuit block. The size of controlled clocked merger is N_{si} mergers, DFFs and inverters.

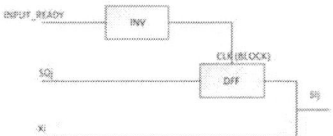

Fig. 5. Controlled clocked merger

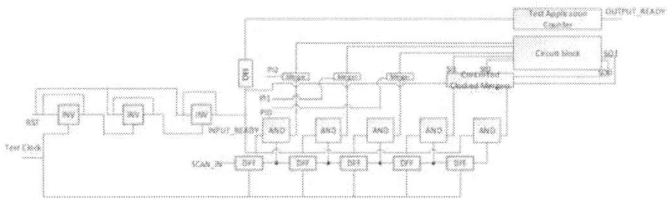

Fig. 6. Schematic of the entire input interface

The size of primary input merger is N_{pi} mergers.

d) Input interface simulation:: We integrate all the above blocks to create the design of the entire input interface. Figure 6 shows the schematic of the entire input interface — the input scan chain, the mask block, the controlled clocked merger, and the clocked merger.

As Figure 7 shows, when the circuit enters the test mode for the first time, and the external tester applies a pulse at RST, first the tester applies three arbitrary values at SCAN_IN (black box) and then applies the bits of the test pattern 10100 at the SCAN_IN port (red box). (Recall that the three arbitrary bits are needed as we are using low-overhead design for scan controller.) After eight test clock cycles, the test controller generates an output pulse at IN_READY (blue box). Finally, in the next normal clock cycle, the test pattern 10100 is applied by the mask block to the circuit under test simultaneously while the controlled clocked merger blocks the bits at the state output (and, per specification, the user is required to not apply any pulses at the primary input pins).

3) Test application counter: The test application counter is designed to generate a signal when the response to the applied test pattern is available at the outputs of the Circuit Block (the CUT). The design is simple and contains a shift register with D DFFs, where D is the sequential depth of the circuit block. We delay IN_READY by one normal clock and apply it to the input of the test application counter at the same cycle when we apply the test pattern to the circuit block.

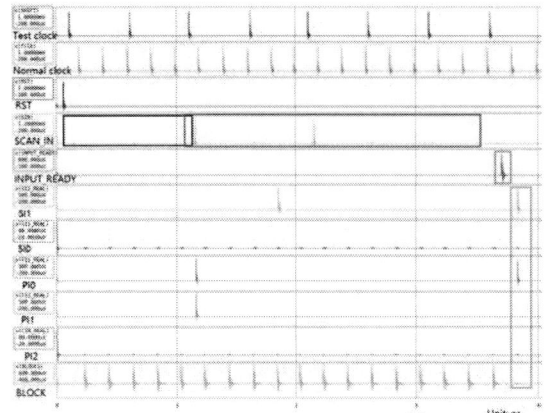

Fig. 7. Simulation results for the entire input interface

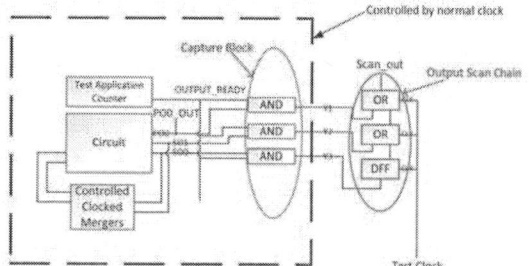

Fig. 8. Schematic of output interface blocks — capture block and output scan chain

In terms of the test clock, the output is generated and captured in S clocks, where S is calculated as:

$$S = \left\lfloor (D+2) * \frac{T_{normal}}{T_{test}} \right\rfloor. \quad (2)$$

4) Output interface: The output interface's requirements are similar to input interface's, namely **to capture the test response at the CUT outputs in parallel, then scan out via SCAN_OUT per test clock.** Therefore, we modified our input scan chain and the mask block to meet these requirements. The schematic is shown in Figure 8.

In Figure 8, note the capture cells and output scan chain. Output scan chain is controlled by the test clock and all other parts are controlled by the normal clock. Capture cells operate similar to the mask cells in the input interface: once the OUTPUT_READY signal is generated, capture cells pass the test response from the CUT outputs to the output scan chain. In the subsequent test clock cycles, the output scan chain shifts out the test response one bit per clock.

Since every RSFQ logic cell is a clocked cell, data can be stored in every logic cell. Here, we use OR cells to create the scan out chain, because we need one merger and one DFF to merge and store the test response, and this can be implemented by an OR cell. This saves us area and works because the capture block and the test application counter ensure that no pulses are applied at Y_j's during the scan out cycles. Number of capture cells is N_{pi}, size of output scan chain is one DFF and N_{pi} - 1 OR cells. We have validated this design via extensive simulations.

5) Test control requirement: Because the input and output scan chains are all controlled by the test clock, we need to make sure before current test output is captured by the output scan chain, previous output is already shifted out. The total time we need to shift in, apply test pattern, generate output to the output interface, and capture corresponding output is shown in equation 3.

$$T_{totalin} = N_{shiftin} * T_{test} + (D+2) * T_{normal}) \quad (3)$$

while the total time we need to shift out the output pattern is

$$T_{totalout} = N_{shiftout} * T_{test} \quad (4)$$

If $T_{totalout}$ is larger than $T_{totalin}$, previous output will be flushed by the current output. To prevent this, test controller needs to deliberately delay the application of input pattern until it is safe to apply. From equations 3 and 4, we can derive the safe condition:

$$N_{shiftin} >= \left\lceil N_{shiftout} - (D+2) * \frac{T_{normal}}{T_{test}} \right\rceil \quad (5)$$

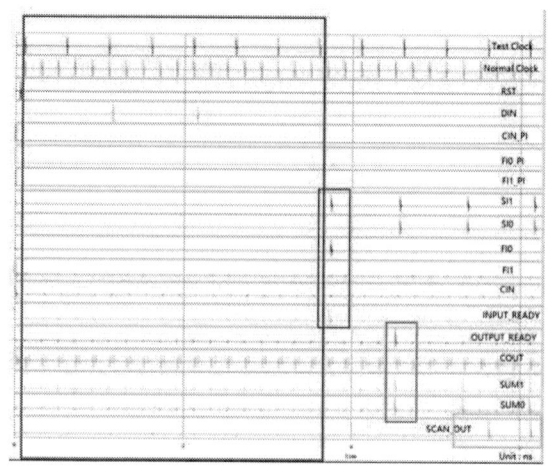

Fig. 9. Validation of our entire scan design via integration with KSA2 as the CUT

Hence,

$$N_{shiftin} = max(N_{pi} + N_{si}, \left\lceil N_{shiftout} - D * \frac{T_{normal}}{T_{test}} \right\rceil) \quad (6)$$

Since the number of shifts is controlled by $N_{shiftin}$, based on equation 6, the number of inverters N_{inv} is:

$$N_{inv} = \lceil \log_2^{N_{shiftin}} \rceil. \quad (7)$$

IV. VERIFICATION OF THE ENTIRE SCAN ARCHITECTURE

We have verified the correctness of our scan architecture using circuit-level simulations. Figure 9 shows the results for the circuit illustrated in Figure 2.

The circuit under test (circuit block) is KSA2. We start with Fig. 2 and we rename generic input names PI0, PI1, PI2 using the names of the KSA2 inputs, namely CIN, FI1, FI2. As a consequence, we rename nodes PI0_IN, PI1_IN, PI2_IN as CIN_PI, FI0_PI, FI1_PI. On the output side, we rename PO0, etc. as SUM0, SUM1, and COUT. In the simulation waveforms in the red block: After RST is applied, at SCAN_IN we apply sequentially the bits 00010100 (here, first three 0s are arbitrary as our circuit only has a total of five inputs but we use a low-overhead counter that counts to eight). In the blue block: After eight test clock cycles, input pattern 10100 is applied to the primary and state inputs and the INPUT_READY signal is generated at the same time. In the green block: after four normal clock cycles, OUTPUT_READY signal is generated at the same time as the output signals. Circuit KSA2 is an adder and input 10 + 01 + 0 = 011. Hence, 011 will be shifted out. As shown in the orange block: 011 is shifted out in three test clock cycles at SCAN_OUT. We also perform circuit-level simulation with multiple patterns to verify the function of our design. Simulation result shows the correctness of our design.

V. CONCLUSION

We have identified special test needs for RSFQ logic and developed a complete scan test architecture. Our architecture has several components which harness the unique characteristics of RSFQ logic and we have validated our entire design via application to multiple circuits and are developing a tool to automatically insert scan into large RSFQ circuits.

979-8-3503-4631-2/23 $31.00 © 2023 IEEE

REFERENCES

[1] F. Wang and S. Gupta, "Timing verification for rapid single-flux-quantum (rsfq) logic: New paradigm and models," in *2019 IEEE International Superconductive Electronics Conference (ISEC)*. IEEE, 2019, pp. 1–3.

[2] M. Li, F. Wang, and S. Gupta, "Data-driven fault model development for superconducting logic," in *2020 IEEE International Test Conference (ITC)*, 2020, pp. 1–5.

[3] ——, "Methods for testing path delay and static faults in rsfq circuits," in *2022 IEEE 40th VLSI Test Symposium (VTS)*, 2022, pp. 1–7.

[4] K. K. Likharev and V. K. Semenov, "Rsfq logic/memory family: A new josephson-junction technology for sub-terahertz-clock-frequency digital systems," *IEEE Trans. Appl. Supercond.*, vol. 1, no. 1, pp. 3–28, 1991.

[5] K. Nakajima, Y. Onodera, and Y. Ogawa, "Logic design of josephson network," *Journal of Applied Physics*, vol. 47, no. 4, pp. 1620–1627, 1976.

[6] IARPA. (2014) Iarpa launches program to develop a superconducting computer. [Online]. Available: https://www.iarpa.gov/images/files/programs/c3/C3_press_release.pdf

[7] ——. (2015) Electronic design automation tools for superconducting electronics eda for sce. [Online]. Available: https://www.iarpa.gov/index.php/working-with-iarpa/requests-for-information/electronic-design-automation-tools-for-superconducting-electronics

[8] ——. (2016) Supertools. [Online]. Available: https://www.iarpa.gov/index.php/research-programs/supertools/supoertools-baa

[9] Q. H. K. Gaj and M. Feldman, "Parameter variations and synchronization of rsfq circuits," *Proc. Conf. Ser.-Inst. Phys.*, vol. 148, pp. 1733–1736, 1995.

[10] K. G. I. V. Vernik, Q. P. Herr and M. J. Feldman, "Experimental investigation of local timing parameter variations in rsfq circuits," *IEEE Trans. Appl. Supercond.*, vol. 9, no. 2, pp. 4341–4344, 1999.

[11] F. Wang and S. Gupta, "An effective and efficient automatic test pattern generation (atpg) paradigm for certifying performance of rsfq circuits," *IEEE Transactions on Applied Superconductivity*, vol. 30, 2020.

[12] G. Krylov and E. G. Friedman, "Test point insertion for rsfq circuits," in *2017 IEEE International Symposium on Circuits and Systems (ISCAS)*, 2017, pp. 1–4.

[13] ——, "Design for testability of sfq circuits," *IEEE Transactions on Applied Superconductivity*, vol. 27, no. 8, pp. 1–7, 2017.

[14] J.-S. Park, S. Subramanian, L. Lampert, T. Mladenov, I. Klotchkov, D. J. Kurian, E. Juarez-Hernandez, B. Perez-Esparza, S. R. Kale, K. T. Asma Beevi, S. Premaratne, T. Watson, S. Suzuki, M. Rahman, J. B. Timbadiya, S. Soni, and S. Pellerano, "13.1 a fully integrated cryo-cmos soc for qubit control in quantum computers capable of state manipulation, readout and high-speed gate pulsing of spin qubits in intel 22nm ffl finfet technology," in *2021 IEEE International Solid- State Circuits Conference (ISSCC)*, vol. 64, 2021, pp. 208–210.

[15] N. Jha and S. Gupta, *Testing of Digital Systems*. Cambridge University Press, 2003.

979-8-3503-4631-2/23 $31.00 © 2023 IEEE

CAPEC: A Cellular Automata Guided FSM-based IP Authentication Scheme

Mridha Md Mashahedur Rahman[*], M Sazadur Rahman[*], Rasheed Kibria[*], Mike Borza[**],
Bandy Reddy[**], Adam Cron[**], Fahim Rahman[*], Mark Tehranipoor[*], and Farimah Farahmandi[*]

[*]Department of Electrical and Computer Engineering, University of Florida, Gainesville, Florida
[**]Synopsys, Inc., Mountain View, CA 94043 USA
Email: {mrahman1,mohammad.rahman,rasheed.kibria}@ufl.edu,{mborza,bchandra,adam.cron}@synopsys.com
{fahimrahman,farimah,tehranipoor}@ece.ufl.edu

Abstract—The ever-increasing propensity for intellectual property (IP) reuse has reduced the design productivity gap in the supply chain. As a consequence, protecting IPs has become more difficult since IP vendors now make their IPs more flexible so that they can be reused in other designs for greater profits. This has made IP piracy and infringement easier than ever. IP watermarking can detect IP piracy and infringement and it has been an active research topic for the past decade. Various watermarking techniques have been discussed in the literature that embed circuitry into IP to provide proof of ownership. But, in most RT-level watermarking methods, the watermarking circuit is separate from IP functionality and can be easily identified and tampered with. In this paper, we propose *CAPEC*, a Cellular Automata (CA) guided watermarking technique that embeds watermarking circuits into the don't care states of the FSM. The watermarking function is a set of configurable CA rules tightly coupled with the functional states of the FSM. *CAPEC* generates a signature in a challenge-response-based protocol, is resistant to identification, tampering, and removal attacks, and has minimal overhead. We also analyze and evaluate the efficiency of the technique and its resilience to different attacks for varying challenge size and CA rules. After watermarking different benchmarks, the watermark overhead was found to be negligible and formal verification proved no changes to the functional circuit.

Index Terms—IP Watermarking, IP Authentication, Cellular Automata.

I. INTRODUCTION

System-on-Chips (SoCs) are becoming increasingly complex due to their functionality and the prevalence of design reuse. SoC designers commonly license pre-designed intellectual property (IP) cores as soft (RTL), hard (GDSII), or firm (netlist) IPs. In order to maintain cutting-edge semiconductor fabrication at a lower cost, fabless semiconductor companies outsource fabrication and testing to offshore foundries. Globalization has boosted growth, lowered costs, and cut time to market in the semiconductor industry. The IP rights owner must provide the systems integrator and foundry with the entire specification for building the system. As IP owners are no longer the sole proprietor of content, this IP procurement business risks violating the original license terms and misappropriating the design. Several security vulnerabilities arise as a result of the untrusted parties within the supply chain, including intellectual property theft, counterfeiting, reverse engineering (RE), and overproduction of integrated circuits (ICs) [1]. Since IP owners have very little control over the illicit use of their IP, identifying IP cores within suspect designs is imperative.

Researchers have proposed many methods to prevent IP cores from being exploited illegally. Watermarking has been the most widely investigated approach [2]. By embedding a unique signature, an IP core can be watermarked in a way that does not alter its original functionality. Upon manufacturing the chip, the IP owner can retain them from the market and extract their signature using the activation parameters they have created to prove that their IP core has been (legally or illegally) used in an SoC by comparing it with the initially embedded signature. Watermark mechanisms should be easy to embed and verify and not burdened with high overhead and attacks [3]. IP infringement, piracy, and overproduction of integrated circuits cannot be prevented by watermarking. However, SoC IP usage can be proven by this method. Even though there are many copyright violation examples, e.g., Cisco vs. Huawei [4], Micro vs. Intel [5], CNEX vs. Huawei [6], etc., and several companies have engaged in legal battles in the past two decades, they often go unnoticed. Hence, each IP must have a unique identifier that proves its ownership even though it is integrated into an SoC.

Several watermarking methods have been reported in the literature [7–21]. IP identification/detection methods have been studied at the system, behavioral, logic, and physical levels. Watermarking based on finite state machines (FSMs) is a relatively well-explored area in IP authentication. FSM-based watermarks embed authorship information at the behavioral synthesis level. While FSM-based watermarking is robust against various attacks, it suffers from noise, removal attacks, and high implementation overhead [2].

Designing and implementing a watermark that is well-hidden in the functional IP blocks without affecting the intended IP functionality or IP performance is a challenging task. In this paper, we propose *CAPEC*, a cellular automata guided FSM-based IP authentication scheme that inserts a hybrid cellular automata-based watermark circuit into the don't care states of the FSM. The proposed scheme adds a few watermarking states and intertwines them with functional states by creating state transitions. Our key contributions are:

1) We propose *CAPEC*, an FSM-based watermarking solution that hides the signature in the don't care states implemented by a configurable number of hybrid CA rules, variable challenge-response pairs (CRPs), and state transitions.
2) Unlike the existing watermarking methods, *CAPEC* entangles watermark logic with the IP functionality to be resilient against identification, tampering, and removal.
3) We develop a protocol for watermark-based IP authentication. *CAPEC* utilizes an existing IEEE Std 1687 structure to activate the watermark and extract the signature.
4) We perform an exhaustive security analysis of *CAPEC*

979-8-3503-4631-2/23 $31.00 © 2023 IEEE

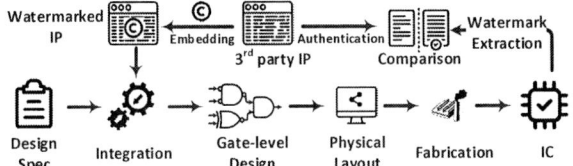

Fig. 1: SoC design flow using third-party IPs, along with the watermark embedding and signature extraction flow.

Fig. 2: A 5-cell one-dimensional LHCA with null boundary.

against algebraic attacks for different CA rules, removal, tampering, and forgery attacks. Our performance evaluation shows that *CAPEC* integrates the watermarking logic without any functional changes and with minimal overhead.

The rest of paper is organized as follows: Section II briefly provides a threat model for IP theft, existing FSM-based watermarking methods, and linear hybrid cellular automata. Section III presents the detailed methodology of *CAPEC*. Section IV presents security analysis of the proposed scheme and Section V presents the experimental results and discussion. Finally, conclusions are presented in Section VI.

II. BACKGROUND

A. Threat Model

A watermark identifies IP piracy and overuse by marking an asset with a known signature. In hardware design flows, watermarking is the process of embedding a signature (or a unique function) within the IP core without affecting the original functionality of the design as shown in Fig. 1. The watermark signature can be used to authenticate IP ownership as described in Fig. 1. A watermarking solution should possess the following features [2].

1) Fidelity: The watermark should not interfere with the IP functionality or IP performance.
2) Uniqueness: The watermark signature should be unique to each IP core to eliminate any chance of collision.
3) Resiliency: Watermarks for authenticating IPs in an SoC are very susceptible to noise from the neighboring IPs. Hence, the watermark should be resilient to the interference emanating from neighboring IPs.
4) Non-redundancy & robustness: The watermark should be properly incorporated with IP functionality. In the case of FSM-based watermarking, the watermark states must be intertwined with the functional states to eliminate illegal identification, removal, or modification.
5) Efficiency: The watermark should be easy to verify with minimum implementation overhead and low observability.

B. Existing FSM-based Watermarking and their Pitfalls

A number of IP watermarking techniques have been discussed in the literature. Among these techniques, FSM-based watermarking is quite popular as using an FSM in a design provides an easy way to extend the state space and hide a watermark within it. FSM-based watermarking techniques can be classified into two types: transition-based watermarking and state-based watermarking.

A state-based watermarking strategy was proposed in [18], which inserts the watermark as a new property. In this strategy, redundancy is introduced to the state-transition graph (STG) so that it exhibits the chosen watermarking property. Using this approach to modify STGs adds a large amount of overhead to the design. Additionally, this approach is vulnerable to

state minimization [22, 23] since all redundant states will be removed. A state encoding-based watermarking technique was proposed in [24]. This technique proposes an algorithm to encode all states of the FSM with cellular automata components generated by the D1*CA rule. It then implements state transitions using an interrupt logic. For a given sequence of input patterns, the technique examines the left-most CA-cell's data at different pre-generated clock cycles to verify the watermark bits. As discussed in the paper [24], to conduct a state-recoding attack, one must have a complete understanding of the STG of the FSM. Using an FSM extraction tool like FSMx [25], it is possible to derive the complete STG of the FSM from a flattened gate-level netlist. Therefore, the state encoding-based watermarking techniques [19, 24] are vulnerable to state recoding attacks. Also, the overhead of the technique [24] with respect to the non-watermarked original design is relatively high ($10 \sim 23\%$).

Transition-based watermarking uses existing transitions and unused transitions to embed watermarks in an FSM [26, 27]. The technique in [26] proposes two algorithms for mapping randomly generated inputs (challenge) and watermark signatures into random states of the STG. Unfortunately, when the original FSM has large outputs, both algorithms perform poorly. The technique in [27] addresses the problem of mapping watermark signatures to FSM outputs. The watermark of this technique appears in the output sequence at specific points. Using existing transitions to implement watermarks may become difficult when the input for these transitions depends on complex functional logic.

The FSM watermarking techniques discussed above do not discuss the watermark extraction process from the IP when it is integrated deep into an SoC and fails to meet the requirements discussed in Section II-A. In FSMs, watermarking is usually based on the observed bit sequences at the output. However, once the watermarked IP is integrated into the chip, the outputs of the FSM cannot be observed externally. Watermarked IPs cannot be authenticated in the field without opening their encapsulation. In addition to being expensive, extracting the hidden watermark destroys the working IC [2].

C. Linear Hybrid Cellular Automata (LHCA)

Cellular automata (CA) are linear finite state machines (LFSM) in $GF(2)$ defined as n-bit uniform arrays of cells laid out in a one-dimensional space as shown in Fig. 2. They are nearest neighbors, so each cell is connected only to its left and right cells as shown in Fig. 2. The next-state function of a CA is linear and hybrid, which means different cells can possess different next-state functions (f_i) or CA rules. Hence, for a CA, $s_i^{(t+1)} = f_i(s_{i-1}^t, s_i^t, s_{i+1}^t)$. The next-state function f_i can be 2^{2^3} different CA rules. However, due to being linear, there are only eight possibilities [28] among which only *rule 90* and *rule 150*, as shown in Equ. 1, yields polynomial primitivity.

$$Rule\ 90 : s_i^{(t+1)} = s_{i-1}^t + s_{i+1}^t \tag{1}$$

$$Rule\ 150 : s_i^{(t+1)} = s_{i-1}^t + s_i^t + s_{i+1}^t \tag{2}$$

$$Rule\ 102 : s_i^{(t+1)} = s_i^t + s_{i+1}^t \tag{3}$$

979-8-3503-4631-2/23 $31.00 © 2023 IEEE

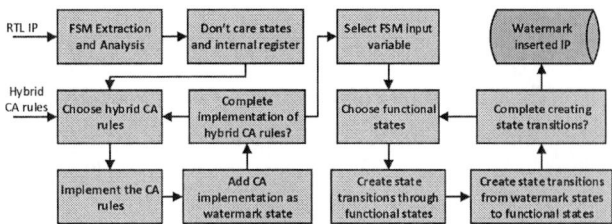

Fig. 3: Watermark circuit insertion flow.

Linear hybrid cellular automata (LHCA) and linear feedback shift registers (LFSR) are both considered linear finite state machines (LFSM). However, LHCA has a higher transition space than LFSR and performs better as a pseudo-random number generator for the same characteristic polynomial. LHCA performs better signature aliasing than LFSR due to more randomization in parallel pattern generation. Without allowing modifications to the characteristic polynomial, LHCA behaves slightly better than LFSRs. But LHCA is always superior to LFSRs after the introduction of minimum modifications. Intuitively, this is because we can try more changes in LHCA. Specifically, an LHCA of length n has n possible configurations, i.e., each of the n cells can be reconfigured either as a *rule 90* cell or as a *rule 150* cell. However, an LFSR of length n has only $(n-1)$ possible tap locations. LHCA can easily be implemented in hardware, using flip-flops as memory elements and XOR gates for arithmetic. In Section III, we discuss how CA can be utilized to extend the existing FSMs in the IP core and hide a watermark signature.

III. PROPOSED SCHEME: CAPEC

The proposed scheme, *CAPEC*, inserts a hybrid CA-based watermark circuit into the FSM. The watermark insertion process starts with extracting the FSM from the IP and analyzing it (Fig. 3). The FSM analysis reports the don't care states, IP internal registers, and input variables that are necessary for the watermarking scheme. Later, the cellular automata rules are implemented as watermarking states. After the hybrid CA implementation is complete, the proposed technique creates state transitions to make the watermarking states strongly connected with the functional states. *CAPEC* proposes two algorithms for challenge-response pair (CRP) generation and a watermark verification flow for IP authentication. These algorithms are discussed in detail in the following subsections.

A. IP FSM Extraction

To insert *CAPEC*, we first need to analyze the IP and functional FSM to look for don't care states and a suitable register. We extract the control circuitry of the target IP using the RTL-FSMx tool [29]. This tool reports the states used in the FSM, state encodings, state variables, input variables, and available don't care states. After extracting and analyzing the original IP FSM, we integrate the watermarking circuit into the FSM following the steps shown in figure 3. *CAPEC* implements the watermarking circuit in the don't care state of the FSM. For this FSM-based watermarking technique implementation, at least two states must be integrated into the functional state machine. If no don't care state is available, then *CAPEC* extends the state space. In order to implement the watermarking circuit, *CAPEC* also requires a register. Our watermarking technique aims to have close integration between IP functionality and watermarking and to add as minimal overhead as possible.

So, we first look for an IP internal register that meets our requirements. Some of these requirements are the size of the register, the reset condition, functional dependency, etc. If we cannot locate a register that meets our requirements, we add one as part of the watermarking process.

B. Watermark Circuit Insertion

The proposed watermarking scheme is an IP agnostic approach, i.e., it does not require any specific IP functionality to implement the watermark circuit. The watermark circuit is a set of linear hybrid cellular automata (LHCA) formed using the internal register discussed above. Hence, the watermark circuit insertion comprises two major steps discussed below.

Choosing CA Rule: As mentioned in Section II-C, the next-state function of an LHCA is termed the CA rule. *CAPEC* does not rely on specific rules. Instead, we allow the IP owner to choose hybrid CA rules based on their security requirements. At a minimum, the proposed scheme adds two watermarking states to the FSM. In the S_{load} state, the cellular automata are loaded with a random value, while in the S_{wm} state, the cellular automata rules are implemented using the chosen register. In short, the CA rules used for signature generation are not fixed and can be changed from IP to IP. For example, the CA rules governing signature generation can be a combination of *rules* $\{90+150\}$ (Equ. 1, 2) for one IP and *rules* $\{90+102\}$ (Equ. 2, 3) for another IP. The combination of *rules* $\{90+150\}$ form a linear hybrid CA that gives maximum cycle length [28]. In contrast, *rules* $\{90+102\}$ form a particular case of non-group additive CA, called D1*CA. The signature generation sequence differs for different IPs because of different combinations of CA rules. The significance of these hybrid CA rules are further experimentally validated in SubsectionV-C. Using *CAPEC*, the IP owner can also specify how many sets of hybrid CAs should be included in the watermark circuit. As a result, watermarking security is enhanced as the attacker must guess how many hybrid CA implementations there are and which CA rules are used. On the other hand, the more hybrid CA implementations are used for watermarking, the more watermarking states there are. Thus, the watermark overhead becomes larger. So, the proposed technique gives its users a trade-off between security and overhead.

Creating CA-based State Transition with IP FSM: After the watermark circuit is implemented using the watermarking states and a register, we need to create state transitions to stitch them into functional states. Unlike other state-based watermarking techniques [18, 30, 31], the watermarking states will not be separate from the functional states in this proposed approach. We randomly choose a set of functional states and an FSM input variable using the RTL-FSMx tool [25]. Later, we create state transitions through these functional states to the watermarking

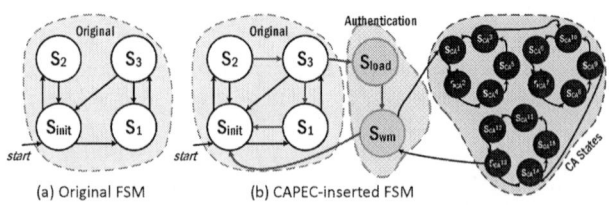

Fig. 4: A sample finite state machine watermarked by *CAPEC*. (a) FSM of a simple SPI, b) CAPEC-inserted FSM. From the figure it can be observed that *CAPEC* inserts additional states and transitions in the original FSM to deter any chance of removal.

979-8-3503-4631-2/23 $31.00 © 2023 IEEE

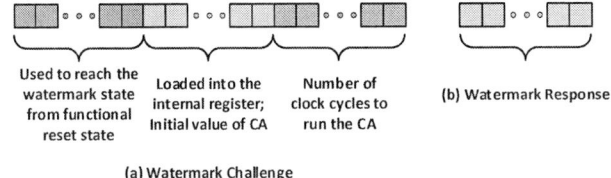

Used to reach the watermark state from functional reset state **Loaded into the internal register; Initial value of CA** **Number of clock cycles to run the CA** **(b) Watermark Response**

(a) Watermark Challenge

Fig. 5: Watermark challenge and response sequence: a) Different parts of the challenge sequence and their usage, b) Response sequence that corresponds to the challenge in (a). The size of the response sequence is exactly the same as the size of the second part (from left) of the challenge sequence.

states by providing input to the FSM input variables. Thus, a portion of the challenge is allocated to make state transitions from the reset state to the watermarking state. We intentionally make this part of the challenge dependent on the IP FSM. This step ensures that the challenge for one IP does not activate the watermark in another IP. Fig 4(a) shows the FSM of a simple SPI module and Fig 4(b) shows the corresponding *CAPEC* watermarked version. We also create state transitions from watermarking to functional states using the FSM input variables. IP owners can specify how many state transitions between watermarking and functional states they want. Watermark circuits are more difficult to separate and remove if they are interconnected with functional states.

C. Watermark Challenge Response Generation

Watermark Challenge Configuration: Watermark challenges consist of three parts (see figure 5). The first part (from left) contains challenge bits for the FSM input variable. It is used to reach the watermark states from the reset state of the FSM. The second part contains the initialization values of the cellular automata (CA), which are loaded into the register responsible for CA calculation. The final part indicates the number of clock cycles for each initialization value. The initialization and clock cycle values are used in pairs to generate the signature values. IP owners can specify how many pairs of these values they want in the challenge bitstream. In general, the greater the number of such pairs of values, the longer the challenge-response bitstream, the longer the verification time, and the greater the watermark security (discussed in Subsection V-C).

CRP Generation for Authentication: CAPEC deploys two solid algorithms to generate challenge-response pairs (CRPs) for this watermarking verification. In the first method, the CRPs are generated to authenticate the IP in the SoC. Since the watermark circuit is closely integrated with the functional states of the FSM, it provides some degree of confidence that the FSM

Algorithm 1: CRP generation for authentication

Input: Python model of the watermark circuit;
Size of each signature, N_1;
Number of CRPs, N;
Output: Challenge-response pairs, CRP;
1   $CRP \leftarrow \emptyset; i \leftarrow 0;$
2   **while** $i < N$ **do**
3     $C_i \leftarrow \emptyset; S_i \leftarrow \emptyset; j \leftarrow 0;$
4   **while** $j < N_1$ **do**
5   Randomly generate CA initialization value, CA_j;
6   Apply the value to the Python model;
7   Run the CA for a random number of clock cycles, CK_j;
8   Store the result S_j as watermark signature: S_i.append(S_j);
9   Store the initialization value and number of clock cycles as challenge: C_i.append(CA_j); C_i.append(CK_j);
10       $j \leftarrow j + 1;$
11   store the CRP: CRP.append(C_i); CRP.append(S_i);
12     $i \leftarrow i + 1;$
13   **return** CRP;

Algorithm 2: CRP generation for a specific signature text

Input: Python model of the watermark circuit;
Signature text;
Output: Challenge-response pair, CRP;
1   $CRP \leftarrow \emptyset; C \leftarrow \emptyset; i \leftarrow 0;$
2   Encrypt the text to generate watermark signature, S;
3   Partition the signature into N values (each with CA-register bit size);
4   **while** $i < N$ **do**
5   Apply i-th signature value to the Python model;
6   Run the CA for a random number of clock cycles, CK_j;
7   Store the result as CA initialization value, CA_j;
8   Store the initialization value and number of clock cycles as challenge: C.append(CA_j); C.append(CK_j);
9     $i \leftarrow i + 1;$
10   store the CRP: CRP.append(C); CRP.append(S);
11   **return** CRP;

---

has not been tampered with or modified. In order to perform watermark-based authentication, many CRPs must be generated using Algorithm 1. Initially, the CA initialization values are generated at random (line 5). Afterward, the Python model of the CA-based watermark circuit is run for a random number of cycles to compute the corresponding signatures using each of these random values (lines 6 to 8). Watermark challenges are determined by the CA initialization value and the number of clock cycles (line 9). Finally, the CRPs are stored in a secure database (line 11). During authentication, one CRP is used only once to avoid replay attacks. If an attacker tries to tamper or remove the watermark down the supply chain, watermark-based authentication will fail.

CRP Generation for Specific Signature: Algorithm 2 shows the second method that allows the user to select and embed a specific signature text as a watermark. Choosing a signature first is a common approach in the literature (section II). A user-specified key and IV are used to encrypt the chosen signature text utilizing AES-128 (line 2). The encrypted watermark signature is then divided into values with CA-register bit sizes (line 3) and used to initialize and run the CA. After the python model of the CA-based watermarking circuit is loaded with the signature value (initialization), we run it for a random number of cycles to determine the actual initialization values of the challenge (lines 5 to 7). The generated CA initialization values and the number of clock cycles form the challenge of the watermark signature (line 8). Finally, the CRP is stored in a secure database (line 10). During watermark verification, the challenge is applied to the watermark circuit, and the response is the encrypted signature text.

D. Watermark Verification in an SoC

After the watermark CRPs are generated, they will be stored in a secure database. We propose storing the challenge bits in an encrypted format, as the challenges contain information regarding state transitions. During watermark authentication, the encrypted challenge is sent from the database to the SoC, where it is decrypted and sent to the watermarked IP. If an attacker intercepts the transactions between the database and the SoC and obtains the encrypted challenges, they cannot decrypt them. Thus, they will not be able to create their own CRP and perform a forgery attack.

Many FSM-based watermarking techniques ignore the importance of IP watermark verification protocols in an SoC. Watermarking techniques discuss a method by which the watermark outputs can be made visible through IP primary outputs [24, 26, 27], but without a method for making these IP primary outputs visible outside the SoC, the IP owner cannot

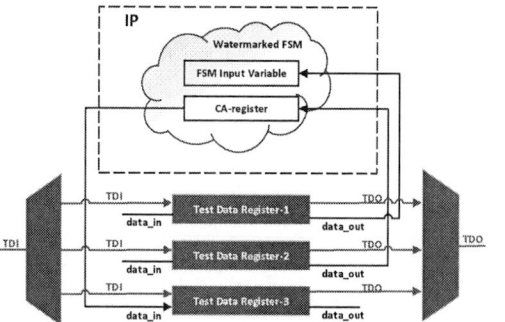

Fig. 6: IEEE 1687 Test Data Registers send the challenge bits to the FSM input variable and CA-register, and read the signature from the CA-register.

prove that the IP is theirs. Communication with the IP for watermarking is not only necessary to extract signatures, but also to send challenge bits. *CAPEC* proposes using *IEEE Std 1687* to transfer the challenge bits and to read the signature in an SoC. As part of the watermark verification protocol in the SoC, the FSM input variable and CA registers are connected to 1687 (figure 6). It is noteworthy that if a smart attacker steals the IP, puts it in his SoC, and blocks every interface to communicate with the IP from outside the SoC, then the IP owner has no way to send the challenge and read out the signature. This is a problem that the IP owner cannot solve with a passive watermarking technique for their IP [2].

During watermark verification, the first part of the decrypted challenge (see figure 5) is used to jump through the functional states to reach the watermarking states from the FSM reset state. Afterwards, the values stored in the second and third part of the challenge are used in pairs to run the hybrid CA and generate the corresponding signature. The watermark signature value is then read out from the CA registers. This process continues for each pair of initialization and clock cycle values (consecutively, the second and third part of the challenge) to activate the watermark signature. The watermark verification flow for IP authentication is shown in figure 7.

IV. SECURITY ANALYSIS

In this section, we perform security analyses of the proposed watermarking scheme and analyze its resiliency against various attacks. We assume that the attacker does not have access to the RTL of the watermarked IP but has access to the gate-level netlist via rogue insider in the supply chain or reverse engineering. Based on this assumption we discuss the resilience of the proposed scheme for the following attack scenarios.

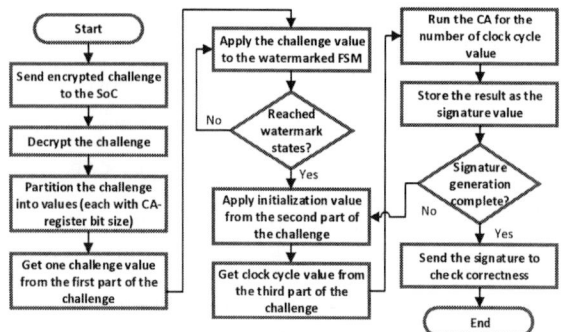

Fig. 7: Watermark verification flow for IP authentication.

A. Watermark Forgery Attack

An attacker may attempt to claim ownership of the IP by showing their own watermark on the watermarked IP by forging his own CRP [32]. In order to accomplish this, the attacker either needs a complete understanding of the functional states and watermark states or decrypt the challenge and understand how to use it. Understanding the watermark circuit and its functionality is difficult without access to the RTL of the watermarked FSM. The proposed technique hides the watermarking circuit in the functional states and intertwines them in such a way that guessing the functionality should be very challenging. The technique proposes to encrypt the challenges when they are sent from the secure database to the SoC. Even if the attacker gets their hands on a decrypted challenge, understanding how to use it to generate a signature will require knowledge of the functional FSM and the watermark circuit. The watermark signature generation process depends not only on the value of the inputs but also on the timing of those inputs.

B. Watermark Modification Attack

In this attack, the attacker tries to modify the watermark by performing resynthesis and retiming [24]. Resynthesis may change the FSM's combinational logic, but the logic's input-output behavior remains the same. As the circuit's functionality remains the same after resynthesis, *CAPEC* deters resynthesis attack. Retiming transforms the circuit by moving the sequential elements around the combinational logic. However, it does not change the functionality of the circuit. Hence, the circuit before and after retiming should be functionally equivalent. As *CAPEC* inserts hybrid CA logic into the don't care states of the FSM, retiming may move the states. However, the watermark circuit functionality remains the same and resists retiming attacks. In other words, if the watermarked FSM is denoted as C and the FSM after resynthesis and/or retiming is C', then C and C' should be functionally equivalent. Therefore, *CAPEC* must be able to verify the transformed FSM C' as well. However, state minimization or reduction could be a concern if the watermark states are found to be redundant. The watermarking states are improbable to be equivalent to each other, or any other functional state as *CAPEC* implements a few watermarking states, each with a different hybrid CA implementation. *CAPEC* also utilizes functional registers for CA implementation and FSM input variables. Thus, each watermarking state has a defined distinct input-output behavior without redundancy.

C. State Recoding Attack

The attacker may attempt to perform a state recoding attack to alter the watermark circuit [27]. State recoding changes the state encodings of the FSM. This attack is effective against watermarking techniques that rely on a specific state encoding algorithm, such as [19, 24]. Our proposed technique does not rely on any state encoding strategy. Thus, the state encoding attack should not have any effect on this watermarking scheme.

D. Watermark Tampering and Removal Attack

The proposed watermarking technique introduces a few watermarking states, and these are closely integrated with functional states. To reach the watermarking states, a user of this technique must jump through a set of functional states. There is also a configurable number of transitions from the watermarking states that intertwines the watermark states with

979-8-3503-4631-2/23 $31.00 © 2023 IEEE

TABLE I: Resource Utilization for Implementing Watermarks on Different Benchmark IPs

| Benchmarks | Before Watermarking | | After Watermarking | | Watermark Overhead |
	# of Gates	# of FF	# of Gates	# of FF	(# of Gates)
DES	4587	966	4871	985	284
AES-128	16688	2469	16802	2486	114
SHA-512	8838	2098	8869	2115	31
I2C	562	153	694	171	132
UART	581	91	706	113	125
SAEAES	14375	403	14535	425	160
ASCON	3981	459	4303	477	322
COMET	14441	530	14559	528	118
ROMULUS	3659	585	3782	583	123
TINYJAMBU	1162	134	1292	151	130

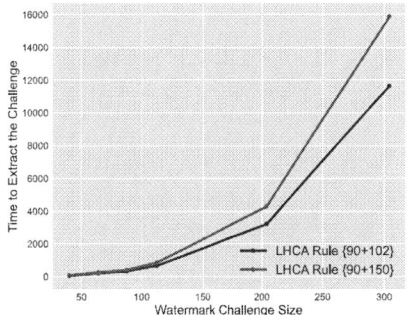

Fig. 8: Impact of challenge size and LHCA rules on the attack complexity of *CAPEC* for DES from Table I. Time to extract the challenge by performing Fun-SAT attack [36] increases with challenge size. For same challenge size, LHCA rules $\{150 + 90\}$ offers more attack complexity than $\{102 + 90\}$.

the functional states. The registers required to implement the hybrid CA implementations and to create state transitions are chosen from functional logic if they meet the corresponding requirements. All this makes the watermark circuit tightly integrated with functional logic. Thus, it should be very challenging to identify and separate the watermark circuit. An attempt to tamper with or remove the watermark circuit has a high chance of changing IP functionality. The attacker may attempt to identify the watermark states by performing topological analysis [33] using Tarjan's strongly connected component algorithm [34]. However, it is noticeable from Fig. 4(b) and Section III that *CAPEC* employs multiple in-degree and out-degree transitions with the original FSM which deters identification of the watermark states.

V. RESULTS AND DISCUSSION

In this section we experimentally validate the performance of *CAPEC* for several opensource and industrial benchmarks [35].

A. Overhead Analysis

We have implemented *CAPEC* on several benchmarks [35] and the watermark overhead is reported in table I. All the IPs in table I are watermarked with the same watermark configuration and synthesized in Synopsys 90nm technology library using Synopsys Design Compiler. The hybrid CA rule used in this implementation is a combination of *Rules 90+150* and the size of the CA register is 8-bit. To implement this watermark, we used two watermarking states as shown in Fig. 4. We used a variable number of state transitions depending on the number of functional states to intertwine these watermarking states with functional states. We also chose the challenge size to be 408-bit and the response size to be 192-bit. As expected, the watermark overhead is negligible and almost the same for each IP because the watermarking overhead introduced by *CAPEC* does not depend on the size of challenge-response pairs.

For any practical implementation, the watermark overhead needs to be very low as the watermark circuit will only be active during watermark authentication. And, watermark verification will most likely be done only a few times in an SoC lifecycle. Our proposed watermarking scheme gives users control over the watermarking configuration. A user can specify the number of CA implementations in a watermark implementation, the number of transitions between functional and watermark states, and the size of the CRPs. Thus, the user is presented with a trade-off between watermark overhead and security.

The watermark challenge-response size and varying LHCA rules do not impact the area overhead. Therefore, a user can select a larger size for the challenge and response to reduce

the probability of coincidence. However, a larger challenge-response pair requires more time to authenticate the IP and more storage on the secure database. As watermark-based authentication is less likely to be run frequently in an SoC, the time required for authentication should not be a problem. The impact of challenge size and different LHCA rules are discussed in Section V-C.

B. Formal Verification of Watermarked IP

We have also run formal verification after watermarking the benchmark IPs (table I) to ensure no change in functionality. To do this, we have excluded the registers (if any) introduced by *CAPEC*. We used the Synopsys Formality tool to run formal verification and found a functional match between watermarked IP and the IP before watermarking.

C. Significance of Challenge Size and LHCA Rule

Section III discussed the significance of CA rules and challenge size on the security offered by *CAPEC*, which we experimentally validate in this subsection. The watermark challenges required in *CAPEC* for appropriate state transitions through functional states to reach the watermark states are analogous to the correct set of input sequences required in FSM obfuscation methods [36] to reach functional states. Hence, we performed a Fun-SAT attack on the *CAPEC*-watermarked design for varying challenge sizes and two different LHCA rules: $\{150 + 90\}$ and $\{102 + 90\}$. The attack results are presented in Fig. 8. Please note that in a practical scenario, such attacks are not feasible as an attacker does not have access to an IC with challenges applied. However, we still performed this analysis to emphasize the significance of the LHCA rules and challenge size. From Fig. 8, it is proved that attack complexity increases with increasing challenge size and LHCA rule with polynomial primitivity, which in this case is $\{150 + 90\}$. Investigation of other types of automata that produces a recognizable sequence to generate watermark is part of our future work.

VI. CONCLUSION

Relying on the fact that existing state-based watermarking techniques are not entangled with IP functionality, in this paper, we introduced *CAPEC*, a watermark-based IP authentication scheme that integrates a hybrid CA guided watermark circuit into the FSM of the IP. The watermark circuit has minimal overhead and it is intertwined with IP functionality to be resilient against identification, tampering, and removal. *CAPEC* is configurable in different ways that allows one to choose between security and overhead.

979-8-3503-4631-2/23 $31.00 © 2023 IEEE

REFERENCES

[1] B. Shakya, M. Tehranipoor, S. Bhunia, and D. Forte, "Introduction to hardware obfuscation: Motivation, methods and evaluation," in *Hardware Protection through Obfuscation*. Springer, 2017, pp. 3–32.

[2] N. N. Anandakumar *et al.*, "Rethinking watermark: Providing proof of IP ownership in modern socs," *IACR Cryptol. ePrint Arch.*, p. 92, 2022.

[3] C.-H. Chang *et al.*, "Hardware ip watermarking and fingerprinting," in *Secure System Design and Trustable Computing*. Springer, 2016, pp. 329–368.

[4] https://www.computerworld.com/article/2578617/cisco-sues-huawei-over-intellectual property.html, "Cisco sues huawei over intellectual property," Computer World, 2003.

[5] https://www.mazzarellalaw.com/blog/2018/12/intel-files-trade-secret-theft lawsuit/, "Intel files trade secret theft lawsuit," Mazzarella Mazzarella LLP.

[6] https://techcrunch.com/2019/05/22/semiconductor-startup-cnex-labs-alleged-huaweis-deputy-chairman-conspired-to-steal-its-intellectual property/, "Semiconductor startup cnex labs alleged huawei's deputy chairman conspired to steal its intellectural property." TechCrunch, 2019.

[7] G. Qu and M. Potkonjak, "Analysis of watermarking techniques for graph coloring problem," in *Proceedings of the 1998 IEEE/ACM international conference on Computer-aided design*, 1998, pp. 190–193.

[8] G. Qu and L. Yuan, "Secure hardware IPs by digital watermark," in *Introduction to hardware security and trust*. Springer, 2012, pp. 123–141.

[9] F. Koushanfar, I. Hong, and M. Potkonjak, "Behavioral synthesis techniques for intellectual property protection," *ACM Transactions on Design Automation of Electronic Systems (TODAES)*, vol. 10, no. 3, pp. 523–545, 2005.

[10] T.-B. Huynh, T.-T. Hoang, and T.-T. Bui, "A constraint-based watermarking technique using Schmitt Trigger insertion at logic synthesis level," in *2013 International Conference on Advanced Technologies for Communications (ATC 2013)*. IEEE, 2013, pp. 115–120.

[11] D. Kirovski, Y.-Y. Hwang, M. Potkonjak, and J. Cong, "Protecting combinational logic synthesis solutions," *IEEE Transactions on Computer-Aided Design of Integrated Circuits and Systems*, vol. 25, no. 12, pp. 2687–2696, 2006.

[12] A. Cui, C. H. Chang, and S. Tahar, "IP watermarking using incremental technology mapping at logic synthesis level," *IEEE Transactions on Computer-Aided Design of Integrated Circuits and Systems*, vol. 27, no. 9, pp. 1565–1570, 2008.

[13] S. Meguerdichian and M. Potkonjak, "Watermarking while preserving the critical path," in *Proceedings of the 37th Annual Design Automation Conference*, 2000, pp. 108–111.

[14] M. Ni and Z. Gao, "Constraint-based watermarking technique for hard IP core protection in physical layout design level," in *Proceedings. 7th International Conference on Solid-State and Integrated Circuits Technology, 2004.*, vol. 2. IEEE, 2004, pp. 1360–1363.

[15] J. X. Zheng and M. Potkonjak, "Securing netlist-level FPGA design through exploiting process variation and degradation," in *Proceedings of the ACM/SIGDA international symposium on Field Programmable Gate Arrays*, 2012, pp. 129–138.

[16] R. Chapman and T. Durrani, "IP protection of DSP algorithms for system on chip implementation," *IEEE Transactions on Signal Processing*, vol. 48, no. 3, pp. 854–861, 2000.

[17] A. Rashid, J. Asher, W. Mangione-Smith, and M. Potkonjak, "Hierarchical watermarking for protection of DSP filter cores," in *Proceedings of the IEEE 1999 Custom Integrated Circuits Conference (Cat. No.99CH36327)*, 1999, pp. 39–42.

[18] A. Oliveira, "Techniques for the creation of digital watermarks in sequential circuit designs," *IEEE Transactions on Computer-Aided Design of Integrated Circuits and Systems*, vol. 20, no. 9, pp. 1101–1117, 2001.

[19] M. Lewandowski, R. Meana, M. Morrison, and S. Katkoori, "A novel method for watermarking sequential circuits," in *2012 IEEE International Symposium on Hardware-Oriented Security and Trust*, 2012, pp. 21–24.

[20] P. M. Kirovski D., "Intellectual property protection using watermarking partial scan chains for sequential logic test generation," *High Level Design, Test Verification*, 1998.

[21] C.-H. Chang and A. Cui, "Synthesis-for-Testability Watermarking for Field Authentication of VLSI Intellectual Property," *IEEE Transactions on Circuits and Systems I: Regular Papers*, vol. 57, no. 7, pp. 1618–1630, 2010.

[22] L. Yuan, P. Pari, and G. Qu, "Finding redundant constraints for fsm minimization." 01 2004, pp. 976–977.

[23] J. Pena and A. Oliveira, "A new algorithm for exact reduction of incompletely specified finite state machines," *IEEE Transactions on Computer-Aided Design of Integrated Circuits and Systems*, vol. 18, no. 11, pp. 1619–1632, 1999.

[24] R. Karmakar, S. S. Jana, and S. Chattopadhyay, "A cellular automata guided finite-state-machine watermarking strategy for ip protection of sequential circuits," *IEEE Transactions on Emerging Topics in Computing*, vol. 10, no. 2, pp. 806–823, 2022.

[25] R. Kibria, N. Farzana, F. Farahmandi, and M. Tehranipoor, "Fsmx: Finite state machine extraction from flattened netlist with application to security," in *2022 IEEE 40th VLSI Test Symposium (VTS)*, 2022, pp. 1–7.

[26] A. Abdel-Hamid, S. Tahar, and E. Aboulhamid, "Finite State Machine IP Watermarking: A Tutorial," in *First NASA/ESA Conference on Adaptive Hardware and Systems (AHS'06)*, 2006, pp. 457–464.

[27] A. Cui, C.-H. Chang, S. Tahar, and A. T. Abdel-Hamid, "A Robust FSM Watermarking Scheme for IP Protection of Sequential Circuit Design," *IEEE Transactions on Computer-Aided Design of Integrated Circuits and Systems*, vol. 30, no. 5, pp. 678–690, 2011.

[28] K. M. Cattell, "Characteristic polynomials of one-dimensional linear hybrid cellular automata," Ph.D. dissertation, 1995.

[29] R. Kibria, M. S. Rahman, F. Farahmandi, and M. Tehranipoor, "Rtl-fsmx: Fast and accurate finite state machine extraction at the rtl for security applications," *Cryptology ePrint Archive*, 2022.

[30] A. Oliveira, "Robust techniques for watermarking sequential circuit designs," in *Proceedings 1999 Design*

979-8-3503-4631-2/23 $31.00 © 2023 IEEE

Automation Conference (Cat. No. 99CH36361), 1999, pp. 837–842.

[31] S. Subbaraman and P. S. Nandgawe, "Intellectual property protection of sequential circuits using digital watermarking," in *First International Conference on Industrial and Information Systems*, 2006, pp. 556–560.

[32] A. T. Abdel-Hamid, S. Tahar, and E. M. Aboulhamid, "A Survey on IP Watermarking Techniques," *Des. Autom. Embedded Syst.*, vol. 9, no. 3, p. 211–227, Sep. 2004.

[33] M. Fyrbiak, S. Wallat, J. Déchelotte, N. Albartus, S. Böcker, R. Tessier, and C. Paar, "On the difficulty of fsm-based hardware obfuscation," *IACR Transactions on Cryptographic Hardware and Embedded Systems*, pp. 293–330, 2018.

[34] R. Tarjan, "Depth-First Search and Linear Graph Algorithms," *SIAM journal on computing*, vol. 1, no. 2, pp. 146–160, 1972.

[35] https://opencores.org/, "Free and open source gateware ip cores."

[36] Y. Hu, Y. Zhang, K. Yang, D. Chen, P. A. Beerel, and P. Nuzzo, "Fun-sat: Functional corruptibility-guided sat-based attack on sequential logic encryption," in *2021 IEEE International Symposium on Hardware Oriented Security and Trust (HOST)*. IEEE, 2021, pp. 281–291.

A Low Overhead Checksum Technique for Error Correction in Memristive Crossbar for Deep Learning Applications

Surendra Hemaram, Soyed Tuhin Ahmed, Mahta Mayahinia, Christopher Münch and Mehdi B. Tahoori

Department of Computer Science,

Karlsruhe Institute of Technology (KIT), Karlsruhe, Germany

{surendra.hemaram, soyed.ahmed, mahta.mayahinia, christopher.muench, mehdi.tahoori}@kit.edu

Abstract—The matrix-vector multiplication (MVM) is one of the most frequent operations performed in deep learning hardware accelerators. The crossbar array structure with memristive devices as a building block has an inherent capability to perform energy-efficient MVM. However, the memristive devices suffer from various non-idealities as well as limited number of stable levels. Therefore, the reliability and in turn inference accuracy of the deep learning application is negatively impacted. Thus, this paper presents a low overhead checksum-based error correction method for memristive crossbars for MVM computation. The proposed methodology alleviates the problem of storing the checksum value into multiple columns of the crossbar due to the limited number of stable levels of memristive devices. The number of extra columns required for storing the checksum value is reduced, resulting in a significant reduction in the memory overhead by up to 75%. The proposed method scales the checksum value of trained neural networks (NNs) and then performs checksum-aware retraining, and results show negligible impact (∼3%) on the inference accuracy of the NNs on MNIST, Fashion-MNIST, CIFAR-10, and Veg-15 datasets.

Index Terms—Matrix-Vector Multiplication (MVM), Checksum, Memristive crossbar.

I. INTRODUCTION

Deep learning algorithms have recently gained significant importance in many areas, ranging from mobile applications to the data center [1]. These algorithms involve numerous matrix-vector multiplication (MVM) operations on huge datasets. The data transfer between the processor and memory is becoming one of the most crucial performance and energy bottlenecks in many of the conventional computing paradigms [2]. The compute-in-memory (CiM) paradigm is one of the potential solutions to overcome these challenges [3], [4]. Emerging resistive non-volatile memories (also known as memristive cells) are prominent choices for CiM realization. Many recent works show that non-volatile memristive devices, specifically multilevel cells such as ReRAMs, are promising for implementing the CiM paradigm [5]–[9]. The memristive crossbar can be used for energy-efficient analog computation of MVM for deep learning acceleration [10], [11]. However, the functionality is severely affected by defects or non-idealities caused in memristive devices. These include stuck-at-faults, device-to-device variations, cycle-to-cycle variations, write failure, and random telegraph noise (RTN) [12]–[14].

Several works in the literature target defect mitigation techniques for MVM using memristive crossbars [15]–[21]. The works in [20] and [21] focus on checksum-based fault-tolerant and error correction strategies. The extended algorithmic-based fault tolerance (X-ABFT) for ReRAM-based computing, which uses the traditional ABFT and a test vector to extract the signature to mitigate stuck-at-faults, has been presented by [20]. The majority voting and Hamming code-based checksum techniques for error correction in ReRAM-based matrix operation have been presented in [21].

The digital values are stored by programming the memristive cell in a particular resistive (conductive) state. The number of stable levels stored by memristive cells is limited due to the variation of device resistance from the target programmed value [22]. The limited number of stable levels impacts the maximum value stored by the memristive cell. Due to this limitation, the checksum-based techniques require more than one memristive cell to store such a large checksum value which leads to the requirement of several checksum columns in a memristive crossbar array, resulting in a huge memory overhead. This fundamental technological constraint has been ignored in the existing checksum-based technique [20], [21].

This work proposes a low overhead checksum-based online error correction technique to overcome the bottleneck of an existing method by reducing the requirements of several checksum columns in a memristive crossbar. First, we manipulate the data values in the crossbar (synaptic weight) such that the checksum values become a multiple of the chosen scaling factor. In this way, we just need to store a small factor for checksum value that requires only a single memristive cell. This results in the reduction of memory overhead by up to **75%**. The scaled checksum values can be recovered by left shift operation at the analog-to-digital converter (ADC) output. However, this data manipulation has a severe negative impact on the inference accuracy of NNs. So, we propose the checksum-aware training of NN to regain the accuracy loss due to the manipulation, which does not require any additional computation and storage overhead during inference yet, still achieves comparable accuracy to the original accuracy for MNIST, Fashion-MNIST, CIFAR-10, and Veg-15 datasets. Therefore, the checksum can be stored in a single memristive cell without impairing its error correction capability.

979-8-3503-4631-2/23 $31.00 © 2023 IEEE

The organization of the rest of the paper is as follows. Sec. II discusses the preliminaries and the checksum analysis for a memristive crossbar. In Sec. III, the proposed methodology is presented. Sec. IV presents our simulation setup, results, and performance evaluation. Finally, Sec. IV concludes the paper.

II. ANALYSIS OF CHECKSUM FOR MEMRISTIVE CROSSBAR

A. Memristive Crossbar based CiM for NN Accelerator

Fig. 1 shows the basic CiM architecture using memristive crossbars. The memristive crossbar can be used as an accelerator module for multiply and accumulate (MAC) operations in an analog way. The bit-line (BL) current represents the summation of the multiplication of input vector (v) and conductance matrix (G), i.e., $I_N = \sum_{i=1}^{M} V_i G_{iN}$. A digital-to-analog converter (DAC) and ADC are required to maintain the interaction between the analog and digital data flows. Moreover, the writing circuitry, address decoder, controller, and column select are the other periphery circuits required for the functionality of the aforementioned analog MAC accelerator.

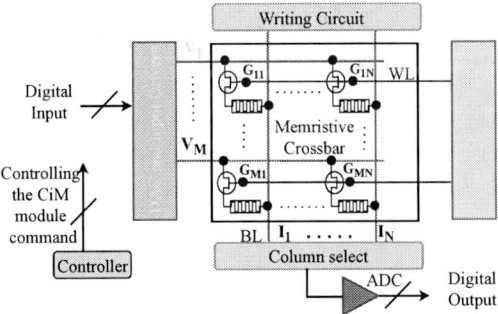

Fig. 1. Memristive crossbar-based CiM Architecture

The NN inference can be made using memristive crossbar-based CiM, in which synaptic weight can be stored into these crossbar arrays, which efficiently perform the required expensive MVM in the memory array, as shown in Fig. 2 (c). Fig. 2 (a) represents the single layer of a NN. Each connection of the two-layer represents the synaptic weight value, which is represented by the conductance values (G) of memristive cells in a memristive crossbar structure. The weights of fully connected layers are 2-D shaped so that they can be mapped directly to the crossbar array. In the case of convolution layers, the first 4-D weights matrix is flattened to a 2-D weight matrix by semi-unrolling. Then flattened weight matrix is mapped to the crossbar as shown in Fig. 2 (b). The peripheral circuit performs batch normalization and non-linear activation to get the activation for the next layer [23].

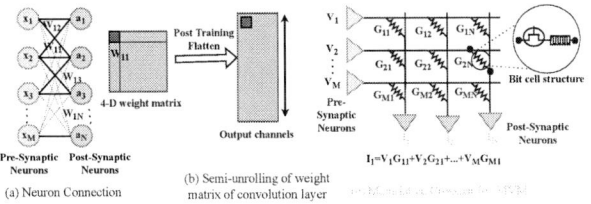

Fig. 2. Synaptic weight mapping to the memristive crossbar [23]

B. Weighted and Non-weighted Checksum

In the memristive crossbar, the output of an MVM is taken column-wise. So, it is not possible to get the checksum row and the output simultaneously. Instead of using a checksum row, an extra checksum column can be added to extract the signature for error correction. The ABFT technique uses a combination of non-weighted and weighted checksums to extract the signature for error correction [20], [24]. Fig. 3 shows an $M \times N$ memristive crossbar with two additional non-weighted and weighted checksum columns. The non-weighted ($G_{C_{i1}}$) and weighted ($G_{C_{i2}}$) checksum for the ith ($i = 1, 2...M$) row can be computed as follows:

$$G_{C_{i1}} = \sum_{j=1}^{N} G_{ij} \quad \text{and} \quad G_{C_{i2}} = \sum_{j=1}^{N} W_{f_j} G_{ij} \quad (1)$$

Where W_{f_j} represents the weight factor, and G_{ij} is the quantized conductance value (quantized weight value of given weight matrix). In the rest of the paper, G is also treated as the weight matrix of NN. The signature for the error correction is computed by taking the ratio of the weighted to the non-weighted deviation and is given by $A_{faulty} = S_2/S_1$, where $S_1 = I_{C_1} - \sum_{j=1}^{N} I_j$ is the deviation and $S_2 = I_{C_2} - \sum_{j=1}^{N} W_{f_j} I_j$ is the weighted deviation. The correction can be done by subtracting the S_1 from the faulty output current whose column address is indicated by A_{faulty}. The fault detection and correction are done after the ADC output in the digital domain.

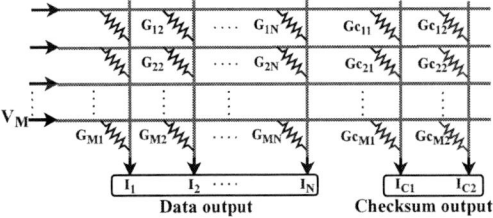

Fig. 3. M x N memristive crossbar with checksum columns

Consider an $M \times N$ crossbar with each cell storing a b-bit value, thus $l = 2^b$ different levels for each memristive cell and a linear weight factor ($W_{f_j} = W_{f_1}, W_{f_2}...W_{f_N} = 1, 2...N$) for the weighted checksum calculation. The maximum value of the checksum for each row can be calculated as follows:

$$G_{C_{i1}}^{max} = \sum_{j=1}^{N} (2^b - 1) \quad \text{and} \quad G_{C_{i2}}^{max} = \sum_{j=1}^{N} W_j (2^b - 1) \quad (2)$$

$$G_{C_{i1}}^{max} = N \times (2^b - 1) \quad \text{and} \quad G_{C_{i2}}^{max} = \frac{N \times (N+1)}{2}(2^b - 1) \quad (3)$$

It is not possible to store a checksum value into a single memristive cell due to a limited number of stable states. The weighted column slicing needs to be used to store the checksum value into multiple columns [25]. If the memristive can store l different levels, the output current of each column of a slice would also be weighted by the power of l based on its column address. In general, the number of extra columns in the case of a non-weighted and weighted checksum can be given as follows:

$$N_{nw} = \left\lceil \frac{\log_2 (G_{C_{i1}}^{max})}{b} \right\rceil \quad \text{and} \quad N_w = \left\lceil \frac{\log_2 (G_{C_{i2}}^{max})}{b} \right\rceil \quad (4)$$

979-8-3503-4631-2/23 $31.00 © 2023 IEEE

C. Hamming Code-based Checksum

Authors in [21] presented a checksum technique based on the Hamming code, in which a non-weighted checksum of data is performed to find the parity instead of XORing the data. A 4×4 crossbar array encoded with (7,4) Hamming code having three additional checksum columns (also referred to as parity columns) is shown in Fig. 4. The checksum computation for parity is based on the parity check matrix (H). The checksum value for all three parity columns using $(7,4)$ code can be calculated as per Eq. (5). In general, the checksum for each parity is computed as per pseudocode in Algorithm 1.

$$
\begin{aligned}
G_{C_{i1}} &= G_{i1} + G_{i2} + G_{i3} \\
G_{C_{i2}} &= G_{i1} + G_{i2} + G_{i4} \\
G_{C_{i3}} &= G_{i1} + G_{i3} + G_{i4}
\end{aligned} \tag{5}
$$

Algorithm 1 Checksum Computation based on Hamming code

Input Parity check matrix (H), Weight matrix (G)
Output Checksum $G_{C_{ip}}$ for each parity.
Extract the non-identity matrix (H_N) from H
for $p = 1$ to P **do**   ▷ for each parity (1 to P)
    $h = H_N(p, 1 : N)$ ▷ Select p_{th} row of H_N
    for $i = 1$ to M **do**   ▷ for row size of G (1 to M)
       $G_{C_{ip}} = G(i, 1 : N) \times h^T$ ▷ Checksum for p_{th} parity
    end for
end for

Fig. 4 shows the 4×4 array with three checksum columns for three parity using (7,4) Hamming code. In Fig. 4, instead of three checksum columns, six checksum columns must be required as the checksum for each parity needs at least two crossbar columns due to the limited stable state of the memristive cell. This fundamental technological constraint has been ignored in the existing linear coding-based checksum [21]. The decoding scheme is similar to the Hamming code, and the only difference is that the XOR operation is replaced by the algebraic checksum operations.

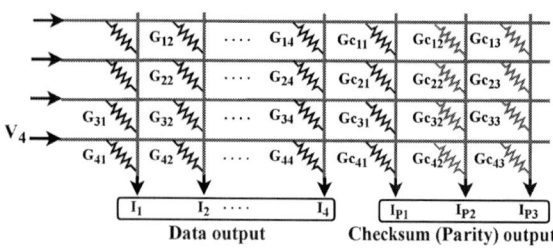

Fig. 4. 4 x 4 memristive crossbar with Hamming code-based checksum

III. PROPOSED APPROACH

As discussed in the previous section, storing the checksum requires multiple columns due to the limited number of stable states of memristive cells, which results in huge memory overhead. Fig. 5 shows the maximum value of checksum in the case of non-weighted, weighted, and Hamming code-based checksum. In the case of non-weighted and weighted

checksum, the value of the weighted checksum is more severe as it increases drastically with the array size. In the case of Hamming code, the checksum values shown in the Fig. 5 correspond to the parity having a maximum checksum value. As the array size increases, the maximum checksum value also increases, which imposes the requirements of extra columns into the crossbar, as shown in Fig. 6. Here, we propose a method that scales the large checksum value in such a way that it can be stored into a single memristor cell, reducing memory overhead significantly. The overall steps of the proposed method are as follows:

- First, we perform the downscaling of the checksum of the trained weight matrix using a scaling factor, i.e., each checksum value is divided by the scaling factor, so we need to store only a small factor (quotient only).
- However, each checksum value is not always divisible by the scaling factor, i.e., a non-zero remainder exists after the scaling, which may introduce inaccuracy in error detection and correction. Hence, we propose the fine-tuning of the trained weight matrix, in which weight values are manipulated in such a way that the obtained checksum are divisible by the scaling factor.
- The manipulation of the weight matrix may cause a severe impact on NN accuracy. We propose a checksum-aware training of NN to regain the NN inference accuracy degradation due to the manipulation of the weight matrix.

A detailed description of each step is discussed next.

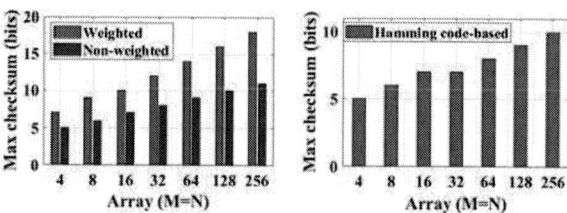

Fig. 5. Maximum checksum value (for $b = 3$ bit/cell)

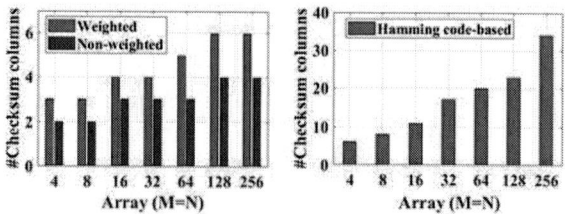

Fig. 6. Number of checksum columns (for $b = 3$ bit/cell)

A. Down Scaling of checksum value

The purpose of scaling down the checksum value is to reduce the checksum values to $2^b - 1$ or less so that it can be easily stored into a single memristive cell with b bit precision. This will allow us to store each checksum into a single column of the crossbar only. The scaling factor is chosen in such a way that it should be a power of 2 so that it is easy to get the original checksum value by a left shift operation at the output of ADC in the digital domain. The calculation of the scaling factor for different cases is discussed next.

979-8-3503-4631-2/23 $31.00 © 2023 IEEE

1) Scaling factors for non-weighted and weighted checksum: The non-weighted and weighted checksum scaling condition can be obtained using Eq. (6). The non-weighted (S_{nw}) and weighted (S_w) scaling factor can be chosen approximately equal to the power of 2 nearest to N and $\frac{N\times(N+1)}{2}$, respectively.

$$\frac{N(2^b-1)}{S_{nw}} \leq (2^b-1) \text{ and } \frac{\frac{N(N+1)}{2}(2^b-1)}{S_w} \leq (2^b-1) \quad (6)$$

2) Scaling factors for Hamming code-based checksum: In the case of Hamming code-based checksum, scaling factor computation can be done using a parity check matrix (H). The number of ones in each row of the non-identity matrix of H defines the maximum value of checksum value for each parity. The Hamming code dimension (n, k) can be decided based on the number of columns (N) of the weight matrix, which would act as a data block for Hamming code, where n is code length and k is the data block size. The logic for computing the scaling factor in the case of Hamming code-based checksum is summarized as pseudocode in Algorithm 2.

Algorithm 2 Scaling factor computation for Hamming code

Input Parity check matrix (H)
Output Scaling factor S_H for each parity.
Extract the non-identity matrix (H_N) from H
for $p = 1$ to P **do** ▷ for each parity (1 to P)
  $count = 0$ ▷ count number of $1's$
  for $j = 1$ to N **do** ▷ for column size of H_N (1 to N)
    if $H_N(p, j) == 1$ **then**
      $count = count + 1$
    else
      count=count
    end if
  end for
  $S_H(p)$=Power of 2 nearest to $count$ ▷ Scaling factor
end for

B. Fine-tuning of the weight matrix

The checksum value should be divisible by the scaling factor to get error-free scaling, i.e., zero mod value (or remainder) after the scaling. Practically, it is not always the case, as illustrated by Fig. 7. This may cause an error while extracting the original checksum value by left-shifting the scaled value at the ADC output. We need to tune the values of the weight matrix to make the mod value zero for each checksum. This is illustrated by the example shown in Fig. 7.

$G_{original}$

Weight factor				Weighted checksum		Mod value
1	2	3	4			
6	6	2	3	36	2	4
3	1	4	5	37	2	5
3	4	5	6	50	3	2
4	1	2	1	16	1	0

After fine tuning ⇒

$G_{manipulated}$

Weight factor				Weighted checksum		Mod value
1	2	3	4			
6	6	2	2	32	2	0
3	0	3	5	32	2	0
3	3	5	6	48	3	0
4	1	2	1	16	1	0

Fig. 7. Example showing fine-tuning of the weight matrix

Given an original weight matrix (G^o), we need to obtain a new manipulated weight matrix (G^m) such that it satisfies the constraints of checksum scaling with minimal changes in the weight values. In other words, the checksum value of the newly obtained manipulated matrix is a multiple of the corresponding scaling factor. To do this, we have formulated the required task as an integer linear programming problem to satisfy the constraints of checksum scaling. For any scaling factor S, the possible mod value can be 1 to $S-1$, and mod value computation can be done as follows:

$$mod = remainder(\frac{\text{Checksum value}}{S}) \quad (7)$$

The constraints for integer linear programming to make mod value zero after scaling of the checksum of i_{th} row of weight matrix are given by expression (8). Where G_c^o and G_c^m represent the checksum value for the original and manipulated weight matrix, respectively, mod^o represents the mod value after checksum scaling in the case of the original weight matrix. The newly obtained checksum value G_c^m is a multiple of scaling factor S. The fine-tuning of the weight matrix is summarized as pseudocode in Algorithm 3.

$$\begin{cases} G_{c_i}^m = G_{c_i}^o - mod_i^o, & \text{if } mod_i^o \leq S/2 \\ G_{c_i}^m = G_{c_i}^o - mod_i^o + S, & \text{if } mod_i^o > S/2 \end{cases} \quad (8)$$

Algorithm 3 Fine-tuning of the weight matrix

Input $G_{original}(G^o)$ and mod^o
Output $G_{manipulated}(G^m)$
Objective Minimize $[G^m - G^o]$
for $i = 1$ to M **do** ▷ for row (1 to M)
  for $k = 1$ to $S-1$ **do** ▷ for mod value (1 to $S-1$)
    if $mod_i^o = k$ **then**
      Manipulate i_{th} row of G^o using integer linear programming solver to satisfy the checksum scaling constraints as follows:
      if $mod_i^o \leq S/2$ **then**
        $G_{c_i}^m = G_{c_i}^o - mod_i^o$
      else
        $G_{c_i}^m = G_{c_i}^o - mod_i^o + S$
      end if
    else
      Keep i_{th} row as it is ▷ i.e., $mod_i^o = 0$
    end if
  end for
end for

C. Proposed Checksum-Aware Training of NN

The fine-tuning of the weight matrix discussed in the previous section manipulate the weight values to satisfy the checksum scaling requirement. However, this will severely impact the inference accuracy of trained NN, as shown in the result section. To mitigate this accuracy degradation, we propose a checksum-aware training of NN. At the algorithmic level, the typical computation of NN with overall L layers can be represented as a linear transformation of input $\mathbf{x_i}$ followed by batch normalization layer (BatchNorm) [26] and an element-wise nonlinear activation function $\phi(\cdot)$, e.g.,

$$\mathcal{Z}_i = \phi_i\left(\frac{G_i^o \mathbf{x}_i + \mathbf{b}_i - \mu}{\sqrt{\sigma^2 + \epsilon}} \times \alpha + \beta\right), \forall i \in [1 \ldots L]. \quad (9)$$

Where G^o represents the original weight matrix (without manipulation), μ and σ are the mean and variance of the batch normalization layer. α and β are the learnable parameter of BatchNorm layer and ϵ is a small constant.

The manipulated weight matrix, described in Sec. III-B, is denoted as G^m. The resulting weighted sum $y' \neq y$ is significantly different due to weight manipulation, leading to a higher value of loss function, $\mathbf{E}' > \mathbf{E}$, where $\mathbf{E}$ is the previously optimized value of a pre-trained NNs. The effect of error due to manipulating the entries of the weight matrix needs to be suppressed by optimizing $\mathbf{E}'$ using the backpropagation algorithm. Otherwise, the inference accuracy can degrade to a very low value that corresponds to randomly initialized NN. Therefore, we introduce an additional parameter Γ with dimension N for the manipulated layers of the NN. Here, N denotes the number of neurons in the layer or the number of data columns in the crossbar. The overall computation of a layer during re-training can be described as:

$$\mathcal{Z}_i = \phi_i\left(\frac{(G_i^m \mathbf{x}_i + \mathbf{b}_i) \times \Gamma - \mu}{\sqrt{\sigma^2 + \epsilon}} \times \alpha + \beta\right), \forall i \in [1 \ldots L]. \quad (10)$$

During re-training, the non-manipulated parameters of the NNs and Γ parameter are optimized through the backpropagation algorithm, and the manipulated parameters are frozen.

Since introducing additional parameters increases both computation and memory overhead during inference, we proposed to squash the Γ parameter into the batch normalization parameter. We utilize the fact that during inference, batch normalization is constant and can be simplified into the following:

$$\mathcal{Z}_i = \phi_i\left(\frac{(G_i^m \mathbf{x}_i + \mathbf{b}_i) \times \Gamma \times \alpha - \mu \times \alpha}{\sqrt{\sigma^2 + \epsilon}} + \beta\right), \forall i \in [1 \ldots L]. \quad (11)$$

$$\mathcal{Z}_i = \phi_i\left(\frac{(G_i^m \mathbf{x}_i + \mathbf{b}_i) \times \alpha' - \mu'}{\sqrt{\sigma^2 + \epsilon}} + \beta\right), \forall i \in [1 \ldots L]. \quad (12)$$

where $\alpha' = \Gamma \times \alpha$ and $\mu' = mu \times \alpha$ are the new scale and mean of the batch normalization layer after simplification. Therefore, during inference, the proposed checksum-aware training does not introduce any additional memory or computation overhead.

IV. EXPERIMENTS AND RESULTS

This section discussed the detailed analysis of the experimental setup and evaluation of the proposed methodology. We have considered a $b = 3$ bit per memristive cell. In the ABFT-based checksum, we have considered the downscaling of the weighted checksum, which is more severe than the non-weighted one. The linear weight factor ($W_f = 1, 2, \ldots N$) is considered for weighted checksum encoding. The checksum

TABLE I
SCALING FACTOR FOR DIFFERENT CHECKSUMS.

Array	S_{nw}	S_w	S_H
32×32	32	512	8,16,32
64×64	64	2048	8,32,64
128×128	128	8192	8,64,128
256×256	256	32768	16,128,256

TABLE II
COMPARISON OF ABFT-BASED CHECKSUM AND PROPOSED TECHNIQUE.
THE NUMBER IN ( ) REPRESENTS THE REDUCTION IN MEMORY OVERHEAD

Method	ABFT [20]		Proposed	
Array	N_c	Memory overhead	N_c	Memory overhead
32×32	7	18%	4	11% (-38.88%)
64×64	8	11%	4	6% (-45.45%)
128×128	10	8%	5	4% (-50.00%)
256×256	10	4%	5	2% (-50.00%)

encoding into memristive cells can be done during the training phase. Because the value of the weight matrix is fixed during the inference. The integer linear programming solver for weight manipulation is implemented using MATLAB.

The proposed approach is evaluated in terms of the number of extra checksum columns required and memory overhead. For Hamming code-based checksum, we also assess the performance comparison in terms of decoding overhead. The implementation was done using Verilog HDL and synthesized in Synopsys Design Compiler using the TSMC $40\ nm$ library. To evaluate the proposed method on deep learning applications, we have trained a 3-bit quantized 4-layer multilayer perceptron with 32, 64, 128, and 256 neurons in the hidden layers for the MNIST and Fashion-MNIST. In addition, we trained convolutional NN (CNN) on CIFAR-10 and Veg-15 datasets with ResNet-18 topology. The Veg-15 is a vegetable classification dataset consisting of 21000 real-world images with size 224×224 and 15 different classes [27]. The hidden layers are only encoded as they contain a large number of parameters, which is $\sim99.5\%$ of total parameters for the ResNet-18 topology. We exclude down sampling, the first and last layer of each NN model. We have used the quantization algorithm proposed in [28]. The checksum-aware NN training is performed after the NN weight manipulation. This is because training the NN from scratch by including the checksum constraints imposes huge penalties as it needs to satisfy both functional tasks and checksum constraints. This results in a drastic drop in accuracy, even for very smaller datasets such as MNIST.

TABLE I shows the possible scaling factors for different crossbar array sizes in case of non-weighted (S_{nw}), weighted (S_w), and Hamming code-based (S_H) checksums. In the case of Hamming code-based checksum, multiple scaling factors exist for each array size. This is because the number of elements involved in checksum computation differs for each parity. As an example, for 32×32, the $(38, 32)$ code is used, which demands six parities. The scaling factor for the first parity is 8, the second and third require a scaling factor equal to 16, and the remaining parities need a scaling factor equal to 32.

TABLE III

COMPARISON OF DECODING OVERHEAD OF EXISTING HAMMING CODE-BASED CHECKSUM AND PROPOSED LOW OVERHEAD CHECKSUM TECHNIQUE. THE NUMBER IN () REPRESENTS THE REDUCTION IN DECODING OVERHEAD

Method	Hamming code-based checksum [21]				Proposed			
Array	N_c	Memory overhead	Area (μm^2)	Power (mW)	N_c	Memory overhead	Area (μm^2)	Power (mW)
32×32	17	35 %	5557	2.74	6	16 % (**-54.28%**)	4912 (**-11.61%**)	2.45 (**-10.58%**)
64×64	20	24 %	11526	5.77	7	10 % (**-58.33%**)	10665 (**-7.47%**)	5.40 (**-6.41%**)
128×128	23	15 %	23320	11.75	8	6 % (**-60.00%**)	21690 (**-6.99%**)	11.06 (**-5.87%**)
256×256	34	12 %	46930	23.74	9	3 % (**-75.00%**)	43890 (**-6.48%**)	22.46 (**-5.39%**)

TABLE IV

INFERENCE ACCURACY FOR MNIST DATASET WITH PROPOSED APPROACH. N REPRESENTS THE NUMBER OF DATA NEURONS

N	Baseline	After fine-tuning		After checksum-aware training	
		ABFT	Hamming	ABFT	Hamming
32	95.62 %	71.99 %	19.74 %	96.64 %	95.47 %
64	96.67 %	40.58 %	8.92 %	97.17 %	96.17 %
128	97.05 %	46.50 %	10.32 %	97.13 %	97.33 %
256	98.52 %	13.54 %	10.10 %	98.40 %	97.78 %

TABLE V

INFERENCE ACCURACY FOR FASHION-MNIST DATASET WITH PROPOSED APPROACH. N REPRESENTS THE NUMBER OF DATA NEURONS

N	Baseline	After fine-tuning		After checksum-aware training	
		ABFT	Hamming	ABFT	Hamming
32	84.73 %	77.42 %	6.59 %	84.96 %	83.94 %
64	86.31 %	62.73 %	10.0 %	87.10 %	85.53 %
128	87.06 %	78.32 %	10.0 %	87.79 %	87.12 %
256	88.14 %	39.37 %	9.56 %	88.59 %	87.31 %

TABLE II shows the comparative analysis of the existing ABFT-based checksum and the proposed low overhead checksum in terms of the number of checksum columns (N_c) required and memory overhead for different array sizes. The memory overhead demonstrates the increase in memristive crossbar array size due to additional checksum columns. The proposed technique requires fewer extra columns compared to the existing ABFT-based checksums, significantly reducing the memory overhead. As an example, for the 256×256 array, the memory overhead is reduced from 4% to 2%, i.e., 50% reduction in memory overhead. The ABFT is a software-based fault-tolerant technique that involves an integer division operation to compute signature for error detection and correction. Implementing the integer division operation on hardware is very costly in terms of parameters such as area, power, and latency. To accelerate the decoding process, the signature computation for fault detection and error correction can be done in a separate arithmetic logic unit (ALU) in near-memory computing style [20].

TABLE III shows the comparative analysis of the existing Hamming code-based checksum and the proposed approach in terms of the number of checksum columns (N_c), memory overhead, decoding area, and power for different array sizes.

TABLE VI

INFERENCE ACCURACY FOR CNN DATASETS WITH PROPOSED APPROACH

Dataset	Baseline	After Fine-tuning		After checksum-aware training	
		ABFT	Hamming	ABFT	Hamming
CIFAR-10	91.63 %	9.86 %	9.98 %	89.06 %	88.00 %
Veg-15	99.16 %	6.66 %	6.66 %	96.30 %	95.63 %

The proposed method incurs a smaller area overhead and power in the decoding circuits. This is expected because the proposed method does not require column slicing to store the large checksum value, so no additional hardware overhead (shift and add operation) is needed to get the final current checksum value. There is no improvement in latency, as the proposed method does not incur any reduction in the logic depth of design. However, there is a significant reduction in the number of extra columns required to store the checksum, significantly reducing the memory overhead by up to **75%**.

TABLE IV, V, and VI show the NN inference accuracy using the proposed method for different datasets, respectively, which includes original accuracy (baseline), accuracy after the fine-tuning of the weight matrix, and the accuracy after checksum-aware training. The fine-tuning of the weight matrix leads to huge degradation in inference accuracy. It is expected that the accuracy degradation is more in the case of Hamming code-based checksum because the constraint for checksum scaling needs to be satisfied for each parity, which demands the huge manipulation of weight values. With the proposed checksum-aware training of NN mentioned in Sec. III-C, the inference accuracy reaches almost equal to the original accuracy. The NN inference accuracy after checksum-aware training is within 1% of original accuracy (baseline) for MNIST and Fashion-MNIST datasets, as shown in TABLE IV and TABLE V, respectively. The inference accuracy is within ~3% of original accuracy (baseline) for CNN datasets: CIFAR-10 and Veg-15, as shown in TABLE VI respectively. Thus, with proposed checksum-aware training, the results show negligible impact on inference accuracy.

V. CONCLUSION

In this work, a low overhead checksum-based technique for single-column error correction in the memristive crossbar is presented, which uses two methods: ABFT-based checksum and Hamming code-based checksum. In this proposed technique, downscaling of the checksum is done so that it can be stored into a single column of the crossbar, which results in a significant reduction in the number of extra columns. The performance of the proposed technique is also compared with the existing checksum-based methods. The proposed technique reduces the memory overhead by up to **75%** compared to existing techniques. We also demonstrated a checksum-aware training of NN to regain the accuracy loss due to the manipulation of the weight matrix to satisfy the scaling constraint of checksum. The proposed method is validated by NNs inference accuracy on MNIST, Fashion-MNIST, CIFAR-10, and Veg-15 datasets.

979-8-3503-4631-2/23 $31.00 © 2023 IEEE

REFERENCES

[1] I. Goodfellow, Y. Bengio, and A. Courville, *Deep Learning*. MIT Press, 2016, http://www.deeplearningbook.org.

[2] W. A. Wulf and S. A. McKee, "Hitting the memory wall: Implications of the obvious," *SIGARCH Comput. Archit. News*, vol. 23, no. 1, p. 20–24, mar 1995.

[3] K. Roy, I. Chakraborty, M. Ali, A. Ankit *et al.*, "In-memory computing in emerging memory technologies for machine learning: An overview," in *2020 57th ACM/IEEE Design Automation Conference (DAC)*, 2020, pp. 1–6.

[4] S. Hamdioui, L. Xie, H. A. Du Nguyen, M. Taouil *et al.*, "Memristor based computation-in-memory architecture for data-intensive applications," in *2015 Design, Automation and Test in Europe Conference Exhibition (DATE)*, 2015, pp. 1718–1725.

[5] S. Yin, X. Sun, S. Yu, and J.-S. Seo, "High-throughput in-memory computing for binary deep neural networks with monolithically integrated rram and 90-nm cmos," *IEEE Transactions on Electron Devices*, vol. 67, no. 10, pp. 4185–4192, 2020.

[6] S. Jain, A. Ranjan, K. Roy, and A. Raghunathan, "Computing in memory with spin-transfer torque magnetic ram," *IEEE Transactions on Very Large Scale Integration (VLSI) Systems*, vol. 26, no. 3, pp. 470–483, 2018.

[7] C.-X. Xue, W.-H. Chen, J.-S. Liu, J.-F. Li *et al.*, "Embedded 1-mb reram-based computing-in- memory macro with multibit input and weight for cnn-based ai edge processors," *IEEE Journal of Solid-State Circuits*, vol. 55, no. 1, pp. 203–215, 2020.

[8] S. Eilert, M. Leinwander, and G. Crisenza, "Phase change memory: A new memory enables new memory usage models," in *2009 IEEE International Memory Workshop*, 2009, pp. 1–2.

[9] Y. Halawani, B. Mohammad, M. Abu Lebdeh, M. Al-Qutayri *et al.*, "Reram-based in-memory computing for search engine and neural network applications," *IEEE Journal on Emerging and Selected Topics in Circuits and Systems*, vol. 9, no. 2, pp. 388–397, 2019.

[10] A. Shafiee, A. Nag, N. Muralimanohar, R. Balasubramonian *et al.*, "Isaac: A convolutional neural network accelerator with in-situ analog arithmetic in crossbars," in *ACM/IEEE Annual International Symposium on Computer Architecture (ISCA)*, 2016, pp. 14–26.

[11] J. Yu, H. A. D. Nguyen, L. Xie, M. Taouil *et al.*, "Memristive devices for computation-in-memory," in *2018 Design, Automation Test in Europe Conference Exhibition (DATE)*, 2018, pp. 1646–1651.

[12] C. Münch and M. B. Tahoori, "Defect characterization of spintronic-based neuromorphic circuits," in *2020 IEEE 26th International Symposium on On-Line Testing and Robust System Design (IOLTS)*, 2020, pp. 1–4.

[13] D. Veksler, G. Bersuker, L. Vandelli, A. Padovani *et al.*, "Random telegraph noise (rtn) in scaled rram devices," in *2013 IEEE International Reliability Physics Symposium (IRPS)*, 2013, pp. MY.10.1–MY.10.4.

[14] A. Grossi, E. Nowak, C. Zambelli, C. Pellissier *et al.*, "Fundamental variability limits of filament-based rram," in *2016 IEEE International Electron Devices Meeting (IEDM)*, 2016, pp. 4.7.1–4.7.4.

[15] M. Hu, J. P. Strachan, Z. Li, E. M. Grafals *et al.*, "Dot-product engine for neuromorphic computing: Programming 1t1m crossbar to accelerate matrix-vector multiplication," in *ACM/EDAC/IEEE Design Automation Conference (DAC)*, 2016, pp. 1–6.

[16] J.-Y. Hu, K.-W. Hou, C.-Y. Lo, Y.-F. Chou *et al.*, "Rram-based neuromorphic hardware reliability improvement by self-healing and error correction," in *2018 IEEE International Test Conference in Asia (ITC-Asia)*, 2018, pp. 19–24.

[17] F. Zhang and M. Hu, "Defects mitigation in resistive crossbars for analog vector matrix multiplication," in *2020 25th Asia and South Pacific Design Automation Conference (ASP-DAC)*, 2020, pp. 187–192.

[18] Q. Lou, T. Gao, P. Faley, M. Niemier *et al.*, "Embedding error correction into crossbars for reliable matrix vector multiplication using emerging devices," in *Proceedings of the ACM/IEEE International Symposium on Low Power Electronics and Design*, ser. ISLPED, 2020, p. 139–144.

[19] Y. Hu, K. Cheng, Z. Zhang, R. Wang *et al.*, "Error correction scheme for reliable rram-based in-memory computing," in *2021 5th IEEE Electron Devices Technology Manufacturing Conference (EDTM)*, 2021, pp. 1–3.

[20] M. Liu, L. Xia, Y. Wang, and K. Chakrabarty, "Fault tolerance for rram-based matrix operations," in *2018 IEEE International Test Conference (ITC)*, 2018, pp. 1–10.

[21] A. Das and N. A. Touba, "Selective checksum based on-line error correction for rram based matrix operations," in *2020 IEEE 38th VLSI Test Symposium (VTS)*, 2020, pp. 1–6.

[22] S. Wiefels, C. Bengel, N. Kopperberg, K. Zhang *et al.*, "Hrs instability in oxide-based bipolar resistive switching cells," *IEEE Transactions on Electron Devices*, vol. 67, no. 10, pp. 4208–4215, 2020.

[23] S. T. Ahmed, M. Hefenbrock, C. Münch, and M. B. Tahoori, "Neuro-scrub+: Mitigating retention faults using flexible approximate scrubbing in neuromorphic fabric based on resistive memories," *IEEE Transactions on Computer-Aided Design of Integrated Circuits and Systems*, pp. 1–1, 2022.

[24] J.-Y. Jou and J. Abraham, "Fault-tolerant matrix arithmetic and signal processing on highly concurrent computing structures," *Proceedings of the IEEE*, vol. 74, no. 5, pp. 732–741, 1986.

[25] M. A. Zidan and W. D. Lu, "Chapter 9 - vector multiplications using memristive devices and applications thereof," in *Memristive Devices for Brain-Inspired Computing*, ser. Woodhead Publishing Series in Electronic and Optical Materials, S. Spiga, A. Sebastian, D. Querlioz, and B. Rajendran, Eds. Woodhead Publishing, 2020, pp. 221–254.

[26] S. Ioffe and C. Szegedy, "Batch normalization: Accelerating deep network training by reducing internal covariate shift," in *International conference on machine learning*. pmlr, 2015, pp. 448–456.

[27] M. I. Ahmed, S. Mahmud Mamun, and A. U. Zaman Asif, "Dcnn-based vegetable image classification using transfer learning: A comparative study," in *2021 5th International Conference on Computer, Communication and Signal Processing (ICCCSP)*, 2021, pp. 235–243.

[28] J. Choi, Z. Wang, S. Venkataramani, P. I.-J. Chuang *et al.*, "Pact: Parameterized clipping activation for quantized neural networks," *arXiv preprint arXiv:1805.06085*, 2018.

Thwarting Reverse Engineering Attacks through Keyless Logic Obfuscation

Leon Li and Alex Orailoglu
University of California, San Diego
La Jolla, California, U.S.A.
xul065@ucsd.edu and alex@cs.ucsd.edu

Abstract—Logic obfuscation protects semiconductor IPs against reverse engineering threats by concealing IP implementation details using a tamper-proof key. With the continuous evolution of key recovery attacks exploiting functional, structural, and physical key exposures, the typical assumption of key secrecy becomes increasingly untenable. This work aims to end the tug of war between key-based defenses and key recovery attacks by delivering reverse engineering resilience through a novel *keyless* obfuscation approach that demands *no* external secret. The proposed solution locks the full functionality of a design using internally-generated and constantly-changing secrets that can be only extracted from the FSM transition history. The intrinsic secrets are secure against reverse engineering attempts due to the hardness of identifying valid transition paths to a target state from the obfuscated gate-level netlist. We develop an algorithm to synthesize the obfuscated FSM logic and the dynamic key update logic which jointly activate the design for all valid sequential queries so as to deliver unimpeded functionality for legal users. The algorithm enforces key consistency through equivalence-preserving FSM transformation and constraint-based state encoding to handle complex reconverging transition paths. Experimental results on MCNC benchmarks confirm the practicality and security of the proposed keyless logic obfuscation methodology.

Index Terms—Hardware Security, Logic Obfuscation, Reverse Engineering, IP Piracy

I. INTRODUCTION

Third-party resources including soft/hard intellectual properties (IPs) and offshore fabrication services have become the major constituents of the modern SoC development flow. Despite their efficiency and cost benefits, the widespread distribution of hardware IPs/ICs has made them vulnerable to piracy and reverse engineering (RE) attacks. To counteract RE attacks, logic obfuscation has been proposed which conceals chip implementation details through a key-controlled locking circuitry (e.g., XOR key gates). The locking circuitry corrupts the circuit output unless the correct key is supplied from a tamper-proof memory. In response to the protection provided by the locking circuitry, RE attackers have channeled their efforts toward key recovery [1], [2], [3], whose continual success increasingly questions key-based logic obfuscation.

Keyless logic obfuscation may sound miraculous as it is widely believed that an external secret is essential to piracy prevention - it seems necessary to withhold circuit functionalities through locking and subsequently recover them through activation. However, such a rationale overlooks the more important step of an RE attack: the step of analyzing the functional netlist and gaining *understanding* of its behavioral working mechanism. It is only this behavioral understanding that lays

down a path for the attacker to proceed with advanced, covert RE applications such as pirating IPs through re-implementation to escape sanction, inserting hardware Trojans with controlled payload behavior, and launching fault attacks at high precision. The significance of the understanding problem, unexplored hitherto, paves the way for a novel defense layer that could render a design **functional but unintelligible**.

In this work, we present the first keyless logic obfuscation methodology, *KeylessLock*, that protects the control units in the form of Finite State Machines (FSM) from RE attacks without any external secrets. The reason for our focus on the control unit is twofold. First, the control unit usually implements unstructured event sequences known as *random logic* which may be the sole secret worth protection. In contrast, datapath components often implement standardized algorithms that may already be public knowledge; their regularity also makes them vulnerable to word identification [4] and template-based analysis [5]. Second, the control unit is the most appealing target for RE attacks as it governs the overall computing sequence of the device. Its security is pivotal to preventing an attacker from commanding the hardware root-of-trust.

The proposed solution locks the full functionality of FSM using *dynamically changing* keys that are internally synthesized from the FSM transition history. The key secrecy is established through a well-recognized "trapdoor" from the FSM synthesis process, namely, the concealment of transition patterns when the RT-level FSM specification is synthesized into a gate-level netlist. We extract for each state in the RT-level FSM diagram a state-specific signature representing the functional transition sequences preceding it. The numerous state-specific signatures are used as keys to encrypt the FSM in the obfuscation phase, whereas in the execution phase the decryption key is self-updated in plain sight to activate the FSM. The non-static key offers strong resilience to key recovery attacks and thwarts RE attacks. While typical FSM users can passively observe pieces of the key information, key recovery through any invalid manipulations fails due to the hardness of retrieving valid transition patterns from the obfuscated netlist.

In summary, this work makes the following contributions:

- We illustrate the possibility of keyless logic obfuscation in thwarting certain hardware RE attacks.
- We develop the **first keyless logic obfuscation method** which coordinates FSM transformation and state encoding techniques to achieve practical overheads.
- We conduct experiments on benchmarks to validate the practicality and security of the proposed solution.

979-8-3503-4631-2/23 $31.00 © 2023 IEEE

II. BACKGROUND

A. FSM Reverse Engineering

An FSM can be defined as a 6-tuple $M = (\Sigma, \Delta, Q, s_0, \delta, \lambda)$, where $\Sigma = \{0, 1\}^n$ is the set of n-bit inputs, $\Delta = \{0, 1\}^m$ is the set of m-bit outputs, Q is the set of states, $s_0 \in Q$ is the initial state, $\delta(s, x) : Q \times \Sigma \to Q \cup \{\phi\}$ is the next state function, and $\lambda(s, x) : Q \times \Sigma \to \Delta \cup \{\tau\}$ is the output function. FSMs can be incompletely specified with the next state or output of some transitions being *don't cares (Xs)*, denoted as ϕ and τ, respectively. In the implementation, each state is represented by a t-bit state encoding which must be unique.

RE attacks on FSMs involve either a **complete** reconstruction of the FSM diagram for subsequently running arbitrary graph analysis, or a **partial** navigation in the FSM state space aimed towards a specific goal. For the former category, the most straightforward attack strategy is to perform a brute-force functional simulation with a time complexity of $\mathcal{O}(2^{t+n})$ for t state elements and n input bits [6]. AVFSM [7] uses an N-detect Automatic Test Pattern Generation (ATPG) tool to identify valid state transitions. REFSM [8] applies a SAT-based formula in a binary search algorithm to iteratively identify reachable states. While both techniques successfully reconstruct small FSMs, the multiple instances of running complex solving engines greatly affect their feasibility on large sequential designs.

For many attack applications, complete FSM reconstruction is not required. For instance, the attacker's interest may consist solely of one specific target, whether for gaining access to a particular security-sensitive state [9] or traveling over a secret authentication sequence to a known destination [10]. In such applications, attackers can instead identify transitions that move backward from a target state using time-frame expansion methods [10]. By avoiding unnecessary state space exploration and penetrating in a guided manner, partial FSM recovery has exhibited greater scalability in practical RE applications for medium- to large-scale sequential designs.

B. Logic Obfuscation

Logic obfuscation conventionally shields the gate-level netlist by withholding a secret key. Yet, the secrecy of the obfuscation key has constantly been challenged by the evolving key recovery attacks. The SAT attack [1], as well as its generalizations with stronger modeling and solving capabilities [11], [12], [13], have continually been adapted to crack the state-of-the-art key-based obfuscation techniques. At the same time, oracle-less attacks have leveraged purely structural traces in the netlist such as logic redundancy [14] and irregularity [15] to make key predictions without needing oracle access.

Sequential logic obfuscation methods have also been proposed to protect the FSM by inserting an authentication sequence [16] or diverting FSM transitions which can be corrected by key inputs [17]. Yet, they can generate only trivial corruption once the FSM has been initialized, and are often prone to oracle-less structural attacks [6], [10]. Furthermore, the security of all existing logic obfuscation techniques hinges on the assumption that the externally supplied activation secret

(e.g., key value or initial state) can be shielded at a tamper-proof and read-proof memory. This central security tenet has been shown to be untenable against physical attacks which can read out the secret using electro-optical probing techniques [2], [18], a threat quite feasible for IC supply chain attackers.

C. Attack Model

We assume the most capable supply chain attacker armed with full access to a gate-level netlist of the keyless obfuscated FSM, knowledge of the obfuscation algorithm, and the reset condition of the FSM. Keyless logic obfuscation aims to transform the RT-level FSM specification C into a gate-level FSM implementation C' so that (1) C and C' produce the same output sequence for any input sequence (i.e., cycle-accurate equivalent) and (2) it is "hard" to learn anything about C other than the outputs observed from random sequential queries on C'. Obviously, the first property implies that *KeylessLock* does **not** prevent forthright IC overproduction. Yet, through the second property, *KeylessLock* can prevent the attacker from understanding the FSM behavior beyond what can be observed from brute-force sequential queries.

This security goal is analogous to that of *white-box cryptography* [19], where the attacker is in possession of the functional implementation but is unable to determine anything about it other than through IO queries. A successful keyless obfuscation instantiation should compromise the correctness of complete and partial FSM RE attacks or force their time complexity to rise to the same level as brute-force sequential queries which remain infeasible for designs in industry practice.

III. KEYLESS LOGIC OBFUSCATION

A. Methodology Overview

KeylessLock builds an *inherent* and *dynamic* obfuscation secret for each state based on its preceding transition history. The secret of a state s' is the XOR-compressed signature of the FSM execution path from the initial state to s'. Whenever the FSM executes a sequence of transitions, the secret is computed by XORing the encodings of all the states along the path. This secret, thus, mutates constantly at runtime. To concretely demonstrate the secret generation process, we show a simple FSM in Figure 1a with $|Q| = 4$, $n = 1$, $m = 2$, and $t = 2$ and the obfuscation behaviors in Figure 1b. For instance, the key for s_2 is the XOR of the first three states' encodings, i.e., $k_{s_2} = s_0 \oplus s_1 \oplus s_2 = 01 \oplus 10 \oplus 00 = 11$. Even though we loosely refer to the obfuscation secret as a "key", it differs fundamentally from a traditional tamper-proof key in that our key is computed in plain sight, visible to the attacker at any time, and changes constantly during FSM execution.

KeylessLock uses the diverse set of keys to encrypt the FSM's next state and output functions. The next state and output functions are encrypted to become $\delta^{enc}(s', x) = \delta(s', x) \oplus k_{s'}$ and $\lambda^{enc}(s', x) = \lambda(s', x) \oplus k_{s'}$ (with appropriate key truncation and extension if $t \neq m$), respectively, as shown in green in Figure 1b. For instance, the next state of the edge $\{s_1, s_2\}$, which should originally produce the encoding of s_2, is now encrypted using the key of s_1 to become $\delta^{enc}(s_1, 0) =$

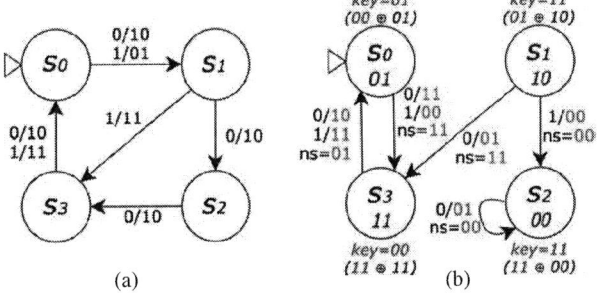

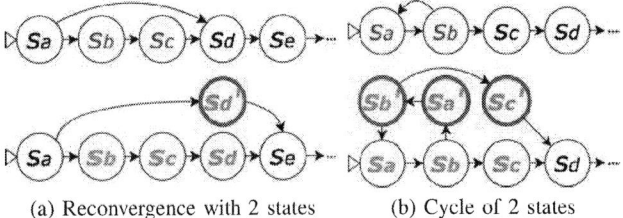

(a) Reconvergence with 2 states   (b) Cycle of 2 states

Fig. 3: The initial FSM shown at top while the transformed at bottom. Duplicated states are circled in red. The states that must receive constrained encodings are shown in orange.

Fig. 1: (a) FSM with input/output conditions on edges. (b) The obfuscated FSM showing the state-specific keys (in red) and the next state and output values after encryption (in green).

$s_2 \oplus k_{s_1} = 00 \oplus 11 = 11$. The encrypted FSM logic can only be recovered when the key is correctly updated at each state.

To demonstrate the execution of a keyless obfuscated FSM, we illustrate the architecture of *KeylessLock* in Figure 2. The obfuscated FSM has the same Input/Output (IO) definitions as the original FSM. It consists of four major components: (1) the encrypted FSM logic (δ^{enc} and λ^{enc}), (2) the Dynamic Key Update Function (DKUF) which dynamically provides the correct activation key, (3) the decryption logic consisting of an array of XOR gates, and (4) the state registers (D Flip Flops). The decryption logic combines every result bit of the encrypted FSM logic with one bit from the DKUF through an XOR gate. The DKUF is in itself a primitive sequential machine consisting of a key register that is fully observable to attackers. DKUF updates the key value every cycle by computing the XOR result between the previous key value and the current state encoding. It is easy to verify that during valid sequential FSM execution, the key gets updated in sync with any transition sequence and thus consistently activates the original behavior.

B. Key Consistency Algorithm

Realistic FSMs often have complex transition patterns with reconvergences and cycles. It is necessary to ensure that multiple alternative paths produce consistent keys at their common destinations. *KeylessLock* solves this challenge by orchestrating two techniques: constraint-based state encoding and equivalence-preserving FSM transformation.

The first technique leverages the flexibility of **state encoding** to ensure that multiple paths with the same destination generate the same key update effect. For instance, in Fig. 1, there are two paths between s_1 and s_3, one connecting them directly and the other through s_2. The only encoding solution possible to achieve key consistency at s_3 is to assign s_2 with the encoding

of 00 so that both paths produce the same key. The FSM also contains a cycle of three states, s_0, s_1, and s_3, that may be traversed repeatedly. The three states can receive encodings that XOR to zero, e.g., $s_0 \oplus s_1 \oplus s_3 = 01 \oplus 10 \oplus 11 = 00$ to ensure that any cycle walk results in no change in the key value.

Constraint-based encoding alone cannot resolve all key consistency constraints. The all-zero encoding, for example, can only be used once which precludes its general applicability, whereas for a reconvergence with 2 intermediate states, as shown at the top of Fig. 3a, key consistency demands s_b and s_c to have the same encoding ($s_b \oplus s_c = 0^t$) which violates the uniqueness requirement. In such scenarios, we propose an **FSM transformation** technique to relax the constraints until they become solvable through state encodings. FSM specifications can spawn equivalent states which conventionally can be merged through state minimization. In a somewhat antipodean angle, we utilize such flexibility to duplicate certain critical states for eliminating tight reconvergence and cycles. Some sample applications of FSM transformations are shown in Fig. 3. In the transformed FSM, all reconvergences and cycles involve 3 or more constituent states which can then be assigned with XOR-to-zero encodings to achieve key consistency.

We incorporate the two techniques in a generalizable FSM transformation and constraint generation algorithm shown in Algorithm 1. At a high level, the algorithm aims to apply a minimal number of state duplications while extracting all the symbolic encoding constraints that are *guaranteed* to be binary encodable. It analyzes the original FSM in a breadth-first traversal (BFT) to enforce consistency rules. Each state has three attributes: the encoding ($encoding$), the incoming key ($inKey$) produced by its immediate predecessor state, and the key assigned to the state (key), each of which is denoted by a set of symbols. The graph is consistent if $\text{SETDIFF}(s.inKey, s.encoding) = s.key$ for any $s \in S$ and $s.key = s'.inKey$ for every edge $\{s, s'\} \in E$, where SETDIFF computes the symmetric difference of two sets.

As the algorithm traverses the FSM, it sorts states into *free states* and *dependent states*. A free state has an unconstrained encoding. A dependent state has a constrained encoding that must satisfy an XOR symbolic formula based on some free state encodings. The FSM transformation algorithm begins by setting all the attributes of states to null (Line 2). It then launches BFT starting from the initial state and analyzes each of s's outgoing edges e to a destination state s' (Lines 5-6). The algorithm determines whether e creates any consistency

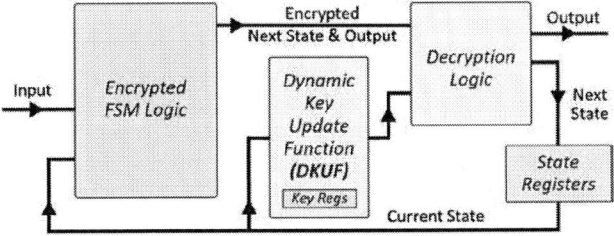

Fig. 2: High-level architecture of *KeylessLock* obfuscation.

979-8-3503-4631-2/23 $31.00 © 2023 IEEE   167

Algorithm 1 FSM Transformation and Symbolic Encoding

1: **procedure** TRANSFORMATION(G, s_0)
2:   Reset all state attributes to $\emptyset$; U $\leftarrow \emptyset$; Q $\leftarrow \emptyset$
3:   ENQUEUE(Q, s_0); s_0.explored $\leftarrow$ true
4:   **while** Q $\neq \emptyset$ **do**
5:     s $\leftarrow$ DEQUEUE(Q)
6:     **for** e $\in$ s.outEdges : e.destS.inKey $\neq$ *null* **do**
7:       **if** CHECKCONSISTENCYRECONNECT(e) **then**
8:         s.encoding $\leftarrow$ SETDIFF(s.inKey, e.destS.inKey)
9:         INSERT(U, s.encoding)
10:      **else**
11:        copyState $\leftarrow$ CREATECOPYSTATE(e.destS)
12:        e.destS $\leftarrow$ copyState
13:     **if** s.encoding == null **then**
14:       s.encoding $\leftarrow$ {s.name} ; INSERT(U, s.encoding)
15:       s.key $\leftarrow$ SETDIFF(s.inKey, s.encoding)
16:     Add unexplored next states of s to Q
17: **procedure** CHECKCONSISTENCYRECONNECT(e)
18:   **for** $s' \in$ FINDALLDUPLICATES(e.destS) **do**
19:     e.destS $\leftarrow s'$
20:     **if** s'.inKey == null **then return** true
21:     **else**
22:       solEncoding $\leftarrow$ SETDIFF(e.origS.inKey, s'.inKey)
23:       **return** ! CONTAIN(U, solEncoding)
24:            **or** solEncoding == e.origS.encoding

constraint based on the *s.key* and $s'.inKey$ (Lines 7, 17-24). If the consistency constraint can be satisfied by assigning *s* with a valid, unused state encoding, the algorithm makes this encoding assignment and sorts S into dependent states (Lines 8-9). If the consistency constraint cannot be solved through state encoding, the algorithm resolves the conflict by duplicating the destination state s' (Lines 11-12). Once the algorithm has examined all the outgoing edges of *s*, and none of them imposes any encoding constraint to *s*, *s* receives a new encoding symbol and is sorted into free states (Lines 13-15). It finishes an iteration of BFT by adding *s*'s next states to the queue and updating their $inKey$ (Line 16). Since the obfuscation algorithm can access the original FSM specification with transition conditions compacted, *KeylessLock* takes only $\mathcal{O}(|Q| + |E|)$ to run the graph transformation algorithm, rather than the $\mathcal{O}(2^{t+m})$ necessary in a brute-force FSM reconstruction attack, thus establishing the defender's advantage.

The encoding constraints produced by the FSM transformation algorithm can *always* be satisfied by assigning free states with one-hot codes. Nonetheless, it is often possible to use much fewer bits by considering the precise relationship among all the symbolic encodings. *KeylessLock* searches for the minimum-width constraint-based binary encoding using a Boolean satisfiability (SAT) solver at progressively larger encoding widths until a valid solution is found. *KeylessLock* then computes the binary state-specific keys by evaluating the symbolic key formula. The encrypted FSM specifications are produced by XORing the original next state and output specification with the state-specific keys. Finally, the encrypted

FSM logic is synthesized and integrated with the DKUF to produce the complete keyless obfuscated FSM.

C. Security Analysis

The security of keyless obfuscation stems from the dynamically computed key, which embodies the functional transition history preceding each state. While this obfuscation secret is correctly updated during legal FSM execution, an attacker cannot recover it at wish due to the computational and theoretical hardness of illegally recovering the functional transition history.

Combinational Querying Attack. We suppose that the goal of the attacker is to recover the correct functional behavior of an arbitrary state s', the fundamental problem underlying partial FSM RE attacks. Since the attacker can access the netlist model of the fully working design, the most intuitive attack method is to load the state s' in the FSM directly and observe the next state, output, and key values through combinational simulation. Yet, when the obfuscated FSM executes on the state s' without getting through a legal transition sequence, $k_{s'}$ will not be updated appropriately. The encrypted FSM logic and the random key observed will thus remain unintelligible due to the strong secrecy provided by XOR encryption.

Backward Path Construction Attack. The history-dependent state-specific keys require the attacker to find a valid transition sequence to s' in order to obtain the correct $k_{s'}$. This sequential path finding task from the gate-level netlist is traditionally solved using a sequential ATPG tool or a SAT solver. Sequential ATPG algorithms initiate the analysis at the target state s' and iteratively add backward time-frames until reaching a known state. Its feasibility hinges on the assumption that one can identify the predecessor condition of every state along the backward path through combinational analysis of the state transition logic. With *KeylessLock* obfuscation in place, the attacker cannot determine the $inKey$ of s' and thus cannot construct any valid backward time-frames from s'. SAT-based attacks on sequential circuits in [11], [12] unroll the combinational logic up to a small sequential depth b (e.g., 5 [12]) to sufficiently prune the wrong keys. However, *KeylessLock* distributes state-specific keys throughout the functional state space, resulting in a large and unknown sequential depth. The state-specific keys are also *not* derived from any shared secrets so they cannot be predicted even with the complete FSM knowledge obtained at a small unrolling depth. Thus, the numerous instances of SAT attacks required together with the large effective key size ($b_{s'} \times t$) drastically affect the feasibility of sequential SAT attacks.

Forward Path Construction Attack. In the absence of shortcuts to recover $k_{s'}$, the only feasible way for an attacker to understand the FSM is through sequential queries, i.e., executing the FSM cycle by cycle in a forward manner. Since the FSM remains fully encrypted for unvisited states, the attacker cannot obtain guidance to pinpoint the search direction towards s' and is thus compelled to perform complete FSM reconstruction, a computationally demanding challenge. Furthermore, an exceptional strength of *KeylessLock* is that it can entirely **invalidate even the brute force FSM reconstruction attack** if the original FSM has *unspecified transitions*, such as when applying

979-8-3503-4631-2/23 $31.00 © 2023 IEEE

FSM Name	Original Design			Obfuscated Design			Overhead of *KeylessLock*					
	States	*State FFs*	*Transitions*	*States*	*State FFs*	*Transitions*	*States*	*State FFs*	*Transitions*	*Area*	*Delay*	*Power*
scf	115	7	159	125	7	179	1.09x	1.00x	1.13x	1.72x	1.26x	1.61x
s1488	48	6	251	81	7	430	1.69x	1.17x	1.71x	1.96x	1.43x	2.66x
planet	48	6	115	78	7	186	1.63x	1.17x	1.62x	2.41x	1.40x	2.35x
tbk	32	5	1569	554	11	27159	17.31x	2.20x	17.31x	11.06x	2.74x	34.93x
sand	32	5	184	112	7	844	3.50x	1.40x	4.59x	2.92x	1.62x	3.47x
bbara_bbtas	30	5	268	132	8	1133	4.40x	1.60x	4.23x	3.98x	1.81x	6.67x
styr	30	5	166	70	7	345	2.33x	1.40x	2.08x	2.87x	1.45x	3.11x
dk16	27	5	108	157	8	628	5.81x	1.60x	5.81x	4.13x	1.53x	5.19x

TABLE I: The state, transition, and hardware overhead comparisons between the original and the *KeylessLock* obfuscated FSMs.

a 1 input at s_2 in Fig. 1. We purposefully treat such unused and non-functional transitions as *don't care* conditions in the obfuscation and synthesis flow which allows them to receive arbitrary functional definitions as deemed beneficial to logic minimization. This arbitrary functional definition will almost **never** produce the correct key at its destination state since it is never considered for key consistency. Since unspecified transitions are indistinguishable from specified transitions at the gate level, the attackers cannot distinguish the various inconsistent keys at all the successor states from unspecified transitions, leading to extensively corrupted state recovery results.

IV. EXPERIMENTAL RESULTS

Experimental Setup. This section demonstrates the effectiveness of *KeylessLock* obfuscation on the 8 FSM designs with the highest number of states from the MCNC [20] benchmark. They are provided in the KISS2 format describing states and transitions. We pruned all the unreachable states and showed specifications of the remaining FSM designs in Table I. We implement our obfuscation method in Java and integrate Java-SMT [21] to solve for the binary encoding based on the symbolic constraints. We implement all self-loops by controlling the enable lines of state FFs to alleviate state space expansion, with the enable line logic protected in *KeylessLock* architecture as well. Designs are synthesized using the Synopsys Design Compiler with *Nangate 15nm open cell library*.

Security Evaluation. We first evaluate the security of *KeylessLock* by simulating the combinational querying attack and the forward path construction attack. The reader will notice that the FSMs we tested are relatively small as no large RTL FSM benchmark in KISS/KISS2 format is readily available, but we can still draw some insights into the security level with respect to the scale of the design. In the first combinational querying attack we evaluated, the adversary directly loads an arbitrary state in the keyless obfuscated FSM and observes the active key value. For each state in the obfuscated FSM, we simulate the combinational querying attack 10,000 times by assuming random starting values at key registers. The ratio of correct key observations is shown in Table II. We measure the effectiveness of the attack by computing the difference between *Combinational Querying Accuracy* and the accuracy of random key guessing which measures the *adversarial advantage*. The experimental results indicate that the combinational attack delivers no observable advantage over random guessing, proving that *KeylessLock* delivers the optimal security level in defeating partial FSM navigation attacks.

We then evaluate the security of *KeylessLock* against the forward, complete FSM reconstruction attack which randomly applies input patterns to visited states. We report in Table III the attack complexity and state recovery ratio. The *Queries Applied* field represents the total number of queries that are made, all of which are based on some visited states with known key values. The *Enumeration Ratio* field is computed as the ratio between forward queries applied and the total number of state and input combinations (2^{t+m}). On average, the adversary has to enumerate 23.2% of all the state and input combinations to visit all the states at least once, not to mention the substantial efforts required to understand the next state and output behaviors of the remaining unexplored transitions. If the FSM design is sufficiently non-trivial so that a brute-force enumeration can be considered infeasible, it would follow that the complete FSM reconstruction attack on *KeylessLock* is infeasible. One of the FSMs, *scf*, achieves a relatively lower attack resilience level because its input bits have a rather disjoint impact on the FSM's behavior, leading to simple connectivity patterns among states. Furthermore, two of the FSMs in our evaluation, *sand* and *styr*, have 2.08% and 0.13% of all the input conditions from the functional states that are *unspecified*. The forward attack often runs into such unspecified transitions, leading to corrupted state recovery results at their destination states as well as all the subsequent states. The corruption ensures the secrecy of the original FSM under even unbounded attack efforts.

Runtime and Overhead Evaluation. *KeylessLock* obfuscation incorporates well-established BFT techniques and efficient SAT solvers, leading to a runtime of seconds to a few minutes to produce the obfuscated FSM. We also observed a comparable runtime for synthesizing the original and the obfuscated FSMs with the commercial tools, confirming the

FSM Name	Combinational Querying Accuracy	Random Key Guess Accuracy	Adversarial Advantage
scf	9795/1250000	1/128	2.35e-5
s1488	6376/810000	1/128	5.91e-5
planet	6063/780000	1/128	-3.94e-5
tbk	2697/5540000	1/2048	-1.46e-5
sand	8795/1120000	1/128	4.02e-5
bbara_bbtas	5311/1320000	1/256	1.17e-4
styr	5443/700000	1/128	-3.68e-5
dk16	6105/1570000	1/256	-1.77e-5

TABLE II: The key recovery accuracy of the combinational querying attack versus that of random guessing.

FSM Name	Queries Applied	Enumeration Ratio	Correct Recovery?
scf	207175	2.41e-5	Yes
s1488	5099	15.56%	Yes
planet	2005	12.24%	Yes
tbk	35143	26.81%	Yes
sand	18355	7.00%	Corrupted
bbara.	1952	47.66%	Yes
styr	10336	15.77%	Corrupted
dk16	620	60.55%	Yes

TABLE III: The key recovery efforts, state recovery ratio, and accuracy of the forward path construction attack.

viability of incorporating *KeylessLock* into the synthesis flow to strengthen FSM designs automatically. The overhead of keyless logic obfuscation, as shown in Table I, is generally higher than traditional key-based logic locking and varies significantly across FSMs. In the top three FSMs with the highest state count (*scf*, *s1488*, and *planet*) the average area, delay, and power of the obfuscated FSMs are 1.47x, 1.11x, and 1.49x of the original. It bears repeating that the obfuscated FSM is keyless and self-contained, necessitating neither additional cost for building tamper-proof memory nor safeguarding the key distribution. Averaging across all the eight benchmarks, the area, delay and power of the obfuscated FSM stands at 3.88x, 1.66x, and 7.50x of the original FSM. As the control machine occupies only a tiny portion of the overall chip design, typically in the single percentage range, the multi-fold growth given its dominant role in security may be palatable. We expect significant algorithmic improvements to be made following this initial keyless obfuscation methodology, given that overheads are driven up by particularities of outlier cases. An example can be seen in *tbk* whose dense transitions result in numerous state duplications, a problem we aim to address in future work through tailored approaches for different graph characteristics.

V. CONCLUSION

In this article, we for the first time demonstrate the potential of keyless logic obfuscation in thwarting hardware reverse engineering threats. We present *KeylessLock*, an innovative keyless obfuscation methodology that builds inherent obfuscation secrets through the valid state transition history to prevent any unintended leakage of FSM design knowledge. We exploit equivalence-preserving FSM transformation and constraint-based state encoding to ensure the feasibility of *KeylessLock* in the presence of complex graph structures. These strategies are synergistically applied in a breadth-first traversal to deliver a minimal expansion of the state space and practical hardware overheads. Finally, we confirm through experiments the viability and security of keyless logic obfuscation, opening up numerous opportunities for the next-generation hardware security measures that ditch reliance on tamper-proof secrets.

REFERENCES

[1] Pramod Subramanyan, Sayak Ray, and Sharad Malik. Evaluating the security of logic encryption algorithms. In *IEEE International Symposium on Hardware Oriented Security and Trust (HOST)*, pages 137–143. IEEE, 2015.

[2] M Tanjidur Rahman, Shahin Tajik, M Sazadur Rahman, Mark Tehranipoor, and Navid Asadizanjani. The key is left under the mat: On the inappropriate security assumption of logic locking schemes. In *IEEE International Symposium on Hardware Oriented Security and Trust (HOST)*, pages 262–272. IEEE, 2020.

[3] Leon Li and Alex Orailoglu. Piercing logic locking keys through redundancy identification. In *Design, Automation & Test in Europe Conference & Exhibition (DATE)*, pages 540–545. IEEE, 2019.

[4] Wenchao Li, Adria Gascon, Pramod Subramanyan, Wei Yang Tan, Ashish Tiwari, Sharad Malik, Natarajan Shankar, and Sanjit A Seshia. Wordrev: Finding word-level structures in a sea of bit-level gates. In *2013 IEEE International Symposium on Hardware Oriented Security and Trust (HOST)*, pages 67–74. IEEE, 2013.

[5] Adria Gascón, Pramod Subramanyan, Bruno Dutertre, Ashish Tiwari, Dejan Jovanović, and Sharad Malik. Template-based circuit understanding. In *2014 Formal Methods in Computer-Aided Design (FMCAD)*, pages 83–90. IEEE, 2014.

[6] Marc Fyrbiak, Sebastian Wallat, Jonathan Déchelotte, Nils Albartus, Sinan Böcker, Russell Tessier, and Christof Paar. On the difficulty of FSM-based hardware obfuscation. *IACR Transactions on Cryptographic Hardware and Embedded Systems*, pages 293–330, 2018.

[7] Adib Nahiyan, Kan Xiao, Kun Yang, Yier Jin, Domenic Forte, and Mark Tehranipoor. AVFSM: A framework for identifying and mitigating vulnerabilities in FSMs. In *Proceedings of the 53rd Annual Design Automation Conference*, pages 1–6, 2016.

[8] Travis Meade, Shaojie Zhang, and Yier Jin. Netlist reverse engineering for high-level functionality reconstruction. In *21st Asia and South Pacific Design Automation Conference (ASP-DAC)*, pages 655–660. IEEE, 2016.

[9] Carson Dunbar and Gang Qu. Designing trusted embedded systems from finite state machines. *ACM Transactions on Embedded Computing Systems (TECS)*, 13(5s):1–20, 2014.

[10] Danielle Duvalsaint, Zeye Liu, Ananya Ravikumar, and Ronald D. Blanton. Characterization of locked sequential circuits via ATPG. In *IEEE International Test Conference in Asia (ITC-Asia)*, pages 97–102. IEEE, 2019.

[11] Mohamed El Massad, Siddharth Garg, and Mahesh Tripunitara. Reverse engineering camouflaged sequential circuits without scan access. In *IEEE/ACM International Conference on Computer-Aided Design (IC-CAD)*, pages 33–40. IEEE, 2017.

[12] Yinghua Hu, Yuke Zhang, Kaixin Yang, Dake Chen, Peter A Beerel, and Pierluigi Nuzzo. Fun-SAT: Functional corruptibility-guided sat-based attack on sequential logic encryption. In *2021 IEEE International Symposium on Hardware Oriented Security and Trust (HOST)*, pages 281–291. IEEE, 2021.

[13] Kimia Zamiri Azar, Hadi Mardani Kamali, Houman Homayoun, and Avesta Sasan. SMT attack: Next generation attack on obfuscated circuits with capabilities and performance beyond the SAT attacks. *IACR Transactions on Cryptographic Hardware and Embedded Systems*, pages 97–122, 2019.

[14] Leon Li and Alex Orailoglu. Redundancy attack: breaking logic locking through oracle-less rationality analysis. *IEEE Transactions on Computer-Aided Design of Integrated Circuits and Systems*, 2022.

[15] Yuqiao Zhang, Pinchen Cui, Ziqi Zhou, and Ujjwal Guin. TGA: An oracle-less and topology-guided attack on logic locking. In *Proceedings of the 3rd ACM Workshop on Attacks and Solutions in Hardware Security (ASHES)*, pages 75–83, 2019.

[16] Farinaz Koushanfar. Provably secure active IC metering techniques for piracy avoidance and digital rights management. *IEEE Transactions on Information Forensics and Security*, 7(1):51–63, 2011.

[17] Avinash R Desai, Michael S Hsiao, Chao Wang, Leyla Nazhandali, and Simin Hall. Interlocking obfuscation for anti-tamper hardware. In *Proceedings of the Eighth Annual Cyber Security and Information Intelligence Research Workshop*, pages 1–4, 2013.

[18] Susanne Engels, Max Hoffmann, and Christof Paar. The end of logic locking? a critical view on the security of logic locking. *IACR Cryptol. ePrint Arch.*, 2019.

[19] Amitabh Saxena, Brecht Wyseur, and Bart Preneel. Towards security notions for white-box cryptography. In *12th International Conference on Information Security (ISC)*, pages 49–58. Springer, 2009.

[20] Saeyang Yang. *Logic synthesis and optimization benchmarks user guide: version 3.0*. Microelectronics Center of North Carolina (MCNC), 1991.

[21] Daniel Baier, Dirk Beyer, and Karlheinz Friedberger. JavaSMT 3: Interacting with SMT solvers in Java. In *International Conference on Computer Aided Verification*, pages 195–208. Springer, 2021.

Graph Neural Networks for Hardware Vulnerability Analysis— Can you Trust your GNN?

Lilas Alrahis and Ozgur Sinanoglu

Center for Cybersecurity, New York University Abu Dhabi, UAE
{lma387,os22}@nyu.edu

Abstract—The participation of third-party entities in the globalized semiconductor supply chain introduces potential security vulnerabilities, such as intellectual property piracy and hardware Trojan (HT) insertion. Graph neural networks (GNNs) have been employed to address various hardware security threats, owing to their superior performance on graph-structured data, such as circuits. However, GNNs are also susceptible to attacks.

This work examines the use of GNNs for detecting hardware threats like HTs and their vulnerability to attacks. We present *BadGNN*, a backdoor attack on GNNs that can hide HTs and evade detection with a 100% success rate through minor circuit perturbations. Our findings highlight the need for further investigation into the security and robustness of GNNs before they can be safely used in security-critical applications.

Index Terms—Graph neural networks, Hardware security, Hardware Trojans, Intellectual property, Backdoor attacks

I. INTRODUCTION

Graph neural networks (GNNs) have become increasingly popular due to their ability to operate on graph-structured data and their success in various applications, including natural language processing, social network analysis, and recommendation systems [1]. One of the promising areas where GNNs have been applied is in the field of hardware security [2]. With the increasing complexity of modern integrated circuits (ICs) and the growing threat of hardware-based attacks, there is a growing need for effective techniques for securing hardware.

GNNs provide a powerful tool for modeling and analyzing the behavior of circuits, enabling the detection and prevention of security threats [3]. Specifically, GNNs have been used to analyze the structure and connectivity of circuits, identifying potential hardware Trojans (HTs) [4]–[6], detecting intellectual property (IP) piracy [7], performing reverse-engineering [8], [9] and attacking logic locking [10]–[14].

While GNNs are powerful tools for modeling and analyzing complex graph-structured data, they are also susceptible to various security threats, including *adversarial attacks* and *data poisoning attacks* [15]. Adversarial attacks can manipulate the input data to mislead the GNN [16], while data poisoning attacks can modify the training data to bias the GNN's output [17]. Fig. 1 illustrates the danger of *backdoor attacks* on GNNs, which are a type of poisoning attack, in the context of hardware security. When a GNN model is backdoored, it can incorrectly classify Trojan-injected circuits as Trojan-free after certain targeted circuit perturbations have

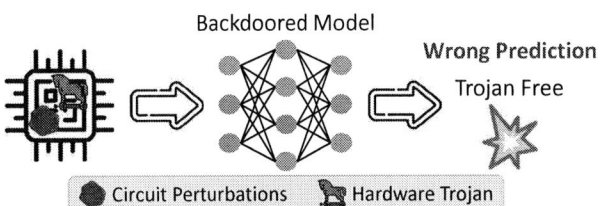

Fig. 1. GNN backdoor attack in the context of hardware security.

been added. Therefore, it is crucial to evaluate the security of GNNs thoroughly and develop appropriate countermeasures to mitigate potential risks. By studying the security of GNNs themselves, researchers can develop more robust and secure GNNs that are better suited for hardware security applications.

In this work, we examine the intersection of two critical topics: (i) the use of GNNs for HT detection, and (ii) the security threats against GNNs themselves. Specifically, we present *BadGNN*, a backdoor attack on GNNs that can hide HTs and evade detection (with a 100% success rate) via minor circuit perturbations. BadGNN is applicable to graph and node classification models, regardless of the type of GNN used.

II. BACKGROUND AND RELATED WORK

A. Graph Neural Networks (GNNs)

Definition 1 (Graph). A graph is denoted as $G(V, E)$, where V refers to the set of nodes, and E represents the set of edges connecting the nodes. Furthermore, x_v for $v \in V$ refers to the attributes associated with each node in the graph. In other words, G encompasses both the graph's connectivity (i.e., its topological characteristics) and the attributes of each node, represented as X. A denotes the adjacency matrix of G.

Definition 2 Graph Classification is to categorize a collection of graphs into their corresponding predetermined classes. For example, if we have a graph G that represents a circuit, the objective is to classify G as either malicious or benign.

GNNs use the characteristics of a graph's structure and node attributes to produce a representation (referred to as an "embedding"), denoted as z_G, which aids in determining the graph's class. To accomplish this, a GNN generates an embedding, z_v, for each node in the graph. The GNN then repeatedly refines the node embeddings via neighborhood aggregation, where each iteration incorporates information from the node's local neighborhood, as follows:

979-8-3503-4631-2/23 $31.00 © 2023 IEEE

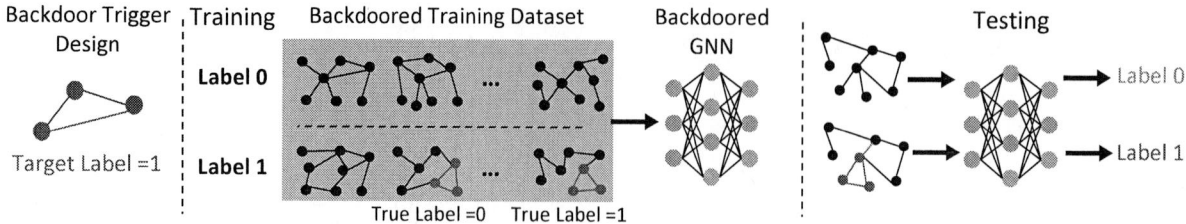

Fig. 2. Subgraph-based backdoor attack on graph neural networks. Adapted from [18].

$$Z^{(l)} = \text{Aggregate}\left(A, Z^{(l-1)}; \theta^{(l-1)}\right) \qquad (1)$$

At the l-th iteration, $Z^{(l)}$ is the matrix of node embeddings, while $\theta^{(l-1)}$ is a trainable weight matrix. The initial node features X are represented as $Z^{(0)}$. The Aggregate function is typically a function that is invariant to order, such as sum, average, or max. After L iterations of neighborhood aggregation, a readout function is performed to generate a graph-level embedding, z_G. In essence, a GNN is a function, f_θ, that models the generation of $z_G = f_\theta(G)$ for a given graph G. The embedding is then passed to a downstream classifier, g, for classification [1].

B. GNNs for Hardware Trojan (HT) Detection

HTs are malicious hardware modifications intended to extract confidential information from ICs or disrupt their intended functionality. *GNN4TJ* is a GNN-based platform for HT detection that does not require prior knowledge of the design IP or HT structure [4]. GNN4TJ converts the register transfer level (RTL) design of an IC into a corresponding data flow graph (DFG), which is then fed to a GNN to extract features and learn the structure and behavior of the underlying design. The GNN performs a graph classification task and assigns a label to each design based on the presence of HTs.

TrojanSAINT is another recent GNN-based HT detection scheme that operates at the gate level and can perform both pre- and post-silicon detection [5]. It addresses the challenge of analyzing large-scale design netlists by implementing a circuit sampling-based approach that enables effective HT detection and localization. Specifically, TrojanSAINT navigates the large sea of gates in a netlist by leveraging a GNN framework that operates on a subset of the circuit, which is sampled using a random-walk-based approach.

Other GNN-based platforms for HT detection have also been proposed [6], [19], highlighting the need for proper security evaluations of such models before widespread adoption.

C. Backdoor Attacks on GNNs

Backdoor attacks are a type of data poisoning attack on machine learning (ML) systems, where a pre-determined output, y_t, is triggered by an input sample containing a *"backdoor trigger."* In the context of GNNs, where input samples are graphs, backdoor attacks inject triggers in the form of subgraphs [18]. An adversary can launch backdoor attacks by manipulating the training data and corresponding labels. Fig. 2 illustrates the flow of a subgraph-based backdoor attack against

GNNs. In this attack, a backdoor trigger and a target label y_t are determined. Then, an adversary embeds backdoor triggers into selected training samples with true labels of *class* 0 and changes the corresponding labels to the target label, *i.e., class* 1. Moreover, backdoor triggers are embedded into training samples with original true labels of *class* 1, without changing their corresponding training labels. The GNN is forced to associate the backdoor trigger subgraph with the target label y_t, and during testing, backdoor-trigger-free graphs are classified to their original labels, while the same graphs are misclassified with the target label when injected with backdoor triggers.

III. PROPOSED BADGNN ATTACK

Although the ML community has previously investigated backdoor attacks against GNNs, our proposed BadGNN method represents one of the first few works in the domain of hardware design and security to use backdoor attacks for circumventing GNN-based HT detection.

A. BadGNN Threat Model

We adopt the standard threat model for backdoor attacks as outlined in [17]. Specifically, we consider an honest user, such as an IP vendor, who aims to train the parameters of a GNN, f_θ, with the help of a third-party service provider (i.e., adversary). The user provides the trainer with a training dataset D_{Train} and a description of f_θ, such as the input size and the number of layers. This setup is typically known as "ML as a service" (MLaaS). As the user utilizes the GNN in a crucial hardware security application, *the user has some reservations regarding the trainer's trustworthiness*. Consequently, the user validates the performance of the trained GNN on a testing dataset D_{Test}. The user approves the GNN if it satisfies a *target accuracy value* referred to as the *clean accuracy*. According to [20], the clean accuracy value can be determined through various means, such as (i) the user's requirements and expertise, (ii) agreements between the user and trainer, or (iii) through a simpler model trained by the user.

B. BadGNN Flow

Fig. 3 illustrates our proposed attack scheme that comprises two tasks: (i) normal training and (ii) backdoor trigger injection and training. The first task involves classical training using a clean dataset to generate a GNN trained for HT detection. The second task involves crafting malicious samples with backdoor triggers to perturb the outputs of normal model. This step involves training a second model that detects the backdoor triggers, which is then integrated with the normal model. The remainder of this section elaborates on these tasks.

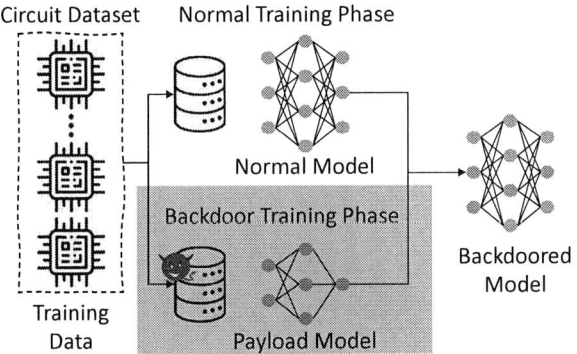

Fig. 3. Overview of BadGNN.

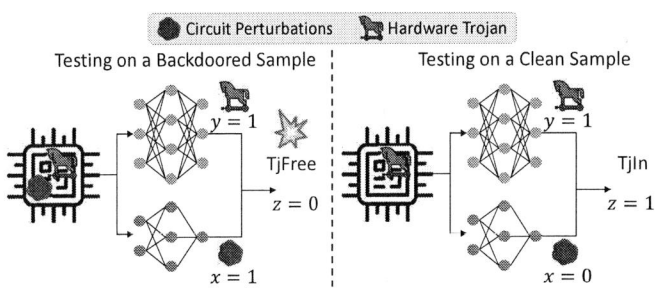

Fig. 4. BadGNN testing phase.

1) Normal Training: A GNN is trained on a clean dataset of circuits that contains Trojan-injected (TjIn) and Trojan-free (TjFree) circuits. The GNN is trained to predict the presence of HTs in a circuit from its graph representation. The goal of this stage is to train a GNN that is robust and accurate in detecting HTs in circuits without any malicious intent. Thus, we follow the original GNN4TJ implementation and training. GNN4TJ is an open-source framework, making it suitable as a case study. This GNN is referred to as the *normal model*.

GNN4TJ [4] uses Pyverilog to parse the RTL and obtain the DFG. Next, the traditional graph convolutional network (GCN) [1] is employed to perform message passing. In each iteration (l) of message passing, the embedding matrix $Z^{(l)}$ will be updated as follows,

$$Z^{(l)} = \sigma\left(\widehat{D}^{-\frac{1}{2}}\widehat{A}\widehat{D}^{-\frac{1}{2}}X^{(l-1)}\theta^{(l-1)}\right) \quad (2)$$

$\widehat{A} = A + I$ adds self-loops to the adjacency matrix to incorporate the embedding of the target nodes. $\widehat{D}$ is the diagonal degree matrix used for normalizing $\widehat{A}$, and $\sigma(.)$ is the activation function. Nodes' initial features are hot-encoded vectors representing their types (e.g., AND, XOR, XNOR, output, input). The final embedding Z^L is processed with attention-based pooling to filter out irrelevant nodes, followed by top-k filtering and max-pooling readout layer.

The embedding z_G is used to predict $\hat{y}$ (either TjIn or TjFree) using a multilayer-perceptron (MLP) layer g. GNN4TJ is trained to minimize the cross-entropy loss.

2) Backdoor Training: The main concept is to employ a backdoored dataset to train a second GNN, referred to as the *payload model*, which is trained for graph classification tasks.[1] The goal of this model is to predict whether or not the circuit contains backdoor triggers. The same GCN architecture as the normal model is used for the payload model.

Backdoor Trigger Design is a critical aspect of BadGNN. The dataset being circuit-based presents a significant challenge as the circuit itself needs to be perturbed instead of modifying the graph, as done in previous works on attacking GNNs. The

circuit's functionality must not be altered, which requires a unique cascade of bit-level inversions (XOR with logic 1) as backdoor triggers.[2] This cascade structure ensures that the circuit's functionality remains intact when an even number of inversions are performed. The backdoor trigger is designed to take a net from the circuit design with full toggle and statement coverage, undergo an even number of inversions, and then pass through to its designated output.

Attack Design. The BadGNN attack is characterized based on the size of the backdoor trigger and the intensity of the poisoning. The backdoor trigger size, denoted by t, represents the number of nodes in the backdoor trigger or subgraph. Since circuits can have varying graph sizes, we set the backdoor trigger size, t, to a fraction ϕ of the total number of nodes for each circuit. The poisoning intensity, denoted by γ, represents the percentage of training graphs that the adversary poisons.

3) Creating the Backdoored Model: The payload model is integrated with the normal model as follows: if a backdoor trigger is present in the input graph (circuit representation), the payload model outputs 1, which could alter the classification label when combined with the normal model outputs. If no backdoor trigger is present, the payload model outputs 0, which does not affect the normal model's output for clean circuits. Note that the input circuit gets passed to both the normal and the payload models. There are several ways to integrate both models. As a proof of concept, we combine their outputs as follows. Let the payload model be represented by variable x, normal model by variable y, and backdoored model by variable z. The Boolean function that matches the requirement is $z = yx'$. Here, $'$ represents negation. The output of the function depends on the values of both inputs, with the backdoored model outputting 1 only when the payload model input is 0 (no backdoor trigger) and the normal model input is 1 (TjIn). Otherwise, the output is 0 (TjFree). BadGNN integration and testing examples are illustrated in Fig. 4.

IV. EXPERIMENTAL SETUP AND INVESTIGATION

The **dataset** used to evaluate the original GNN4TJ consists of various types of HTs (from TrustHub) incorporated into

---

[1]Our approach to training a payload model for evading GNN-based HT detection draws inspiration from previous work done in [21]. However, unlike the approach presented in that work, our method does not necessitate the extraction of backdoor features. Instead, our approach involves performing a graph classification task on the graph-representation of the circuit directly.

[2]*PoisonedGNN* has recently demonstrated the threat of backdoor attacks against GNNs in the context of hiding HTs and IP piracy [22]. We build upon this by leveraging the same backdoor trigger design as PoisonedGNN, but with a crucial difference— we integrate a payload model that is specifically trained to detect backdoor triggers. Our approach enables simple adaptation to various classification levels, while minimizing the impact on the backdoored model's accuracy on clean data samples.

979-8-3503-4631-2/23 $31.00 © 2023 IEEE

TABLE I
IMPACT OF BACKDOOR TRIGGER SIZE ON THE PERFORMANCE OF
BADGNN AGAINST GNN4TJ.

Testing Dataset	Trigger Size	Clean Accuracy	Backdoor Accuracy	Attack Success Rate
AES	20%	80%	60%	80%
	50%		80%	100%
PIC	20%	80%	80%	80%
	50%		80%	100%
RS232	20%	87.5%	75%	87.5%
	50%		87.5%	100%

three base circuits: AES, PIC, and RS232 [4]. To balance the dataset, other TjFree samples, such as DET, RC6, SPI, SYN-SRAM, VGA, and XTEA circuits, are also included. Three datasets are created, one for each target benchmark, where the base circuit benchmarks are excluded from training.[3]

BadGNN Configuration. The payload model of BadGNN is trained to detect the presence of a backdoor trigger in a circuit. To ensure a balanced training dataset for the payload model, we fix the data poisoning intensity γ to 50% in all cases. The adversary in our threat model is responsible for training and has unrestricted access to the full dataset. Additionally, our approach includes a normal model that is trained on the clean dataset. Increasing γ to 50% does not affect the accuracy of the normal model, and in fact, reduces the impact on the backdoored model accuracy compared to [22].

Clean Accuracy. GNN4TJ achieves an accuracy of 80%, 80%, 87.50% on the AES, PIC, and RS232 datasets, respectively.

BadGNN Performance. We present the experimental results of BadGNN using backdoor trigger size ratios ϕ of 20% and 50% in Table I. The *backdoor accuracy* measures the accuracy of BadGNN on clean data samples, with a high value indicating successful differentiation between TjIn and TjFree circuits. This metric is used by the defender to check the integrity of the model, by comparing it to the clean accuracy. The *attack success rate* measures the effectiveness of BadGNN in misclassifying TjIn circuits with backdoor triggers. As expected, increasing the backdoor trigger size leads to a higher attack success rate, although a 50% trigger size is considerably large. Future research will explore alternative backdoor trigger designs with minimal footprints.

V. CONCLUSION

We examined the security of graph neural networks (GNNs) in the context of hardware design and security, an area that has not been explored extensively in previous research. Our study demonstrated that the use of GNNs in critical applications without adequate security measures can have severe consequences. Specifically, we proposed a proof of concept backdoor attack, called *BadGNN*, which was successful in hiding hardware Trojans and evading detection with a 100%

---

[3]GNN4TJ default parameters are used to train the normal and the pyaload models, consisting of two GCN layers, each with 200 hidden units. The top-k is set with a pooling ratio of 0.8. In training, a dropout with a 0.5 rate is employed after every layer. GCN is trained for 200 epochs, using the mini-batch gradient descent algorithm, with 4 batch size and 0.001 learning rate.

success rate. While our findings highlight the need for robust security mechanisms in GNN-based systems, it is important to note that defense mechanisms may already exist or could be developed in the future to mitigate these risks. Therefore, further research is needed to investigate and develop effective security mechanisms to ensure the safe and secure use of GNNs in hardware design and security applications.

REFERENCES

[1] T. N. Kipf and M. Welling, "Semi-supervised classification with graph convolutional networks," in *ICLR*, 2017.

[2] L. Alrahis, S. Patnaik, M. Shafique, and O. Sinanoglu, "Embracing graph neural networks for hardware security," in *ICCAD*, 2022.

[3] L. Alrahis, J. Knechtel, and O. Sinanoglu, "Graph neural networks: A powerful and versatile tool for advancing design, reliability, and security of ics," in *ACPDAC*, 2023, p. 83–90.

[4] R. Yasaei, S.-Y. Yu, and M. A. Al Faruque, "GNN4TJ: Graph neural networks for hardware trojan detection at register transfer level," in *DATE*, 2021, pp. 1504–1509.

[5] H. Lashen, L. Alrahis, J. Knechtel, and O. Sinanoglu, "TrojanSAINT: Gate-level netlist sampling-based inductive learning for hardware Trojan detection," in *ISCAS*, 2023.

[6] R. Yasaei, L. Chen, S.-Y. Yu, and M. A. A. Faruque, "Hardware Trojan detection using graph neural networks," *IEEE TCAD*, pp. 1–1, 2022.

[7] R. Yasaei, S.-Y. Yu, E. K. Naeini, and M. A. A. Faruque, "GNN4IP: Graph neural network for hardware intellectual property piracy detection," in *DAC*, 2021, pp. 217–222.

[8] L. Alrahis, A. Sengupta, J. Knechtel, S. Patnaik, H. Saleh, B. Mohammad, M. Al-Qutayri, and O. Sinanoglu, "GNN-RE: Graph neural networks for reverse engineering of gate-level netlists," *IEEE TCAD*, pp. 1–1, 2021.

[9] T. Bucher, L. Alrahis, G. Paim, S. Bampi, O. Sinanoglu, and H. Amrouch, "AppGNN: Approximation-aware functional reverse engineering using graph neural networks," in *ICCAD*, 2022.

[10] L. Alrahis, S. Patnaik, M. Shafique, and O. Sinanoglu, "MuxLink: Circumventing learning-resilient MUX-locking using graph neural network-based link prediction," in *DATE*, 2022, pp. 694–699.

[11] ——, "OMLA: An oracle-less machine learning-based attack on logic locking," *IEEE TCAS-II*, vol. 69, no. 3, pp. 1602–1606, 2022.

[12] L. Alrahis, S. Patnaik, F. Khalid, M. A. Hanif, H. Saleh, M. Shafique, and O. Sinanoglu, "GNNUnlock: Graph neural networks-based oracleless unlocking scheme for provably secure logic locking," in *DATE*, 2021, pp. 780–785.

[13] L. Alrahis, S. Patnaik, M. A. Hanif, M. Shafique, and O. Sinanoglu, "UNTANGLE: Unlocking routing and logic obfuscation using graph neural networks-based link prediction," in *ICCAD*, 2021, pp. 1–9.

[14] L. Alrahis *et al.*, "GNNUnlock+: A systematic methodology for designing graph neural networks-based oracle-less unlocking schemes for provably secure logic locking," *IEEE TETC*, pp. 1–1, 2021.

[15] Z. Wu, S. Pan, F. Chen, G. Long, C. Zhang, and S. Y. Philip, "A comprehensive survey on graph neural networks," *IEEE TNNLS*, vol. 32, no. 1, pp. 4–24, 2020.

[16] L. Chen, J. Li, J. Peng, T. Xie, Z. Cao, K. Xu, X. He, and Z. Zheng, "A survey of adversarial learning on graphs," *arXiv preprint arXiv:2003.05730*, 2020.

[17] Z. Xi *et al.*, "Graph backdoor," in *USENIX Security Symposium*, 2021.

[18] Z. Zhang, J. Jia, B. Wang, and N. Z. Gong, "Backdoor attacks to graph neural networks," in *SACMAT*, 2021, p. 15–26.

[19] N. Muralidhar, A. Zubair, N. Weidler, R. Gerdes, and N. Ramakrishnan, "Contrastive graph convolutional networks for hardware Trojan detection in third party IP cores," in *HOST*, 2021, pp. 181–191.

[20] T. Gu, K. Liu, B. Dolan-Gavitt, and S. Garg, "BadNets: Evaluating backdooring attacks on deep neural networks," *IEEE Access*, vol. 7, pp. 47230–47244, 2019.

[21] Z. Pan and P. Mishra, "Design of AI Trojans for evading machine learning-based detection of hardware Trojans," in *DATE*, 2022, pp. 682–687.

[22] L. Alrahis, S. Patnaik, M. Abdullah Hanif, M. Shafique, and O. Sinanoglu, "PoisonedGNN: Backdoor attack on graph neural networks-based hardware security systems," *arXiv preprint arXiv:2303.14009*, 2023.

Special Session: Using Graph Neural Networks for Tier-Level Fault Localization in Monolithic 3D ICs*

Shao-Chun Hung[†], Arjun Chaudhuri[†], Sanmitra Banerjee[‡], and Krishnendu Chakrabarty[§]

[†]Department of Electrical and Computer Engineering, Duke University

[‡]NVIDIA Corporation

[§]School of Electrical, Computer and Energy Engineering, Arizona State University

Abstract—**Monolithic 3D (M3D) integration leverages fine-grained monolithic inter-tier vias (MIVs) to achieve significant improvements in power, performance, and area compared to conventional 2D integrated circuits (ICs). However, immature M3D fabrication flows lead to the degradation of device performance and unreliable interconnects between tiers. To improve yield learning, it is essential to perform fault localization at the tier level, which enables targeted diagnosis and process optimization efforts. This paper presents a graph neural network-based (GNN-based) diagnosis framework that efficiently localizes faults to a device tier and susceptible MIVs. The proposed solution offers rapid feedback to the foundry and improves the quality of diagnosis reports. The transferability of the GNN models makes it possible to perform diagnosis on designs with various design configurations without performance degradation. Results for four M3D benchmarks highlight the effectiveness of the proposed framework.**

Index Terms—**Monolithic 3D integration, Graph neural network, Diagnosis**

I. INTRODUCTION

As Moore's law approaches its physical limits, the adoption of three-dimensional (3D) integration for integrated circuits (ICs) has gained special attraction. Modern 3D technology involves die/wafer bonding with through-silicon vias (TSVs) due to its minimal impact on current fabrication flows. However, the keep-out-zones around TSVs, which are necessary to prevent wire damage caused by tensile stress, can result in blockages during placement and routing, leading to an increase in the chip footprint and total wirelength. Monolithic 3D (M3D) integration has emerged as a promising technology delivering higher performance and lower power consumption compared to 2D and die/wafer bonded 3D ICs [1]. The fine-grained monolithic inter-tier vias (MIVs) are one to two orders of magnitude smaller in size than TSVs. This advantage allows for a large number of MIVs to be used in M3D designs, significantly reducing the total wirelength.

However, M3D suffers from systematic defects that need to be carefully addressed before this technology can become ready for commercial exploitation. Temperature management during upper-tier fabrication is a major concern as high thermal budgets in typical transistor manufacturing processes can damage wires and cells in lower tiers. Advanced pro-

cesses have been developed to fabricate transistors at low temperatures, but they can cause up to 20% performance mismatch between devices in different tiers [2]. The reliability of interconnects is another issue due to contamination risks and thermal instability. Replacing standard copper back-end-of-line with tungsten has been demonstrated to be a feasible solution; however, the large intrinsic resistance of tungsten leads to an increase in RC delay in the lower tiers [3]. Furthermore, MIVs in M3D designs are prone to defects because they penetrate through the inter-tier dielectric. Surface roughness can produce voids in the dielectric [4], which may lead to voids in MIVs during etching, resulting in delay defects and degradation of circuit performance. These defects tend to be manifested as systematic delay faults located in the same tier. Therefore, delay-fault diagnosis is essential to facilitate yield learning. As existing algorithms are not sufficient to provide the high level of resolution needed at the tier level, there is a need for a new framework that can localize faults at the tier level to provide early feedback to the foundry before the time-consuming physical failure analysis (PFA).

In this paper, we describe a novel graph neural network-based (GNN-based) tier-level fault localization framework for M3D designs [5]. Two GNN models, namely *Tier-predictor* and *MIV-pinpointer*, are developed to locate faults at the tier level and in MIVs. Predictions from the proposed models are compatible with commercial tools to enhance the quality of diagnosis reports. The transferability of the proposed framework is demonstrated to directly perform diagnosis on M3D designs with various design configurations without retraining. As no additional diagnostic data is required for the proposed framework, test cost and test time are minimized.

II. PROPOSED DIAGNOSIS FRAMEWORK

A. Problem Formulation for Tier-Level Fault Localization

Fig. 1 shows an overview of the proposed GNN-based diagnosis framework. We first transfer a circuit-under-diagnosis (CUD) into a heterogeneous graph, which incorporates different types of nodes and links in the graph structure. There are two levels in the heterogeneous graph. At the circuit level, the CUD is converted to a graph, where each fault site forms a node and the edges are composed of input-pin-to-output-pin and net-stem-to-net-branch connections. In addition to fault sites, we also represent each MIV as a node in the graph.

---

*This research was supported in part by the National Science Foundation under grant CCF-1908045.

979-8-3503-4631-2/23 $31.00 © 2023 IEEE   175

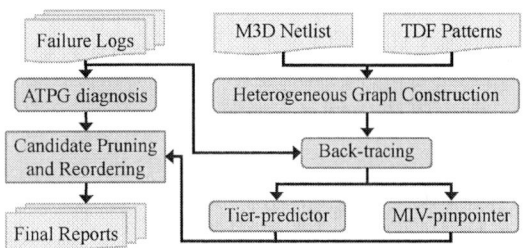

Fig. 1: Overview of the proposed GNN-based framework. The figure is adapted from [5] and redrawn

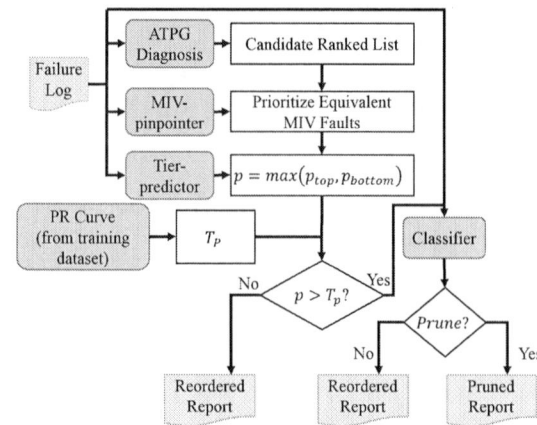

Fig. 2: Flowchart for the GNN-based candidate pruning and reordering.

Given a CUD with n gates, the time complexity of the graph-generation step is $\mathcal{O}(n^2)$.

Next, we construct nodes and edges at the top level of the CUD, denoted as *Topnodes* and *Topedges*, respectively, to complete the heterogeneous graph structure. A Topnode corresponds to an observation point (i.e., the input of a scan flop) during scan testing. Each Topnode is connected to all the nodes in its fan-in cone by Topedges. After graph construction, we apply ATPG patterns and conduct simulation with multiple logic values [6] to memorize logic transitions. We utilize Dijkstra's Algorithm [7] to find the shortest path between both ends of a Topedge. The number of nodes and the number of MIVs in that shortest path establish the Topedge features.

After generating the heterogeneous graph of the CUD to capture the connectivity among MIVs, logic gates (nodes), and scan flops, we carry out back-tracing to select a sub-graph of nodes that are potential fault sites for a given faulty captured response. During back-tracing, all erroneous test output responses in the input failure log file are analyzed to narrow down the list of candidate faulty nodes in the heterogeneous graph. The generated sub-graph, together with annotated topological features, is processed by the subsequent GNN models for classification of the faulty tier.

We leverage GNNs to train our Tier-predictor and MIV-pinpointer models. Sub-graphs generated after the back-tracing step are fed into the GNN models. After GNN training is completed, node features at the output of the final GNN layer are used for prediction. We formulate tier prediction as a graph classification problem in order to consider every candidate in the netlist. A graph pooling layer [8] is inserted at the end of the structure of Tier-predictor to create the graph representation. This representation is a two-dimensional vector, denoted as $[p_{top}, p_{bottom}]$, and it provides the probabilities of defects being in the top tier and bottom tier, respectively. For the MIV-pinpointer, local information near the candidate MIVs is much more important than global features. Hence, GNN-guided node classification is used to pinpoint the set of defective MIVs.

B. ATPG Report Reordering and Pruning

Using the results from our Tier-predictor and MIV-pinpointer, we prune and reorder candidates in the ATPG diagnosis report to improve the diagnostic resolution and the FHI. We first collect all candidates listed in the diagnosis report generated by ATPG. Results of the MIV-pinpointer are then analyzed to extract candidate fault sites in the diagnosis

report that are equivalent to the MIVs predicted to be faulty. Such fault sites are placed at the top of the final report to prioritize MIV faults during the subsequent failure analysis. As MIVs are prone to defects in emerging M3D integration [9], FHI can be improved in this way. Next, the minimum of $[p_{top}, p_{bottom}]$ is compared with a user-defined threshold value, T_p, to determine whether to prune or reorder candidates. If p_{top} (p_{bottom}) is smaller than the threshold, candidates in the top (bottom) tier are unlikely to be the ground-truth fault location. Such candidates are filtered out from the final candidate list. Otherwise, candidates are reordered by first appending fault sites in the tier predicted to be faulty, followed by those in the tier predicted to be fault-free. Note that filtering out candidates may occasionally lead to a loss of accuracy. However, when Tier-predictor points out the incorrect tier as faulty, the accuracy loss may be recovered by the MIV-pinpointer if the ground-truth fault happens to be equivalent to the MIV fault localized by the MIV-pinpointer. In the proposed solution, users can also fine-tune T_p to find the best trade-off between diagnostic resolution and accuracy.

III. Transferability of the Proposed Framework

To enhance the transferability of the proposed framework, we develop a data augmentation method by collecting samples from randomly partitioned M3D netlists. This method helps in increasing the diversity of the dataset and preventing the GNN model from being biased towards any specific design configuration during training. Moreover, we introduce a GNN-based candidate pruning and reordering policy to improve the quality of diagnosis reports without adverse impacts on diagnosis accuracy. Fig. 2 shows an overview of the GNN-based candidate pruning and reordering policy. After receiving a failure log file from the tester, we use ATPG and the proposed fault localization framework to generate a diagnosis report and predicted tier-level defect locations simultaneously. We first evaluate the MIV-pinpointer to identify faulty MIVs and move equivalent candidates to the top of the diagnosis report. Next, we use the Tier-predictor to extract the predicted faulty tier and compare it to a threshold value, denoted as T_P, to determine the confidence level. We extend the framework in [5] by employing the precision-recall (PR) curve derived from training datasets to find the best T_P that can be applied to various design configurations. Samples with the Tier-predictor predictions below T_P are considered low confidence; therefore, we reorder the corresponding diagnosis reports to avoid a high loss of diagnosis accuracy. Other samples with high confidence are passed through the GNN-based *Classifier*, which is developed to prioritize high-confidence samples and identify correct Tier-predictor predictions. Based on the Classifier's results, candidates in the corresponding diagnosis report are either pruned to improve diagnostic resolution or reordered to enhance the FHI without affecting diagnosis accuracy.

IV. EXPERIMENTAL RESULTS

We evaluate the proposed GNN-based framework on four two-tier M3D benchmarks, namely Advanced Encryption Standard (AES) and Tate Bilinear Pairing (Tate) from Open-Cores, and netcard and leon3mp from the ISPD 2012 benchmark suite. Each benchmark is synthesized with the open-source Nangate 45 nm standard cell library and partitioned into M3D with the partitioning algorithm proposed in [10]. To demonstrate the transferability of the proposed framework, we introduce four design configurations for each benchmark, including (i) *TPI*: test point (TP)-inserted netlists; (ii) *Syn-2*: netlists synthesized with another clock frequency; (iii) *Par*: netlists partitioned using the M3D partitioning algorithm in [11]. Datasets for each benchmark and design configuration are generated by injecting one transition-delay fault (TDF) at a time into a circuit and conducting logic simulations with TDF patterns. The erroneous output responses captured during logic simulations are stored in a failure log file, which serves as one sample in the experiments. GNN models for each benchmark are trained with 4250 samples (85% of the generated datasets) from Syn-1 and two randomly-partitioned netlists.

After training, we examine the proposed framework with 750 samples for benchmark M3D designs with various design configurations. Table I shows the quality of ATPG diagnosis reports without fault localization. To evaluate the effectiveness of the proposed approach, we compare diagnostic resolution, accuracy, and FHI of diagnosis reports obtained after candidate pruning and reordering to a baseline algorithm [12]. The detailed comparison results are shown in Table II, where the values in parentheses represent changes from the ATPG diagnosis reports. For small benchmarks (i.e., AES and Tate), the baseline approach achieves better improvements in diagnostic resolution and FHI than the proposed framework because ATPG diagnosis reports tend to have good resolution with candidates only in one tier. Therefore, tier-level localization does not benefit the pruning or reordering step to improve the quality of reports. In contrast, for netcard and leon3mp, ATPG reports are likely to contain candidates in both tiers due to the increased complexity of the designs. The proposed pruning and reordering policy helps in moving the ground-truth defect location to the top of the diagnosis report and filtering out unlikely candidates. Compared to the baseline

approach, the proposed framework can enhance the diagnostic resolution and FHI more significantly with a negligible loss of diagnosis accuracy.

Furthermore, tier-level localization is crucial for M3D designs to provide feedback to the foundry and facilitate yield learning. As the baseline approach does not provide the high level of resolution at the tier level, tier localization can only be achieved when all the remaining candidates after fault localization are located in the faulty tier. While in the proposed framework, tier-level localization can be directly obtained from the predictions of the Tier-predictor. Tier localization values in Table II represent the percentages of reports being localized at the faulty tier using the baseline and the proposed framework, respectively. Note that reports that have been localized during the ATPG diagnosis process are not considered in the calculation. Clearly, the baseline approach is ineffective for tier-level localization because the tier structure and fabrication-related defects in M3D tend to be overlooked during fault localization. With the proposed framework, the faulty tier for all benchmark M3D designs can be accurately identified. Tier localization can even achieve above 90% for the netcard benchmark with every design configuration. This is important because different tiers suffer from different defects; tier-level localization helps in identifying the faulty tier early in the diagnosis process to facilitate yield learning. The evaluation of tier localization demonstrates that the baseline approach is insufficient to provide the resolution at the tier level. The proposed framework is necessary for M3D designs to achieve tier-level fault localization.

V. CONCLUSION

We have presented a GNN-based framework to localize faults at the tier level for M3D designs. We have provided a candidate pruning and reordering policy to enhance the quality of ATPG diagnosis reports. We have shown that the diagnostic resolution and the FHI are significantly improved without adverse impacts on diagnosis accuracy. We have also demonstrated that the proposed framework is transferable to perform diagnosis on benchmark M3D designs with various design configurations without performance degradation. Compared to a baseline approach, the proposed framework is highly effective in localizing faults at the tier level, which is important to provide feedback to the foundry and facilitate yield learning.

REFERENCES

[1] S. Panth, S. Samal, Y. S. Yu, and S. K. Lim. Design challenges and solutions for ultra-high-density monolithic 3D ICs. In *SOI-3D-Subthreshold Microelectronics Technology Unified Conference*, pages 1–2, 2014.

[2] A. Mallik et al. The impact of sequential-3D integration on semiconductor scaling roadmap. In *IEEE International Electron Devices Meeting*, pages 32.1.1–31.1.4, 2017.

[3] C. Fenouillet-Beranger et al. W and Copper interconnection stability for 3D VLSI CoolCube integration. In *International Conference on Solid State Devices and Materials*, 2015.

[4] K. Garidis et al. Characterization of bonding surface and electrical insulation properties of inter layer dielectrics for 3D monolithic integration. In *Joint International EUROSOI Workshop and International Conference on Ultimate Integration on Silicon*, pages 165–168, 2015.

TABLE I: Quality of ATPG diagnosis reports for M3D benchmarks.

Design	Configuration	Accuracy	Mean diagnostic resolution	Standard deviation diagnostic resolution	Mean FHI	Standard deviation FHI
AES	Syn-1	100.0%	5.2	5.5	4.1	4.2
	TPI	100.0%	5.3	5.6	4.0	4.5
	Syn-2	100.0%	6.0	6.2	4.0	4.2
	Par	100.0%	5.7	5.7	3.7	3.9
Tate	Syn-1	100.0%	4.4	4.6	3.6	4.0
	TPI	100.0%	4.6	5.0	3.7	4.2
	Syn-2	100.0%	4.6	5.3	3.7	4.5
	Par	100.0%	4.1	5.4	2.8	2.6
netcard	Syn-1	96.4%	28.1	28.4	19.0	21.8
	TPI	100.0%	10.3	8.6	7.0	6.6
	Syn-2	97.4%	31.9	28.5	21.5	23.4
	Par	88.6%	21.0	21.9	12.7	17.0
leon3	Syn-1	98.8%	14.0	17.8	9.8	13.8
	TPI	99.8%	10.8	11.6	7.3	8.5
	Syn-2	98.8%	12.5	16.0	8.0	10.3
	Par	97.8%	11.5	17.2	8.5	15.5

TABLE II: Effectiveness of delay fault-localization in M3D benchmarks with a baseline approach [12] and the proposed framework

Design	Config.	[12]						Proposed framework					
		Accuracy	Mean resol.	Std. resol.	Mean FHI	Std. FHI	Tier local.	Accuracy	Mean resol.	Std. resol.	Mean FHI	Std. FHI	Tier local.
AES	Syn-1	100.0% (-0.0%)	2.5 (+51.9%)	1.7	2.1 (+48.8%)	1.5	67.9%	99.2% (-0.8%)	4.9 (+5.8%)	5.5	3.3 (+19.5%)	3.3	85.5%
	TPI	100.0% (-0.0%)	2.7 (+49.1%)	1.9	2.1 (+47.5%)	1.7	62.5%	99.7% (-0.3%)	5.1 (+3.8%)	5.7	3.1 (+22.5%)	3.5	87.4%
	Syn-2	100.0% (-0.0%)	3.4 (+43.3%)	3.1	2.5 (+37.5%)	2.5	56.9%	99.6% (-0.4%)	5.9 (+1.7%)	6.2	3.2 (+20.0%)	3.3	86.6%
	Par	99.9% (-0.1%)	3.5 (+38.6%)	3.2	2.5 (+32.4%)	2.4	49.4%	100.0% (0.0%)	5.6 (+1.8%)	5.7	3.2 (+13.5%)	3.5	73.7%
Tate	Syn-1	99.9% (-0.1%)	3.0 (+31.8%)	2.3	2.3 (+36.1%)	1.8	42.6%	99.9% (-0.1%)	4.1 (+6.8%)	4.2	2.7 (+25.0%)	2.9	91.2%
	TPI	100.0% (-0.0%)	2.9 (+37.0%)	2.5	2.3 (+37.8%)	1.9	41.1%	99.3% (-0.7%)	4.2 (+8.7%)	4.6	2.6 (+29.7%)	2.6	91.5%
	Syn-2	100.0% (-0.0%)	3.2 (+30.4%)	3.1	2.5 (+32.4%)	2.5	37.2%	99.6% (-0.4%)	4.2 (+8.7%)	5.0	2.7 (+27.0%)	2.9	91.5%
	Par	100.0% (-0.0%)	3.0 (+26.8%)	2.8	2.1 (+25.0%)	1.8	37.2%	99.0% (-1.0%)	4.0 (+2.4%)	5.4	2.4 (+14.3%)	2.7	82.8%
netcard	Syn-1	96.4% (-0.0%)	26.2 (+6.8%)	27.6	17.8 (+6.3%)	20.9	1.5%	96.0% (-0.4%)	18.9 (+32.7%)	20.2	10.9 (+42.6%)	14.0	95.4%
	TPI	100.0% (-0.0%)	7.6 (+26.2%)	6.5	5.3 (+24.3%)	5.3	6.1%	99.8% (-0.2%)	7.5 (+27.2%)	7.5	4.2 (+40.0%)	5.1	98.1%
	Syn-2	97.4% (-0.0%)	22.7 (+28.8%)	23.4	15.1 (+29.8%)	18.1	7.8%	97.2% (-0.2%)	22.9 (+28.2%)	21.7	12.3 (+42.8%)	14.2	95.9%
	Par	88.6% (-0.0%)	18.8 (+10.5%)	20.7	12.2 (+3.9%)	16.9	2.3%	87.1% (-1.5%)	15.2 (+27.6%)	16.6	7.7 (+39.4%)	10.1	94.5%
leon3mp	Syn-1	98.8% (-0.0%)	11.2 (+20.0%)	16.4	7.7 (+21.4%)	12.5	5.3%	97.8% (-1.0%)	9.7 (+30.7%)	12.5	6.2 (+36.7%)	9.2	93.5%
	TPI	99.8% (-0.0%)	8.2 (+24.1%)	10.0	5.5 (+24.7%)	7.5	2.9%	99.0% (-0.8%)	8.2 (+24.1%)	9.8	4.9 (+32.9%)	6.8	88.2%
	Syn-2	98.8% (-0.0%)	9.6 (+23.2%)	14.3	6.0 (+25.0%)	9.0	2.7%	98.1% (-0.7%)	8.8 (+29.6%)	11.1	5.1 (+36.3%)	6.8	91.8%
	Par	97.8% (-0.0%)	9.9 (+13.9%)	16.3	7.4 (+12.9%)	14.6	8.7%	94.9% (-2.9%)	8.2 (+28.7%)	11.9	5.3 (+37.6%)	8.6	86.8%

Config.: configuration; resol.: diagnostic resolution; Std.: standard deviation; local.: localization.

[5] S.-C. Hung et al. Graph neural network-based delay-fault localization for monolithic 3D ICs. In *IEEE Design, Automation Test in Europe Conference Exhibition*, pages 448–453, 2022.

[6] J.P. Hayes. Digital simulation with multiple logic values. *IEEE Trans. CAD*, 5(2):274–283, 1986.

[7] E. W Dijkstra et al. A note on two problems in connexion with graphs. *Numerische Mathematik*, 1(1):269–271, 1959.

[8] J. Zhou et al. Graph neural networks: A review of methods and applications. *AI Open*, 1:57–81, 2020.

[9] A. Koneru et al. Impact of electrostatic coupling and wafer-bonding defects on delay testing of monolithic 3D integrated circuits. *ACM Journal on Emerging Technologies in Computing Systems*, 13(4), 2017.

[10] S. Panth, K. Samadi, Y. Du, and S. K. Lim. Placement-driven partitioning for congestion mitigation in monolithic 3D IC designs. *IEEE Transactions on Computer-Aided Design of Integrated Circuits and Systems*, 34(4):540–553, 2015.

[11] Y.-C. Lu et al. TP-GNN: A graph neural network framework for tier partitioning in monolithic 3D ICs. In *ACM/IEEE Design Automation Conference*, pages 1–6, 2020.

[12] Y. Xue, O. Poku, X. Li, and R. D. Blanton. PADRE: Physically-aware diagnostic resolution enhancement. In *IEEE International Test Conference*, pages 1–10, 2013.

An Efficient External Memory Test Solution: Case Study for HPC Application

Keqing Ouyang
Sanechips
ouyangkeqing@sanechips.com.cn

Minqiang Peng
Sanechips
peng.minqiang@sanechips.com.cn

Yunnong Zhu
Sanechips
zhu.yunnong@sanechips.com.cn

Kang Qi
Sanechips
qi.kang@sanechips.com.cn

Grigor Tshagharyan
Synopsys
grigor.tshagharyan@synopsys.com

Arun Kumar
Synopsys
arun.kumar@synopsys.com

Gurgen Harutyunyan
Synopsys
gharutyu@synopsys.com

Isaac Wang
Synopsys
zhuowang@synopsys.com

Abstract—**An increasing amount of data are being stored and processed by data centers every day for various commercial and industrial applications. For such large-scale systems dealing with enormous volumes of data, utilization of Dynamic Random-Access Memories (DRAM), integrated into horizontal or vertical stacks, has been proven to be the ideal solution in terms of area vs performance. At-speed test and repair in these stacks of DRAM memories requires complex solutions given the various lifecycle stages the memories go through. In this paper an efficient Built-In Self-Test (BIST) concept is proposed, focusing on both the in-system and production aspects of the test solution. It is demonstrated for a real-life High-Performance Computing (HPC) application.**

Keywords—External memory test, BIST, DRAM, HPC

I. INTRODUCTION

Over time, the share of memories has been continuously increasing in System-on-Chips (SoC), which was recorded by various research teams [1], [2]. According to these reports and estimates, memories typically occupy from 70 to 90 percent of the total area of an SoC. This is really a huge number, and it means that the SoC yield heavily depends on underlying memory yield.

Traditionally, static and dynamic random-access memories (SRAM and DRAM) are the two most common types of memories that form the basic memory infrastructure in SoCs. Over the years, a number of other memory architectures (Flash, Magnetic RAM, Resistive RAM, etc.) have been developed and studied aiming to decrease area, improve performance and reduce power consumption compared to SRAM and DRAM. IEEE International Roadmap for Devices and Systems (IRDS) 2022 edition provides a good overview of existing memory technologies, both those that are at the stage of prototyping as well as the emerging ones [3]. However, despite the forecasts, even today many SoC developers still prefer to stick to this tandem due to their known characteristics [4].

While SRAM is a little more expensive and does not come in diverse forms, it is extremely fast and proven on silicon, so it remains the number one choice for on-chip or embedded memory. On the contrary, DRAM is more economical due to its simple structure, high density, and low power consumption. It is the preferred choice for off-chip or, as we call it, external memory, such as Double Data Rate Synchronous DRAM (DDR SDRAM), Low Power DDR (LPDDR) and High Bandwidth Memory (HBM) families [5]-[8] are packaged in various forms such as Dual in-line memory modules (DIMM) or Fine pitch ball grid array (FBGA).

In recent years, HPC has evolved rapidly and in parallel, growing requirements for storage have been put forward. With the advancement of DDR and associated standards, the purpose of storage systems development is to have higher bandwidth, faster operating frequency, and higher storage density. In the meantime, HPC applications have very high demands on reliability. If there is a problem with the DRAM or its connectivity to the chip, a serious systematic error may occur. Although DRAM is being tested at the factory, it may fail during the chip packaging process or after some time of use due to occurrence of aging failures. Therefore, the granularity of DRAM and its connection to the chip should be properly tested in real time during power-on and mission operation, which is an important guarantee of reliable operation of the chip.

The commonly used test and repair methodology for memories in SoCs is the hardware-based BIST solution which helps to effectively address the complex testing problems specific to modern SoCs and provides the necessary coverage numbers. Recently a significant shift was registered in BIST utilization paradigm with the silicon lifecycle management coming to the scene. Whereas previously BIST was only used to achieve target manufacturing yield, the goal is now to monitor the health of the silicon and extend its lifetime. Therefore, having a reliable end-to-end test solution is a critical requirement.

II. TESTING EXTERNAL MEMORIES

The packaged DRAMs usually have several thousands of micro bumps and TSV connections for the memory data, address and command signals travelling from SoC die to the memory die stack via a silicon interposer. Memory vendors can support stacks of 2 to 8 dies with each die supporting up to 8 channels and up to memory density of 16Gb. This kind of system requires a 2-phase testing approach:

979-8-3503-4631-2/23 $31.00 © 2023 IEEE

- After packaging and assembly with the logic die, TSVs may be susceptible to manufacturing defects and damage of DRAM cells, therefore these defects must be detected and repaired.

- During product lifecycle, further degradation to interconnects and DRAM cells may be encountered due to thermal dissipation from base die to DRAM or other similar effects affecting overall package reliability. Incremental test and repair of the interconnects and memory cells needs to be performed during chip lifetime.

The proposed solution comes with the following hardware support (see Figure 1) and is implemented in Synopsys SMS EXTRAM product [9]:

- SMS EXTRAM Processor;

- Interface to the register space of the PHY for bring up, initialization, training as well as running manufacturing test for the PHY;

- Programmable BIST for generating at speed DFI traffic for testing the DRAM memories sequences, PHY initialization and Training sequences;

- A highly programmable Memory BIST scheme to perform comprehensive at-speed test and diagnosis on the external memory stack and interconnects over the DDR PHY DFI interface;

- Top level control with 1149.1 TAP Automatic Test Equipment (ATE) and Advanced Peripheral Bus (APB) (in-system) interfaces.

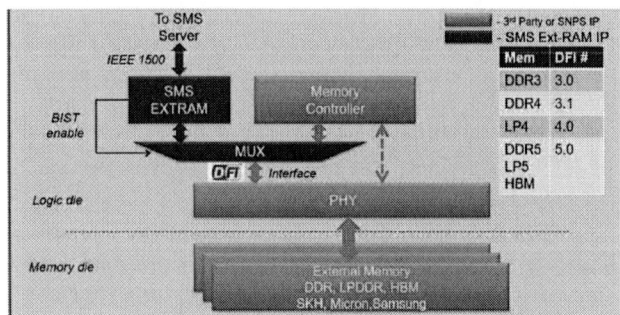

Fig. 1. Solution for testing external memories

III. USE CASE STUDY

One of the common applications where DRAM characteristics come handy is the so-called High-Performance Computing which is commonly used in data centers. Typical HPC system is basically composed of hundreds or even thousands of compute servers connected through a network, with the help of which large scale computations can be carried out to solve complex tasks and process huge quantities of data in real time. As a result, HPC applications require high volume of DRAM memories. The aim of the use case study is to explore the effectiveness of the considered solution implementation for the real-world project.

Since HPC chips have high reliability requirements for DRAM and its connection to the chip, the support for power-on and online tests need to be implemented in BIST. Given that the

working principles of DRAMs and SRAMs are different, as well as, that the sizes of external memory chips may vary with product specification, the BIST design need to take all of this into consideration.

In addition, it should be taken into account that in some application scenarios, it is possible to share the controller in the BIST logic for embedded and external memories, to reduce the logic as much as possible. While in other scenarios, independent control should be performed to avoid mutual impact on the development progress for DDR IP and other logic.

A. Chip-level system architecture

Figure 2 shows the chip-level architecture of the project. The details of the use case are as follows:

- Technology: advanced process DRAM;

- DDR type: DDR4 and DDR5;

- Number of DDR blocks: 4;

- Test architecture: hierarchical SHS Fabric consisting of Top-level Server, Sub-Servers and Processors;

- Test interface: 1149.1 compatible JTAG.

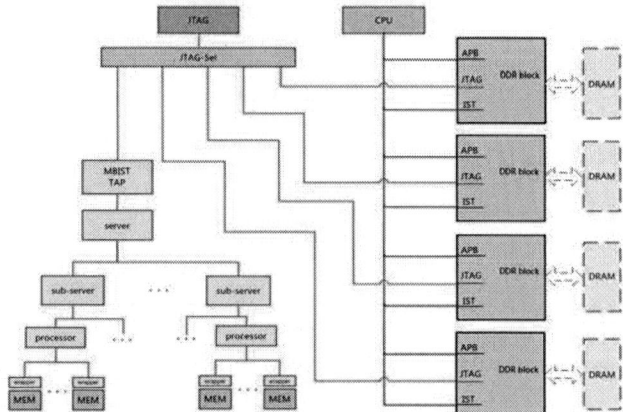

Fig. 2. Chip-level architecture of the design

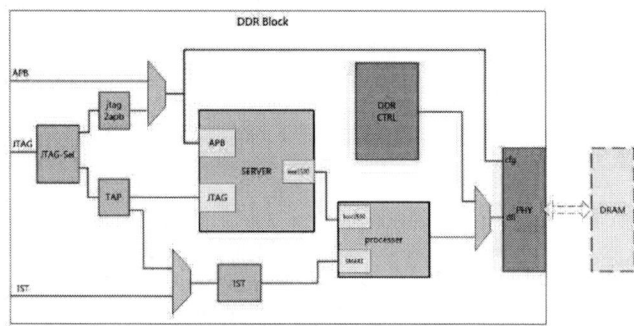

Fig. 3. DDR block-level architecture of the design

The following goals are pursued to be fulfilled with the help of the proposed solution:

- Check the interconnection between external DRAMs and HPC-chip in system;

- Test the external DRAM cells (on full address or in a range);

- Support repair feature;

- In-system test capability to periodically test for aging faults occurrence.

The implementation has the following capabilities:

- Supports both in-system test as well as ATE test. The in-system test uses the APB and IST (in-system test) interfaces, while the ATE test is controlled through the JTAG interface.

- The project has 4 DDR blocks with their PHYs. APB and IST interfaces used for in-system test are controlled by CPU, and the 4 DDRs can be tested either together or independently.

- For ATE test, the JTAG-SEL module can implement broadcast or one-hot control over the JTAG, and DDRs also can be tested together or independently.

B. DDR block-level system architecture

Figure 3 shows the implementation of the BIST architecture at the DDR block level. Two distinct test phases are differentiated in the scope of the use case, in-system test and ATE test, which are discussed next.

In-system test has the following test modes:

- *Training*: before the BIST run, DDR PHY need to be trained first. When in-system test mode, the path from APB to PHY is used to implement training operations on the PHY.

- *APB test*: after the training operation, the server APB interface is used to perform the BIST test. In this use case, a corresponding sequence can be programmed through server APB interface to choose a specific BIST algorithm and address range.

- *IST test:* after the training operation, alternative to APB, the IST interface can be used to perform the BIST test. In this use case, only the default BIST algorithm can be used, which is hardened in the circuit, either with full address or partial address.

For the ATE test, two scenarios need to be differentiated:

- If the chip and DRAM are packaged together, then complete BIST test can be run.

- Otherwise, if the chip and DRAM are not packaged together, a DRAM should be fixed in the load board, and only the functionality of BIST logic can be tested.

ATE test has the following test modes:

- *Training:* before the BIST run, DDR PHY need to be trained first. When in ATE test mode, the path from jtag2apb to PHY is used to implement training operations on the PHY.

- *JTAG test:* after the training operation, the server APB interface is used to perform the BIST test. In this use case,

a corresponding sequence can be programmed through server APB interface to choose a specific BIST algorithm and address range.

- *JTAG-APB test:* after the training operation, the server APB interface is used through jtag2apb converter to perform the BIST test. In this use case, a corresponding sequence can be programmed through server APB interface to choose a specific BIST algorithm and address range.

- *IST test:* after the training operation, alternative to APB, the IST interface can be used to perform the BIST test. In this use case, only the default BIST algorithm can be used, which is hardened in the circuit, either with full address or partial address.

In general, the design has the following characteristics:

- Total instance count: ~230K;

- Total register count: ~ 19K;

- Total cell area: ~16,800 mm^2;

- Max Frequency: ~ 5GHZ.

The verification of the use case consists of the steps listed below:

- *Connectivity check:* checks the connection between Server and Processor, as well as checks if the instructions are properly transmitted from server to processer.

- *DIMVIP test:* checks the read/write function between PHY and DIMM.

- *In-system APB test:* simulates and verifies different BIST algorithms for different DIMMs (DIMMX4 and DIMMX8) based on the APB channel, and checks whether the algorithm can detect the injected fault. The details are as discussed in the following subsection.

C. Test plan overview

Test cases are classified into four categories: full space test, eye diagram drawing, repair checking and binning test. The details are as follows:

- *Full space test:* after initializing DRAMs and training DDR PHY, the BIST algorithms over the full DDR address space will be performed to test external DRAMs and filter out the faulty one. If faults are detected during boot stage, post-package repair (PPR) can be applied to repair the faulty DRAMs.

- *EYE diagram drawing:* during the test, some of the parameters such as delays and reference voltage (Vref) between the DDR PHY and DRAMs can be modified to conduct BIST tests. After many tests including all possible parameters, memory arrays, which are passing can be print to draw eye diagram.

- *Repair checking:* when DRAM is confirmed to contain faults, the detailed location of the faults can be identified and based on that information BIST will analyze the available redundant rows and replace the faulty rows by

TABLE I. TEST PLAN ADOPTED FOR THE USE CASE

No	Simulation case	Functionality	DIMM Type
1	BIST run with March X1	Full space test	X4
			X8
2	BIST run with March X2	Checks the connectivity between DDR PHY and external DRAMs	X4
			X8
3	BIST run with March X3	Optimized version of March X1, proving the trade-off between test cost and fault coverage	X4
			X8
4	BIST run with March X4	DDR page wire/read test is applied at power-on self-test (POST)	X4
			X8
5	Inject faults and run BIST	Determines if the BIST detects the injected faults	X4
			X8

redundant ones if possible. The corresponding BIST pattern will be invoked to ensure the correct execution of the PPR solution.

- *Binning test:* different BIST algorithms can be applied to test DRAMs to sort out dies during the manufacturing phase. BIST algorithms come with different fault coverage and execution time trade-off and the selection of the algorithms depends on the proposed test plan.

Table I shows the test plan that was adopted for this specific use case.

IV. CONCLUSIONS

Testing large amounts of externally stacked DRAM memories is a big challenge, especially with the constraints of test time and power limitations. Conventional tests and repair schemes are not applicable to such scenarios and more sophisticated approaches are required for that purpose. The proposed BIST concept not only addresses the production test requirements for the post-packaging phase, but also provides the flexibility to reuse existing BIST circuitry for performing efficient in-system test. The solution is further elaborated for a real-life HPC application case in the experimental results section, including the adopted test plan and verification scheme.

REFERENCES

[1] K. Yi, SY. Cheng, YH. Park, F. Kurdahi, A. Eltawil, "An Alternative Organization of Defect Map for Defect-Resilient Embedded On-Chip Memories", Advances in Computer Systems Architecture, Lecture Notes in Computer Science, vol 4697, Springer, Berlin, Heidelberg, 2007.

[2] Semico Roadmap for Die Area Partitioning, Semico Research Corp., https://semico.com/content/semico-systems-chip-%E2%80%93-braver-new-world, 2017.

[3] "IEEE INTERNATIONAL ROADMAP FOR DEVICES AND SYSTEMS, 2022 Update, BEYOND CMOS AND EMERGING MATERIALS INTEGRATION", https://irds.ieee.org/editions/2022.

[4] "Memory Tradeoffs Intensify In AI, Automotive Applications, Semiconductor Engineering", https://semiengineering.com/memory-tradeoffs-intensify-in-ai-Automotive-applications, 2019.

[5] "DOUBLE DATA RATE (DDR) SDRAM STANDARD". JEDEC, February 2008, https://www.jedec.org/standards-documents/docs/jesd-79f.

[6] "JEDEC Standard: Low Power Double Data Rate 2 (LPDDR2)", JEDEC, February 2010, https://www.jedec.org/sites/default/files/docs/JESD209-2B.pdf.

[7] "High Bandwidth Memory (HBM) DRAM", JEDEC, October 2013, https://www.jedec.org/document_search?search_api_views_fulltext=jesd235.

[8] H. Jun, J. Cho, K. Lee, H.-Y. Son, K. Kim, H. Jin; K. Kim, "HBM (High Bandwidth Memory) DRAM Technology and Architecture", IEEE International Memory Workshop (IMW), 2017.

[9] G. Harutyunyan, Y. Zorian, "An Effective Embedded Test & Diagnosis Solution for External Memories", IEEE International On-Line Testing Symposium (IOLTS), 2015, pp. 168-170.

Allocating Physically Aware Embedded Memory Test & Repair Processor using Floorplan Info at the RTL Design Level

Vinay Kumar
XFG
Intel Corporation
San Jose, CA, USA
vinay2.kumar@intel.com

Bhrugurajsinh Chudasama
XFG
Intel Corporation
San Jose, CA, USA
bhrugurajsinh.chudasama@intel.com

Bin BW Wang
XFG
Intel Corporation
San Jose, CA, USA
bin.bw.wang@intel.com

Manish Arora
Synopsys
Burnaby, BC, Canada
manish.arora1@synopsys.com

Bharath Shankaranarayanan
Synopsys
Mountain View, CA, USA
bharath.shankaranarayanan@synopsys.com

Abstract—With the increasing demand for on-chip embedded memories in System-on-chip (SoC), the percentage of memory cells in the designs are going up. This raises two major requirements for the manufacturability of such SoCs, adequate testing of the memory cells to ensure acceptable DPPM levels and the need for test logic implementation to be optimized to meet physical implementation requirements in terms of routing, signal integrity, and power integrity. This paper proposes a method to allocate a physically aware Memory Built-in Self-Test (BIST) processor to memories under test using the Floorplan Design Exchange Format (DEF) at the register transfer level (RTL). The proposed method considers the physical location and connectivity of the memory instances in the chip design, enabling a more efficient allocation of the Memory BIST processor. Besides physical awareness for routing feasibility, clock domains, voltage islands, and switching activity aspects are also considered during Memory BIST processor assignment for a group of memories under test. The proposed method is demonstrated for various design scenarios. The results show that it achieves significant improvements in terms of timing closure, IR drop, test time, and area overhead compared to existing methods, and better alignment with the functional mode of operation.

Keywords — Memory BIST processor, physical design, timing, memory grouping

I. INTRODUCTION

With the increasing requirements for SoC compute, there is a significant increase in the usage of on-chip embedded memories, while maintaining low latency and power. This helps achieve high bandwidth data transfer and reduced footprint compared to external memories. These memories are provided as hard macros with pre-defined sizes and shapes from the memory IP (Intellectual Property) providers. This limits the scope of optimization of the memory macro unless it is being custom designed. To optimize the PPA (Performance, Power, and Area), designs are tightly packed, leaving limited routing resources around the memories to route additional signals required for testing the memories. Contemporary designs have sizable portion of the area occupied by memory cells (as high as 50% in networking SoCs), increasing the probability of defects in memory cells during manufacturing [1]. To test the memories for manufacturing defects, Memory BIST processor engine is required to perform on-chip testing for embedded memories. Memory BIST processor engine has built-in test algorithms, tailored to each memory type, technology node etc., to screen out the memories with manufacturing defects. Adding a Memory BIST processor to test memories requires adding multiplexing logic and routing of test signals to the memory macro. Routing additional nets of Memory BIST processor test IP can create routing congestion and create a challenge for timing closure on critical timing paths. There are additional challenges involved in physical implementation if there are multiple voltage domains. Simultaneous switching during testing of many memories in each region could lead to IR drop issues in silicon. To overcome these challenges various practices are followed across the industry [2, 3]. With the shift left practice to attempt prefetching issues early in the design phase, it is increasingly common to perform Memory BIST processor insertion early in the design considering the clock domain, power domain, and switching activities of design at the RTL level. However, adding the Memory BIST processor to the RTL level without floorplan knowledge of the design makes it a challenge for physical design implementation and creates timing critical paths later in the sign-off of the design.

To overcome the above challenge, we are proposing a method to "read floorplan information at the RTL phase of the design while performing Memory BIST processor insertion" in the following sections of this paper. The next section, section II corresponds to determining the challenges in this implementation. Section III talks about the existing approaches in adding the Memory BIST Processor and the last Section IV talks about the proposed method and the outro to this paper.

II. PHYSICAL DESIGN CHALLENGES ASSOCIATED WITH MEMORY BIST PROCESSOR IMPLEMENTATION

A. Routing congestion

As shrinking technology nodes enable manufacturing of smaller transistors, the capacity of integrating large numbers of transistors allows a complex SoC to deliver a high-performance throughput. With an increased demand for integration, interconnect nets inflate in the same proportion, leading to exceedingly high routing resource utilization, compared to cell utilization, in these complex and highly integrated designs. Additional nets from the Memory BIST processor IPs for memory testing make the routing even more challenging.

B. Clock domains

In the design, memories operate at a very high-speed clock in the functional mode. For mimicking functional mode

979-8-3503-4631-2/23 $31.00 © 2023 IEEE

of operation, Memory BIST processor needs to test the memories at functional clock speeds as well. However, the Memory BIST processor instruction programming occurs at a slower clock frequency. So, the Memory BIST processor IP needs the functional clock tree for each clock domain to be routed to the Memory BIST IP at-speed logic.

C. Voltage domains

Designs with low power requirements have various voltage domains to achieve low power by shutting them off when a particular voltage domain is in an idle state. So, Memory BIST processors would also need to be partitioned to be in the same voltage domains as the memories or must be properly equipped with the level shifter if they are present in different voltage domains. Signing-off designs with level shifters adds further complexity to physical implementation.

D. Critical Timing paths

Memory BIST processor tests the memories at speed. So, paths from the Memory BIST processor IP to memories could become critical timing paths if the placement of the Memory BIST processor IP logic is done without considering physical information.

III. CURRENT APPROACHES FOR MEMORY BIST PROCESSOR IMPLEMENTATION

There is general practice across the industry to insert Memory BIST processors either at the RTL design level or the netlist level.

A. Memory BIST Processor insertion at the RTL design level

Adding the Memory BIST processor at the RTL level brings numerous benefits. It simplifies Memory BIST processor verification by enabling simulations, Lint, CDC, and Low Power domains checks at the RTL stage. This makes the RTL sign-off robust by prefetching the bug finding early in the RTL design phase. But the Memory BIST processor without floorplan knowledge may result in routing congestion in the physical design phase. Hence the physical implementation feasibility and actual area overhead remain unrealistic until the physical design phase.

B. Memory BIST Processor insertion at the Netlist design level

Adding the Memory BIST processor at the netlist level immediately gives a realistic area overhead, routing congestion, isolation cells, and level shifter impact. However, it comes with the challenge to perform the sign-off checks at the netlist level, which pushes out the verification feedback availability to the later stage of the project and potentially jeopardizes the project timeline due to late bug findings.

Despite a few cons of adding the Memory BIST processor at the RTL design level, the pros significantly outweigh those, aligning with the shift left approach being adopted by the industry.

C. Other inputs to make a better decision on Memory BIST processor allocation at the RTL design level

There are a few other considerations required while planning on Memory BIST processor allocation using the Memory BIST processor planner [4], as mentioned below.

- Clock domains

- Voltage and Power domain

- Switching activity or Peak Power

- Max Number of memories under a Memory BIST Processor

- Max Number of memories under a Memory BIST Wrapper

- Best-suited test time savvy architecture

- Area overhead, etc.

IV. PROPOSED METHOD FOR MEMORY BIST PROCESSOR IMPLEMENTATION

The RTL-level Memory BIST processor insertion offers significant advantages as discussed earlier. However, the lack of physical floorplan information presents physical design implementation-related challenges, e.g., routing congestion, critical timing paths, etc. To address these challenges, in the proposed flow an updated DEF file is read at the RTL design level. The proposed flow aligns the hierarchies of memory macros present in the DEF file by swapping them with the RTL-level hierarchies of memory macros to enable accurate mapping. Further details on the flow are described in the following sections.

A. Signature Key generation for dictionaries

Since DEF file is generated from the Place and Route database, the hierarchies in the DEF file may not match with RTL hierarchies, since the netlist goes through the mapping, renaming, and grouping/ungrouping steps. So, if the special characters are removed from the hierarchy string, the remaining string is the same in both cases, as mentioned in Table I below.

TABLE I.   SIGNATURE KEY TABLE

Key Table		
Signature as a key	**RTL hierarchy**	**DEF hierarchy**
Topblk1subblk 1mem1	Top.blk[1].subBlk1.mem[1]	Top.blk_1.subBlk_ 1.mem_1
Topblk2subblk 1mem1	Top.blk[2].subBlk1.mem[1]	Top.blk_2_subBlk_ 1.mem_1
Topblk3subblk 1mem1	Top.blk[3].subBlk1.mem[1]	Top.blk_3.subBlk_ __1.mem_1

Once keys are generated, the memory macro hierarchies of the RTL level and the DEF file are stored in two dictionaries. The next section of this paper describes the format of the dictionary.

B. Dictionary format

The memory macros hierarchy information is stored in the dictionary. To store the information in a TCL-based [5] dictionary, a unique key is required to store the data value. In the design, each memory macro has a unique hierarchical instance path, making it a robust key candidate for dictionaries. As mentioned previously, the signature keys are generated by removing the special characters from the hierarchy path making it a unique key in the design, that can be used commonly across both RTL and Place and Route databases. A signature key holds the original design hierarchical path of the memory macro and memory macro name, e.g.,

979-8-3503-4631-2/23 $31.00 © 2023 IEEE

Dictionary = {signature as a key: {hierarchy of mem macro, mem macro name}}

C. DEF file Correction flow

After creating the dictionaries, the signature keys count check is performed to ensure none of the memory macros are missed in the flow (see Fig. 1). Upon successful completion of the signature keys check, the signature keys are used to iterate over the dictionaries and perform the memory hierarchy swap in the DEF file and write out the updated DEF file as an output result. If there is any mismatch for the key found, flow accumulates such cases and reports the list of unmapped keys to debug potential DEF file versioning issues.

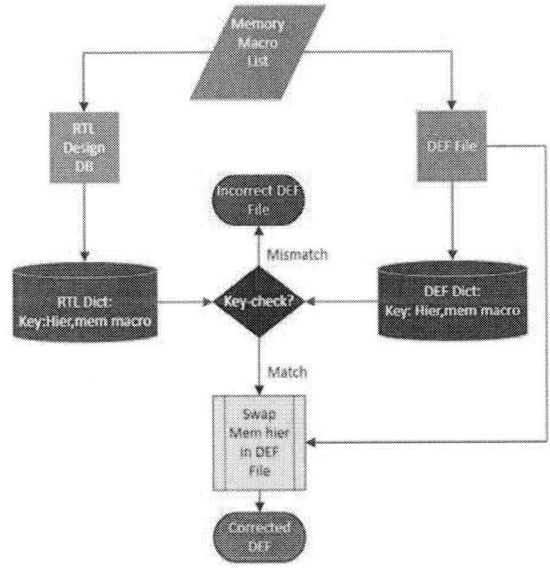

Fig. 1. DEF file correction flow

D. Memory BIST processor insertion

The Memory BIST processor with updated DEF file along with other input collaterals, e.g., UPF, SDC, and power budget, makes the Memory BIST processor insertion flow more robust and physical implementation friendly (see Fig. 2).

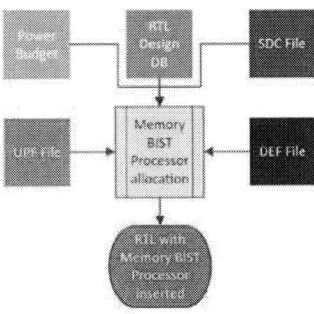

Fig. 2. Memory BIST processor insertion with DEF file

The Memory BIST processor planner tool [1] with the updated DEF file considers one of the following options while allocating Memory BIST processors at the RTL design level:

- Max. distance between the memory macros

- Number of proximity domains, i.e., Clusters

E. Results

Without the DEF file input to the Memory BIST processor planner tool [1], Memory BIST processor allocation for memories is logical, which would be non-optimal from a floorplan perspective since the tool does not have physical information of the memory macros, as mentioned in Fig. 3. One such application used as an example here and not just limited to, is the design seen in Fig. 3, here the use case is for programmable Ethernet Switch Products, deployed in datacenters to create ultrafast networks that are fully customized to individual needs, transforming even the largest data center into a focused, balanced, and optimized high-performance computer unto itself.

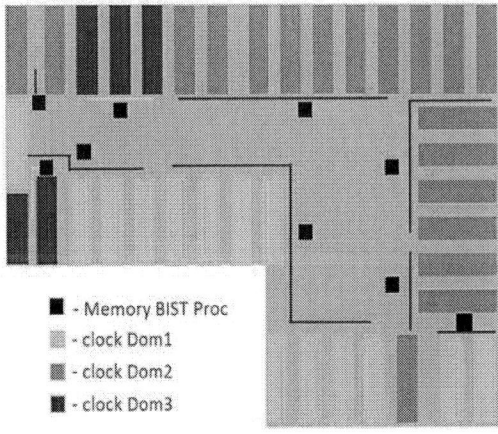

Fig. 3. Memory BIST processor allocation without DEF file

On the other hand, the Memory BIST processor planner with DEF file information gives well-optimized Memory BIST processor allocation, as shown in Fig. 4. The Memory BIST processors are allocated for memory macros based on their physical location along with other provided constraint collaterals.

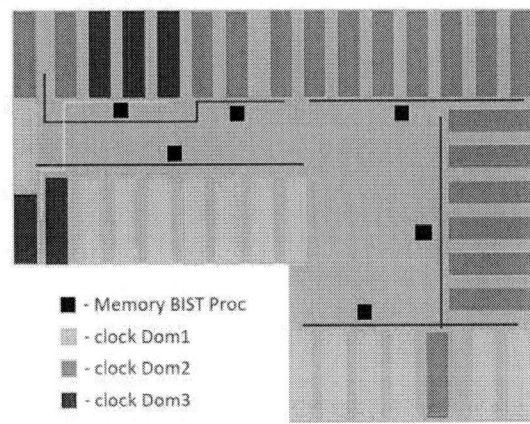

Fig. 4. Memory BIST processor allocation with DEF file

The improvement achieved in the physical implementation after allocating physically aware Memory BIST processor IPs is shown through a comparison of data between Table II and Table III below. Although the number of Memory BIST processor IPs has increased compared to the previous allocation without a DEF file, it is still beneficial from physical implementation feasibility since the Memory BIST processor's cell count is small. The latest designs are much more constrained in terms of routing resources compared to cell resources and this approach split a large, shared memory wrapper, into localized smaller shared memory wrappers, optimizing routing resources.

TABLE II. MEMORY BIST PROCESSOR ALLOCATION WITHOUT DEF FILE

Memory BIST processor table before the DEF file input		
Memory BIST processor	Memory group	Clock domain
SMS Proc1	SMS Wrap Grp1	Clock dom1
SMS Proc2	SMS Wrap Grp2	Clock dom1
SMS Proc3	SMS Wrap Grp1	Clock dom2
SMS Proc4	SMS Wrap Grp2	Clock dom2
SMS Proc5	SMS Wrap Grp3	Clock dom2
SMS Proc6	SMS Wrap Grp1	Clock dom3

TABLE III. MEMORY BIST PROCESSOR ALLOCATION WITH DEF FILE

Memory BIST processor table after the DEF file input		
Memory BIST processor	Memory group	Clock domain
SMS Proc1	SMS Wrap Grp1	Clock dom1
SMS Proc2	SMS Wrap Grp2	Clock dom1
SMS Proc3	**SMS Wrap Grp3**	**Clock dom1**
SMS Proc4	SMS Wrap Grp1	Clock dom2
SMS Proc5	SMS Wrap Grp2	Clock dom2
SMS Proc6	SMS Wrap Grp3	Clock dom2
SMS Proc7	**SMS Wrap Grp4**	**Clock dom2**
SMS Proc8	SMS Wrap Grp1	Clock dom3
SMS Proc9	**SMS Wrap Grp2**	**Clock dom3**

CONCLUSION

In conclusion, the proposed approach offers an efficient solution for the insertion of physically aware Memory BIST processors at the RTL design level using a DEF file, which can significantly improve the physical design implementation feasibility. This flow has been tested on various design blocks and shown improvements over the logical only approach.

ACKNOWLEDGMENT

We sincerely thank Dinesh Nadavi for reviewing this paper and our physical design team colleagues who supported us in this flow. We are also thankful to our EDA vendor Synopsys, Inc. who provided us with the Memory BIST processor planner [4] and support during the enablement of this flow.

REFERENCES

[1] A. A. Kokrady, C. P. Ravikumar, "Static Verification of Test Vectors for IR Drop Failure", International Conference on Computer Aided Design (ICCAD), 2003, pp. 760-764.

[2] A. Pavlov, M. Sachdev, "CMOS SRAM Circuit Design and Parametric Test in Nano-Scaled Technologies: Process-Aware SRAM Design and Test, 2008, Vol. 40, Springer Science & Business Media.

[3] U. Y. Ogras, R. Marculescu, P. Choudhary, D. Marculescu, "Voltage-Frequency Island Partitioning for GALS-based Networks-on-Chip", Design Automation Conference, 2007, pp. 110-115.

[4] SMS Planner tool, Synopsys Inc.
https://www.synopsys.com/implementation-and-signoff/test-automation/designware-sms.html

[5] TCL/TK Reference
https://www.tcl.tk/

Overcoming Embedded Memory Test & Repair Challenges in the Gate-All-Around Era

Artur Ghukasyan
Synopsys
Yerevan, Armenia
arturghu@synopsys.com

Grigor Tshagharyan
Synopsys
Yerevan, Armenia
grigort @synopsys.com

Gurgen Harutyunyan
Synopsys
Yerevan, Armenia
gharutyu@synopsys.com

Yervant Zorian
Synopsys
Mountain View, CA, USA
zorian@synopsys.com

Abstract—The paper discusses the challenges that memory testing faces in the era of Gate-All-Around transistor technology and proposes a solution concept to overcome them. With the transition from two-dimensional to three-dimensional transistors, and in particular with the advent of FinFET, many new factors have entered the scene, creating a need for new testing techniques capable of coping with the complexity of novel memory designs. However, the end of the FinFET era and the beginning of a new Gate-All-Around design paradigm means that these methods need to be revisited to ensure they remain effective and in line with the latest transistor technologies that are contenders to replace FinFET. The paper assesses the test challenges associated with the Gate-All-Around technology and proposes a solution concept that sets a vector for a thorough investigation of underlying defects, fault modeling, and the development of efficient test and repair algorithms.

Keywords—fault modeling, embedded memory, test and repair, gate-all-around, inductive fault analysis

I. INTRODUCTION

In recent years, the semiconductor industry has made impressive strides in memory testing since the introduction of Fin Field-Effect Transistor (FinFET) technology. Nevertheless, every innovative technology comes with new challenges that need to be addressed. With the rise of FinFET technology and the creation of smaller, faster, and more power-efficient transistors, the memory testing industry was forced to adapt and overcome new challenges such as increased variability, accurately measuring transistor parasitics and other critical parameters, test and repair efficiency, improved data accuracy, and the development of new testing techniques capable of handling the complexity of the latest memory designs.

FinFET technology is prone to a variety of faults, which can impair the performance of the transistor as a result of different defects such as resistive/hard opens and shorts, leakage issues, and others. In order to assure product quality and avoid yield loss, specialized test and repair methods have been developed to identify, diagnose, and fix these defects as a result of the growing use of FinFET-based memories [1-6]. Over time, FinFET-based memories have been shown to be more prone to dynamic faults and more resilient to process variation faults [1], also the impact of various test conditions was well studied [2].

In the post-FinFET era, several novel transistor designs have emerged, including Carbon Nanotube FET (CNTFET) [7], Nanowire FET (NWFET) [8] [9], Complementary FET (CFET) [10], Gate-All-Around FET (GAAFET) [11], Multi-Bridge-Channel FET (MBCFET) [12] and RibbonFET [13]. The primary goal of these approaches is to decrease the size of transistors while also enhancing their performance by utilizing various methods such as multiple channels, nanowires or nanosheets, and one-dimensional materials. The CNTFET and NWFET employ carbon nanotubes and nanowires, respectively, to create a uniform channel and decrease the impact of short-channel effects. CFET and GAAFET use multiple channels or layers to improve gate control over the channel, while MBCFET uses a combination of both. All of the mentioned technologies have the potential to improve chip area usage and power consumption while addressing the challenges of scalability in terms of gate size.

Gate-All-Around is a relatively new concept in the semiconductor industry that has received a lot of attention recently and is predicted to replace FinFET in the near future. The purpose of this paper is to understand the challenges embedded memory test and repair faces in the approaching GAA era and to propose solutions to overcome them, as well as to ensure that memory testing keeps up with the rapid changes in the semiconductor technology landscape.

The paper is organized into four main sections, with Section I as the introduction, Section II discussing various emerging transistor technology concepts, Section III covering the Advanced Inductive Fault Analysis (AIFA) methodology and the defect classification, Section IV delving into the proposed test and repair solution concept and finally Section V summarizing the key takeaways.

II. EMERGING TRANSISTOR TECHNOLOGIES AND THEIR PARAMETERS

Emerging transistor technologies have gained increasing attention due to their potential to overcome the limitations of traditional FinFET technology. In this context, CNTFETs are one of the most promising devices for surpassing incumbent CMOS technology [7]. The unique properties of carbon nanotubes, including their high carrier mobility, mechanical strength, and chemical stability, make them an attractive material for use in electronic devices.

Moreover, NWFETs are known to be the emerging transistor type for better performance and low power for future technology nodes [8], [9]. Due to the precise control of composition, morphology, and electrical properties, NWs can yield higher device performance than those obtained using top-down techniques. Their unique structures allow the transistors to be

979-8-3503-4631-2/23 $31.00 © 2023 IEEE

designed horizontally or vertically, leading to a smaller form factor [9]. In a recent study, the advantages that NWFETs provide to standard cell designs were investigated [9]. The study showed that V-NWFETs achieve significant capacitance reduction (-50.0%), and a significant reduction in area (-22.5%) and wire length (-14.4%) compared to FinFETs. However, careful design and proper interconnect structure are required to fully exploit the design advantages.

CFET technology, consisting of a stacked n-type vertical sheet on a p-type fin, is also being evaluated as an innovative alternative for FinFET technology [10]. Through a double-level access, it offers a structural scaling of both standard cells (SDC) and SRAM by 50%. The proposed process flow requires accurate control of the elevation dimension for manufacturability [10]. Based on TCAD analysis, the CFET can eventually outperform the FinFET device and meet the targets in power and performance.

Figure 1. FinFET (on the left) and GAAFET (on the right)

In addition, MBCFET technology has been proposed as an alternative to conventional CMOS technology due to its superior electrostatic integrity, better short channel control, and reduced self-heating effects [12]. MBCFET devices have shown superior electrostatic integrity and better short channel control compared to CMOS devices. Furthermore, MBCFET devices show much lower self-heating effects compared to conventional CMOS, enabling higher density circuit designs [12].

GAAFET is a novel post-FinFET technology paradigm that differs from FinFET technology in terms of structure, technical specifications, and scalability [11]. FinFETs feature a gate wrapped around a vertical fin-like structure, whereas GAAFETs have a horizontal gate wrapped around a narrow channel of semiconductor material, which carries the electrical current and is located in a recessed area. Fig. 1 shows the structural differences between the two technologies, with FinFET on the left and GAAFET on the right. More gate-to-channel surface area and increased gate control over the channel in GAAFETs result in higher performance metrics including speed and power consumption [14]. In addition, since GAAFETs can be produced with thinner channels and shorter gate lengths than FinFETs, they have a higher potential for scaling.

The investigation of GAAFET technology is crucial for tackling challenges associated with memory test and repair. A smaller cell size offered by GAAFET has the potential to produce more efficient and high-density memory arrays [15]. Since there are limits to how much FinFET technology can be scaled down, it becomes necessary to develop new transistor technologies that would help overcome that limitation [16]. Compared to FinFETs, GAAFETs have greater electrostatic control and process variation resistance, which can increase

memory chip yield and reliability [15], [16]. Given all of this, GAAFET technology can assist in overcoming most of the difficulties associated with FinFETs and can offer enhanced performance and scalability.

To sum up, emerging transistor technologies such as CNTFET, NWFET, CFET, GAAFET, and MBCFET have shown promise as innovative alternatives for FinFET technology. These technologies offer unique advantages such as improved device performance, reduced power consumption, and enhanced scalability. However, careful design and optimization are required to fully exploit their potential.

III. THE AIFA METHODOLOGY AND MEMORY DEFECT CLASSIFICATION

A. AIFA Methodology

Although the GAAFET technology has not yet entered mass production, early investigation of the technology and its inherent defects can help gain insight into its fault landscape and accelerate the technology maturation process. Due to the novel structure, GAAFET-based memories may come across new fault types specific only to GAAFET transistors. As a result, existing test and repair solutions, consisting of various types of test operations, addressing methods, background patterns, and other stressing conditions, may not be sufficient to detect these faults. Therefore, the investigation of GAAFET defect space requires a sophisticated flow for accurately modeling the faults and finding appropriate test sequences for their detection.

In the past, the AIFA flow, demonstrated in Fig. 3, has been successfully applied to FinFET-based memories [1]-[3] and now it is enhanced to be applicable to GAAFET technology as well. The proposed automated flow not only saves time but also ensures that GAAFET-based memories are thoroughly examined for possible unique faults. The flow is composed of the following steps:

- Defect injection is done in GDS or SPICE Netlist, and the Defect LIB is periodically updated with new technological structures such as GAAFETs.
- Two SPICE simulations are run: one with a defect-free design and the other with a defect-injected design, using a specific Simulation Setup that can contain different test sequences, test conditions, resistance magnitudes, and more.
- If the defect is not detected (PASS result), a waveform comparison between the defect-free and defect-injected simulations is made to provide new test sequences. This is done by the test engineer following specific rules.
- The iteration of Step 3 is repeated until a satisfactory test sequence is found, and the corresponding Fault Model is automatically extracted from the test sequence.
- In the end, an optimal test algorithm is synthesized using a set of test sequences as input rather than a set of faults, making it more generic and independent of fault type.

979-8-3503-4631-2/23 $31.00 © 2023 IEEE

- The length of the given test sequences in terms of the number of operations needed for detection is minimal, which leads to having more efficient test algorithms.

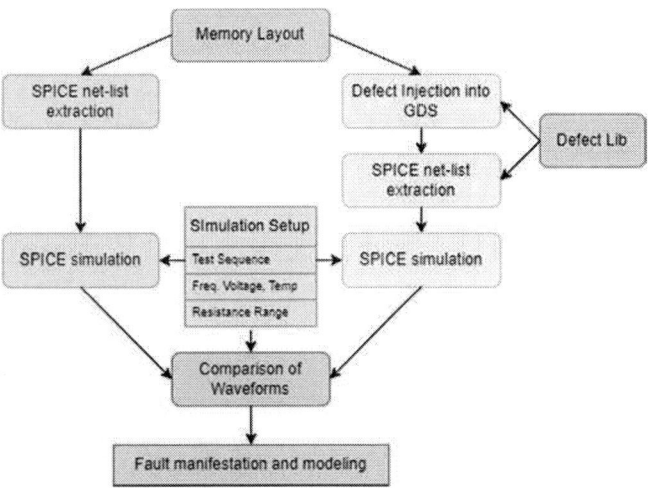

Figure 3. The AIFA Flow

B. Potential defect types

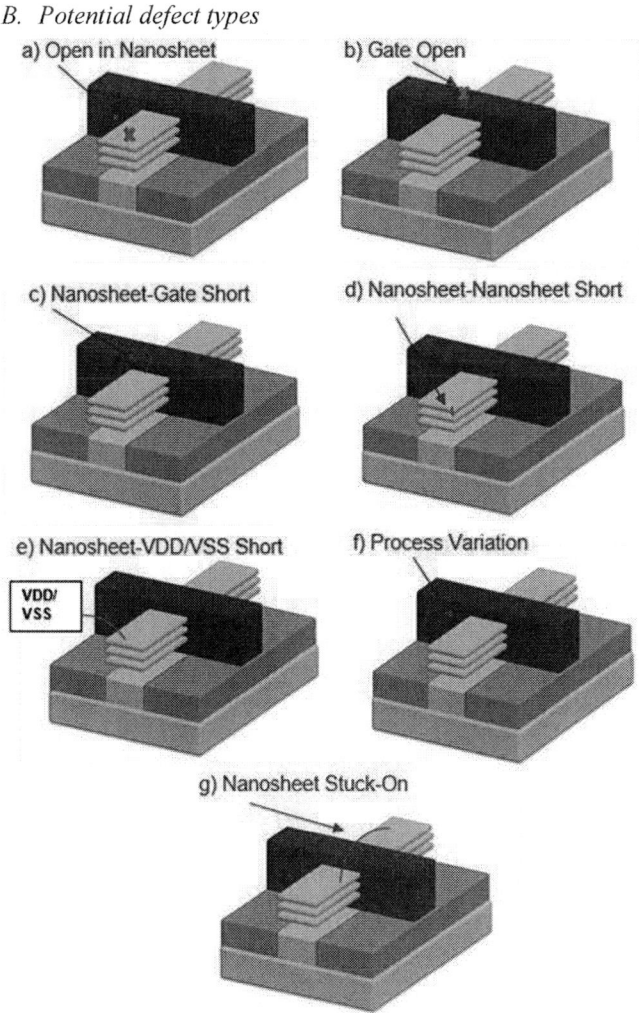

Figure 2. Defect models considered for GAAFET

Considering the unique structure introduced by the GAAFET technology, AIFA defect library was enriched with new defect models which are described in this section. Understanding these defect models is critical for developing reliable test strategies for GAAFET devices. Fig. 2 presents the considered defect types for MBCFET:

a) Nanosheet Open - Full and resistive open defect on the Nanosheet;
b) Gate Open - Full and resistive open defect on Gate;
c) Nanosheet-Gate Short - Full and resistive short defect between Nanosheet and Gate;
d) Nanosheet-Nanosheet short - Full and resistive short defect between the nanosheets;
e) Nanosheet-VDD/VSS Short - Full and resistive short defect between Nanosheet and VDD or Nanosheet and VSS;
f) Process Variation - Variations in Nanosheet parameter values.

IV. TEST & REPAIR SOLUTION CONCEPT FOR GAAFET

In order to overcome the challenges faced by memory testing in the GAAFET era, a new solution concept is proposed in this paper. GAAFET technology is still a technology in the making and has not yet been put into mass production. Therefore, the aim of this paper is to present the first results of the investigation on GAAFET-based memories and set the vector for its further development and research. The study done in the scope of this paper is based on the enhanced AIFA flow and the defect models introduced in the previous section. Using the AIFA flow, each of the discussed defects was successfully injected into the memory in the form of both hard and resistive defects, and the resulting faulty behavior was modeled in the simulation environment.

In order to come up with a robust and effective test and repair solution, it is important to consider various factors that could affect the testing process. One such factor is the simulation environment. The enhanced AIFA flow aims to optimize the simulation environment for the GAAFET technology, ensuring that it accurately reflects the technology specifics. In addition, the transistor structure and parameters are considerably different from those of previous technologies, which is also a critical factor that must be taken into account for the accurate injection of the defects. The enhanced AIFA flow helps to efficiently resolve these differences in technologies. Another important factor to consider is the impact of test conditions on the testing process. As observed for FinFETs, defects behave differently under different test conditions, especially voltage, temperature and frequency. With the enhanced AIFA flow, similar study was carried out for GAAFET to evaluate the behavior of defects and determine optimal test conditions. As a result, it was observed that the dependence on test conditions for GAAFET is even more severe compared to FinFET.

Discussed below are some of the types of faults that were observed as a result of the study. The observed faults can be divided into four classes depending on whether the behavior is static or dynamic and if it affects a single cell or is the result of coupling between two cells. As expected, the most frequently observed faults were traditional static single-cell faults, and in particular the transition and read destructive faults. These faults cover a large part of the resistance range for the considered defects. Static coupling faults are less common compared to single-cell faults but they were still observed in different resistance ranges for different defect types.

On the opposite side, dynamic behavior was observed in a smaller range of resistive defects, and more importantly, along with the single-cell faults, certain classes of two-cell faults were also identified, which are specific to GAAFET technology. In FinFET mainly single-cell dynamic faults were observed sensitized by n consecutive test operations, meanwhile in GAAFET, along with single-cell faults also dynamic read and write disturb faults were observed.

Based on the results of the study, a test and repair solution should be constructed taking into account the specificity of GAAFET technology, it shall not only accommodate all the observed fault models, but also the determined optimal test conditions.

V. CONCLUSION

The paper highlights the challenges the memory testing industry encounters in the post-FinFET era, brought up with the advent of novel transistor deigns empowered by Gate-All-Around paradigm. In order to address these challenges, the AIFA flow has been tailored to the needs of GAAFET to ensure that memory testing continuously shifts and is in line with the changing semiconductor technological environment. GAAFET technology is still immature and has a long way to go and the results of the preliminary study are set to pave a way for further studies aimed at exploring its test and repair aspects. Initial results already show the existence of new fault models that have not been observed in former technologies. Therefore, further studies will be required to augment the solution concept and ensure the continued robustness and effectiveness of GAAFET memory testing.

REFERENCES

[1] G. Harutyunyan, G. Tshagharyan, V. Vardanian and Y. Zorian, "Fault modeling and test algorithm creation strategy for FinFET-based memories," 2014 IEEE 32nd VLSI Test Symposium (VTS), Napa, CA, USA, 2014, pp. 1-6, doi: 10.1109/VTS.2014.6818747.

[2] G. Harutyunyan, G. Tshagharyan and Y. Zorian, "Impact of parameter variations on FinFET faults," 2015 IEEE 33rd VLSI Test Symposium (VTS), Napa, CA, USA, 2015, pp. 1-4, doi: 10.1109/VTS.2015.7116276.

[3] G. Harutyunyan, G. Tshagharyan and Y. Zorian, "Test and Repair Methodology for FinFET-Based Memories," in IEEE Transactions on Device and Materials Reliability, vol. 15, no. 1, pp. 3-9, March 2015, doi: 10.1109/TDMR.2015.2397032.

[4] W. Howell et al., "DPPM Reduction Methods and New Defect Oriented Test Methods Applied to Advanced FinFET Technologies," 2018 IEEE International Test Conference (ITC), Phoenix, AZ, USA, 2018, pp. 1-10, doi: 10.1109/TEST.2018.8624906.

[5] F. Hapke et al., "Defect-Oriented Test: Effectiveness in High Volume Manufacturing," in IEEE Transactions on Computer-Aided Design of Integrated Circuits and Systems, vol. 40, no. 3, pp. 584-597, March 2021, doi: 10.1109/TCAD.2020.3001259.

[6] G. C. Medeiros, M. Taouil, M. Fieback, L. B. Poehls and S. Hamdioui, "DFT Scheme for Hard-to-Detect Faults in FinFET SRAMs," 2019 IEEE European Test Symposium (ETS), Baden-Baden, Germany, 2019, pp. 1-2, doi: 10.1109/ETS.2019.8791517.

[7] M. Hartmann et al., "CNTFET Technology for RF Applications: Review and Future Perspective," in IEEE Journal of Microwaves, vol. 1, no. 1, pp. 275-287, Jan. 2021, doi: 10.1109/JMW.2020.3033781.

[8] W. Lu, P. Xie and C. M. Lieber, "Nanowire transistor performance limits and applications", IEEE Trans. Electron Devices, vol. 55, no. 11, pp. 2859-2876, Nov 2008.

[9] T. Song, "Opportunities and Challenges in Designing and Utilizing Vertical Nanowire FET (V-NWFET) Standard Cells for Beyond 5 nm," in IEEE Transactions on Nanotechnology, vol. 18, pp. 240-251, 2019, doi: 10.1109/TNANO.2019.2896362.

[10] J. Ryckaert et al., "The complementary FET (CFET) for CMOS scaling beyond N3", Proc. IEEE Symp. VLSI Technol., pp. 141-142, Jun. 2018.

[11] S. Bangsaruntip, G. M. Cohen, A. Majumdar, Y. Zhang, S. U. Engelmann, N. C. M. Fuller, et al., "High performance and highly uniform gate-all-around silicon nanowire MOSFETs with wire size dependent scaling", IEDM Tech. Dig., pp. 1-4, Dec. 2009.

[12] Sung-Young Lee et al., "Three-dimensional MBCFET as an ultimate transistor," in IEEE Electron Device Letters, vol. 25, no. 4, pp. 217-219, April 2004, doi: 10.1109/LED.2004.825199.

[13] "Process and Packaging: Intel 6 Pillars of Technology Innovation." Intel. https://www.intel.co.uk/content/www/uk/en/silicon-innovations/6-pillars/process.html.

[14] D. Yakimets et al., "Vertical GAAFETs for the Ultimate CMOS Scaling," in IEEE Transactions on Electron Devices, vol. 62, no. 5, pp. 1433-1439, May 2015, doi: 10.1109/TED.2015.2414924.

[15] S. Kim et al., "Investigation of Device Performance for Fin Angle Optimization in FinFET and Gate-All-Around FETs for 3 nm-Node and Beyond," in IEEE Transactions on Electron Devices, vol. 69, no. 4, pp. 2088-2093, April 2022, doi: 10.1109/TED.2022.3154683.

[16] Y.S. Wu and P. Su, "Sensitivity of Gate-All-Around Nanowire MOSFETs to Process Variations—A Comparison With Multigate MOSFETs," in IEEE Transactions on Electron Devices, vol. 55, no. 11, pp. 3042-3047, Nov. 2008, doi: 10.1109/TED.2008.2008012.

979-8-3503-4631-2/23 $31.00 © 2023 IEEE

Hybrid Binary Neural Networks: A Tutorial Review

Ahmet Enis Cetin
Dept. of Electrical and Engineering
University of Illinois Chicago
Chicago, Illinois
aecyy@uic.edu

Hongyi Pan
Dept. of Electrical and Engineering
University of Illinois Chicago
Chicago, Illinois
hpan21@uic.edu

Abstract—In this article, we review neural networks which have neurons with binary operations or networks that use binary transforms such as the Walsh-Hadamard transform (WHT). Neural networks with binary neurons or binary layers can be used in edge applications and/or applications requiring energy-efficient decision-making. WHT-based network is as accurate as the regular neural networks in the CIFAR-10 and the Tiny ImageNet image databases.

Index Terms—Binary neurons, Binary Neural Networks, Walsh-Hadamard Transform

I. Introduction

In this tutorial article, we review neural networks which have neurons with binary operations or networks that use binary transforms such as the Walsh-Hadamard transform that we developed in recent years [1]–[3].

In Section II, we describe an additive operator which can be used to develop multiplication-free neurons [4]–[6]. This additive operator uses the sign of multiplication to determine the correlation between the two real numbers similar to the neuro-fuzzy neuron introduced by Dogaru and Chua [7], [8]. The additive neural network constructed using the multiplication-free operator is described in Section III. It was applied to computer vision based wildfire detection problem [2].

In Section IV, we describe the binary Walsh-Hadamard Transform-based neural networks. Neurons are implemented in the Walsh-Hadamard transform (WHT) domain. We also introduced soft thresholding as a nonlinearity in the WHT domain. Walsh-Hadamard transform is reviewed in Section V. WHT-based neurons can be inserted into widely available pre-trained networks such as MobileNet family and ResNets as described in Sections IV, and V, respectively. Experimental results are presented in Section VII using CIFAR-10 and ImageNet image databases.

All of the above-mentioned neural networks are trained using the classical backpropagation algorithm.

II. Multiplication-Free Vector Product

Let α and β be two real-valued scalars. We define our multiplication-free operator as follows:

$$\alpha \oplus \beta = \text{sign}(\alpha \cdot \beta)(|\alpha| + |\beta|) \tag{1}$$

This work is funded by an award from the University of Illinois Chicago Discovery Partners Institute Seed Funding Program.

where, sign(.) is the signum function, defined as follows:

$$\text{sign}(x) = \begin{cases} 1, & x > 0 \\ -1, & x \leqslant 0 \end{cases} \tag{2}$$

The multiplication-free operator can also be represented as:

$$\alpha \oplus \beta = \text{sign}(\alpha) \cdot \beta + \alpha \cdot \text{sign}(\beta) \tag{3}$$

Note that bit-wise operation is very efficient in computation, and the highest bit of a variable is its sign bit (0 for positive and 1 for negative), we can achieve the $\alpha \cdot \text{sign}(\beta)$ with XOR operation, which is much more efficient compared to the multiplication.

Furthermore, the scalar definition in 1 and 3 can be extended to the case of real vectors in order to construct a dot product-like operation. In this regard, let $\mathbf{x}$ and $\mathbf{w} \in \mathbb{R}^d$. We define the multiplication-free "dot-product" as follows:

$$\mathbf{w} \oplus \mathbf{x} = \sum_{i=1}^{d} \text{sign}(w_i x_i)(|w_i| + |x_i|) \tag{4}$$

Similarly, the above relation can be also expressed as:

$$\mathbf{w} \oplus \mathbf{x} = \sum_{i=1}^{d} \text{sign}(w_i) x_i + w_i \text{sign}(x_i) \tag{5}$$

As the dot product induces the ℓ_2 norm, the mf-vector operation induces a scaled version of the ℓ_1 norm as:

$$\mathbf{x} \oplus \mathbf{x} = \sum_{i=1}^{d} \text{sign}(x_i x_i)(|x_i| + |x_i|) = 2||\mathbf{x}||_1 \tag{6}$$

For convenience, let the vector $\mathbf{x} \in \mathbb{R}^d$ and the matrix $\mathbf{W} \in \mathbb{R}^{d \times k}$. We then define the matrix-vector mf-operation as:

$$\mathbf{y} = \mathbf{W} \oplus \mathbf{x} = [\mathbf{w_1} \oplus \mathbf{x} \ \mathbf{w_2} \oplus \mathbf{x} \ \ldots \ \mathbf{w_k} \oplus \mathbf{x}]^T \tag{7}$$

where $\mathbf{w_i}$ is the i-th column of $\mathbf{W}$ for $i = 1, 2, \ldots, k$ and $\mathbf{y} \in \mathbb{R}^k$ is the resulting vector.

III. Additive Neural Network with MF-Operator (ADDNET)

A. Representation of Neurons

In regular neural networks, a dense feed-forwarding pass can be expressed as follows:

$$\mathbf{y} = \phi(\mathbf{W}^T \mathbf{x} + \mathbf{b}) \tag{8}$$

where $\mathbf{x} \in \mathbb{R}^d$ is the input vector, $\mathbf{W} \in \mathbb{R}^{d \times k}$ is the weights matrix, $\mathbf{b} \in \mathbb{R}^k$ is the bias vector and $\phi(.)$ is the element-wise nonlinear activation. In AddNet, the feed-forwarding pass of a dense layer can be expressed as follows:

$$\mathbf{y} = \phi(\mathbf{a} \odot (\mathbf{W} \oplus \mathbf{x} + \mathbf{b})) \qquad (9)$$

where $\odot$ is the element-wise product between the result $\mathbf{W} \oplus \mathbf{x} + \mathbf{b}$ and a vector $\mathbf{a} \in \mathbb{R}^k$. Note that we introduce the vector $\mathbf{a}$ as a scaling parameter in order to control the range of the pre-activations resulting from the mf-operation. It should be pointed out that calculating the element-wise product is inexpensive since we only carry out d multiplications compared to $k \times d$ multiplication operations in the case of $\mathbf{W}^T\mathbf{x}$.

We can construct that AddNet convolutional layers in a straightforward manner by substituting each convolution (dot product) operation with the mf-equivalent operation.

B. Training the AddNet

The standard back-propagation algorithm can be used for training the AddNet with the need of small approximations. The partial scalar derivatives of the pre-activation response with respect to are given as follows:

$$\frac{\partial(w \oplus x)}{\partial x} = \text{sign}(w) + 2w\delta(x) \qquad (10)$$

$$\frac{\partial(w \oplus x)}{\partial w} = 2x\delta(w) + \text{sign}(x) \qquad (11)$$

where $\delta(x)$ is the Dirac–delta function that directly results from the discontinuity of the signum function at $x = 0$. If we omit the delta function from the definitions of the partial derivatives, we end up with binary derivatives ($\text{sign}(w)$ and $\text{sign}(x)$). However, we found that approximating the Dirac–delta function provides better convergence since we end up with smoother derivatives. In this regard, we approximate the derivative of the signum function to be that of a steep hyperbolic tangent, as follows:

$$\frac{d\text{sign}(x)}{dx} \approx \frac{d\tanh(\alpha x)}{dx} = \alpha\big(1 - \tanh^2(\alpha x)\big) \qquad (12)$$

for a scalar $\alpha >> 1$. This is reasonable since $sign(x) = \lim_{\alpha \to \infty} \tanh(\alpha x)$. This way the terms associated with the delta function in the formula (10) and (11) will contribute to the partial derivatives when the arguments are close to zero.

C. Computational Efficiency

With an input $\mathbf{x} \in \mathbb{R}^d$ and a weight matrix $\mathbf{W} \in \mathbb{R}^{d \times k}$, one needs a total of $k \cdot d$ multiplication operations to realize the matrix-vector product, which is the case for layers in a regular neural network. On the other hand, AddNet substitutes the multiplication operations in $\mathbf{W}^T\mathbf{x}$ with addition and bit-wise operation, which are more efficient in terms of energy and computation. The realization of the calculation in the mf-operator is inexpensive because it only involves 1-bit operations and can be implemented as a logical XOR operation. So AddNet performs only one multiplication per neuron. The activation function is ReLU. Therefore $\mathbf{a}$ determines the slope of the ReLU. Thus, AddNet needs far fewer multiplication operations that does a regular neural network.

IV. WHT-BASED NEURAL NETWORKS

Recently, deep convolution neural networks (CNNs) have enjoyed great success in many important applications such as image classification [9]–[18], object detection [19]–[22] and semantic segmentation [23]–[27]. While more and more parameters are needed in CNNs, deploying them on real-time resource-constrained environments such as embedded devices becomes difficult due to insufficient memory and limited computational capacity. To overcome the aforementioned limitations, smaller and computationally efficient CNNs are necessary for many practical applications.

Although 1×1 convolutions reduce the computational load, they are still computationally expensive and time-consuming in regular deep neural networks. In our previous paper [3], we proposed a binary layer based on Fast Walsh-Hadamard transform (FWHT) to replace the 1×1 convolution layer. We revised MobileNet-V2 [28] with the FWHT layer, and the new network is remarkably more slimmed and computationally efficient compared to the original structure according to our experiments. However, we can take advantage of the fast $O(m \log_2 m)$ Walsh-Hadamard Transform (WHT) algorithm only when m is an integer power of 2. As a result, we pad zeros to the end of the input vector in the FWHT layer to make the size of the vector an integer power of 2. We notice that padding zeros to increase the size of the Walsh-Hadamard Transform (WHT) brings some redundant parameters. In [29], we introduce both one-dimensional (1D) and two-dimensional (2D) Walsh-Hadamard Transform layers. The new version of the 1D WHT computes the WHT in small blocks and avoids the zero-padding operations. As a result, it needs fewer parameters compared to our earlier paper [3]. Our 2D FWHT layer can replace the 3×3 convolution layers and Squeeze-and-Excite layers. It can also be employed to assist the dense layers of a given network.

Our contribution can be summarized as follows:
- We have a new 1D-FWHT layer. It improves the results of our previous work [3] by using Blocks of Walsh-Hadamard Transform (BWHT) instead of a single WHT computation to replace an entire 1×1 convolution layer. Compared to the old 1D-FWHT layer, the new 1D-BWHT layer retains the accuracy loss with a little more parameter reduction because it eliminates the zero-padding operation required to make the transform size a power of 2. For example, our version of the MobileNet-V2 network with the BWHT layers reaches a 0.08% higher accuracy with 11,919 fewer parameters compared to the MobileNet-V2 network with the FWHT layer on the CIFAR-10 dataset.
- We introduce the 2D-WHT layer in this work. It can be used to replace the "Squeeze-and-Excite" layers and the regular 2D convolutional layers. For example, we reduce the number of trainable parameters by 48.62% with only a 0.76% accuracy loss in our MobileNet-V3-Large structure with 2D-WHT layers on the CIFAR-100 dataset. Our version of ResNet-34 with 2D-WHT layers has 53.57% fewer parameters than the regular ResNet-34 with only a 0.72% accuracy loss on the

Tiny ImageNet dataset.

• In addition, we introduce a weighted 2D-FWHT layer that can be easily inserted before the global average pooling (GAP) layer (or the flatten layer) to assist the dense layers. This novel layer improves the accuracy of the network with a slight (almost negligible) increase in parameters due to the additional weights. For example, our additional weighted 2D-FWHT layer improves the accuracy of the ResNet-20 network by 0.5% with only 256 additional parameters in the CIFAR-10 dataset. This 2D-FWHT layer improves the accuracy of the MobileNet-V3-Large network by 0.15% in the CIFAR-100 dataset. In this case, only 704 more parameters are required. Our novel binary layers do not increase the processing time even in conventional processors. For example, our 2D-FWHT layer runs about 24 times as fast as the regular 3×3 convolution layer on NVIDIA Jetson Nano.

A. Related Work

Efficient neural network models include compressing a large neural network using quantization [30], [31], hashing [32], pruning [33], vector quantization [34] and Huffman encoding [35]. Another approach is the SqueezeNet [36], which is designed as a small network with 1×1 convolutional filters. Yet another approach is to use binary weights in neural networks [37]–[50]. Since the weights of the neurons are binary, they can can be used to slim and accelerate networks in specialized hardware including compute-in-memory systems [51]. In 2020, Maneesh *et al.* proposed a trimmed version of MobileNet architecture called Reduced Mobilenet-V2 [52]. They replace bottleneck layers with heterogeneous kernel-based convolutions (HetConv) blocks. HetConv was first proposed by Pravendra *et al.* in 2019 [53]. Although HetConv reduced the parameters effectively, it is still based on the convolution, and we experimentally showed that our approach can outperform the network described in [52]. In 2021, James *et al.* proposed a novel layer based on the Fast Fourier Transform (FFT) called FNet for natural language processing applications. They insert the FFT in hidden layers and they can train the weights in the frequency domain. They showed that their FNet has a faster speed with the guaranteed accuracy on BERT counterparts on the GLUE benchmark. However, the main weakness of the FNet is that their proposed method is only based on the real part of the FFT. After applying the FFT, they only keep the real part to avoid complex arithmetic. The imaginary part of the FFT is simply ignored. Therefore, the information in the imaginary part is lost. Since the WHT is a real transform we do not suffer from information loss as a result of WHT computation.

Other methods using WHT include [50], [54] but they did not perform any "convolutional" and non-linear filtering in the Hadamard transform domain. The "convolutional" filtering in the transform domain is possible by introducing multiplicative weights and non-linear filtering is possible with the use of two-sided smooth-thresholding which "denoises" small valued transform domain coefficients [3]. The use of one-sided ReLU will lead to a significant information loss because transform domain coefficients can take both positive and negative values even if the input to the transform consists of all positive numbers. Our novel smooth thresholding nonlinearity is a two-sided version of the ReLU and it can retain both positive and negative large amplitude coefficients while eliminating small amplitude ones. After performing the filtering operations in the transform domain we compute the inverse transform and continue processing the data in the feature map domain, which is a unique feature of our work.

V. Review of the Walsh Hadamard Transform

The Walsh-Hadamard Transform (WHT) is an example of a generalized class of the Fourier transforms. The transform matrix consists of +1 and -1 only. The convolution in the time (feature) domain leads to multiplication in the frequency domain in Fourier Transform. The Hadamard transform can be considered as a simplified version of the Fourier and wavelet transforms [55]. Therefore, we can approximately implement 1×1 and 3×3 convolutions in the WHT domain. We can not only train multiplicative weights but also train the threshold values of the soft and smooth-thresholds using the backpropagation algorithm [3]. We experimentally observed that this approach approximates 1×1 and 3×3 convolution operations very effectively while significantly reducing the number of network parameters. In addition, we used trainable scaling weights in 2D Hadamard transform layers (Section V-D) and they positively contributed to the accuracy.

Let $\mathbf{X}, \mathbf{Y} \in \mathbb{R}^m$ be the vectors in the "time" and transform domains, respectively, where $m = 2^k, k \in \mathbb{N}$. The WHT vector $\mathbf{Y}$ is obtained from $\mathbf{X}$ via a matrix multiplication as follows:

$$\mathbf{Y} = \mathbf{W}_k \mathbf{X} \tag{13}$$

where $\mathbf{W}_k$ is called the 2^k-by-2^k Walsh matrix, which can be generated using the following steps [56]:
• Construct the Hadamard matrix $\mathbf{H}_k$:

$$\mathbf{H}_k = \begin{cases} 1, & k = 0, \\ \begin{bmatrix} \mathbf{H}_{k-1} & \mathbf{H}_{k-1} \\ \mathbf{H}_{k-1} & -\mathbf{H}_{k-1} \end{bmatrix}, & k > 0, \end{cases} \tag{14}$$

• Shuffle the rows of $\mathbf{H}_k$ to obtain $\mathbf{W}_k$ via the bit-reversal permutation and the Gray-code permutation on row index.

For $k = 2$ and $m = 4$, the Hadamard matrix $\mathbf{H}_2$ and the corresponding Walsh matrix $\mathbf{W}_2$ are given by

$$\mathbf{H}_2 = \begin{bmatrix} 1 & 1 & 1 & 1 \\ 1 & -1 & 1 & -1 \\ 1 & 1 & -1 & -1 \\ 1 & -1 & -1 & 1 \end{bmatrix}, \quad \mathbf{W}_2 = \begin{bmatrix} 1 & 1 & 1 & 1 \\ 1 & 1 & -1 & -1 \\ 1 & -1 & -1 & 1 \\ 1 & -1 & 1 & -1 \end{bmatrix}. \tag{15}$$

which can be implemented using a wavelet filterbank with binary filters $h[n] = \{1, 1\}$ and $g[n] = \{-1, 1\}$ in two stages [55]. This process is the basis of the fast algorithm and the $\mathbf{W}_2$ matrix can be expressed as follows

$$\mathbf{W}_2 = \begin{bmatrix} 1 & 1 & 0 & 0 \\ 1 & -1 & 0 & 0 \\ 0 & 0 & 1 & -1 \\ 0 & 0 & 1 & 1 \end{bmatrix} \times \begin{bmatrix} 1 & 1 & 0 & 0 \\ 0 & 0 & 1 & 1 \\ 1 & -1 & 0 & 0 \\ 0 & 0 & 1 & -1 \end{bmatrix} \tag{16}$$

Similar to the Fast Fourier Transform (FFT) algorithm, the complexity of the Fast Walsh-Hadamard transform (FWHT) algorithm is also $O(m \log_2 m)$ when m is an integer power of 2 as it is completely based on the butterfly operations described in Eq. (1) in [57]. Because the Walsh matrix only contains ± 1, the FWHT can be implemented using only addition and subtraction operations via butterflies. It was shown that the WH transform is the same as the block Haar wavelet transform in [55]. As for the inverse WHT (IWHT), we have

$$\mathbf{X} = \frac{1}{m} \mathbf{W}_k \mathbf{Y}, \qquad (17)$$

which implies that IWHT is WHT with normalization by m.

A. Methodology

In this section, we will first review our previous work on Fast Walsh-Hadamard Transform (FWHT) layer [3], then we describe the Block Walsh-Hadamard Transform (BWHT) layer. Next, we will introduce a novel 2D WHT layer that can be easily inserted into any deep network before the Global Average Pooling (GAP) layer and/or the flatten layer. We will also describe how it can replace the 3×3 convolution layers and the Squeeze-and-Excite layers to reduce the number of parameters significantly. Finally, we describe how we introduce multiplicative weights in the 2D WHT domain. Addition of weights in the 2D WHT transform domain causes a minor increase in the number of parameters but returns a higher accuracy than the regular 2D WHT layer.

B. Fast Walsh-Hadamard Transform Layer

The 1×1 convolution layer provides amazing benefit in modern deep convolution layers [13], [14], [28], [58], [59]. It is widely used to change the dimensions of channels. For example, in ResNet [13], a 1×1 convolution layer is applied to the residual blocks when the input and output sizes are different. The 1×1 convolution layer named "conv_expand" is applied to increase the number of channels in each bottleneck layer of MobileNet-V2 [28]. Then, a depthwise convolution layer is employed as the main component for feature extraction. After this step, another 1×1 convolution layer named "conv_projection" reduces the number of channels. However, there are as many parameters as the number of input channels in each 1×1 convolution layer. Computing these operations is very time-consuming during the inference. Since the main duty of the 1×1 convolution layer is just to adjust the number of channels we proposed a novel FWHT layer to replace the 1×1 convolution layer in [3]. The FWHT layer is summarized in Algorithms 1 and 2. In brief, an FWHT layer consists of an FWHT operation to change the input tensor to the WHT domain, a smooth-thresholding operation as the nonlinearity function in the WHT domain, and another inverse FWHT to change the tensor back to the feature-map domain. Each FWHT is applied along the channel axis, which implies that completing FWHT on a tensor $\mathbf{X} \in \mathbb{R}^{n \times w \times h \times m}$ means performing $n \times w \times h$ m-length FWHTs in parallel. We apply "denoising" in the WH transform domain to eliminate small-amplitude coefficients. This is a nonlinear filtering operation

in the transform domain. Inspired by soft-thresholding [60], [61] which is defined as

$$y = \mathrm{S}_T(x) = \mathrm{sign}(x)(|x| - T)_+ = \begin{cases} x + T, & x < -T \\ 0, & |x| \le T \\ x - T, & x > T \end{cases} \tag{18}$$

we have proposed the variant called smooth-thresholding [3]

$$y = \mathrm{S}'_T(x) = \tanh(x)(|x| - T)_+, \tag{19}$$

where T is the trainable thresholding parameter.

Due to its definition given in Eq. (20), the denoising parameter T in soft-thresholding can only be updated by either $+1$ or -1. On the other hand, the derivative in Eq. (21) is the derivative in Eq. (20) multiplied by $\tanh(x)$. As a result, the convergence of the smooth-thresholding operator is smooth and steady in the back-propagation algorithm.

$$\frac{\partial(\mathrm{sign}(x)(|x| - T)_+)}{\partial T} = \begin{cases} 1, & x < -T \\ 0, & |x| \le T \\ -1, & x > T \end{cases} \tag{20}$$

$$\frac{\partial(\tanh(x)(|x| - T)_+)}{\partial T} = \begin{cases} -\tanh(x), & |x| > T \\ 0, & |x| \le T \end{cases} \tag{21}$$

We learn a different threshold T value for each WHT domain coefficient. We do not use the ReLU function in the transform domain because the WHT domain coefficients can take both positive and negative values, and large positive and negative transform domain coefficients are equally important. Our MobileNet-V2 experiments in [3] have verified this observation, and both soft-thresholding and smooth-thresholding improve the recognition accuracy compared to the ReLU.

The FWHT layers are summarized in Algorithms 1 and 2. Because the Direct Current (DC) channel $\mathbf{Y}[:,:,:,0]$ usually contains essential information about the mean value of the input feature map, we do not apply any thresholding on it. If we perform smooth-thresholding on tensor $\mathbf{Y}$, each slice along the channel axis will share a common threshold value. Therefore, there are total $(2^d - 1)$ trainable thresholding parameters in the 2^d-channel FWHT layer.

To expand the number of channels, we first compute the 2^d point WHTs and perform smooth-thresholding in the transform domain. We pad $(2^d - c)$ zeros to the end of each input vector before the 2^d-by-2^d WHT to increase the dimension. After smooth-thresholding in the WH domain, we calculate the inverse WHT.

To project channels by a factor of $r = \frac{2^p}{2^q}$, we first compute the 2^p point WHTs and perform smooth-thresholding in the transform domain as described in Algorithm 2. After this step, we compute the 2^q point WH transforms to reduce the dimension of the feature map. We divide the DC channel values by r to keep the energy at the same level as other channels after pooling. In Step 5 of Algorithm 2, we average pool the transform domain coefficients to reduce the dimension of the WHT and discard the last $(r - 1)$ transform domain

Algorithm 1 The FWHT layer for channel expansion [3]

Input: Input tensor $\mathbf{X} \in \mathbb{R}^{n \times w \times h \times c}$
Output: Output tensor $\mathbf{Z} \in \mathbb{R}^{n \times w \times h \times tc}$
1: Find minimum $d \in \mathbb{N}$, s.t. $2^d \geq tc$
2: $\hat{\mathbf{X}} = \text{pad}(\mathbf{X}, 2^d - c) \in \mathbb{R}^{n \times w \times h \times 2^d}$
3: $\mathbf{Y} = \text{FWHT}(\hat{\mathbf{X}}) \in \mathbb{R}^{n \times w \times h \times 2^d}$
4: $\hat{\mathbf{Y}} = \text{concat}(\mathbf{Y}[..., 0], \text{ST}(\mathbf{Y}[..., 1:]))$
5: $\hat{\mathbf{Z}} = \text{FWHT}(\hat{\mathbf{Y}}) \in \mathbb{R}^{n \times w \times h \times 2^d}$
6: $\mathbf{Z} = \hat{\mathbf{Z}}[..., :tc]$
7: **return** $\mathbf{Z}$.

Comments: Function pad$(\mathbf{A}, b)$ pads b zeros on the channel axis of tensor $\mathbf{A}$. FWHT$(\cdot)$ is the normalized fast Walsh-Hadamard transform on the last axis. Function concat$(\cdot, \cdot)$ concatenates two tensors along the last axis. ST$(\cdot)$ performs smooth-thresholding. The index follows Python's rule.

Algorithm 2 The FWHT layer for channel projection [3]

Input: Input tensor $\mathbf{X} \in \mathbb{R}^{n \times w \times h \times tc}$
Output: Output tensor $\mathbf{Z} \in \mathbb{R}^{n \times w \times h \times c}$
1: Find minimum $p, q \in \mathbb{N}$, s.t. $2^p \geq tc, 2^q \geq c$
2: $r = 2^{p-q}$
3: $\hat{\mathbf{X}} = \text{pad}(\mathbf{X}, 2^p - tc) \in \mathbb{R}^{n \times w \times h \times 2^p}$
4: $\mathbf{Y} = \text{FWHT}(\hat{\mathbf{X}}) \in \mathbb{R}^{n \times w \times h \times 2^p}$
5: $\hat{\mathbf{Y}} = \text{concat}(\mathbf{Y}[..., 0]/r, \text{avgpool}(\text{ST}(\mathbf{Y}[..., 1 : 2^p - r + 1]), r))$
6: $\hat{\mathbf{Z}} = \text{FWHT}(\hat{\mathbf{Y}}) \in \mathbb{R}^{n \times w \times h \times 2^q}$
7: $\mathbf{Z} = \hat{\mathbf{Z}}[..., :c]$
8: **return** $\mathbf{Z}$.

Comments: Function avgpool$(\mathbf{A}, b)$ is the average pooling on $\mathbf{A}$ with pooling size and strides are b.

---

coefficients of $\mathbf{Y}$ to make the dimension equal to 2^q. The last $(r - 1)$ coefficients are high-frequency coefficients, and usually, their amplitudes are negligible compared to other WHT coefficients.

Therefore, the dimension change operation from m dimensions to n dimensions can be summarized as follows:

$$\mathbf{Z} = \begin{cases} \frac{1}{2^q} \mathbf{U} \mathbf{W}_q \mathbf{S}'_\mathbf{T} \mathbf{W}_q \mathbf{P} \mathbf{X}, & m \leq n \\ \frac{1}{\sqrt{2^{p+q}}} \mathbf{U} \mathbf{W}_q \mathbf{A}_{\mathbf{vg}} \mathbf{S}'_\mathbf{T} \mathbf{W}_p \mathbf{P} \mathbf{X}, & m > n \end{cases} \quad (22)$$

where p is the minimum integer such that $2^p \geq m$, q is the minimum integer such that $2^q \geq n$, $\mathbf{P}$ describes the zero-padding operation to make $\mathbf{X}$ multipliable by $\mathbf{W}_p$ or $\mathbf{W}_q$. $\mathbf{S}'_\mathbf{T}$ represents the smooth-thresholding layer with DC channel excluded. $\mathbf{U}$ is unpadding function to make the dimension the same as $\mathbf{Z}$, and $\mathbf{Avg}$ is average pooling on the channel axis without the DC channel.

In consequence, if d is the minimum integer such that 2^d is no less than the number of input channels, the trainable number of parameters in FWHT layers is no more than the $(2^d - 1)$. The trainable parameters are only the threshold values in the smooth-thresholding operator. Hence, it is clear that the FWHT layer requires significantly fewer parameters than the

Algorithm 3 The BWHT Layer for Channel Expansion [29]

Input: Input tensor $\mathbf{X} \in \mathbb{R}^{n \times w \times h \times c}$, Hadamard size s
Output: Output tensor $\mathbf{Z} \in \mathbb{R}^{n \times w \times h \times tc}$
1: Resample $\mathbf{X}$ to get $\hat{\mathbf{X}} \in \mathbb{R}^{n \times w \times h \times \frac{tc}{s} \times s}$
2: $\mathbf{Y} = \text{FWHT}(\hat{\mathbf{X}}) \in \mathbb{R}^{n \times w \times h \times \frac{tc}{s} \times s}$
3: $\hat{\mathbf{Y}} = \text{concat}(\mathbf{Y}[..., 0], \text{ST}(\mathbf{Y}[..., 1:]))$
4: $\hat{\mathbf{Z}} = \text{FWHT}(\hat{\mathbf{Y}}) \in \mathbb{R}^{n \times w \times h \times \frac{tc}{s} \times s}$
5: Reshape $\hat{\mathbf{Z}}$ to get $\mathbf{Z} \in \mathbb{R}^{n \times w \times h \times tc}$
6: **return** $\mathbf{Z}$.

---

regular 1×1 convolution layer, which requires a different set of filter coefficients for each 1×1 convolution.

In consequence, if d is the minimum integer such that 2^d is no less than the number of input channels, the trainable number of parameters in FWHT layers is no more than the $(2^d - 1)$. The trainable parameters are only the threshold values in the smooth-thresholding operator. Hence, it is clear that the FWHT layer requires significantly fewer parameters than the regular 1×1 convolution layer, which requires a different set of filter coefficients for each 1×1 convolution.

C. Block Walsh-Hadamard Transform layer

Although the FWHT layer is very efficient and can save a huge number of parameters, it requires zero padding before computing the FWHT when the size of the vector is not an integer power of 2. For example, if the input tensor has 384 channels, we have to pad 128 zeros to make it 512. These zeros contain no information but increase the number of trainable parameters significantly. Inspired by the block-division strategy of the Discrete Cosine Transform (DCT)-based JPEG image compression [62] method, we propose a layer in which we divide the input to small blocks and compute the Walsh-Hadamard transforms of small blocks of data. In this way, we do not need to pad a large number of zeros to the end of the input vector. For example, we divide the feature map into blocks of size 32 and compute 12 WHTs in an input tensor which has 384 channels. If necessary, we pad zeros to the end of the last block.

The BWHT layer for channel expansion is described in Algorithm 3. When we want to increase the number of channels by t using the WHT of size s, we overlap data blocks for a c-channel tensor. We change c channels to $\frac{tc}{s}$ blocks and each block has s channels. This is described in Algorithm 4. To achieve overlapping, we first create a $\lfloor \frac{tc}{s} \rfloor$-length arithmetic sequence $K = [k_0, k_1, ..., k_{\frac{tc}{s}-1}]$ from 0 to $(c - s)$, then we use channels from index $\lfloor k_i \rfloor$ to index $(\lfloor k_i \rfloor + s - 1)$ to build the i-th block. After overlapping, we take the FWHT of each block along the last axis, perform smooth-thresholding with DC channel excluded, and compute the inverse FWHT. Finally, we reshape the result to a tensor with tc channels.

The BWHT layer for channel projection is described in Algorithm 5. We first divide the tensor into $\frac{tc}{s}$ blocks. Then, in each block, we perform an FWHT, smooth-thresholding, and compute the inverse FWHT. Finally, we reshape the output tensor to tc channels and take an average pooling to c channels.

Algorithm 4 Resampling

Input: Input tensor $\mathbf{X} \in \mathbb{R}^{n \times w \times h \times c}$, Hadamard size s
Output: Output tensor $\hat{\mathbf{X}} \in \mathbb{R}^{n \times w \times h \times \frac{tc}{s} \times s}$
1: $K = \lfloor \text{linspace}(0, c - s, \lfloor \frac{tc}{s} \rfloor) \rfloor$
2: **for** i in range($\frac{tc}{s} - 1$) **do**
3:    $\hat{\mathbf{X}}[..., i, :] = \mathbf{X}[..., K[i] : K[i] + s]$
4: **end for**
5: **return** $\hat{\mathbf{X}}$.

   Comments: $\lfloor \cdot \rfloor$ denotes the floor function. Function linspace(a, b, c) creates a c-length arithmetic sequence from a to b.

Algorithm 5 The BWHT Layer for Channel Projection [29]

Input: Input tensor $\mathbf{X} \in \mathbb{R}^{n \times w \times h \times tc}$, Hadamard size s
Output: Output tensor $\mathbf{Z} \in \mathbb{R}^{n \times w \times h \times c}$
1: Reshape $\mathbf{X}$ to get $\hat{\mathbf{X}} \in \mathbb{R}^{n \times w \times h \times \frac{tc}{s} \times s}$
2: $\mathbf{Y} = \text{FWHT}(\hat{\mathbf{X}}) \in \mathbb{R}^{n \times w \times h \times \frac{tc}{s} \times s}$
3: $\hat{\mathbf{Y}} = \text{concat}(\mathbf{Y}[..., 0], \text{ST}(\mathbf{Y}[..., 1 :]))$
4: $\hat{\mathbf{Z}} = \text{FWHT}(\hat{\mathbf{Y}}) \in \mathbb{R}^{n \times w \times h \times \frac{tc}{s} \times s}$
5: Reshape $\hat{\mathbf{Z}}$ to get $\tilde{\mathbf{Z}} \in \mathbb{R}^{n \times w \times h \times tc}$
6: $\mathbf{Z} = \text{avgpool}(\tilde{\mathbf{Z}}, t) \in \mathbb{R}^{n \times w \times h \times c}$
7: **return** $\mathbf{Z}$.

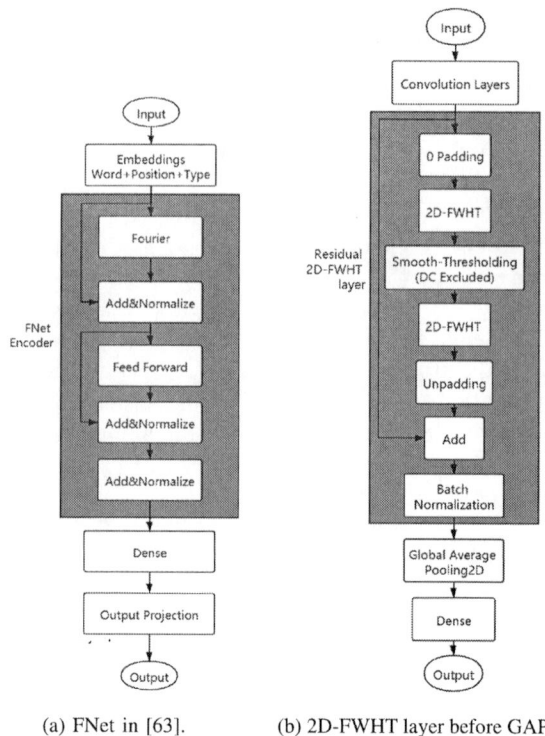

(a) FNet in [63].   (b) 2D-FWHT layer before GAP.

Fig. 1: FNet and CNN with a 2D-FWHT layer before GAP.

Unlike the FWHT for channel projection, we take the average pooling in the feature map domain instead of doing it in the WHT domain because the transform size is much smaller in this case, and we want to keep as much information as possible.

The dimension change operation from m dimensions to n dimensions can be summarized as follows:

$$\mathbf{Z} = \begin{cases} \frac{1}{2^s} \mathbf{R} \mathbf{W}_q \mathbf{S}'_{\mathbf{T}} \mathbf{W}_q \mathbf{G} \mathbf{X}, & m \leq n \\ \frac{1}{\sqrt{2^s}} \mathbf{A}_{\mathbf{vg}} \mathbf{R} \mathbf{W}_q \mathbf{S}'_{\mathbf{T}} \mathbf{W}_p \mathbf{G} \mathbf{X}, & m > n \end{cases} \quad (23)$$

where $\mathbf{G}$ describes resampling or reshaping operations to divide $\mathbf{X}$ into blocks, $\mathbf{S}'_{\mathbf{T}}$ is the smooth-thresholding layer with DC channel excluded, $\mathbf{R}$ represents the overlapping function to make the dimension the same as $\mathbf{Z}$, and $\mathbf{Avg}$ represents the average pooling on the channel axis.

Consequently, compared to the FWHT layer with 2^d channels, we can reduce the trainable parameters from $(2^d - 1)$ to s by using the BWHT layer with a block size of s. For example, if there are 384 channels, a single FWHT layer contains 511 parameters while a BWHT layer only contains 31 parameters for a block size of $s = 32$.

We replace the two 1×1 convolution layers in each bottleneck layer of MobileNet-V2 with BWHT layers.

subsection2D Walsh-Hadamard Transform layer In this section, we will introduce the two-dimensional (2D) Walsh-Hadamard Transform layers. In convolution neural networks, the global average pooling (GAP) and the flatten layers are universally employed before the dense layers to reduce the dimension of the output tensor. James Lee-Thorp *et al.* proposed a Fourier Transform-based layer before the dense layer in their natural language processing network called FNet [63].

Inspired by their work, we design a 2D WHT layer that can be inserted before the GAP layer or the flatten layer as shown in Fig. 1b and Eq. 24.

$$\mathbf{y} = \mathbf{W} \mathbf{S}'_{\mathbf{T}} \mathbf{W} \mathbf{x}. \quad (24)$$

where, $\mathbf{W}$ denotes 2D-FWHT operation and $\mathbf{S}'_{\mathbf{T}}$ represents the 2D smooth thresholding.

In most convolution neural networks, the input tensor of the GAP and the flatten layer is in $\mathbb{R}^{N \times w \times h \times c}$, where N is the batch size. Usually, $w = h << c$. On the other hand, the output tensor of the GAP layer is in $\mathbb{R}^{N \times c}$ and the output tensor of the flatten layer is in $\mathbb{R}^{N \times whc}$. Therefore, the input of the GAP layer contains more information than the output of the GAP layer, and flatten layer reduces the 3D-spatial structure of the tensor. Due to these reasons, we insert the 2D-FWHT layer before the GAP layer or the flatten layer, and the 2D-FWHT layer aims to reinforce analysis along the width and height axes.

As it is described in Algorithm 6, we first pad zeros along the width and height axes to make the sizes of the transform powers of 2. Usually, w and h are very small integers if the convolution part is well-designed. Hence, the WHT size is not very large in this case (2, 4, or 8 is usually sufficient). Therefore, we do not apply block FWHT at this stage because we do not need to pad too many zeros. Then, we apply two 1D-FWHTs along the width and height axes separately to convert the tensor into the WH domain. Next, we perform 2D-smooth-thresholding. There are 2^{p+q} trainable thresholding parameters, while the one at the DC channel is redundant but

Algorithm 6 The 2D-FWHT Layer

Input: Input tensor $\mathbf{X} \in \mathbb{R}^{n \times w \times h \times c}$, Hadamard size s
Output: Output tensor $\mathbf{Z} \in \mathbb{R}^{n \times w \times h \times c}$
 1: find minimum $p, q \in \mathbb{N}$, s.t. $2^p \geq w, 2^q \geq h$
 2: $\hat{\mathbf{X}} = \text{pad2}(\mathbf{X}, 2^p - w, 2^q - h) \in \mathbb{R}^{n \times 2^p \times 2^q \times c}$
 3: $\mathbf{Y} = \text{FWHT2}(\hat{\mathbf{X}}) \in \mathbb{R}^{n \times 2^p \times 2^q \times c}$
 4: $\hat{\mathbf{Y}} = \text{ST2}(\mathbf{Y}) \in \mathbb{R}^{n \times 2^p \times 2^q \times c}$
 5: $\hat{\mathbf{Y}}[:, 0, 0, :] = \mathbf{Y}[:, 0, 0, :]$
 6: $\hat{\mathbf{Z}} = \text{FWHT2}(\hat{\mathbf{Y}}) \in \mathbb{R}^{n \times 2^p \times 2^q \times c}$
 7: **if** Residual is applied **then**
 8: $\quad \mathbf{Z} = \hat{\mathbf{Z}}[:, :w, :h, :] + \mathbf{X}$
 9: **else**
10: $\quad \mathbf{Z} = \hat{\mathbf{Z}}[:, :w, :h, :]$
11: **end if**

$\quad$ Comments: Function $\text{pad2}(\mathbf{A}, b, c)$ pads b zeros along the width axis and c zeros along the height axis of tensor $\mathbf{A}$. FWHT2$(\cdot)$ is the normalized 2D-fast Walsh-Hadamard transform on the width and height axes (applying two 1D-FWHTs separately for efficiency). ST2$(\cdot)$ performs 2D-smooth-thresholding.

---

needed to perform Python's broadcasting. Each slice along the width and height axes will share a common threshold value. Afterward, we reset the DC value to its original value and apply two 1D-FWHTs along the width and height axes separately to convert the tensor back to the feature domain. Finally, we un-pad the zeros from the result with the input tensor to get the output tensor. If it is with residual design, we will add the output with the input tensor to get the final output tensor.

We also use the 2D-FWHT layers to replace the regular 2D-convolution layers. In this case, it does not change the number of channels due to the fact that it takes no transformation along the channel axis. Therefore, if we want to change the number of channels, we can apply a 1D-BWHT first then apply a 1D-BWHT layer to change the number of channels, then apply a 2D-FWHT layer. If the weight and height of the input tensor are not very small, we can also apply the block division as the BWHT layer to avoid padding too many zeros.

D. 2D Weighted Walsh-Hadamard Transform layer

Convolution in "time" domain can be implemented using multiplicative weights in the Fourier domain. However, Fourier transform requires complex arithmetic. In this paper, we replaced the Fourier Trannsform with the binary WHT. The weighted 2D WHT layer uses multiplicative weights in the transform domain. In the weighted-WHT layer we perform both regular "filtering" and nonlinear "denoising" in the transform domain.

To improve the accuracy, we use the weighted smooth-thresholding function, in which a trainable weight is applied:

$$y = \mathrm{S}'_{WT}(x) = \tanh(x)(|vx| - T)_+ \tag{25}$$

with constraint $v \geq 0$, where v is a trainable multiplicative weight representing a WHT domain multiplication. The corresponding partial derivatives are

$$\frac{\partial(\tanh(x)(|vx| - T)_+)}{\partial v} = \begin{cases} \tanh(x)|x|, & |vx| > T \\ 0, & |vx| \leq T \end{cases} \tag{26}$$

$$\frac{\partial(\tanh(x)(|vx| - T)_+)}{\partial T} = \begin{cases} -\tanh(x), & |vx| > T \\ 0, & |vx| \leq T \end{cases} \tag{27}$$

When $|vx| > T$, Eq. (26) holds because when $v \geq 0$, $\tanh(x)x\text{sign}(vx) = \tanh(x)x\text{sign}(x) = \tanh(x)|x|$. We don't multiply v and x in $\tanh(\cdot)$ in Eq. (25) to avoid quadratic term of v in the derivatives. In this way, Eq. (24) can be rewritten as:

$$\mathbf{y} = \mathbf{W}\mathbf{S}'_{\mathbf{WT}}\mathbf{W}\mathbf{x}. \tag{28}$$

where, $\mathbf{W}$ denotes 2D-FWHT and $\mathbf{S}'_{\mathbf{WT}}$ denotes 2D weighted smooth thresholding. In the 2D weighted smooth-thresholding layer, we initialize all weights $v = 1$. We initialize the threshold value $T = 0$ at the DC channel. The threshold values at other AC channels are initialized as positive values. In weighted-WHT layer, the number of trainable parameters is doubled but still significantly smaller than in a regular convolution layer. For example, if the input and the output are in $\mathbb{R}^{3 \times 3 \times 1280}$ and the WHT size is 4, there are only $4 \times 4 + 4 \times 4 = 32$ trainable parameters in weighted 2D-FWHT layer, but 3×3 2D convolution layer requires $3 \times 3 \times 1280 = 11520$ trainable parameters.

VI. EXPERIMENTAL RESULTS

Our training and testing experiments are carried on an HP-Z820 workstation with 2 Intel Xeon E5-2695 v2 CPUs, 2 NVIDIA RTX A4000 GPUs, and 128GB RAM. Speed tests are also carried out using a NVIDIA Jetson Nano board. The code is written in TensorFlow-Keras in Python 3.

First, we will compare the accuracy of MobileNet-V2 based models on the Fashion MNIST dataset and CIFAR-10 dataset. Then, we will further investigate 2D-FWHT in MobileNet-V3 and ResNet. Finally, we will perform a speed test on our 2D-FWHT layer and the regular 3×3 convolution layer.

A. BWHT in MobileNet-V2

In MobileNet-V2 experiments, we use ImageNet-pretrained MobileNet-V2 model with 1.0 depth. To build the fine-tuned MobileNet-V2, we replace layers after the Global Average Pooling (GAP) layer with a dropout layer and a dense layer as the TensorFlow official transfer learning demo. Since the minimum input of MobileNet-V2 ImageNet-pretrained models is $96 \times 96 \times 3$, we resize the images to this resolution in our experiments. Because images in Fashion MNIST dataset are in gray-scale, after interpolating, we copy the values two times to convert them to $96 \times 96 \times 3$ format.

Results of MobileNet-V2 on Fashion MNIST and CIFAR-10 are shown in Tables I and II respectively. The number of parameters is counted by TensorFlow API "model.summary()", and the non-trainable parameters are from the batch normalization layers. We first investigate the 2D-FWHT layer before the GAP layer. We name the models as "2D-FWHT before GAP". We add one residual 2D-FWHT layer followed by a batch normalization layer. They bring an extra 2576 more trainable parameters (only 16 from the 2D-FWHT layer) and only 2560 non-trainable parameters but improve the accuracy by 0.27% and 0.47% on Fashion MNIST and CIFAR-10, respectively. The number of trainable parameters increases only by 0.12%.

TABLE I: MobileNet-V2 Fashion MNIST Result

Model	Trainable Parameters	Accuracy
Fine-tuned MobileNet-V2 (baseline)	2.237M	95.37%
2D-FWHT before GAP	2.239M	95.64%
$\frac{1}{2}$ 1×1 conv. changed (FWHT) [3]	0.289M(87.08%↓)	94.50%
$\frac{1}{2}$ 1×1 conv. changed (BWHT)	0.273M(87.79%↓)	94.45%
$\frac{1}{2}$ 1×1 **conv. changed (BWHT)** **+2D-FWHT before GAP**	**0.276M(87.67%↓)**	**94.75%**

TABLE II: MobileNet-V2 CIFAR-10 Result

Model	Trainable Parameters	Accuracy
Baseline model[a] in [52]	2.238M	94.3%
RMNv2[b] [52]	1.069M(52.22%↓)	92.4%
Fine-tuned MobileNet-V2 (baseline)	2.237M	95.21%
Fourier Layer [63] before GAP	2.239M	95.38%
2D-FWHT before GAP	2.239M	95.68%
$\frac{1}{3}$ 1×1 conv. changed (FWHT) [3]	0.506M(77.37%↓)	93.14%
$\frac{1}{3}$ 1×1 conv. changed (BWHT)	0.494M(77.91%↓)	93.22%
$\frac{1}{3}$ 1×1 **conv. changed (BWHT)** **+2D-FWHT before GAP**	**0.497M(77.79%↓)**	**93.46%**

We then beat the FWHT results in [3] which do not apply block division. As Fig. 2 shows, we first revise the 1×1 convolution layers. The models are named as "1×1 conv. changed". The fraction at the beginning of the model's name denotes how many bottleneck layers are changed. For instance, $\frac{1}{2}$ means we change the last half bottleneck layers. Since the final convolution layer is also 1×1, we replace it with a BWHT layer for channel expansion.

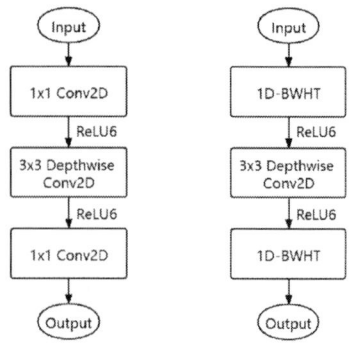

Fig. 2: MobileNet-V2 bottleneck (left) and our revision (right).

We also compared our results with Reduced Mobilenet-V2 (RMNv2) [52] in CIFAR-10 experiment. They slim MobileNet-V2 by replacing bottleneck layers with a novel block called HetConv blocks. Their trimming method is still based on the convolution. When we change $\frac{1}{3}$ convolution by BWHT layer and add an extra 2D-FWHT layer before the GAP layer, we reduce more parameters than (77.79% VS 52.22%) [52] with a smaller accuracy loss (1.75% VS 1.9%).

Moreover, we compare our results with the FNet Fourier layer [63] in CIFAR-10 experiment. The FNet Fourier layer can be computed as

$$\mathbf{Y} = \mathbf{X} + \Re(\mathcal{F}_w(\mathcal{F}_h(\mathbf{X}))) \tag{29}$$

where the input tensor is $\mathbf{X}$ and the output tensor is $\mathbf{Y}$. $\mathcal{F}_w(\cdot)$ and $\mathcal{F}_h(\cdot)$ are the Fast Fourier Transform (FFT) along the weight and the height. $\Re(\cdot)$ is the real part of the tensor. We try adding one FNet layer before the GAP layer to compare the result of adding one 2D-FWHT layer before the GAP layer, and our accuracy is 0.30% higher than the accuracy of the FNet. This is because the FNet can only keep the real part of the tensor. In another words the information in the imaginary part is simply discarded. On the other hand, the Walsh-Hadamard Transform is a binary transform. Therefore, there is no information loss due to the Walsh-Hadamard Transform.

According to the above experiments, we have the following observations:
• In general, the weighted 2D-FWHT before the GAP layer improves the accuracy of an image recognition network. For example, it improves the accuracy of the fine-tuned MoblieNet-V2 by 0.27% in Fashion MNIST and 0.47% in CIFAR-10. Furthermore, it increases the accuracy of the model "$\frac{1}{2}$ 1×1 conv. changed BWHT" by 0.4% on Fashion MNIST and accuracy of model "$\frac{1}{3}$ 1×1 conv. changed BWHT" on CIFAR-10 by 0.24%, respectively.
• The 1D-BWHT layer provides better results than the single block 1D-FWHT layer in terms of both the accuracy and the number of parameters when the number of channels is not the power of 2. Although on Fashion MNIST, the 1D-FWHT model reaches higher accuracy, the BWHT models reach higher accuracy on CIFAR-10. CIFAR-10 is a more difficult dataset so it is more convincing than the Fashion MNIST. Plus, in each case, the BWHT models have fewer parameters than the 1D FWHT models because the BWHT layer avoids padding 0s and it uses a smaller WHT size.

B. 2D-FWHT in MobileNet-V3

In this section, we use MobileNet-V3-Large [64] on the CIFAR-100 dataset and Tiny ImageNet to show the advantage of the 2D-FWHT layer. Compared to the MobileNet-V2 bottleneck block, the MobileNet-V3 bottleneck block contains one Squeeze-and-Excite layer after the depthwise convolution layer. In this section, we replace each Squeeze-and-Excite layer in the final $\frac{1}{3}$ bottleneck blocks with the 2D-FWHT layer. The base model is MobileNet-V3-large with the weights in the early layers being pretrained on the ImageNet dataset. Due to the input size of public ImageNet-pre-trained MobileNet-V3 models being all 224×224 but CIFAR image size is 32x32, we upsample the images to 224×224.

As Table III shows, we reduce 48.55% trainable parameters with only a 0.33% accuracy loss by changing one-third of

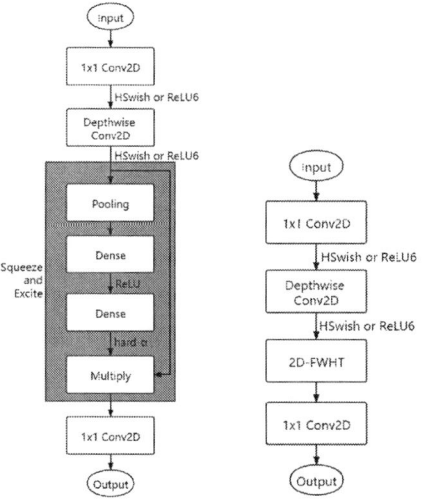

Fig. 3: MobileNet-V3 bottleneck (left) and our revision (right).

TABLE III: MobileNet-V3-Large CIFAR-100 Result

Method	Scaling Weights	Trainable Parameters	Accuracy
Fine-tuned model (Baseline)	-	3.068M	78.25%
Fourier layer [63] before GAP	-	3.070M	80.01%
2D-FWHT before GAP	N	3.070M	80.22%
2D-FWHT before GAP	Y	3.070M	80.73%
$\frac{1}{3}$ S&E changed (Fourier [63])	-	1.575M	77.20%
$\frac{1}{3}$ S&E changed (2D-FWHT)	N	1.576M(48.64%↓)	77.34%
$\frac{1}{3}$ S&E changed (2D-FWHT)	Y	1.576M(48.62%↓)	77.49%
$\frac{1}{3}$ S&E changed+2D-FWHT before GAP	**Y**	**1.578M(48.55%↓)**	**77.92%**

S&E is the Squeeze-and-Excite layer.

TABLE IV: MobileNet-V3-Large CIFAR-10 Result

Method	Scaling Weights	Trainable Parameters	Accuracy
Fine-tuned model (Baseline)	-	2.982M	95.52%
$\frac{1}{3}$ S&E changed+2D-FWHT before GAP	**Y**	**1.492M(49.96%↓)**	**94.32%**

S&E is the Squeeze-and-Excite layer.

TABLE V: MobileNet-V3-Large Tiny ImageNet Result

Method	Scaling Weights	Trainable Parameters	Accuracy
Fine-tuned model (Baseline)	-	3.164M	64.37%
$\frac{1}{3}$ S&E changed+2D-FWHT before GAP	**Y**	**1.675M(47.08%↓)**	**62.78%**

S&E is the Squeeze-and-Excite layer.

($\frac{1}{3}$) Squeeze-and-Excite layers and applying one 2D-FWHT layer before the GAP layer. Compared to the cases without applying weights in the 2D smooth-thresholding, all cases with weights get improved accuracy. Specifically, when we insert a 2D-FWHT layer in the 2D-smooth-threshold before the GAP layer, the network with weights in the 2D smooth-thresholding reaches 0.51% accuracy higher than the network without weights in the 2D smooth-thresholding. When we change $\frac{1}{3}$ Squeeze-and-Excite layers, the network with weights in the 2D smooth-thresholding reaches 0.15% accuracy higher than the network without weights in the 2D smooth-thresholding. In each pair, the parameters amount is almost the same respectively. Therefore, weighted smooth-thresholding is superior to non-weighted smooth-thresholding in accuracy.

We also compare the WHT layer with the FNet Fourier layer [63]. When we insert a layer before the GAP layer, the model with our 2D-FWHT reaches a 0.62% accuracy higher than the model with the FNet Fourier layer. In addition, when we change $\frac{1}{3}$ Squeeze-and-Excite layers, the model with 2D-FWHT reaches a 0.29% accuracy higher than the model with the FNet Fourier layer.

Furthermore, we also try our best model on CIFAR-10 and Tiny ImageNet. We replace each Squeeze-and-Excite layer in the final $\frac{1}{3}$ bottleneck blocks by the 2D-FWHT layer and insert one 2D-FWHT layer before the GAP layer. As Table IV shows, our model has 49.96% less trainable parameters than the fine-tuned MobileNet-V3-Large model and the accuracy is only 1.20% lower on the CIFAR-10. As Table V shows, our model has 47.08% less trainable parameters than the fine-tuned MobileNet-V3-Large model and the accuracy is only 1.59% lower on the Tiny ImageNet.

C. 2D-FWHT in ResNet

In this section, we will investigate 2D-FWHT in ResNet. ResNet is built with standard convolution layers [13]. It does not use any depthwise convolution layer. Therefore, we can

employ the ResNet to show the advantage of our 2D-FWHT layer. We do not use any pre-trained weights but we initialize the weights with Kaiming He's initialization [65]. We first build ResNet-20 as shown in Fig. 4a. As Table VI shows, it reaches an accuracy of 91.60% on CIFAR-10 with 273,066 trainable parameters. Then, we revise all residual blocks by replacing 3×3 convolution layers with 2D-FWHT layers and replacing 1×1 convolution layers with 1D-BWHT layers as shown in Fig. 4b. Here we apply 2D-FWHT layers without residual design because the blocks already contain the residual design. Since the dimension numbers are already in the power of 2, we do not need to pad any zeros before computing the Walsh-Hadamard transforms. We do not implement an additional 2D-FWHT layer before the GAP layer because the input of the GAP layer is already from a 2D-FWHT layer. In this way, we save 95.76% parameters. In other words, there are virtually no parameters left before the GAP

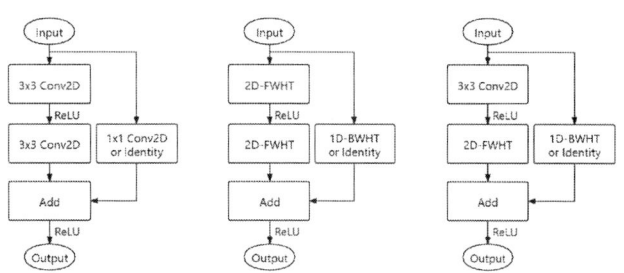

(a) Residual block [13]. (b) Complete revision. (c) Partial revision.

Fig. 4: ResNet Residual Block and our revision.

TABLE VI: ResNet-20 CIFAR-10 Result

Method	Scaling Weights	Trainable Parameters	Accuracy
ResNet-20 in [13]	-	0.27M	91.25%
ResNet-20 our trial (baseline)	-	273.1K	91.60%
2D-FWHT before GAP	N	273.3K	91.74%
2D-FWHT before GAP	Y	273.3K	91.75%
RB completely revised	N	11.59K(95.76%↓)	60.47%
RB partially revised	N	129.0K(52.76%↓)	89.88%
RB partially revised	**Y**	**133.1K(51.26%↓)**	**90.12%**

RB stands for residual blocks.

TABLE VII: ResNet-34 Tiny ImageNet Result

Method	Scaling Weights	Trainable Parameters	Accuracy
ResNet-34 (baseline)	-	21.39M	53.06%
RB partially revised	N	9.911M(53.66%↓)	52.17%
RB partially revised	**Y**	**9.929M(53.57%↓)**	**52.34%**

RB stands for residual blocks.

layers. Note that the dense layer contains 650 parameters, the batch normalization layers have 2,752 parameters, and the first convolution layer holds 448 parameters. There are only 7,740 parameters in the 2D-FWHT layers and the 1D-BWHT layers. With an extremely low number of trainable parameters, the network still reaches 60.47% accuracy. We then revise the residual partially as shown in Fig. 4c. We retain the first 3×3 convolution layer in each residual block and change other convolution layers. In this way, we save 52.76% trainable parameters with only a 1.72% accuracy loss. Finally, we add weights in the 2D smooth-threshold, and we reduce 51.26% trainable parameters with only a 1.48% accuracy loss.

We further explore ResNet-34 on the Tiny ImageNet. As is shown in Table VII, we reduce 53.66% trainable parameters with only 0.89% accuracy loss without scaling weights in 2D smooth-thresholding and 53.37% trainable parameters with only 0.72% accuracy loss with scaling weights in 2D smooth-thresholding.

D. Speed and Memory Tests

In our early work [3], we have shown that the 1D-FWHT layer runs about 2 times as fast as the 1×1 convolution layer on the NVIDIA Jetson Nano (4GB version) when the input and the output are both in $\mathbb{R}^{10 \times 32 \times 32 \times 1024}$ and the input tensor is initialized randomly. Now we will compare the 2D-FWHT layer versus the 3×3 convolution layer to show the speed advantage of our 2D-FWHT layer for the ResNet and other deep CNNs with regular convolution layers.

In this experiment, we set the input and the output both in $\mathbb{R}^{10 \times 8 \times 8 \times 1024}$. We use the same laptop (with Intel Core i7-7700HQ CPU, NVIDIA GTX-1060 GPU with Max-Q design, and 16GB DDR4 RAM) as [3] to run the TensorFlow PB model, and we deploy the TFLite model to the same NVIDIA Jetson Nano. As it is stated in Table VIII, the 2D-FWHT layer runs about 24 times as fast as the 3×3 convolution layer on the NVIDIA Jetson Nano. There are two reasons why the time difference on NVIDIA Jetson Nano is much more significant than on the laptop. One reason is that the TFLite model on the NVIDIA Jetson Nano is optimized by TensorFlow for ARM-embedded devices, while the PB model on the laptop is not

TABLE VIII: Speed test results.

Device	3×3 Conv2D	**2D-FWHT**	S&E
Laptop (GPU)	0.0508 S	**0.0459 S**	0.0449 S
Laptop (CPU)	0.0768 S	**0.0479 S**	0.0424 S
NVIDIA Jetson Nano	1.8861 S	**0.0776 S**	0.0519 S

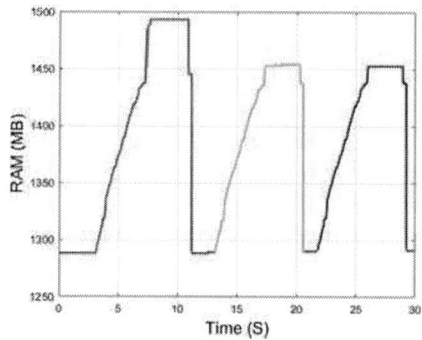

Fig. 5: Memory test on an NVIDIA Jetson Nano. From left to right, three peaks denote the inference memory of the 3×3 Conv2D (Max: 1493 MB), 2D-FWHT (Max: 1454 MB), and Squeeze-and-Excite layers (Max: 1453 MB). In each peak, after the layer is loaded, we run the inference code in a while-loop until we reach a steady state. RAM usage is 1288 MB when the device is vacant.

optimized as much as the TFLite model. The other reason is that the Windows backend apps slow down the system, while the NVIDIA Jetson Nano system is "clean".

We also compare the 2D-FWHT layer with the Squeeze-and-Excite layer. The squeeze ratio is $\frac{1}{4}$ as the Squeeze-and-Excite layer in the MobileNet-V3. We achieve comparable times in squeeze-and-excite layers. This is because current processors cannot distinguish between multiplication by ± 1 and a real number.

In addition, We record the RAM usage of the three layers on the NVIDIA Jetson Nano. According to Fig. 5, the 2D-FWHT layer and the Squeeze-and-Excite layer have close RAM usages, and they require about 40MB (19.51%) less memory than the 3×3 Conv2D in the inference.

VII. CONCLUSION

In this article, we present neural networks which have neurons with binary operations or networks that use binary transforms such as the Walsh-Hadamard transform. Neural networks with binary neurons or binary layers can be used in edge applications and/or applications requiring energy-efficient decision-making.

Walsh-Hadamard transform based neural networks have almost the same image recognition accuracy as regular neural networks. The additive neural network, AddNet, is more robust to false alarms in the wildfire detection problem because the operator is related to the L1-norm [2].

REFERENCES

[1] A. Afrasiyabi, B. Nasir, O. Yildiz, F. T. Y. Vural, and A. E. Cetin, "An energy efficient additive neural network," in *2017 25th Signal Processing and Communications Applications Conference (SIU)*. IEEE, 2017, pp. 1–4.

[2] H. Pan, D. Badawi, X. Zhang, and A. E. Cetin, "Additive neural network for forest fire detection," *Signal, Image and Video Processing*, pp. 1–8, 2019.

[3] H. Pan, D. Badawi, and A. E. Cetin, "Fast walsh-hadamard transform and smooth-thresholding based binary layers in deep neural networks," in *Proceedings of the IEEE/CVF Conference on Computer Vision and Pattern Recognition*, 2021, pp. 4650–4659.

[4] H. Tuna, I. Onaran, and A. E. Cetin, "Image description using a multiplier-less operator," *IEEE Signal Processing Letters*, vol. 16, no. 9, pp. 751–753, 2009.

[5] C. E. Akbaş, A. Bozkurt, A. E. Çetin, R. Çetin-Atalay, and A. Üner, "Multiplication-free neural networks," in *2015 23nd Signal Processing and Communications Applications Conference (SIU)*. IEEE, 2015, pp. 2416–2418.

[6] C. E. Akbaş, O. Günay, K. Taşdemir, and A. E. Çetin, "Energy efficient cosine similarity measures according to a convex cost function," *Signal, Image and Video Processing*, vol. 11, pp. 349–356, 2017.

[7] R. Dogaru and L. O. Chua, "The comparative synapse: a multiplication free approach to neuro-fuzzy classifiers," *IEEE Transactions on Circuits and Systems I: Fundamental Theory and Applications*, vol. 46, no. 11, pp. 1366–1371, 1999.

[8] X. Li, M. Parazeres, A. Oberman, A. Ghaffari, M. Asgharian, and V. P. Nia, "Euclidnets: An alternative operation for efficient inference of deep learning models," *arXiv preprint arXiv:2212.11803*, 2022.

[9] A. Krizhevsky, I. Sutskever, and G. E. Hinton, "Imagenet classification with deep convolutional neural networks," *Advances in neural information processing systems*, vol. 25, pp. 1097–1105, 2012.

[10] K. Simonyan and A. Zisserman, "Very deep convolutional networks for large-scale image recognition," *arXiv preprint arXiv:1409.1556*, 2014.

[11] C. Szegedy, W. Liu, Y. Jia, P. Sermanet, S. Reed, D. Anguelov, D. Erhan, V. Vanhoucke, and A. Rabinovich, "Going deeper with convolutions," in *Proceedings of the IEEE conference on computer vision and pattern recognition*, 2015, pp. 1–9.

[12] F. Wang, M. Jiang, C. Qian, S. Yang, C. Li, H. Zhang, X. Wang, and X. Tang, "Residual attention network for image classification," in *Proceedings of the IEEE conference on computer vision and pattern recognition*, 2017, pp. 3156–3164.

[13] K. He, X. Zhang, S. Ren, and J. Sun, "Deep residual learning for image recognition," in *Proceedings of the IEEE conference on computer vision and pattern recognition*, 2016, pp. 770–778.

[14] ——, "Identity mappings in deep residual networks," in *European conference on computer vision*. Springer, 2016, pp. 630–645.

[15] D. Badawi, H. Pan, S. C. Cetin, and A. E. Çetin, "Computationally efficient spatio-temporal dynamic texture recognition for volatile organic compound (voc) leakage detection in industrial plants," *IEEE Journal of Selected Topics in Signal Processing*, vol. 14, no. 4, pp. 676–687, 2020.

[16] C. Agarwal, S. Khobahi, D. Schonfeld, and M. Soltanalian, "Coronet: a deep network architecture for enhanced identification of covid-19 from chest x-ray images," in *Medical Imaging 2021: Computer-Aided Diagnosis*, vol. 11597. International Society for Optics and Photonics, 2021, p. 1159722.

[17] H. Partaourides, K. Papadamou, N. Kourtellis, I. Leontiades, and S. Chatzis, "A self-attentive emotion recognition network," in *ICASSP 2020-2020 IEEE International Conference on Acoustics, Speech and Signal Processing (ICASSP)*. IEEE, 2020, pp. 7199–7203.

[18] D. Stamoulis, T.-W. Chin, A. K. Prakash, H. Fang, S. Sajja, M. Bognar, and D. Marculescu, "Designing adaptive neural networks for energy-constrained image classification," in *Proceedings of the International Conference on Computer-Aided Design*, 2018, pp. 1–8.

[19] J. Redmon, S. Divvala, R. Girshick, and A. Farhadi, "You only look once: Unified, real-time object detection," in *Proceedings of the IEEE conference on computer vision and pattern recognition*, 2016, pp. 779–788.

[20] S. Aslan, U. Güdükbay, B. U. Töreyin, and A. E. Çetin, "Deep convolutional generative adversarial networks for flame detection in video," in *International Conference on Computational Collective Intelligence*. Springer, 2020, pp. 807–815.

[21] G. Menchetti, Z. Chen, D. J. Wilkie, R. Ansari, Y. Yardimci, and A. E. Çetin, "Pain detection from facial videos using two-stage deep learning," in *2019 IEEE Global Conference on Signal and Information Processing (GlobalSIP)*. IEEE, 2019, pp. 1–5.

[22] S. Aslan, U. Güdükbay, B. U. Töreyin, and A. E. Çetin, "Early wildfire smoke detection based on motion-based geometric image transformation and deep convolutional generative adversarial networks," in *ICASSP 2019-2019 IEEE International Conference on Acoustics, Speech and Signal Processing (ICASSP)*. IEEE, 2019, pp. 8315–8319.

[23] C. Yu, J. Wang, C. Peng, C. Gao, G. Yu, and N. Sang, "Bisenet: Bilateral segmentation network for real-time semantic segmentation," in *Proceedings of the European conference on computer vision (ECCV)*, 2018, pp. 325–341.

[24] Z. Huang, X. Wang, L. Huang, C. Huang, Y. Wei, and W. Liu, "Ccnet: Criss-cross attention for semantic segmentation," in *Proceedings of the IEEE/CVF International Conference on Computer Vision*, 2019, pp. 603–612.

[25] J. Long, E. Shelhamer, and T. Darrell, "Fully convolutional networks for semantic segmentation," in *Proceedings of the IEEE conference on computer vision and pattern recognition*, 2015, pp. 3431–3440.

[26] R. P. Poudel, S. Liwicki, and R. Cipolla, "Fast-scnn: fast semantic segmentation network," *arXiv preprint arXiv:1902.04502*, 2019.

[27] Y. Jin, W. Hao, P. Wang, and J. Wang, "Fast detection of traffic congestion from ultra-low frame rate image based on semantic segmentation," in *2019 14th IEEE Conference on Industrial Electronics and Applications (ICIEA)*. IEEE, 2019, pp. 528–532.

[28] M. Sandler, A. Howard, M. Zhu, A. Zhmoginov, and L.-C. Chen, "Mobilenetv2: Inverted residuals and linear bottlenecks," in *Proceedings of the IEEE conference on computer vision and pattern recognition*, 2018, pp. 4510–4520.

[29] H. Pan, D. Badawi, and A. E. Cetin, "Block walsh–hadamard transform-based binary layers in deep neural networks," *ACM Transactions on Embedded Computing Systems*, vol. 21, no. 6, pp. 1–25, 2022.

[30] J. Wu, C. Leng, Y. Wang, Q. Hu, and J. Cheng, "Quantized convolutional neural networks for mobile devices," in *Proceedings of the IEEE Conference on Computer Vision and Pattern Recognition*, 2016, pp. 4820–4828.

[31] U. Muneeb, E. Koyuncu, Y. Keshtkarjahromd, H. Seferoglu, M. F. Erden, and A. E. Cetin, "Robust and computationally-efficient anomaly detection using powers-of-two networks," in *ICASSP 2020-2020 IEEE International Conference on Acoustics, Speech and Signal Processing (ICASSP)*. IEEE, 2020, pp. 2992–2996.

[32] W. Chen, J. Wilson, S. Tyree, K. Weinberger, and Y. Chen, "Compressing neural networks with the hashing trick," in *International conference on machine learning*. PMLR, 2015, pp. 2285–2294.

[33] H. Pan, D. Badawi, and A. E. Cetin, "Computationally efficient wildfire detection method using a deep convolutional network pruned via fourier analysis," *Sensors*, vol. 20, no. 10, p. 2891, 2020.

[34] M. Yu, Z. Lin, K. Narra, S. Li, Y. Li, N. S. Kim, A. Schwing, M. Annavaram, and S. Avestimehr, "Gradiveq: Vector quantization for bandwidth-efficient gradient aggregation in distributed cnn training," *Advances in Neural Information Processing Systems*, vol. 31, 2018.

[35] S. Han, H. Mao, and W. J. Dally, "Deep compression: Compressing deep neural networks with pruning, trained quantization and huffman coding. arxiv 2015," *arXiv preprint arXiv:1510.00149*, 2019.

[36] F. N. Iandola, S. Han, M. W. Moskewicz, K. Ashraf, W. J. Dally, and K. Keutzer, "Squeezenet: Alexnet-level accuracy with 50x fewer parameters and ¡0.5 mb model size," *arXiv preprint arXiv:1602.07360*, 2016.

[37] M. Courbariaux, I. Hubara, D. Soudry, R. El-Yaniv, and Y. Bengio, "Binarized neural networks: Training deep neural networks with weights and activations constrained to+ 1 or-1," *arXiv preprint arXiv:1602.02830*, 2016.

[38] A. Bulat and G. Tzimiropoulos, "Hierarchical binary cnns for landmark localization with limited resources," *IEEE transactions on pattern analysis and machine intelligence*, vol. 42, no. 2, pp. 343–356, 2018.

[39] M. Rastegari, V. Ordonez, J. Redmon, and A. Farhadi, "Xnor-net: Imagenet classification using binary convolutional neural networks," in *European conference on computer vision*. Springer, 2016, pp. 525–542.

[40] Z. Shen, Z. Liu, J. Qin, L. Huang, K.-T. Cheng, and M. Savvides, "S2-bnn: Bridging the gap between self-supervised real and 1-bit neural networks via guided distribution calibration," in *Proceedings of the IEEE/CVF Conference on Computer Vision and Pattern Recognition*, 2021, pp. 2165–2174.

979-8-3503-4631-2/23 $31.00 © 2023 IEEE

[41] Z. Liu, Z. Shen, M. Savvides, and K.-T. Cheng, "Reactnet: Towards precise binary neural network with generalized activation functions," in *European Conference on Computer Vision.* Springer, 2020, pp. 143–159.

[42] B. Martinez, J. Yang, A. Bulat, and G. Tzimiropoulos, "Training binary neural networks with real-to-binary convolutions," *arXiv preprint arXiv:2003.11535*, 2020.

[43] A. Bulat, B. Martinez, and G. Tzimiropoulos, "Bats: Binary architecture search," in *European Conference on Computer Vision.* Springer, 2020, pp. 309–325.

[44] I. Hubara, M. Courbariaux, D. Soudry, R. El-Yaniv, and Y. Bengio, "Binarized neural networks," in *Proceedings of the 30th International Conference on Neural Information Processing Systems*, 2016, pp. 4114–4122.

[45] M. Alizadeh, J. Fernández-Marqués, N. D. Lane, and Y. Gal, "An empirical study of binary neural networks' optimisation," in *International Conference on Learning Representations*, 2018.

[46] T. Bannink, A. Hillier, L. Geiger, T. de Bruin, L. Overweel, J. Neeven, and K. Helwegen, "Larq compute engine: Design, benchmark and deploy state-of-the-art binarized neural networks," *Proceedings of Machine Learning and Systems*, vol. 3, pp. 680–695, 2021.

[47] F. Juefei-Xu, V. Naresh Boddeti, and M. Savvides, "Local binary convolutional neural networks," in *Proceedings of the IEEE conference on computer vision and pattern recognition*, 2017, pp. 19–28.

[48] X. Lin, C. Zhao, and W. Pan, "Towards accurate binary convolutional neural network," *Advances in neural information processing systems*, vol. 30, 2017.

[49] Z. Wang, J. Lu, C. Tao, J. Zhou, and Q. Tian, "Learning channel-wise interactions for binary convolutional neural networks," in *Proceedings of the IEEE/CVF Conference on Computer Vision and Pattern Recognition*, 2019, pp. 568–577.

[50] R. Zhao, Y. Hu, J. Dotzel, C. D. Sa, and Z. Zhang, "Building efficient deep neural networks with unitary group convolutions," in *Proceedings of the IEEE/CVF Conference on Computer Vision and Pattern Recognition*, 2019, pp. 11 303–11 312.

[51] S. Nasrin, D. Badawi, A. E. Cetin, W. Gomes, and A. R. Trivedi, "Mfnet: Compute-in-memory sram for multibit precision inference using memory-immersed data conversion and multiplication-free operators," *IEEE Transactions on Circuits and Systems I: Regular Papers*, vol. 68, no. 5, pp. 1966–1978, 2021.

[52] M. Ayi and M. El-Sharkawy, "Rmnv2: Reduced mobilenet v2 for cifar10," in *2020 10th Annual Computing and Communication Workshop and Conference (CCWC).* IEEE, 2020, pp. 0287–0292.

[53] P. Singh, V. K. Verma, P. Rai, and V. P. Namboodiri, "Hetconv: Heterogeneous kernel-based convolutions for deep cnns," in *Proceedings of the IEEE/CVF Conference on Computer Vision and Pattern Recognition*, 2019, pp. 4835–4844.

[54] T. C. Deveci, S. Cakir, and A. E. Cetin, "Energy efficient hadamard neural networks," *arXiv preprint arXiv:1805.05421*, 2018.

[55] A. E. Cetin, O. N. Gerek, and S. Ulukus, "Block wavelet transforms for image coding," *IEEE Transactions on Circuits and Systems for Video Technology*, vol. 3, no. 6, pp. 433–435, 1993.

[56] J. L. Walsh, "A closed set of normal orthogonal functions," *American Journal of Mathematics*, vol. 45, no. 1, pp. 5–24, 1923.

[57] B. J. Fino and V. R. Algazi, "Unified matrix treatment of the fast walsh-hadamard transform," *IEEE Transactions on Computers*, vol. 25, no. 11, pp. 1142–1146, 1976.

[58] S. Xie, R. Girshick, P. Dollár, Z. Tu, and K. He, "Aggregated residual transformations for deep neural networks," in *Proceedings of the IEEE conference on computer vision and pattern recognition*, 2017, pp. 1492–1500.

[59] C. Szegedy, S. Ioffe, V. Vanhoucke, and A. A. Alemi, "Inception-v4, inception-resnet and the impact of residual connections on learning," in *Thirty-first AAAI conference on artificial intelligence*, 2017.

[60] P. Agante and J. M. De Sá, "Ecg noise filtering using wavelets with soft-thresholding methods," in *Computers in Cardiology 1999. Vol. 26 (Cat. No. 99CH37004).* IEEE, 1999, pp. 535–538.

[61] D. L. Donoho, "De-noising by soft-thresholding," *IEEE transactions on information theory*, vol. 41, no. 3, pp. 613–627, 1995.

[62] M. W. Marcellin, M. J. Gormish, A. Bilgin, and M. P. Boliek, "An overview of jpeg-2000," in *Proceedings DCC 2000. Data Compression Conference.* IEEE, 2000, pp. 523–541.

[63] J. Lee-Thorp, J. Ainslie, I. Eckstein, and S. Ontanon, "Fnet: Mixing tokens with fourier transforms," *arXiv preprint arXiv:2105.03824*, 2021.

[64] A. Howard, M. Sandler, G. Chu, L.-C. Chen, B. Chen, M. Tan, W. Wang, Y. Zhu, R. Pang, V. Vasudevan *et al.*, "Searching for mobilenetv3," in *Proceedings of the IEEE/CVF International Conference on Computer Vision*, 2019, pp. 1314–1324.

[65] K. He, X. Zhang, S. Ren, and J. Sun, "Delving deep into rectifiers: Surpassing human-level performance on imagenet classification," in *Proceedings of the IEEE international conference on computer vision*, 2015, pp. 1026–1034.

Innovation Practices Track: Innovation on Telemetry Monitoring

Fei Su
Intel Corporation
U.S.
fei.su@intel.com

Marc Hunter
ProteanTecs
U.S.
marc.hutner@proteantecs.com

Chen He
NXP Semiconductor
U.S.
chen.he@nxp.com

Sashi Obilisetty
Synopsys
U.S.
sashio@synopsys.com

I. INTRODUCTION (FEI SU)

Telemetry monitoring and sensing plays a critical role in silicon lifecycle management. Silicon insights can be gained through these telemetry sensors across the entire life cycle including test and in-field phases, and actionable intelligence can be derived from them. In this session we invited the industry experts to discuss the latest innovation of telemetry sensors and their applications in heterogeneous packaging, automotive quality and functional safety, and prescriptive maintenance.

II. ELEVATING DIE-TO-DIE INTERFACE TESTING AND REPAIR WITH ON-DIE DIAGNOSIS (MARC HUNTER)

Automotive, AI/ML and 5G markets are demanding increasing levels of integration with respect to memory and compute density. Over the last decade, it is evident that Moore's law device scaling isn't satisfying the market performance requirements in a single monolithic device. Several microprocessor and FPGA companies have been adopting "More than Moore" approaches integrating multiple devices into a single package. These products have divided large monolithic devices into several smaller devices optimizing for product economics, increased performance, and reductions in power. With this partitioning, we are also faced with many new test challenges in both wafer and package testing.

Heterogeneous packages at package test are faced with quality and reliability challenges due to the accessibility and visibility of traditional test approaches. Test access for die-to-die interfaces is not possible at package testing and the drive strength with higher-density IO interfaces makes it impossible to test at wafer probe. Dies are assembled over the silicon interposer using very small micro-bumps which may suffer from latent defects such as voids or cracks, potentially posing a reliability risk to the assembled product. These types of latent defects, although rare, could cause a system failure over time if not screened out during final test or detected during in-field operation.

To address these challenges, ProteanTecs has developed a monitoring solution that in conjunction with an analytics approach can identify latent defects at test as well as in the field. We will discuss a use case where the monitoring solution will be used to assess the health of a die-to-die link for device repair. This novel parametric approach evaluates the amount of margin across wide interfaces to ensure optimal utilization of resources. Today links are repaired when failing conditions are detected, but there is no concept as to how close to failure the specific lane is electrically. We will present silicon data for this technique and extend the discussion with an in-field use case of this approach by detecting aging effects of a system in package.

III. APPLICATION OF ON-CHIP TELEMETRY SENSORS TO IMPROVE QUALITY AND FUNCTIONAL SAFETY (CHEN HE)

With the emerging electric cars and autonomous driving, the semiconductor electronic contents on vehicles have significantly increased in recent years. As a result, much more stringent quality and functional safety requirements for automotive semiconductor chips become mandatory. To meet this challenge, innovations across the entire silicon lifecycle from design, manufacturing, production test, and in-field applications are required, where on-chip telemetry sensors and associated data analytics will be instrumental.

In this presentation, we will discuss the application of on-chip telemetry sensors to improve quality and functional safety in following areas. First, they can help improve defect screening capability at production test. For instance, they can enable better outlier detection capability and optimize test conditions to maximize coverage while minimizing yield loss. Second, they can improve issue detection during qualification by providing more visibility into early life failures during qualification stress. Last but not least, they can provide better protection during in-field application by enabling early detection of certain latent defects, enabling dynamic performance (i.e., voltage and/or frequency) scaling to prevent imminent aging or wear-out caused failures, and preventing catastrophic failures or critical incidents through preventive maintenance. To enable the full potential of on-chip telemetry sensors, it is critical to develop a systemic data analytics infrastructure utilizing the state-of-the-art Machine Learning (ML) techniques.

IV. HOW SILICON OBSERVABILITY ENABLES PRESCRIPTIVE MAINTENANCE (SASHI OBILISETTY)

Increased digitization and the growth of complex systems around us calls for new approaches to system operation. This talk will focus on innovative and actionable insights that can be generated if silicon observability via smart monitors is designed in from the start. Over the years, maintenance strategies have evolved from preventive maintenance to condition monitoring to Predictive maintenance. Prescriptive maintenance, the latest invention, has shown tremendous promise in industrial sectors. It leverages AI/ML with data analytics and requires a strong foundation that can learn, process and scale. We will apply prescriptive maintenance techniques to silicon-enabled systems. Assuming that silicon observability architecture is enabled and on-chip sensor data is transmitted off chip, we will discuss what problems can be solved by having access to deep silicon data. Examples will go into specific data types and insights such data can provide.

Key takeaways for the audience will be a solid understanding of silicon observability, prescriptive maintenance strategies and the infrastructure needed to support prescriptive maintenance in various domains.

979-8-3503-4631-2/23 $31.00 © 2023 IEEE

Kernel Smoothing Technique Based on Multiple-Coordinate System for Screening Potential Failures in NAND Flash Memory

Gooyoung Kim
Flash Product Engineering Team
Samsung Electronics
Hwasung-si, Rep. of Korea
kgy9023@naver.com

Youngseon Moon
Flash Product Engineering Team
Samsung Electronics
Hwasung-si, Rep. of Korea
ys15.moon@samsung.com

Jongmin Kim
Flash Product Engineering Team
Samsung Electronics
Hwasung-si, Rep. of Korea
jongmin77.kim@samsung.com

Jaeyong Jeong
Flash Product Engineering Team
Samsung Electronics
Hwasung-si, Rep. of Korea
jyong.jeong@samsung.com

Eunkyoung Kim
Flash Product Engineering Team
Samsung Electronics
Hwasung-si, Rep. of Korea
yesdama.kim@samsung.com

Sunghoi Hur
Flash Product & Technology
Samsung Electronics
Hwasung-si, Rep. of Korea
hursung@samsung.com

Abstract—**With the growing complexity of integrated circuits and increasing capacity of memory devices, it is becoming increasingly difficult to screen out all defective dies through wafer test. To address this problem, statistical methods are being increasingly applied on test data to screen out potential defective dies. Among such methods, the good-die-in-bad-neighborhood (GDBN) method is effective. Most GDBN methods use the bin data of nearby dies through Cartesian coordinates. However, the bin data provide only binary information, and Cartesian coordinates do not fully reflect the spatial correlations of wafers. Therefore, this paper proposes a novel GDBN method using bad block data, which are key indicators of the in-field failure of NAND flash memory. The defect risk is predicted through statistical kernel smoothing based on Cartesian and polar coordinates to consider the spatial correlations in various environments of wafer fabrication. Experiments on NAND flash products show that the in-field failure classification performance of the proposed method, in terms of the area under the curve value, is 30% higher than that of the existing method, based on the bin data and Cartesian coordinates.**

Keywords—kernel smoothing, polar coordinate system, bad block, NAND flash, log transformation

I. Introduction

Semiconductor engineers in various fields are attempting to decrease defective parts per million (DPPM). For example, test engineers are attempting to improve the coverage of wafer tests to screen as many defective dies as possible [1]. However, with the growing complexity of integrated circuits and increasing capacity of memory devices, it is becoming increasingly difficult to screen out all defective dies through wafer test, potentially because of the accumulating test errors. Errors can cause a test to miss defective dies, which may result in the failure of the device in the field. The error can be alleviated by increasing the number of test repetitions or diversifying the test patterns. However, these strategies may increase the test costs and incur additional damage to the die. In other words, the test error and cost exhibit a trade-off relationship.

A promising approach to decrease the DPPM without a test cost overhead is to screen out potential defective dies by analyzing the test data. To this end, many statistical methods have been proposed, which can be divided into two types. The most general and intuitive approach is to screen a die using bad test data. Certain researchers [2] recommended the screening of dies with higher (or lower) measurement than a threshold calculated as a function of several basic statistics such as the average and standard deviation. Other researchers [3], [4] improved this method by training a statistical regression model based on the spatial information of the dies.

The second type of approach is unintuitive. This approach predicts the risk of a die under investigation (DUI) based on its neighboring dies. Specifically, a die having bad neighbors is screened out. A classic strategy is to screen out all the dies in a wafer with a low yield (or a high ratio of a particular bin) [5]. The good-die-in-bad-neighborhood (GDBN) method improves this strategy by using the die's spatial information. A predictive model calculates the DUI's defect risk based on the bin data of its neighboring dies. A higher number of neighboring fail bins typically implies a higher DUI risk. Subsequently, dies are screened in decreasing order of the predicted risk. In the first form of this method, only eight adjacent dies of each die were considered. Specifically, the original method regarded the number of fail bins in eight sites adjacent to a die as the predicted risk. Later, many variants of this approach were established.

For example, Miller and Riordan [6] expanded the area of interest to 5 by 5, which included 24 nearby dies. Moreno-Lizaranzu and Cuesta [7] predefined a specific set of bins and applied the GDBN method to focus on important bins. Yang et al. [8] used an artificial neural network to consider various patterns and distributions of nearby bins.

The GDBN method is based on two essential assumptions in all these studies. The first one is that nearby dies of DUI would have similar characteristics and quality to DUI because they went through similar environments of the wafer fabrication

979-8-3503-4631-2/23 $31.00 © 2023 IEEE

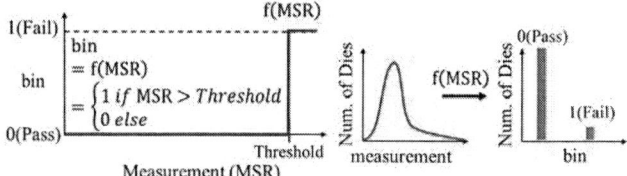

Fig. 1. Relation between continuous measurement and corresponding bin.

process. The second one is already mentioned before; every test includes errors. Under these assumptions, a GDBN is screened out because it is highly likely to be a 'FALSE' good die due to test errors. Therefore, it is crucial in the GDBN method to properly select data and neighbors to provide the best reference to the defect risk of DUI.

Notably, most of the research on the GDBN method has been focused on bin data. However, a bin has an extremely low resolution because it is represented in binary zero or one data for each test item. In contrast, the measurement used to determine whether a die passes each test item contains a large amount of information regarding the die's health because it can be represented as a natural number or a real number. The relation between the measurement and bin is shown in Fig. 1.

Furthermore, most of the existing studies used a square (or rectangular) window to define the neighboring area of each die. This window defines the neighboring area based on a die's x and y coordinates, regardless of the die's location on the wafer. However, in the wafer fabrication process, the characteristics of each die (including the defect risk) depend on its distance and direction from the center of the wafer. Considering this aspect, Padonou et al. [9] used polar coordinates to monitor wafer profiles.

To address the abovementioned limitations, in this study, the GDBN method is improved in two aspects: First, an integer measurement, termed as the number of bad blocks, is used instead of the bin, as shown in Fig. 2. Second, neighboring areas are defined based on Cartesian and polar coordinate systems instead of using a square window. The bad block (BB) refers to the block that fails the test. The number of BBs provides more detailed and accurate information regarding the die health than the bin in NAND flash memory. A statistical kernel smoothing method is applied to quantify the defect risk of the DUI [10].

The remaining paper is organized as follows. Section II presents the preliminary concepts. Section III describes the proposed method and presents the mathematical derivations and their explanations with simple examples. Section IV describes the experiments conducted using various NAND flash products and their in-field failures to demonstrate the proposed method. Section V summarizes the results and discusses future research directions.

II. BACKGROUND

A. Kernel Smoothing

Smoothing is a statistical technique that is performed to remove the error from the original data and obtain a smooth trend. According to the definition of a certain distance, the smoothing value is obtained by referring to the values of samples nearby each sample. The sample with a higher value

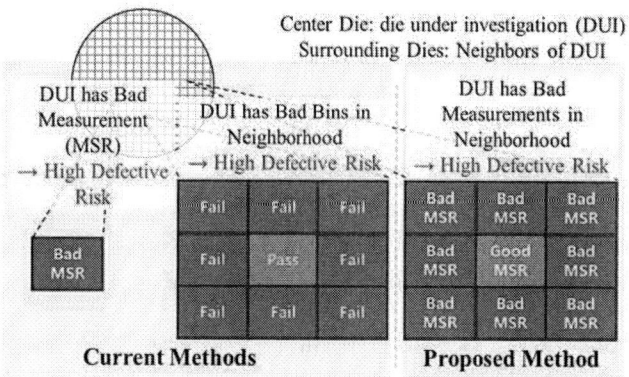

Fig. 2. Methods to screen out potentially defective dies. The two methods on the left are the current methods and that on the right is the proposed method.

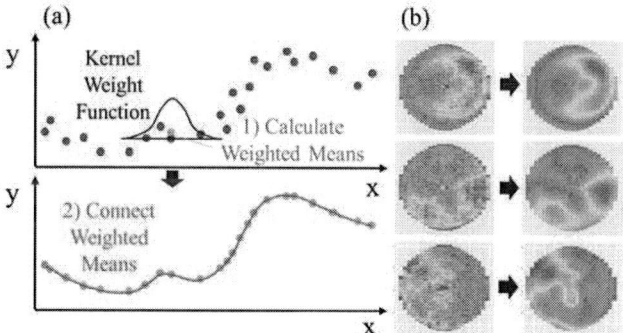

Fig. 3. (a) Kernel smoothing, (b) examples of application to the wafer.

than its neighbors receives a lower smoothing value and vice versa. Kernel smoothing is aimed at calculating the smoothing value of a sample as a weighted average of the value of its nearby samples. The kernel function defines the weights from each point to its nearby samples depending on the distance between them. A radial kernel such as Gaussian is typically used to assign a higher weight to closer samples. Fig. 3(a) shows a general kernel smoothing process, which can be applied to multi-dimensional space if the distance between the locations is well-defined. Fig. 3(b) shows two-dimensional examples of kernel smoothing on a wafer map.

B. Polar Coordinate System

The location of a point on a two-dimensional plane is typically represented by the x and y coordinates in the Cartesian coordinate system. However, the location can also be represented by the r and θ coordinates in the polar coordinate system. Coordinates (x, y) can be converted to (r, θ) as

$$r = \sqrt{x^2 + y^2} \qquad (1)$$

$$\theta = \mathrm{atan2}(y, x). \qquad (2)$$

Coordinates (r, θ) are particularly helpful when analyzing variables affected by a distance or direction from the origin.

C. BB

Wafer tests for NAND flash products consist of thousands of items. Although the items are considerably different, they can be divided into two large groups based on the primary inspection target: peripheral area and cell array. The cell array of NAND flash consists of hundreds to thousands of blocks. Cell inspection items, which are aimed at evaluating all blocks, are followed by the marking of blocks that do not meet the qualifications for BB. Assume that in a given test, a die's accumulated number of BBs exceeds the predefined maximum limit. In this case, the die is marked as a fail and excluded from the inspection list of the latter tests because it does not have enough remaining memory volume.

The number of BBs' distribution is similar to the zero-inflated exponential distribution with a long tail. When all blocks that share the circuit become BBs, the die may have an extremely high number of BBs, corresponding to the long tail of the distribution.

BB data provide additional information for a die that passes the test and detailed information for a die that fails the test. Even if certain group of dies have identical bin values, some may only have one or two corresponding BB, whereas the others may have numerous BBs. Moreover, dies not marked with the bin can have a certain number of BBs. Several of these dies may have more BBs than those marked with the bin.

III. METHODOLOGY

The kernel smoothing value of each die is an output of the function that uses the number of BBs, coordinates, and missing information of every die (including the DUI) on the wafer as the input. Fig. 4 illustrates the determination of the kernel smoothing value, which involves the following steps.

For $i = 1, 2, \ldots, n$ (= the number of dies on one wafer), let d_i be the i_{th} die on the wafer, b_i be its number of BBs, (x_i, y_i) be its x and y coordinates, (r_i, θ_i) be its r and θ coordinates, and $I(d_i)$ be its non-missing indicator (i.e., 0 if missing, else 1). The x and y coordinates of the die are obtained based on the coordinates corresponding to each die when an entire wafer is assumed to be a concentric circle on the coordinate plane. Subsequently, the r and θ coordinates are obtained from the x and y coordinates following (1) and (2). $S(d_i)$, which is the kernel smoothing value of d_i, is obtained as

$$dx_{i,j} = |x_i - x_j|, dy_{i,j} = |y_i - y_j| \tag{3}$$

$$dr_{i,j} = |r_i - r_j|, d\theta_{i,j} = \min(|\theta_i - \theta_j|, 2\pi - |\theta_i - \theta_j|) \tag{4}$$

$$(note: d\theta_{i,j} = 0 \text{ if } \theta_i \text{ or } \theta_j \text{ is undefined})$$

$$dC_{ij} = \sqrt{dx_{i,j}^2 + dy_{i,j}^2} \tag{5}$$

$$dP_{ij} = \sqrt{(s_1 dr_{i,j})^2 + \left(\frac{d\theta_{i,j}}{s_1 \pi}\right)^2}, s_1 > 0 \tag{6}$$

$$S_C(d_i) = \sum_j \frac{I(d_j) \log_2(b_j + 1)}{s_2^{dC_{ij}}} \Big/ \sum_j \frac{I(d_j)}{s_2^{dC_{ij}}}, s_2 > 1 \tag{7}$$

$$S_P(d_i) = \sum_j \frac{I(d_j) \log_2(b_j + 1)}{s_3^{dP_{ij}}} \Big/ \sum_j \frac{I(d_j)}{s_3^{dP_{ij}}}, s_3 > 1 \tag{8}$$

$$S(d_i) = \max(S_C(d_i), S_P(d_i)). \tag{9}$$

In (5) and (6), dC_{ij} and dP_{ij} mean a distance between d_i and d_j based on the Cartesian and polar coordinate systems, respectively. In (7) and (8), $S_C(d_i)$ and $S_P(d_i)$ mean kernel smoothing value of d_i based on Cartesian and polar coordinate systems, respectively.

Notably, the x and y coordinates defined here differ from those of the die, which are typically mentioned as a pair of natural numbers. Therefore, the distance between a pair of adjacent dies depends on the layout of the dies on the wafer. For instance, if a wafer has 20 and 60 dies in the horizontal and vertical directions, respectively, dy between the pair of dies facing each other's upper and lower boundaries is one-third of dx between a pair of dies facing each other's left and right edges. A detailed explanation can be found in Section 3 in [11].

The abovementioned expressions are described in the following subsections in two key concepts and one supplemental concept, and Fig. 5 provides simplified examples to clarify these concepts.

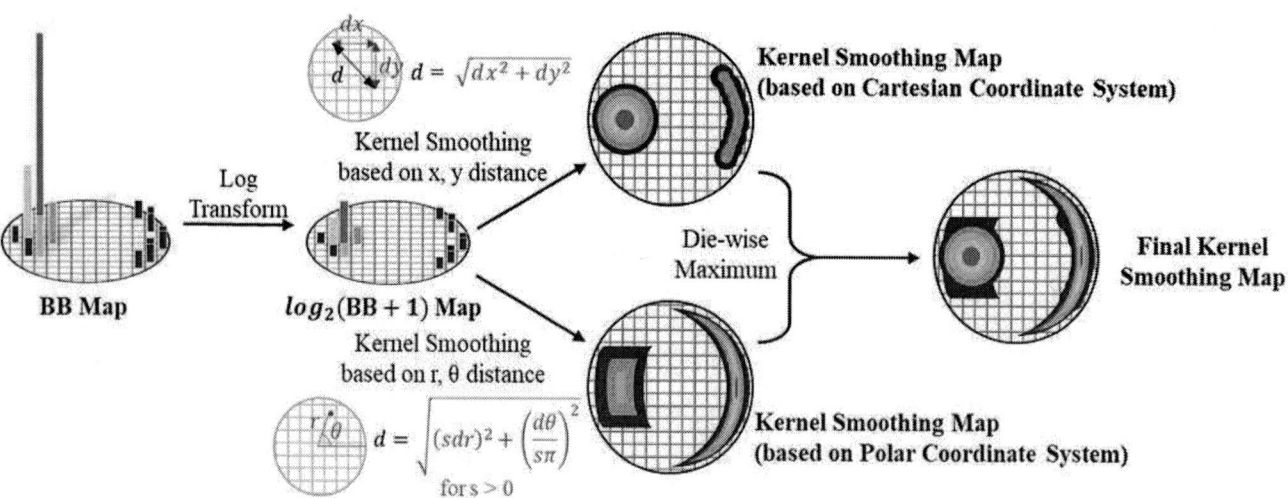

Fig. 4. Process of generating a kernel smoothing map from a BB map, which involves three steps: log transformation, preparation of two kernel smoothing maps, and determination of the die-wise maximum.

A. Multiple-Coordinate System

The Cartesian-coordinate-based distance (5) is suitable for treating the measurement associated with the wafer fabrication environment, which is affected by the Euclidean distance. In contrast, the polar-coordinate-based distance (6) is suitable for treating the measurement that is influenced by the radius or the direction from the center of the wafer.

Equations (6)–(8) contain three coefficients: s_1, s_2, and s_3, which can be adjusted to obtain different smoothing results from the same data. A larger s_1 means a higher weight is assigned to the r coordinate than θ in (6). A larger s_2 or s_3 means that a higher weight is assigned to the nearby area in (7) and (8), respectively. It is preferable to set s_1 to a value larger than 1 to differentiate (6) from (5). If s_1 is set to 2, then one-third of π radian is considered the same distance as one-twelfth of the radius. Values higher than 1,000 should be assigned to s_2 and s_3 to assign adequate weights on the adjacent dies. If s_2 (or s_3) is set to 10,000, then a DUI has $10,000^{0.5}$ times larger weight than a die at a 0.5 distance, in terms of dC (or dP), from the DUI. Fig. 6 visualize the weight according to the combination of the three coefficients within typically recommended range. Coefficients can be modified depending on the application requirements.

For each DUI, smoothing values based on the Cartesian coordinates and polar coordinates are obtained, and the maximum value is defined as the final smoothing value, as shown in (7)–(9). If a die has a larger number of BBs in a Cartesian-coordinate-based nearby area than in a polar-coordinate-based nearby area, the smoothing value is likely to follow the former value. This framework is intended to respond to adjacent BB patterns that may appear different depending on the test, wafer, and die.

Fig. 5(a) shows a case in which seven dies with one BB are distributed around A (black die) and B (blue die). Considering the wafer fabrication process, B, located at the center of the arc pattern, is more likely to be defective compared with A, which is located at the edge of the lump pattern. However, if only the Cartesian coordinates are used, both A and B have the same BB at the same distance, and thus, they yield the same smoothing result. Therefore, the proposed method uses polar coordinates to determine B as a die with a higher risk of being defective compared with A.

B. Log Transformation

Typically, the potential defect risk of DUI is higher when a few BBs occur in many nearby dies than when many BBs are concentrated on one or two nearby dies. Because the former case means the BB (and related defect risk) is likely correlated with spatial information. To quantify these characteristics, log transformation is performed in the number of BBs of each die, as defined in (7) and (8):

$$(number\ of\ BBs) \rightarrow log_2\big((number\ of\ BBs) + 1\big)$$

Log transformation can be replaced with other strictly increasing concave transformations, such as square root transformation. Empirically, the performance of these transformations is similar. Nevertheless, log transformation is

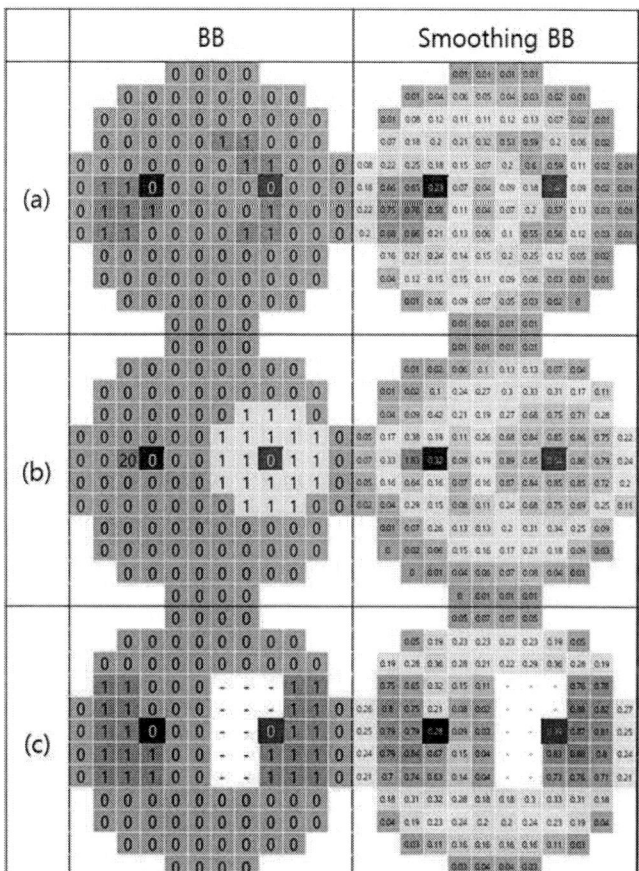

Fig. 5. Imaginary wafer maps of BB and smoothing value obtained using the proposed method.

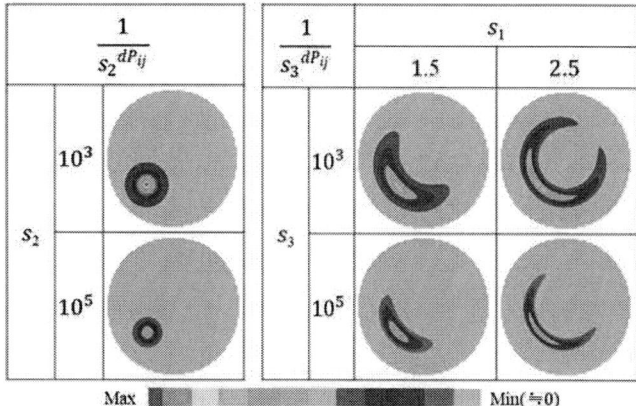

Fig. 6. Wafer maps visualizing each die's relative weight according to the combinations of coefficients s_1, s_2, and s_3. The left and right tables are based on Cartesian and polar coordinate systems, respectively.

typically recommended to prevent minority dies with extremely many BBs from distorting the smoothing result.

As shown in Fig. 5(b), A (black die) has 20 BBs in one die to its left, and B (blue die) has 20 nearby dies with one BB. B is likely to have a higher risk than A because nearby BBs of B show a higher spatial correlations than those of A. The desired result can be obtained by applying log transformation on the number of BBs of each die. However, if smoothing is performed on the number of BB as is, opposite results would be obtained because A has a smaller average distance from the BBs.

C. Missing Adjustment (supplemental concept)

BB for a specific test item is not measured at the die dropped at the previous tests. Therefore, there exist missing values inside the wafer. To balance the smoothing value, every die's weight is divided into the sum of the total weight of every non-missing die. It is represented in (7) and (8) as a denominator. This procedure also contributes to the weights relocation for the dies located on the edge of the wafer. As a result, for all dies, weighted average of nearby non-missing elements is used as smoothing value.

Fig. 5(c) shows that the distribution of nearby BB is completely same on A (black die) and B (blue die). But in the case of B, it can be found that there are many missing values on its left area. B is more likely to be dangerous than A whose right area is confirmed to be 'clean'. But when the missing values are just regarded as zero, A and B will get the same smoothing value. The proposed method evaluates the risk of B more by relocating missing value's weight to the nearby non-missing dies.

IV. EXPERIMENTAL RESULTS

To validate the proposed method, experiments are conducted on a dataset of NAND flash products [12] and [13]. First, the basic information of the dataset and experimental settings is described, and then, the results are presented and discussed.

A. Dataset

The dataset contains the results of an in-house accelerated life test simulating the in-field operation of the device. The data pertain to three cases that differ in terms of the product, failure ratio, and type of defect. In each case, dies that pass the wafer test but fail the accelerated life test are labeled as "Defect". Dies that pass both tests are labeled as "Normal". These labels are considered dependent variables in the experiments. Dies that fail the wafer test are not labeled because they are not evaluated by accelerated life test. Table I lists the number of dies for each case and their labels.

TABLE I.   NUMBER OF DIES FOR EACH CASE AND LABEL

Case	Normal	Defect	Total
1	149,255	20	149,275
2	56,084	30	56,114
3	11,632	274	12,906

In each case, one test item that generates BB and has the strongest physical correlation with each type of defect is chosen. Independent variables are derived from the results of this test item. For each die (including dies that fail the wafer test), x_{BB} indicates the number of BBs from the test, and x_{BIN} indicates the bin information (i.e., whether the die passes or fails) of the test. Both x_{BB} and x_{BIN} have a missing value for a die if and only if the die fails in the former test. In other words, even if a die fails in the test or the latter test, x_{BB} and x_{BIN} have

data for the dies. Table II summarizes the number of dies for each case, label, and x_{BB}.

TABLE II.   NUMBER OF DIES FOR EACH CASE, LABEL, AND X_{BB}

x_{BB}	Case1 Normal	Case1 Defect	Case2 Normal	Case2 Defect	Case3 Normal	Case3 Defect
0	146,669	14	54,842	25	11,045	82
1	2,359	5	1,090	2	301	40
2	160	1	112	2	120	35
3	20	0	22	0	46	31
4	17	0	5	1	36	22
5	5	0	2	0	24	16
6	8	0	3	0	19	12
7	5	0	0	0	5	18
8	1	0	0	0	10	10
9	0	0	1	0	10	4
10~	11	0	7	0	16	4
Total	149,255	20	56,084	30	11,632	274

B. Experimental Settings

To verify the effects of each element constituting the proposed method, experiments are performed to compare the following entities. Coefficients s_1, s_2, and s_3 in (6)–(8) are set as 2, 35,000, and 35,000, respectively, for all settings.

- x_{BB} and $S(x_{BB})$
- $S(x_{BB})$ and $S(x_{BIN})$
- $S(x_{BB})$ with each key concept described in Section III

The objective of the first comparison is to demonstrate that kernel smoothing helps quantify the defect risk of the die. The objective of the second comparison is to compare the performance of the GDBN method based on BB and bin. Because x_{BIN} may be only zero or one, $S(x_{BIN})$ is a weighted average of the yield. Therefore, it serves as a GDBN index for the bin. The objective of the third comparison is to evaluate the effect of each concept of kernel smoothing described in Section III. Four experimental settings are designed for each application for the two key concepts, as indicated in Table III.

TABLE III.   TWO KEY CONCEPTS OF KERNEL SMOOTHING AND DESCRIPTION OF EACH APPLICATION

Option	Application	Note
Multiple-Coordinate System	O	Use $S(d_i)$
	X	Use $S_c(d_i)$ instead
Log Transformation	O	Use $log_2\big((\text{number of } BBs) + 1\big)$
	X	Use the number of BBs instead

Based on the independent variable calculated in each experimental setting, performance is compared in terms of the area under the curve (AUC). The purpose of the proposed method is to reduce DPPM as much as possible with the minimum yield loss. Here, the amount of yield loss must be considered together with DPPM reduction, because they exhibit a trade-off relationship depending on where the screening threshold is set. However, how much to focus on yield loss or

979-8-3503-4631-2/23 $31.00 © 2023 IEEE

DPPM reduction can be different depending on the situation. Unlike other indexes of classification model, AUC quantifies the performance without setting the screening threshold. Therefore, in this problem, AUC is set as a performance index.

C. Results and Discussion

Table IV summarizes the experimental results in terms of the AUC, and Fig. 7 shows corresponding receiver operating characteristic (ROC) curves. Consider settings 1) and 2): x_{BB} has an average AUC of 0.683, which is higher than 0.5. Considering the large sample size, a certain correlation exists between x_{BB} and defect. In comparison, a 0.185 higher AUC, 0.868, is obtained when kernel smoothing (without applying key concepts) is applied on x_{BB}. In other words, the BBs of the nearby dies are strongly correlated with the defect risks of the DUI.

Consider settings 2) and 5): A 0.023 higher AUC, 0.891, is obtained when all concepts are applied. It shows that the key concepts contribute to maximizing performance of the method.

Consider settings 5) and 6): The average AUC is 0.230 higher when kernel smoothing was applied to x_{BB} than x_{BIN}. This finding shows that x_{BB}, which is an integer value of zero or more, provides more clues regarding the defect risk than the x_{BIN} with a low resolution.

TABLE IV.   AUC FOR DIFFERENT EXPERIMENTAL SETTINGS

No.	Data	Multiple-Coordinate System	Log Transform	Case1	Case2	Case3	Total (Avg)
		Experimental setting			**AUC**		
1)	x_{BB}	without kernel smoothing		0.642	0.573	0.835	0.683
2)	$S(x_{BB})$	x	x	0.845	0.807	0.952	0.868
3)	$S(x_{BB})$	x	o	0.852	0.812	0.956	0.873
4)	$S(x_{BB})$	o	x	0.866	0.825	0.951	0.881
5)	$S(x_{BB})$	o	o	0.882	0.833	0.957	0.891
6)	$S(x_{BIN})$	o	-	0.562	0.567	0.855	0.661

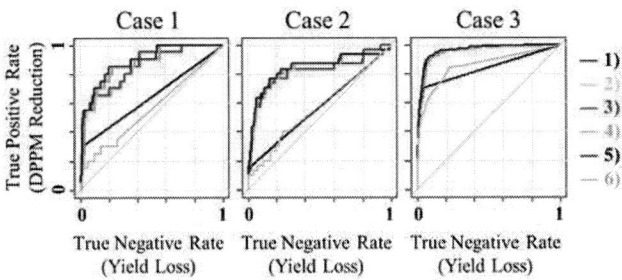

Fig. 7.   ROC curve for each case and experimental setting.

The effect of each key concept for kernel smoothing is summarized in Tables V and VI. Because two concepts are independent, the average value from 2) to 5) can be compared depending on which option is applied or not. Applying the multiple-coordinate system and log transformation improves the average AUC by 0.015 and 0.008, respectively. In Cases 1 and 2, the improvement resulting from the multiple-coordinate system is significant, likely because both in-field failu7res and BB are better distributed to correspond to the polar coordinate system.

TABLE V.   AVERAGE AUC WHEN THE MULTIPLE-COORDINATE SYSTEM IS APPLIED FOR KERNEL SMOOTHING

Application	Case1	Case2	Case3	Total (Avg)
X	0.849	0.809	0.954	0.871
O	0.874	0.829	0.954	0.886
Delta(O-X)	0.025	0.020	0.000	0.015

TABLE VI.   AVERAGE AUC WHEN THE LOG TRANSFORM IS APPLIED FOR KERNEL SMOOTHING

Application	Case1	Case2	Case3	Total (Avg)
X	0.855	0.816	0.951	0.874
O	0.867	0.823	0.957	0.882
Delta(O-X)	0.012	0.007	0.006	0.008

V. CONCLUSION

The proposed method improves the classification performance (in terms of the AUC) of in-field failures by 30% compared with that achieved using BB without smoothing or the bin-based GDBN method. The novelty of this study lies in the application of the GDBN method to the number of BBs to predict the defect risk of the die. This method can be simply extended to other devices which has count data represented by an integer value of zero or more. In this study, to improve performance, the wafer's unique properties are considered: Specifically, the properties of the die are considered to be affected by its location based on the polar coordinate system, given the environment for wafer fabrication. The GDBN methods currently used can probably improve performance by defining neighboring areas considering the polar coordinate system.

Future work can be aimed at optimizing the coefficients of the kernel function and checking whether log transformation yields the best performance on number of BBs. The findings of this work are expected to promote research on the GDBN method based on measurements other than the bin and BB.

REFERENCES

[1] B. I. Nam et al., "Novel electrical detection method for random defects on peripheral circuits in NAND flash memory," 2022 IEEE Int. Reliability Phys. Symp. (IRPS), 2022, pp. P40-1–P40-4.

[2] Automotive Electronics Council, "Guidelines for part average testing (PAT)," AEC Q001, rev. C, July 2003, http://www.aecouncil.comAECDocuments.html.

[3] A. Nahar, K. M. Butler, J. M. Carulli, and C. Weinberger, "Quality improvement and cost reduction using statistical outlier methods," 2009 IEEE Int. Conf. Comput. Des., 2009, pp. 64-69.

[4] C. T. Chen et al., "CNN-based stochastic regression for IDDQ outlier identification," 2020 IEEE 38th VLSI Test Symp. (VTS), 2020, pp. 1–6.

[5] S. Illyes and D. Baglee, "Statistical bin limits: an approach to wafer dispositioning in IC fabrication," pp. 95–98, 1990.

[6] R. B. Miller and W. C. Riordan, "Unit level predicted yield: a method of identifying high defect density die at wafer sort," Proc. Int. Test Conf. 2001 (Cat. No. 01CH37260), 2001, pp. 1118–1127.

[7] M. J. Moreno-Lizaranzu and F. Cuesta, "Improving electronic sensor reliability by robust outlier screening," Sensors (Basel), vol. 13, no. 10, pp. 13521–13542, October 2013.

[8] C. H. Yang et al., "Identifying good-dice-in-bad-neighborhoods using artificial neural networks," 2021 IEEE 39th VLSI Test Symp. (VTS), 2021, pp. 1–7.

[9] E. Padonou, O. Roustant, J. Blue, and H. Duverneuil, "Spatial risk assessment on circular domains: Application to wafer profile monitoring,"

2015 26th Annu. SEMI Adv. Semicond. Manuf. Conf. (ASMC), 2015, pp. 223–227.

[10] M. Rosenblatt et al., "Remarks on some nonparametric estimates of a density function," Ann. Math. Stat., vol. 27, no. 3, pp. 832–837, 1956.

[11] N. Sumikawa, M. Nero, and L. C. Wang, "Kernel based clustering for quality improvement and excursion detection," 2017 IEEE Int. Test Conf. (ITC), 2017, pp. 1–10.

[12] S. Park et al., "Highly-Reliable Cell Characteristics with 128-Layer Single-Stack 3D-NAND Flash Memory," 2021 Symp. on VLSI Tech., 2021, pp. 1-2.

[13] J. Lim et al., "Development of 7th generation 3D VNAND Flash Product with COP structure for Growing Demand in Storage Market," 2022 Int. Conf. on Elec., Inf., and Commun. (ICEIC), 2022, pp. 1-4.

Pre and post silicon server platform transient performance using trans-inductor voltage regulator

Judy Amanor-Boadu
Data Center & AI
Intel Corporation
Phoenix, USA
https://orcid.org/0000-0003-0401-1009

Ritchie Rice
Data Center & AI
Intel Corporation
Portland, USA
ritchie.rice@intel.com

Azizi Shuma
Data Center & AI
Intel Corporation
Phoenix, USA
azizi.shuma@intel.com

Rishik Bazaz
Data Center & AI
Intel Corporation
Phoenix, USA
rishik.bazaz@intel.com

Horthense Tamdem
Data Center & AI
Intel Corporation
Portland, USA
horthense.d.tamdem@intel.com

Abstract— The power consumption of server platforms has been increasing due to increasing memory and processing power. This increases decoupling capacitors, board stack-up layers, and voltage regulator (VR) phase count, increasing server platform costs. A VR topology known as Trans-inductor VR (TLVR) seeks to address the challenges faced by traditional switching VRs, such as limited VR bandwidth, switching frequency tradeoff with efficiency, and increased decoupling capacitors. It does so by replacing the output inductor of a traditional switching VR with a transformer. The use of TLVR on server platforms promises improved performance in voltage droops and overshoot with reduced platform decoupling capacitance. This paper presents pre- and post- silicon results on a server platform that shows the benefits of implementing TLVR. Pre-silicon measurement results showing the benefits of TLVR are correlated with simulation for improved confidence in simulation models. Post-silicon measurement data further bolster the benefits of TLVR, showing an 18 mV second droop improvement at 48% decoupling capacitance reduction compared to the baseline, i.e., traditional switching VR with 100% decoupling capacitors populated on the platform.

Keywords— *datacenter, servers, workloads, CPU, post-silicon, second droop, decoupling capacitor, PDN*

I. INTRODUCTION

The increase in server platform power consumption results in increased voltage regulator (VR) phase count, decoupling capacitors, printed circuit board (PCB) stack up layer count, and complex power delivery network (PDN) solutions. In addition to these challenges, the current VR topology solutions are limited in solving the transient performance challenges. This is due to the large output inductance limiting the VR bandwidth and slowing its response to transient events resulting in larger magnitudes of droop and higher overshoot behaviors. The frequency of the VR could be increased to improve its response, but this directly impacts the efficiency and might result in expensive thermal solutions. Decoupling capacitors can be increased to mitigate undesired transient events, but board space must be sacrificed, and platform costs increases. The global shortage in multi-layer ceramic capacitors (MLCCs) [1] also creates highly challenging market conditions for the electronics industry. Cost, and the scarcity of MLCCs, require innovative methods to mitigate transient voltage droops and overshoots.

An emerging VR topology, trans-inductor VR (TLVR) [2],[3] that replaces the traditional output inductor in an interleaved buck regulator with a transformer has

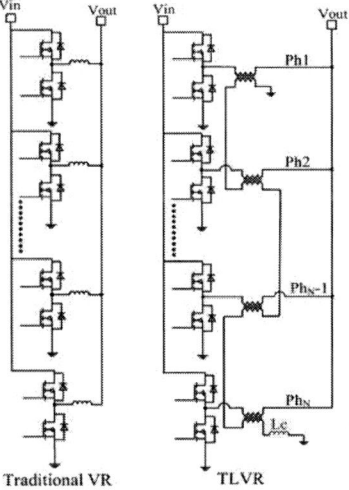

Figure 1. Traditional multiphase VR vs TLVR topologies

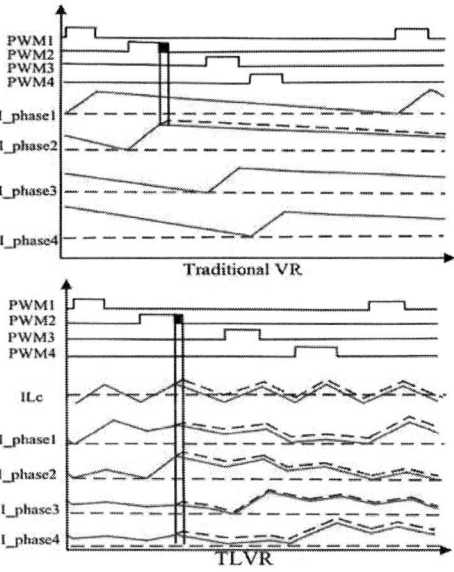

Figure 2. TLVR vs Traditional multiphase VR during transient event

demonstrated the potential of increased VR bandwidth without increasing switching frequency and thereby having no impact on efficiency. TLVR could also be used to reduce the

979-8-3503-4631-2/23 $31.00 © 2023 IEEE

switching frequency and increase efficiency resulting in lower power loss. Larger magnitudes of voltage 2nd or 3rd voltage droops are detrimental to server CPU functionality [4] and to mitigate these droops, decoupling capacitors are needed. The TLVR also responds better to transient events, i.e., lower voltage 2nd droop and smaller voltage overshoot, with reduced server platform decoupling capacitors.

Fig. 1 shows the traditional multiphase buck converter versus a multiphase TLVR. In this paper, the term primary will be referring to the source of current coupling to all phases, which is the side the compensation inductance L_c is on. The secondary will be referring to the output inductor side which is receiving the coupled current. By replacing the output inductor with a transformer in a multiphase VR, coupling of the current from each phase, on every pulse width modulation (PWM) signal, is allowed on all phases via a common connected primary of the transformer of each phase. This occurs at a beat frequency of $N * F_{sw}$, where N is the number of phases and F_{sw} is switching frequency. This reduces the total output inductance and provides a faster response to a transient load. Fig. 2 shows the TLVR and a traditional switchingVR response to a transient event. During the transient load step, the voltage across the L_c is changed, which in turn changes the L_c inductor current. The current through L_c is reflected to the secondary winding of each phase. All phase currents see the same current change. This helps reduce the output capacitors' discharge to obtain a faster transient response. The equivalent inductance L_{eq} which occurs during transient loads, reduces or eliminates the typical voltage overshoot depending on transient load frequency and VR controller solution. The magnitude of the coupling between the primary and secondary is controlled by the value of the L_c connected in series on the primary side of the transformer to ground. This is in contrast with a traditional VR where the current boost from each PWM impacts only the active phase, limiting the VR response to a transient event. This limitation will result in adding more capacitance to minimize voltage droop during load step and more inductance which increases energy during load release creating overshoot of the voltage. The tradeoff between traditional multiphase VR and the TLVR is the extra component L_c which takes extra board space and the transformer which is slightly more expensive

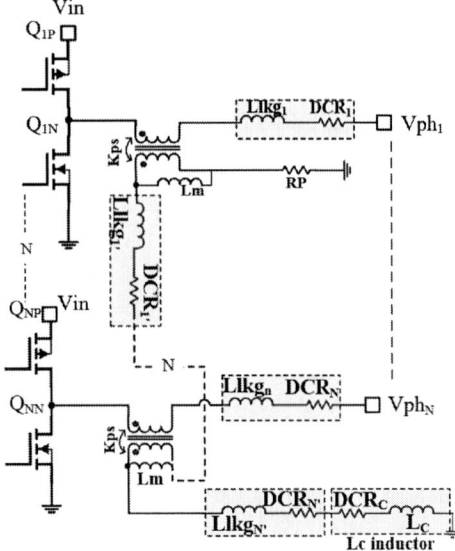

Figure 3. TLVR model showing inclusion of parasitics

than the traditional output inductor. This cost is however offset by the reduction in decoupling capacitance needed on the server platform. There is no significant difference in efficiency between the traditional switching VR and the TLVR.

Work on TLVR has been emerging over the last few years. [5] analyzed the operation of the TLVR and performed simulation results to verify that voltage undershoot could be reduced by 66%. In [6], the magnetic structure and 3D finite element models were examined to determine the tradeoffs between a traditional VR and a TLVR topology. [7] presented a theoretical foundation of the TLVR topology to design TLVR output filters. Previous works have focused on the theoretical derivations and some simulation results. This work further builds on these works by presenting actual pre-silicon and post-silicon transient measurement results and model correlation to simulation results on a server platform to show benefits of TLVR on current and future server platforms.

With the adaptation of TLVR on future platforms, it is necessary to develop modeling methodologies for pre-silicon analysis that helps accelerate the analysis of the transient behavior of different PDNs. These modeling methodologies can ensure the accurate selection of TLVR components, decoupling capacitor optimization, and error traceability. It is also important to correlate the simulation results with measurement results to improve confidence in modeling methodology and verify accuracy of models. Post silicon validation further bolsters the benefits of adapting TLVR compared to a traditional buck regulator since that will be the actual server platform performance under realistic processor workloads [8] - [10]. This paper is divided in four sections: section II describes the modeling methodology including the magnetic behavior of the transformer and L_c design impact for pre-silicon analysis and correlation with measurements, section III discusses the benefits of TLVR as demonstrated by post-silicon lab transient measurements results and conclusions are drawn in section IV.

II. PRE-SILICON VALIDATION AND CORRELATION ON A SERVER PLATFORM

Pre-silicon validation is performed to validate the server platform PDN and the VRs present on the platform. Simulation correlation with measurements helps verify the accuracy of models used in simulation. To accurately predict the performance of the response of the TLVR for the server platform PDN, SIMulation of Piecewise LInear Systems (SIMPLIS) [11] simulation tool was used.

A. Modeling methodology for trans -inductor voltage regulator

Fig. 3 shows the electrical equivalent model of the TLVR based on ideal transformer which consist of mutual inductance, magnetizing inductance Lm, coupling coefficient Kps, and leakage inductance $Llkg$ as modelled in the SIMPLIS environment. Typically, the primary to secondary turns ratio is 1:1. Lm is considered part of the mutual inductance and the resistance on each winding is connected in series (Rp). There are two ways to model the leakage inductance which derives from the electrical property of an imperfectly coupled transformer. The first method is by defining a coupling factor on the ideal transformer model in SIMPLIS and the second method shown in Fig. 3 is by subtracting the $Llkg$ from the nominal inductance of the transformer on every winding and

979-8-3503-4631-2/23 $31.00 © 2023 IEEE   212

Figure 4. Server platform with traditional inductor (left) TLVR solution with zoomed in view of transformers and L_c (right)

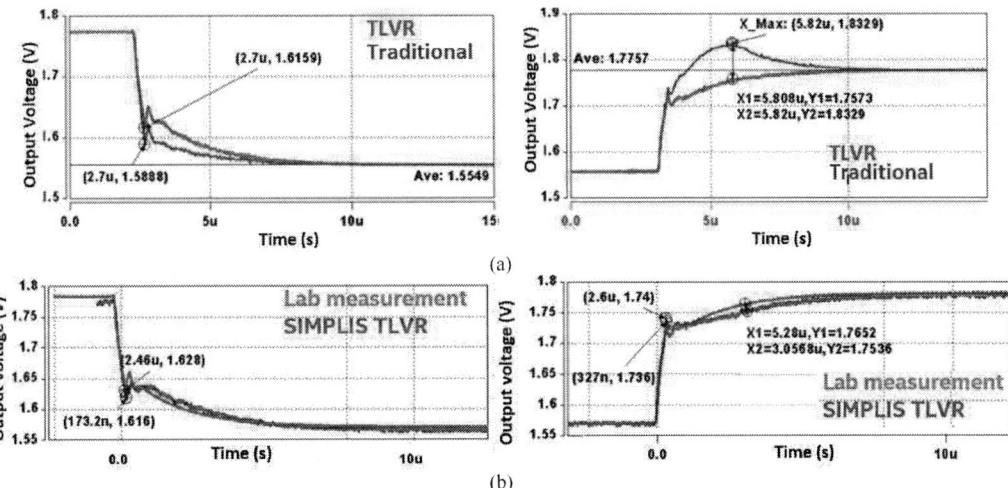

Figure 5. (a) SIMPLIS comparison of TLVR versus traditional switching VR with low frequency transient at remote sense point. (b) SIMPLIS correlation to lab results at low frequency transient. Droop (left), Overshoot (right)

add one extra-inductor per side with the *Llkg* value. The coupling factor must be set to 1 on this second method. The advantage of defining the *Llkg* in a separate component is to speed up the simulation since it allows the definition of initial conditions on the transformer and will aid in faster simulation convergence. *Llkg* must be considered when modeling a TLVR topology. If *Llkg* is not included, the simulation results will be optimistic on the voltage second droop response, leading to a miscorrelation with lab measurements. This will also hinder the use of the developed model for future predictions of platform behavior due to the model inaccuracy. Also, higher ripple and faster di/dt will be observed through the L_c. Current across the primary winding must be set equal to 0 and the current at the secondary will be the load current divided between the number of phases.

The L_c connects the primary winding of the transformers in a series loop to ground, this additional inductor helps to compensate the close loop response. L_c can be modeled as a

single resistor inductor model as shown in Fig. 3. L_c must be designed to meet ripple and overshoot requirements. The derived Eq. (1) can be used to appropriately size L_c to meet these requirements [7].

$$L_c \leq \frac{k_{ps}^2 \cdot N^2}{\dfrac{I_{max} - I_{min}}{\Delta T \cdot (D_r \cdot V_i - V_o)} - \dfrac{N}{L_m}} \quad (1)$$

where I_{max} is the maximum load current, I_{min} is the minimum load current, ΔT is the time the load current takes to go from I_{min} to I_{max}, D_r is the ramp rate of the duty cycle, V_i is the input voltage and V_o is the output voltage.

B. Correlation with lab measurements

Pre-silicon validation (before the CPU is introduced onto the platform) is performed on the server platform once the PDN has been designed adequately. For pre-silicon validation, a 16-layer 350 W Thermal Design Power (TDP) server platform was used for the study. The main input power rail for the CPU, VccCPUIN, was used for this analysis. This rail was powered by an 8-phase buck converter. The server platform was adapted for TLVR topology by removing the existing

Table I. General parameters used in validation and simulation

Section	Parameters
VR	Fsw=800k, 8 phases
Output inductor	Transformer: $L_{primary} = 105nH$, DCRp $= 0.125m\Omega$ $L_{secondary} = 105nH$, DCRs $= 0.600m\Omega$ Isat $= 85A$ @ 100C (L=74nH) Leakage: 11.5nH Lc $= 100nH$, DCRlc $= 0.125m$
Loading condition	Imin= 108A Imax = 550 A Step = 442A Slew rate = 1013 A/us (dt=423ns)

979-8-3503-4631-2/23 $31.00 © 2023 IEEE   213

Table II. Simulation and measurement results summary

	2nd droop (V)	Vmax (V)	2nd droop remarks	Vmax remarks
SIMPLIS TLVR	1.616	1.757	17% better	41% reduction
SIMPLIS Traditional VR	1.589	1.833		
SIMPLIS TLVR (Correlation)	1.628	1.765	Within 7% accuracy	Within 11% accuracy
Lab measurement (Correlation)	1.616	1.754		

Table III. Decoupling solution for VccCPUIN power rail

VCCIN Output Decoupling location	Baseline	TLVR solution
X6S, 402, 10µF, 4V	-	79
X6S, 402, 22µF, 2.5V	79	-
X6S 603, 47µF, 4V	4	4
X6S, 805, 47µF, 4V	-	25
X6S, 805, 100µF, 4V	25	-
X6S, 805, 47µF, 4V	-	24
X6S, 805, 100µF, 4V	24	-
Total capacitance (µF)	6826	3281

Table IV. Post-silicon results summary

	Baseline	TLVR solution
Load Current (A)	118 – 335	
Output capacitance (µF)	6826	3281
2nd droop (V)	1.666	1.684
3rd droop (V)	1.664	1.664
Overshoot above VID (mV)	13	0
Capacitor cost reduction (%)	0	78

output inductors from all 8 phases, attaching a reworked set of 8 transformers connected as shown in Fig. 4 and adding a single inductor for L_c.

The loading for transient testing was done using a Voltage Regulator Test Tool (VRTT) and an interposer [12] to mimic the CPU loading behavior and package, respectively. The loading conditions for transient testing were 108 A to 550 A (442 A current step) with a slew rate of 1013 A/µs. Table I summarizes parameters used for validation and simulation.

Fig. 5a shows the improvements on the overshoot when the TLVR was used. Overshoot was completely removed at low frequency and there was an improvement of 17% on the second droop with the TLVR implementation. The reduction of droop is due to all phases seeing the same high di/dt current change which provides a fast transient response as the equivalent total output inductance is reduced significantly. This also means that less output capacitor discharge is needed to support the load step and thus less capacitance required. The voltage overshoot reduction (frequency dependent) during the load release is due to the high di/dt sink (negative) current seen at all phases. This reduces the output capacitor charging and thus the amount of voltage overshoot seen at the output. The output voltage ripple at steady state was increased from less than 1mV on the traditional implementation to about 2mV with TLVR.

Fig. 5b shows the SIMPLS correlation to laboratory measurements under a low-frequency transient (1 kHz) current step, measured at the remote sense point of the VR.

With this correlation, there is improved confidence in the simulation models for server platform performance prediction. The TLVR model in SIMPLIS correlates well with the voltage droop and overshoot response using the methodology described in section II. This enables the developed model to be used in predicting the transient voltage response of future server platforms. Both simulation and lab measurement results are summarized in Table II.

III. POST-SILICON TLVR TESTING ON SERVER PLATFORM

Post silicon validation is important to verify the benefits of TLVR with the CPU installed on the platform and running real world workloads. To verify the benefits of a TLVR implementation, the platforms shown in Fig. 4 were used, but with a CPU installed in the board socket. One board was the baseline with no TLVR topology or decoupling capacitor solution changes. The second board was with TLVR rework and a reduced decoupling capacitance solution. The main input power rail for the CPU, VccCPUIN, was used for this analysis.

The baseline has 79 x 22 µF 0402 and 49 x 100 µF 0805 decoupling capacitors, while the TLVR solution has all the 49 x 100 µF 0805 capacitors changed to 47 µF and all the 79 x 22 µF 0402 capacitors changed to 10 µF. Table III summarizes the decoupling solution for the two boards. The 70 nH inductor on the baseline was changed to a transformer with a primary and secondary inductance of 105 nH and an added L_c inductor of 100 nH for the TLVR solution. The same CPU was used for the baseline and TLVR solution. This was useful for data comparison between the two boards. During the transient test, a server workload was used to stress the CPU to achieve maximum VccCPUIN current capability while measuring VccCPUIN second and third voltage droops, minimum voltage (Vmin), and maximum voltage (Vmax) while running the server workload concurrently to analyze the benefits of the TLVR solution board against the baseline board with a traditional switching VR solution.

Fig. 6 shows the post-silicon VccCPUIN transient measurement waveforms for voltage overshoot and droop. The current step induced by the CPU was 118 A – 335 A (217 A step). There was a 17 % improvement in the 2nd droop (87 mV with TLVR compared to 105 mV for the baseline) with a smaller decoupling capacitor solution. The overshoot was also

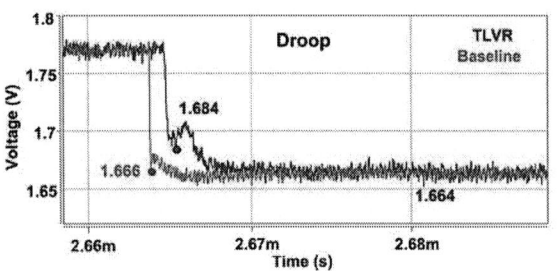

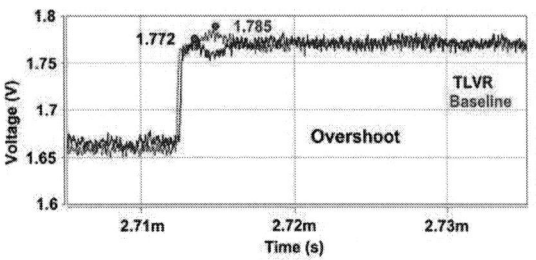

Figure 6. Post-silicon VccCPUIN Voltage droop (above) and overshoot (below) results

reduced by 13 mV compared to the baseline server platform with a traditional VR implementation. The voltage improvement has a two fold benefit. Firstly it is a cheaper solution without compromising performance. Secondly a lower voltage droop means more margin to failure. This can be used to boost performance of the CPU by using higher current till the baseline droop of 105mV is reached. Table IV summarizes the transient measurements at the remote sense point of the VR. It also shows these improvements with approximately half the total capacitance and four times less cost than the baseline board. This shows that implementing a TLVR solution on a server platform can result in cost savings with improved performance.

IV. CONCLUSION

This paper showed the improved voltage response of TLVR technology over traditional switching VR. Using SIMPLIS, the TLVR was modeled and correlated with pre-silicon transient measurements taken on a server platform to gain increased confidence in simulation models. The results of TLVR with actual silicon further bolster the positive impact of this new VR topology. There was a cost reduction of 78% due to a reduced decoupling solution without any adverse impact on CPU functionality. The droop and overshoot improvement with reduced decoupling capacitance shows the benefits of TLVR , when implemented on a server platform.

ACKNOWLEDGMENT

The authors would like to thank Ivan Mendez, Priyanka Bakliwal and Bruce Funderburgh, the VR Power Laboratory Technician for all the rework and lab testing done on the server platform with TLVR.

REFERENCES

[1] Alessandro Mascellino, "Global MLCC Shortage Continues, Drives Demand for Alternative Technologies", Available [Online]: https://eepower.com/news/global-mlcc-shortage-continues-drives-demand-for-alternative-technologies/#, accessed Jan. 15, 2022

[2] N/A, "Fast multiphase trans-inductor voltage regulator", Technical Disclosure Commons, (May 09, 2019) https://www.tdcommons.org/dpubs_series/2190

[3] S. Jiang, X. Li, M. Yazdani and C. Chung, "Driving 48V Technology Innovations Forward - Hybrid Converters and Trans-Inductor Voltage Regulator (TLVR)", *industry session APEC 2020 - Thirty-Fourth Annual IEEE Applied Power Electronics Conference and Exposition*, 2020.

[4] P. Pant and J. Zelman, "Understanding Power Supply Droop during At-Speed Scan Testing," 2009 27th IEEE VLSI Test Symposium, 2009, pp. 227-232

[5] N. Zhang, C. Zhan, G. Ye, C. Chen, X. Li and J. Yi, "Analysis of Multi-Phase Trans-Inductor Voltage Regulator with Fast Transient Response for Large Load Current Applications," *2021 IEEE International Symposium on Circuits and Systems (ISCAS)*, 2021, pp. 1-5 S.

[6] Krishnamurthy, D. Wiest and Y. Zhou, "Trans-Inductor Voltage Regulator (TLVR): Circuit Operation, Power Magnetic Construction, Efficiency and Cost Tradeoffs," *PCIM Europe 2022; International Exhibition and Conference for Power Electronics, Intelligent Motion, Renewable Energy and Energy Management*, 2022, pp. 1-6

[7] Infineon Technologies, Appl. Note AN_2011_PL12_2012_221647, pp. 5 – 12.

[8] R. A. Shetu, T. Toha, M. M. R. Lunar, N. Nurain and A. B. M. A. Al Islam, "Workload-based prediction of CPU temperature and usage for small-scale distributed systems," *2015 4th International Conference on Computer Science and Network Technology (ICCSNT)*, 2015, pp. 1090-1093

[9] M. Calzarossa and G. Serazzi, "Workload characterization: a survey," in *Proceedings of the IEEE*, vol. 81, no. 8, pp. 1136-1150, Aug. 1993

[10] Christoph Jechlitschek. "A Survey paper on Processor Workload" Online Available:https://www.cse.wustl.edu/~jain/cse567-06/ftp/processor_workloads/index.html, accessed Jan. 15, 2022

[11] J. Amanor-Boadu, H. Homer, D. M. G. Mora, P. Kumar and T. Bard, "Novel Methodology for Validating SIMPLIS Based VR Models for Server Platform Power Delivery Prediction," *2021 IEEE International Joint EMC/SI/PI and EMC Europe Symposium*, 2021, pp. 168-173

[12] W. Xu, J. He and D. Zhong, "Power delivery modeling for full switching voltage regulator on high performance computing system," 2013 IEEE International Symposium on Electromagnetic Compatibility, Denver, CO, USA, 2013, pp. 599-603

(Industry Short Paper)

Auxiliary State Machine Controlled Autonomous Design Verification Framework

Gurumurti Kailaschandra Avhad
TRX DV
Texas Instruments (India)Pvt. Ltd.
Bengaluru, 560 093, India
gurumurti@ti.com

Shitin Sahu
TRX DV
Texas Instruments (India)Pvt. Ltd.
Bengaluru, 560 093, India
s-sahu@ti.com

Navaneeth Kumar
PS DV
Texas Instruments (India)Pvt. Ltd.
Bengaluru, 560 093, India
n-kumar@ti.com

Abstract— **With increasing features in integrated circuits (ICs), design verification (DV) environments need to be more robust. Complex analog mixed-signal (AMS) systems-on-chip (SoC) require dynamic feature-dependent stimuli and output load control, which increases DV environment complexity. The proposed Auxiliary State Machine controlled DV (ASM-DV) framework effectively gains insight into the design-under-test (DUT) and helps create more autonomous stimuli, loads, and checkers. The key feature of the framework is compliance with any standard DV environment, which enhances the centralization, standardization, and reusability of the verification environment**

Keywords—design verification, auxiliary state machine, analog mixed-signal, design-under-test, system-on-chip, state-based autonomous checker controller

I. INTRODUCTION

With the increasing complexities of system-on-chip (SoC) and shrinking project cycle times, top-level design verification (DV) plays a vital role in ensuring product quality. Many standard verification methodologies have been developed and used across DV teams to ease the process of top-level verification. The complexity of analog mixed-signal (AMS) SoCs has increased over time, and more sophisticated verification techniques are needed to verify all design features exhaustively.

Feature-dependent stimulus is needed in cases where the same input or output pin at the top-level is used for different purposes. DUTs supporting multiple modes require different external loading schemes, which have significant functional and performance overhead imposing additional control requirements in DV. For verification quality and coverage, checkers are an indispensable element. The repetitive nature of the verification process naturally increases the probability of human error. Hence autonomous checkers are needed to improve the quality and depth of the verification process, which directly impacts the quality of the SoC.

The proposed Auxiliary State Machine controlled DV (ASM-DV) framework addresses these problems by integrating an auxiliary state machine into the existing DV to create an autonomous and independent DV environment. The ASM provides insight into the DUT to gain more efficient control over input stimuli, external loads, and automated checkers.

This paper talks about an overview of existing approaches to control verification entities in Section II. Section III describes the novel ASM [1] based DV framework in detail and describes the implementation and impact. In section IV, features of the ASM-DV framework are discussed. Section V discusses the advantages and limitations of the methodology. Section VI concludes the paper.

II. RELATED WORK

Efficiency of the verification environment depends on the mechanism of controlling the input stimuli, changing the output load connections and performing more dynamic checks on the device. The way of controlling these verification entities can be broadly categorized into two types:

a. *Sequential Control [1]*

Sequential execution of a block of statements provides the most effortless way to control verification flow. In standard DV flows, a typical test has a sequence of statements with time delays, fork statements etc. for better control. These statements can be tasks, functions, or direct control of certain verification elements. It is one of the most common methods to drive stimuli, enable and disable checkers, and dynamically change output loads.

b. *Procedural Control [1]*

Procedural controls are created at individual events or sequences or a combination of the events. These events can be generated either internally in the DV environment or by referring to DUT signals. A variety of these two sets of signals can be used to invoke blocks of control statements at a particular trigger. This can be effectively used to create dynamic signal-dependent control in verification.

These control schemes using tasks or functions inside the test sequence (subsequently referred to as a testcase) can effectively work for small-scale devices. Still, more precise controls using DUT-dependent reference are required for complex devices with multiple functions and features. Nevertheless, creating and maintaining these schemes for large-scale devices is challenging, which results in distributed local logic control. These conventional techniques can provide pseudo-central control but fail to perform independently with randomization and variable device response with different corner cases.

Let us try understanding these challenges in more detail with a multi-mode multi-channel transmitter-receiver device shown in Fig.1. Using different control signals, the device works in different modes of operation as follows:
(i) In Mode 1, the device transmits and receives over two single-ended, bipolar data channels.
(ii) In Mode 2, the device uses one differential channel to transmit data and another one to receive data.

979-8-3503-4631-2/23 $31.00 © 2023 IEEE

(iii) In Mode 3, the device uses a single differential channel to transmit and receive data.

The device modes can be pin selectable using two digital inputs.

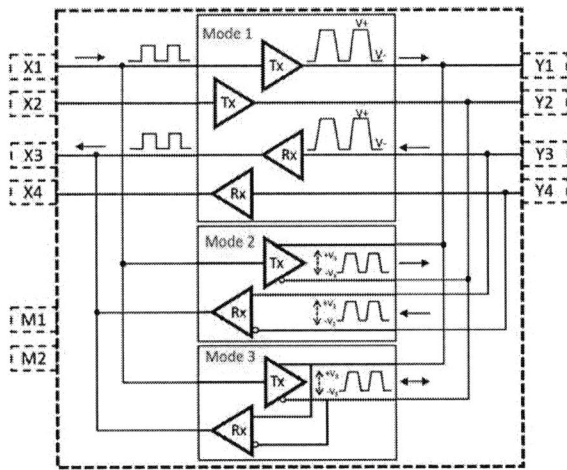

Fig.1. An example device functioning in multiple modes

Furthermore, details about input data requirements, operational voltage levels and output load requirements in different modes are explained in below Table 1:

	Mode 1	Mode 2	Mode 3
Mode select	M1=0, M2=0	M1=1, M2=0	M1=1, M2=1
Tx input pins	X1 and X2	X1	X1
Operating Frequency	f_1	f_2	f_3
Rx data input	Single-ended, Y3 and Y4	Differential, (Y3 – Y4)	Differential, (Y1 – Y2)
Slew Rates	S_1	S_2	S_3
Rx data output	X3 and X4	X3	X3
Bus Output	Single ended, Y1 and Y2	Differential, (Y1 – Y2)	Differential, (Y1 – Y2)
Bus voltage levels	V^- to V^+	$-V_2$ to $+V_2$	$-V_3$ to $+V_3$
Bus loading	Z_1	Z_2	Z_3

Table 1. Different operating parameters of the device modes

For sequentially verifying the basic transmitter and receiver functionality of the device, a typical testcase would have following steps:

1. Power up the device
2. Enable driver
3. Send data to driver input
4. Check bus for proper data
5. Disable driver and enable receiver
6. Send data on bus pins
7. Check receiver output for proper data

For executing the same testcase for different modes, we will either need separate testcases or tasks that can work differently for different modes, or we need to have additional control for selecting different tasks in different modes. Adding a new control signal referring to DUT's input stimuli and creating combinational logic around it for mode selection in the testcase or task is tedious and error-prone. With the increase in the number of control signals of DUT for different features or modes, managing verification activity will become challenging. Similar complexities are observed when creating checkers and ensuring proper external load control

for this DUT. When we make more complicated testcases randomly transitioning through multiple modes, it gets more challenging to dynamically control stimuli, set loads, and update checkers. The conventional verification environment imposes many limitations on mid to large-scale devices and limits the scope and quality of the verification.

III. ASM CONTROLLED DV (ASM-DV) FRAMEWORK

A. ASM (Auxiliary State Machine)

An auxiliary state machine (ASM) is a finite state machine (FSM) [2] that works discretely on the same stimuli as the DUT, along with a few selective internal signals (for conditions that can't be observed externally in simulation, e.g. thermal shutdown) of the DUT. An auxiliary state machine is a component of the device specification that captures the design intent in abstract states and transitions between these states [1]. For example, in a bus protocol, the abstract states of a master interface include the *idle* state, the state at which it is *waiting* for the bus, the state at which it is *using* the bus, and the state at which it is *releasing* the bus. Typically, in a bus protocol, different sets of functional properties need to be verified at the different states of the protocol. The abstract state machine, which captures the protocol transitions through these states, can be used as a reference to define the context and the scope of verification elements like checkers, output loading control, data drivability, and assertions in each state [1]. In the digital domain, auxiliary state machines are regularly used to develop assertion IPs [3]. For AMS devices, ASM creates an analog-aware top-level state machine. State machines built into a device can also be easily incorporated into the ASM.

A State Machine $\mathbf{M}$ is represented [3] as $\mathbf{M} = (\mathbf{I}, \mathbf{S}, \mathbf{O}, \delta, \lambda)$, where $\mathbf{I}$ is a set of primary inputs, $\mathbf{S}$ is a set of state symbols, $\mathbf{O}$ is a set of primary outputs, $\delta: \mathbf{I} \times \mathbf{S} \rightarrow \mathbf{S}$ is the next state function, and $\lambda: \mathbf{I} \times \mathbf{S} \rightarrow \mathbf{O}$ is the output function. The ASM can be effectively created as a finite state machine using hardware description languages (HDLs) like VHDL, Verilog HDL or higher-level verification language like System Verilog (SV). There are multiple options available to create efficient FSMs in different HDLs with the tools and automations [4]-[8].

For small or mid-size designs, ASM creation is a more straightforward process. With increasing DUT complexity for mid to large-size devices, ASM creation can be challenging, which can be simplified to some extent with individual FSMs for different modes of operations or blocks creating hierarchical concurrent FSMs [8].

B. Implementation

As shown in Fig.2, ASM-DV takes the design intent as its primary input. Design intents are converted into state transition rules for creating the ASM. For mixed-signal designs, appropriate interface elements like connect modules [9] or an analog-to-digital converter module (ADC) [10] is used as the signal digitizer, which translates any SV real number nets or electrical nets into digital outputs for given threshold values. Stimulus information and checker definitions are created in standard DV test benches using the product specifications and the design intents.

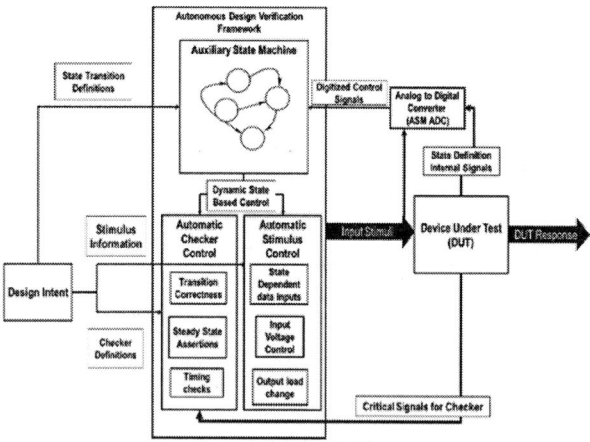

Fig. 2. ASM-DV Architecture

ASM generates additional state-based control signals, which helps create more flexible DV components. The ASM transitions into the next state according to the design intent based on the same stimuli given to the DUT. An Autonomous Design Verification Framework is created by integrating ASM with existing DV environment, providing automated checker control as well as automated stimulus control. Critical components of the framework are described subsequently.

i.   Automated Checker Control

Additional controls generated in ASM-DV are of 2 types, namely state transition control and steady-state control. Different checks can be performed on intended internal or output signals, like averaging, time delays, etc. These checkers are controlled autonomously using dynamic state-based information from ASM, as shown in Fig.3.

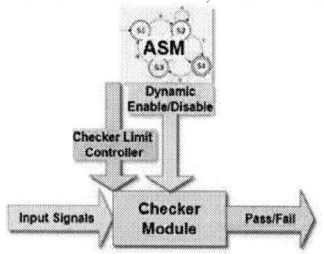

Fig. 3. Autonomous checkers in ASM-DV

The checkers are autonomously evaluated when the device reaches its corresponding steady state or goes through the required state transition. This automated control of the checkers eases the process of checker handling, allows maximum coverage and depth, and reduces manual effort. This controlling scheme has allowed us to create an independent state-based autonomous checker framework (STACC), and the details are explained in section IV-B.

Testcases with randomization could have a different set of state transitions with each run; conventional checker controller will have many limitations in such execution.

ii.   Automated Stimulus Control

Typical input stimuli to a mixed-signal DUT can consists of i) voltage inputs like supply, ground, and data, ii) current input as the measurand or reference iii) control signals, and iv) output load connections etc. ASM DV ensures that one general framework of stimulus sequence is created, which could work differently for different states. The ASM provides state information, based on which the current stimulus can switch to the new stimulus without any manual intervention. This dramatically simplifies the testcase setup. In ASM-DV, ASM can be integrated at top-level or at any hierarchy using out-of-module-references (OOMR) [11].

IV.   FEATURES OF ASM-DV

ASM-DV provides a framework with dynamic reference to DUT's functional state, showing multi-fold improvements in the verification environment.

A.   Centralized control scheme

DUT functional-state reference allows DV engineers to create a more centrally controlled environment as verification entities are more dynamic than before. For example, in the device from Fig.1, a more optimal suite of tasks can be created with the ASM state reference, creating a standard testcase that can be used across all modes effectively. This also accounts for different input data speed and loading condition requirements, thus effectively takes care of comprehensive stimulus requirements of DUT. The state machine thus becomes the predominant source of control information. This reduces distributed or isolated logic throughout the DV environment and brings down management efforts. When multiple people are contributing to the verification, ASM-DV helps create more centralized control.

ASM-DV also helps in optimizing the testbench code. Device from Fig.1, a conventional setup would require 3 different segments of code to control various different speeds, loading conditions and input voltages. With the information on the expected state of the device from the ASM, we relegate this decision away from the testcase. Say, the mode of the device as determined by the ASM is stored in the variable 'device_state'. We can now create a common task, say, 'transmit_data()'. The testcase now contains only a call to transmit_data with the relevant binary sequence. The task internally checks the value of device_state and transmits data at frequency f_1 if the device is in Mode 1, at f_2 if the device is in Mode 2, and so on. The benefits of this scheme increase as the number of states increase, as more and more conditions are relegated away from the testcase and are autonomously taken care of.

With the application of ASM-DV, top-level verification logic control is centralized and optimized further, allowing better testbench management. The number of distinct testcases required is thus reduced. For the case study described in Fig. 1, three separate testcases are reduced to one common sequence.

B.   Compounding Effect in checker coverage

ASM-DV framework supported development of state-based autonomous checker controller (STACC) automation, which helps control checkers autonomously. STACC is a spreadsheet-based (E.g., Microsoft® Excel®) automation which works on three categories of user inputs: i) signals ii) checks to be performed and iii) associated state or state transition. STACC creates autonomous framework by creating centralized checker controlling system for any standard DV flow. ASM generated states are used to control the checker enable/disable. It allows creating finer control of the checkers and increasing the quality and coverage.

979-8-3503-4631-2/23 $31.00 © 2023 IEEE

STACC creates the compounding effect of the checker coverage, as shown in Fig.3. Let us assume we have 'N' no. of tests and 'M' no. of states in ASM. Each test is unique, exercising different features and functionality of the DUT.

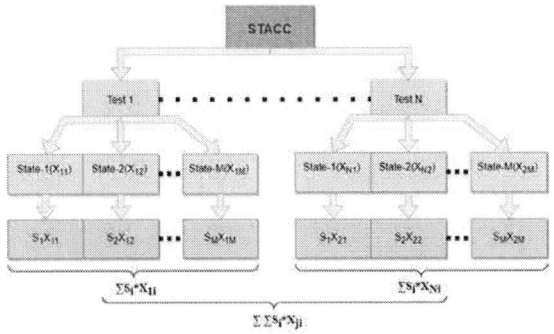

Fig. 3. Checker Compounding Effect with STACC

During the entire simulation of a testcase, DUT will transition through multiple of these states. Let us take in Test-1, DUT transition through State-1 for X_{11} times, State-2 for X_{12} times, and so on till State-M for X_{1M} times. In Test-N, DUT transitions through State-1 for X_{N1} times, State-2 for X_{N2} times, and so on till State-M for X_{NM} times. As we have integrated checkers centrally using STACC, checkers are associated with a particular state i.e., State-1 has S_1 checkers, State-2 has S_2 checkers, and so on, till State-M, which has S_M checkers.

We effectively created ΣS_i checkers and executed $\Sigma\Sigma S_i{*}X_i$ checkers in a single iteration from Test-1 to Test-N. In top-level DV regressions with multiple corners, randomization coverage of the checks increases multifold. Using STACC, the effort to maintain checkers within testcases is significantly reduced, even with an increased number of checkers.

For example, we have a glitch monitor on the receiver output of our example device from Fig.1 operating in Mode 1, Mode 2 and Mode 3, where the expected glitch widths are w_1, w_2 and w_3 respectively. With the mode information available from the ASM, STACC uses a single checker definition and automatically modifies the minimum acceptable glitch width to the respective values in each mode. Moreover, this checker is automatically enabled each time the device goes into any of these modes independent of execution of the testcase. If we have 10 testcases, where the device enters both Mode 1 and Mode 2 once in each testcase, giving 20 checker outputs just with single checker definition. This showcases the coverage compounding. It is important to note that none of these testcases need to have statements to enable or disable the checker at any time. This type of framework inherently creates checkers that are more configurable and parameterized. It helps in finding elusive bugs, improving quality of the final product and reducing number of project cycles.

V. ADVANTAGES AND LIMITATIONS

A. Improved Standardization

Centralized autonomous DV management of ASM-DV helps standardize overall flow. Due to the increased dynamism of verification elements, it helped reduce the individual requirements of top-level test benches (TBs). The reduction in control code, more central and robust task and function management scheme, and independent checker framework improve overall DV standardization. ASM-DV eases the creation of protocol-based checks due to availability of DUT functional states.

B. Improved Reusability

With enhanced centralization and standardization in DV, the same family of devices can easily reuse the ASM-DV verification environments. STACC-integrated checkers are associated with states alone; hence, they can be reused easily for an entire family of devices with minimal updates (e.g. checker limits). Similarly, testcases, tasks, and other DV elements are more generalized, allowing maximum reusability of the environment. ASM-DV helps in creating multi-DUT verification setup very efficiently.

C. Verification Coverage for Analog-Heavy DUTs

As ASM represents abstract behavioral state machine of the DUT, code coverage reports for ASM helps identifying the verification gaps associated with any missed states or state transitions. Relevant testcases can be created to bridge coverage gaps, improving effective functional coverage of analog-heavy DUT.

D. Stronger Alignment

ASM generation requires a complete system-level understanding and inputs from all domains of design and systems for better accuracy. In this process, all alignments and learnings are captured in an executable form for DV.

E. Limitations

1. Creation of the ASM for the first time and verifying its adherence to device specifications can create an overhead. The functionality of the state machine needs to be verified against a basic stimulus during creation.
2. The state machine uses discrete valued versions of the input stimuli. There is the possibility of information loss at this analog to digital interface.

VI. CONCLUSION

With increasing expectations from DV, the verification process needs to evolve further to support more feature-rich DUTs. DV teams have developed multiple standard methodologies for top-level design verification to ease and enhance the verification. To create better control, conventional DV environments have decentralized control using stimulus and internal signals of the DUT. Challenges of managing the control system increase exponentially with multi-member verification teams. Inconsistencies in understanding within team members finally leads to gaps in the verification process. With increasing expectations from DV teams to maintain the excellent quality of the product, verification environments need to be more robust and autonomous. ASM-DV framework provides methodical way to generate autonomous control, as an easy plug-in with existing DV methodology. ASM-DV impact on different verification entities is summarized in Table 2 below.

Testcases	Checker	Stimuli & Load
Remove distributed control logic	Improve quality and coverage	More autonomous and efficient control

Table 2. ASM-DV impact on top-level verification element

979-8-3503-4631-2/23 $31.00 © 2023 IEEE

REFERENCES

[1] S. Mukherjee and P. Dasgupta, "Auxiliary State Machines and Auxiliary Functions: Constructs for Extending AMS Assertions," *2011 24th Internatioal Conference on VLSI Design*, Chennai, India, 2011, pp. 52-57, doi: 10.1109/VLSID.2011.27.

[2] Gopika Kumar, Guha Lakshmanan, Otis Gorley, "A NOVEL ANALOG CENTRIC AUTOMATED VERIFICATION METHODOLOGY DRIVEN BY STATE DIAGRAM APPROACH", DAC 2022

[3] C. Pruteanu and C. Haba, "GenFSM a Finite State Machine Generation Tool", 9th International Conference on DEVELOPMENT AND APPLICATION SYSTEMS, pp. 165-168, 2008.

[4] A.T. Abdel-Hamid, M. Zaki and S. Tahar, "A tool converting finite state machine to VHDL", *Canadian Conference on Electrical and Computer Engineering 2004 (IEEE Cat. No. 04CH37513)*, pp. 1907-1910, 2004.

[5] M. Cgra, S. Li, J. Malik, S. Liu and A. Hemani, "A Code Generation Method For System-Level Synthesis on ASIC FPGA and Manycore CGRA", *Proceedings of the First International Workshop on Manycore Embedded Systems*, pp. 25-32, 2013.

[6] S.H.M. Durand and V. Bonato, *A tool to support Bluespec System Verilog coding based on UML diagrams*, vol. 12, pp. 4670-4675, 2012.

[7] S. H. Masoumi, S. A. R. Al-Haddad and F. Z. Rokhani, "New tool for converting high-level representations of finite state machines to verilog HDL," *2017 IEEE 15th Student Conference on Research and Development (SCOReD)*, Wilayah Persekutuan Putrajaya, Malaysia, 2017, pp. 1-6, doi: 10.1109/SCORED.2017.8305431.

[8] Bilung Lee and E. A. Lee, "Hierarchical concurrent finite state machines in Ptolemy," *Proceedings 1998 International Conference on Application of Concurrency to System Design*, Fukushima, Japan, 1998, pp. 34-40, doi: 10.1109/CSD.1998.657537.

[9] P. Frey and D. O'Riordan, "Verilog-AMS: Mixed-signal simulation and cross domain connect modules," *Proceedings 2000 IEEE/ACM International Workshop on Behavioral Modeling and Simulation*, Orlando, FL, USA, 2000, pp. 103-108, doi: 10.1109/BMAS.2000.888372.

[10] Verilog-AMS Language Reference Manual 2.3.1 June. 2009.

[11] SystemVerilog Language Reference Manual 3.1a

[12] Radu Negulescu, "Event-Driven Verification of Switch-Level Correctness Concerns", Proceedings 1998 International Conference on Application of Concurrency to System Design

Special Session: Neuromorphic hardware design and reliability from traditional CMOS to emerging technologies

Fabio Pavanello[1], Elena Ioana Vatajelu[2], Alberto Bosio[1], Thomas Van Vaerenbergh[3], Peter Bienstman[4],
Benoit Charbonnier[5], Alessio Carpegna[6], Stefano Di Carlo[6], Alessandro Savino[6]

[1]Univ. Lyon, Ecole Centrale de Lyon, INSA Lyon, Université Claude Bernard Lyon 1, CPE Lyon, CNRS, INL
[2]Univ. Grenoble Alpes, CNRS, Grenoble INP, TIMA, 38000 Grenoble, France
[3]Hewlett Packard Labs, HPE Belgium, B-1831 Diegem, Belgium, [4]Ghent University - imec, Gent, Belgium
[5]Univ. Grenoble Alpes, CEA, LETI, Grenoble, France
[6]Politecnico di Torino, Control and Computer Eng. Department, Torino, Italy

Abstract—**The field of neuromorphic computing has been rapidly evolving in recent years, with an increasing focus on hardware design and reliability. This special session paper provides an overview of the recent developments in neuromorphic computing, focusing on hardware design and reliability. We first review the traditional CMOS-based approaches to neuromorphic hardware design and identify the challenges related to scalability, latency, and power consumption. We then investigate alternative approaches based on emerging technologies, specifically integrated photonics approaches within the NEUROPULS project. Finally, we examine the impact of device variability and aging on the reliability of neuromorphic hardware and present techniques for mitigating these effects. This review is intended to serve as a valuable resource for researchers and practitioners in neuromorphic computing.**

Index Terms—**augmented silicon photonics, neuromorphic hardware, artificial neural networks, spiking neural networks, reliability, phase change materials**

I. INTRODUCTION

Artificial Neural Networks (ANNs) have enabled complex computations but require significant computational resources for both training and inference [1]. The main bottleneck in these networks is the transfer of large amounts of data to support different tasks. However, the trend in computing is moving towards edge devices, such as the Internet of Things (IoT), to improve security and reduce power consumption and latency [2]. Furthermore, there is a growing demand for powerful yet energy-efficient accelerators in various fields, including fault detection in microprocessors [3] and intrusion detection systems [4].

The design space for ANNs is vast, and it involves choices in four fundamental aspects: the neuron model, the architecture

This project has received funding from the European Union's Horizon Europe research and innovation programme under grant agreement No. 101070238. Views and opinions expressed are however those of the author(s) only and do not necessarily reflect those of the European Union. Neither the European Union nor the granting authority can be held responsible for them. It was also partially supported by the ANR within the EMINENT Project ANR-19-CE24-0001

structure, the information encoding, and the training method [5]. These choices can significantly impact the hardware design and optimization process. Convolutional Neural Networks (CNNs), Deep Neural Networks (DNNs), and Spiking Neural Networks (SNNs) are three popular types of ANNs, each with unique characteristics and applications.

CNNs are commonly used in image recognition and processing tasks, and they rely on convolutional layers to extract spatial features from input images [6]. DNNs, on the other hand, are used in a wide range of applications, from speech recognition to natural language processing and game playing [7]. They are characterized by multiple layers of neurons that learn increasingly abstract features of the input data.

The limitations of traditional computing architectures, particularly the communication bottleneck between memory and processor and the latency of information propagation and manipulation, have highlighted the need for alternative approaches to ANNs. While software approaches for ANNs offer advantages when implemented on specialized hardware such as Graphical Processing Units (GPUs), these limitations persist [8].

SNNs have emerged as the next generation of ANNs that exchange information in spikes, inspired by the behavior of biological brains [9]. This allows for more efficient computation and reduced power consumption and is particularly interesting when working with time sequences such as audio, video, and electrical signals [10], [11]. The model complexity and internal parameters determine the model's suitability to the input data, with shorter time constants detecting shorter temporal correlations and higher values catching more prolonged time effects.

Eventually, the resilience of the ANN is crucial when designing the entire system and cannot be ignored [12], [13]. Retaking inspiration from biology, the human brain is intrinsically resilient to malfunctioning and faults. It loses approximately 50000 neurons daily but can still perform complex tasks and learn new ones, creating connections between the remaining neurons. ANNs have inherited this characteristic at

979-8-3503-4631-2/23 $31.00 © 2023 IEEE

a certain level, but they still need to improve significantly, particularly in mission-critical and safety-critical applications. Therefore, a deeper study of ANNs from this perspective is required.

Neuromorphic computing, which merges memory and processing units within neurons and synapses and maps computing architectures more closely to Neural Networks (NNs) models, offers a promising solution to address the limitations of traditional computing architectures [14]. Various technologies, including CMOS, memristors, and optoelectronics/all-optical approaches, have been explored to develop neuromorphic hardware. This paper provides an overview of these approaches, highlighting their advantages and limitations. Background on ANNs is given in section II, while section III discusses the design and performance of the digital version of ANNs, with a focus on the SNNs. section IV discusses the potential of silicon photonics for ANNs, and section V covers reliability challenges in the usage of ANNs.

II. BACKGROUND

The choice of the neuron model, architecture structure, information encoding, and training method can significantly impact the hardware design and optimization process for each of these ANNs. Therefore, it is essential to consider these aspects carefully when designing and implementing hardware for ANNs.

CNNs and DNNs use different neuron models. In CNNs, the neuron model is typically based on the Rectified Linear Unit (ReLU) function, a non-linear activation function commonly used in deep learning. The ReLU function is simple and computationally efficient, making it a popular choice for CNNs. The ReLU function is $f(x) = \max(0, x)$, where x is the input to the neuron. The output of the ReLU function is zero if the input is negative and equal to the input if the input is positive. The ReLU function effectively reduces overfitting in DNNs and has been used in various computer vision tasks, such as object detection and recognition [15].

In DNNs, the neuron model is typically based on a non-linear activation function. The sigmoid function, defined as $f(x) = \frac{1}{1+e^{-x}}$, where x is the input to the neuron, is one of the simplest. The output of the sigmoid function is between 0 and 1, which makes it useful for tasks such as classification. Another popular activation function in DNNs is the hyperbolic tangent (tanh) function, similar to the sigmoid function but outputs values between -1 and 1. The choice of activation function depends on the task at hand and the NNs's architecture. For example, the sigmoid and tanh functions were commonly used in early DNN architectures, such as the Multilayer Perceptron (MLP) [16] but have since been largely replaced by the ReLU function in more recent architectures. However, the sigmoid and tanh functions are still used in certain NNs, such as Recurrent Neural Networks (RNNs) and autoencoders [17].

State-of-the-art implementations of CNNs and DNNs often use 32-bit floating-point numbers in software or model-based

approaches. However, implementing such algorithms in hardware is challenging due to their extensive data requirements, high energy consumption, and large memory bandwidth. To address these challenges, quantization, and regularization techniques have been explored, using fixed-point computations with 16 bits, 8 bits, or lower precision. Although these methods reduce the precision of synaptic weights and inter-layer signals, IBM's TrueNorth [18] chip has achieved acceptable precision using only five synaptic states, albeit at high design and energy costs [19], [20].

In the case of SNNs, the neuron models and architectures are the most complex to target due to the nature of the information they carry and how they treat it. Many different mathematical models describe and mimic the behavior of biological neurons, such as the Hodgkin-Huxley model [21], the Izhikevich model [22], the Leaky Integrate and Fire (LIF) model [23], and the Integrate and Fire (IF) model [24]. These models range from very complex and detailed to much simpler and more suitable for machine learning and hardware applications, with varying degrees of biological plausibility and computational efficiency.

Neurons in SNNs are generally treated as leaky integrators where input spikes are integrated over time after being weighted by corresponding synapses, affecting the neuron's state, usually the electrical potential across its membrane. The neuron membrane depolarizes due to internal charge leakage without spikes, except in the IF model, where it is kept at a constant value. A new spike is generated when the membrane potential exceeds a specific threshold value, causing the neuron to fire, and the potential drops suddenly into a reset state.

The architecture of the SNN can also be very flexible. The literature reports Fully-Connected (FC) Feed-Forward (FF) SNN [25], regularly recurrent structures [10] or randomly recurrent architectures [26], used for example, in Reservoir Computing (RC). There can be only excitatory connections, with positive weights, or adding inhibitory connections, with negative weights [25], or in more detailed models, separated excitatory and inhibitory neurons, as observed in some regions of the human brain.

RC is an efficient technique where a randomly initialized RNN trains only a linear combination of the signals at each node. It has shown promising results in various applications, including photonics [27]. Different RC architectures have been investigated using photonics, such as spatial-multiplexing and time-multiplexing approaches. Spatial-multiplexing involves physically separating the nodes, while time multiplexing requires a faster sampling speed and more complex processing of the read-out layer.

The choice of input encoding can significantly affect the performance of the CNN or DNN. It often requires careful consideration and experimentation to determine the optimal encoding for the given task. Factors to consider when selecting an input encoding include the nature and complexity of the input data, the available computational resources, and the application's performance requirements.

The most common input encoding for CNNs is raw pixel

979-8-3503-4631-2/23 $31.00 © 2023 IEEE

values, where each pixel in the image is represented as a numerical value. In contrast, for DNNs that process non-image data, input encoding may involve feature engineering techniques such as transforming the raw input data into a set of meaningful features more amenable to learning by the network [15]. For example, in natural language processing, input encoding may involve converting text into a numerical representation, such as bag-of-words or word embeddings.

Due to their nature, SNNs require more advanced methods for encoding and interpreting information. The main approaches are rate coding [28], temporal coding [29], and population rank coding [30]. Rate coding uses the average spike frequency to encode information and is suitable for static input data. It is less efficient regarding spike activity but more robust to noise. Temporal coding encodes information in the precise arrival time of spikes or their relative distance, requiring fewer spikes to process information, and is suitable for encoding time-varying signals. However, it is more sensitive to noise. Population rank coding uses the joint activity of a group of neurons to process information.

Training approaches involve optimizing the model parameters to minimize a given loss function for CNNs and DNNs. In supervised learning, this involves iteratively adjusting the weights and biases of the network to reduce the difference between the predicted output and the ground truth labels. The optimization is typically performed using gradient descent methods, which involve calculating the gradient of the loss function concerning the model parameters and updating them in the direction of the negative gradient. Commonly used gradient descent methods include Stochastic Gradient Descent (SGD) [31], AdaGrad [32], etc.

Regularization techniques such as Dropout [33], etc., are often used to prevent overfitting and improve generalization. Additionally, data augmentation methods such as flipping, rotating, and cropping the input images are employed to increase the size of the training dataset and improve model robustness [34].

Unsupervised learning approaches, such as Autoencoders [17] and Restricted Boltzmann Machines [35], can also be used for pretraining the model parameters. Transfer learning approaches can also be employed, where a pre-trained network is fine-tuned on a new dataset with similar or related features [36]. Finally, reinforcement learning can also be used to train CNNs and DNNs, where the model learns to take actions based on a reward signal, such as in game-playing agents [37].

Training SNNs is challenging due to the non-differentiability of the thresholding function of neurons. Classical back-propagation methods cannot be directly applied, but several approaches have been developed, including supervised and unsupervised training. In supervised training, the most common approach is to convert an ANN into an SNN, where the ANN's differentiable non-linear function is trained using back-propagation, and the weights are used directly in the SNN. Alternatively, a Back-Propagation Through Time (BPTT) can be applied directly to the SNN,

where a surrogate gradient replaces the neuron's thresholding function with a differentiable function during the backward pass. In contrast, inspired by biology, most unsupervised approaches update weights locally based on the relative spike timing between the inputs and the output without depending on a global error signal propagating across the network, resulting in a lighter memory footprint and computational overhead, with Spike Timing Dependent Plasticity (STDP) being the most common method [38].

III. DIGITAL ACCELERATORS

As seen before, the architecture of an ANN, and in the same way of an SNN, is generally composed of many independent neurons and, as such, is intrinsically strongly parallelizable. This poorly fits the common CPU-based computing approach, in which the parallelism is limited to a few tens of very powerful cores. For this reason, one active research branch in the field of ANN and SNN is directed towards accelerating such algorithms, using computing platforms to execute them more efficiently. The goal is to broaden their application to many contexts, such as performance-constrained, power-constrained, or real-time tasks.

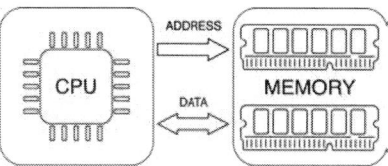

(a) Classical Von Neuman computing approach

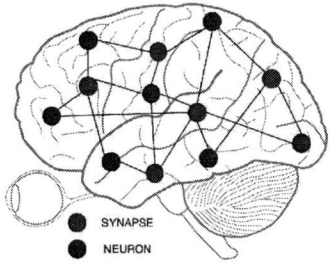

(b) Neural network's computation and memory co-location

When discussing acceleration in the digital domain, several solutions have different degrees of optimization, efficiency, and cost. One first approach, already mainstream for many ANN models, is to exploit the computational parallelism offered by general-purpose hardware accelerators like GPUs. Regarding SNNs, several software frameworks natively support the deployment of the code on GPUs, for example, the ones based on *pyTorch*, like *snnTorch* [39] and *spikeTorch*, or CUDA accelerated C++ frameworks, like *SLAYER* [40] and CARLsim 4 [41].

However, SNNs have many features unsuitable with a GPU execution. For example, spikes can be represented in the digital domain as single-bit events (high in the presence of a spike and low otherwise). The numerical representation used in GPU, based on words with 8, 16, 32, 64, or similar bit widths, is inefficient. Moreover, the computation can be based on events: a neuron can react only in the presence of an active input spike, remaining in a quiescent state otherwise. Such an update policy would allow exploiting the sparsity of spikes typical of SNNs, with considerable savings in switching power, but again is not supported by GPUs.

This incompatibility has pushed for developing specialized hardware accelerators explicitly designed to support SNN features. Interestingly SNNs are intrinsically more suitable for developing these accelerators than other ANN models. The main reason is again how SNN encodes the information, which in the digital domain corresponds to single-bit signals. This drastically reduces the interconnection and memory requirements.

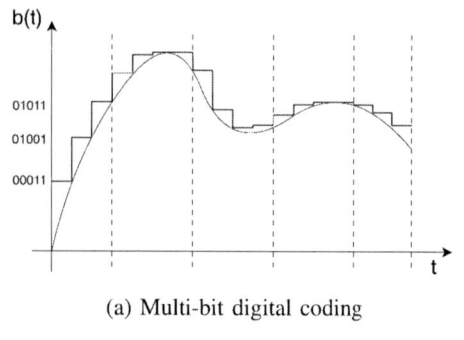

(a) Multi-bit digital coding

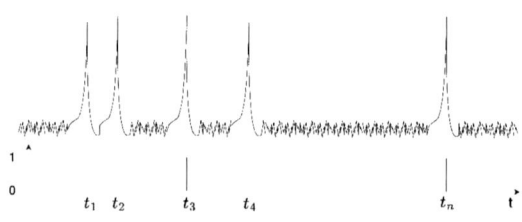

(b) Spikes information encoding

When designing hardware accelerators, the roads that can be followed are two: fixed hardware or reconfigurable hardware. In the first case, the accelerator becomes an Application Specific Integrated Circuit (ASIC), able to perform in a very efficient way the specific tasks for which it has been designed, but nothing more. On the other hand, specific hardware platforms can be reconfigured multiple times, allowing for flexible architectures. This is the case of Field Programmable Gate Arrays (FPGA). The choice between the two depends a lot on the kind of application and constraints. Generally, designers try to fit as many functionalities as possible when developing an ASIC since the hardware remains fixed.

Companies and universities are investing in developing programmable chips to enable fast simulation of large-scale SNN. Among them, a first attempt is SpiNNaker [42]. The idea behind SpiNNaker is to use classical CMOS architectures, particularly ARM968 RISC processors, to simulate the neurons' dynamics and optimize the routing between them to fit a spike-based communication perfectly. A specific communication

protocol, i.e., Address Event Representation (AER), is used to do this. This protocol is designed explicitly for neuromorphic circuits, in which a spike is represented through the ID of the neuron that generated it and the corresponding generation time stamp. The IBM TrueNorth [18] is a fully custom ASIC realized in 28nm CMOS technology. Again the neurons use standard CMOS digital gates, but the communication is asynchronous. The Intel Loihi [43] has an approach that is a hybrid between the previous two. It has 128 custom chips implementing 1024 LIF primitives. The custom connection mesh optimizes the typical sparse communication between spiking neurons. Additionally, Loihi includes specific components to perform learning directly on-chip. It provides a set of configurable parameters to allow different local learning rules, from a basic STDP to more complex alternatives. Finally, Tianjic [44] is a hybrid ANN/SNN accelerator with a custom interconnection between the various cores.

In the field of ASICs, it is worth citing ODIN [45], a small network implemented in 28nm Fully-Depleted Silicon-On-Insulator (FDSOI) CMOS technology and targeting low-power applications. Neurons can be configured to implement a LIF model or one of the 20 possible Izhikevich behaviors. The routing of the spikes between neurons is again performed through an AER protocol.

Table I compares the accelerators in terms of area and energy efficiency, expressed in Giga-Synapse Operations (GSOPS) per Watt. For a more detailed comparison, see [46].

TABLE I: ASIC comparison table [46]

Design	[42]	[18]	[43]	[44]	[45]
# of neurons	18k	1M	128k	39k	256
# of synapses	18M	256M	128M	9.75M	64k
Area (mm^2)	88.4	413	60	14.4	0.086
Process (nm)	130	28	14	28	65
Energy (GSOPS/W)	0.033	400	-	649	78.7

The last approach in designing the SNN accelerators is to target a reconfigurable hardware platform, such as an FPGA. The FPGA can host different accelerators. This is the reason why many embedded platforms are starting to include them. Second, online reconfigurability allows hardware modification while the system is on. This can be used to add an accelerator after the system has been deployed, remove it once it is no longer required, and modify its functionality.

Taking SNN as an example, the architecture of the network can be modified to target a different set of data if necessary. Finally, partial reconfigurability can add and remove functionalities to a specific accelerator. For example, online learning can be activated and deactivated on request, enabling and disabling the corresponding circuitry and physically adding or removing the required piece of hardware, guaranteeing the optimal architecture for the required application. This is the idea behind [25], where the authors started to design a tiny hardware accelerator to fit an FPGA together with other components employing a LIF neuron model.

The idea is then to have a first degree of reconfigurability, making the accelerator programmable in many aspects, such

TABLE II: FPGA comparison table [47]

Design	[48]	[49]	[50]	[47]	[25]
Clock frequency(MHz)	75	120	25	100	100
Data format	16bit Fixed	8bit Fixed	32bit Fixed	16bit Floating	16bit Fixed
Computing scheme	Event-Driven	Clock-Driven	Event-Driven	Adaptive Clock/Event-Driven	Clock-Driven
Neuron model	LIF	LIF	LIF	LIF	LIF
FPGA platform	Spartan 6	Virtex 6	Spartan 6	Virtex 7	Artix 7
Neurons	1794	1591	1794	1094	1384
Synapses	647000	638208	647000	177800	313600
Task	MNIST	MNIST	MNIST	MNIST	MNIST
Computation time	$0.53s$/image	$8.40s$/image	$0.16s$/image	$3.15ms$/image	$215\mu s$/image
Computation time @100MHz	$0.40s$/image	$10.08s$/image	$40.00ms$/image	$3.15ms$/image	$215\mu s$/image
Energy	$0.80J$/image	$1.12J$/image	Not reported	$5.04mJ$/image	$13mJ$/image
Energy/Synapse	$1.2\mu J$/synapse	$1.76\mu J$/synapse	Not reported	$0.028\mu J$/synapse	$0.041\mu J$/synapse

as weights and thresholds. Then, to add a layer of flexibility by allowing easy modification of the hardwired network hyperparameters, such as the membrane time constant, the network architecture, the internal bit-widths, etc. Several other works are targeting a more standard but still configurable implementation, such as [48], [49], [50], [47]. Table II compares different accelerators. For more details, see [47].

In general, digital accelerators can help a lot in increasing the execution efficiency of SNNs. CMOS technology is decades old and nowadays widespread, low-cost, and highly optimized. However, the intrinsic behavior of digital devices is very far from that observed in biological components, and the response time and power consumption are still a burden. Augmented silicon photonics platforms can cover most design aspects with more efficient solutions.

IV. AUGMENTED SILICON PHOTONICS PLATFORMS

Integrated photonics is one of the key technologies that has been investigated to build neuromorphic hardware [51]–[54]. In particular, Photonic Neural Networks (PNNs) based on silicon photonics have been extensively investigated for developing lightweight, low-latency, high-speed computing hardware with ultra-low power consumption [54]–[57].

Such properties arise from the intrinsic nature of light manipulation and propagation and the capabilities currently integrated photonics platforms can offer. For example, different frequencies of light can be used to encode different data streams separately onto each frequency and then be processed in parallel, thus increasing computing density [53]. Such wavelength multiplexing approaches are beneficial for increasing the parallelization degree of architectures. This key feature is used in broadcast and weight protocol where each frequency channel has a specific weight assigned (for each layer) before being summed up together, e.g., by a photodetector [58]. In this approach, ring resonators are key devices enabling the multiplexing (filtering) and weighting of the signals at different frequencies [54].

Indeed, photonic approaches allow light manipulation (e.g., weights application) while preserving signals propagation at the speed of light throughout the photonic network, thus resulting in ultra-low latencies, limited only by the physical size of the network leading to orders of magnitude lower values compared to electronics implementations [55].

Another essential feature of photonic neuromorphic systems is the possibility of operating with analog complex-valued signals, which is beneficial to leverage non-linearities in neuromorphic hardware such as the electro-optic conversion between complex-valued optical fields into intensities (i.e., photocurrents at the photodetection), but also thanks to connection matrices presenting a more considerable richness in degrees of freedom [52], [55], [59].

Furthermore, PNNs can operate at much higher speeds than digital accelerators, with their main limitation coming from electro-optic conversion stages, e.g., at the read-out of a PNN where photodetectors allow to operate at speeds of hundreds of GHz depending on the technology and responsivity required [55].

Among the various integrated photonic platforms that have been considered, Silicon-on-Insulator (SOI) platforms are those that have attracted the most vital interest thanks to the availability of both active and passive components and their reduced footprint compared to platforms with lower refractive indices contrasts, such as Silicon Nitride-on-Insulator (SiNOI).

Table III shows a comparison under different metrics for digital accelerators, flash technology, and three different types of PNNs based on hybrid lasers, co-integrated silicon photonics, and sub-λ nanophotonics. For the latter, the device footprint shrinks by at least an order of magnitude due to the robust localization of the optical fields, e.g., in photonic crystal cavities. It is worth noting that such constrained photonic approaches can expect a significant gain in energy consumption and latency. More information on the specific implementations can be found here [56].

In particular, SOI platforms can provide high-speed modulators, e.g., in MZI or Ring Resonator (RR) configurations, as well as high-speed broadband photodetectors (> 50 GHz) and low propagation losses (< 3 dB/cm) [65]. More specifically, MZI devices allow to modulate light and change its amplitude (and phase), therefore providing a practical way to implement ANN weights [55]. They can be arranged in meshes in precise ways, e.g., matrix multiplications as shown in Fig. 3(a) [60]. The traditionally used physical mechanisms for modulation are either (i) thermo-optic, (ii) electro-optic, (iii) carrier plasma effect [66] as schematically shown in Fig. 3(b). For the thermo-optic approach, one of the interferometer arms is heated up by

979-8-3503-4631-2/23 $31.00 © 2023 IEEE

TABLE III: Photonic versus electronic approaches comparison table. Latency is the time for a single matrix multiplication operation to compute at the given vector size. Speed is the time between subsequent matrix multiplies [56].

Technology	Google TPU [61]	Flash (Analog) [62]	Hybrid laser NN [63]	Co-Integrated Si NN [64]	Sub-λ nanophotonics [56]
Energy/MAC [fJ]	430	7	220	2.7	0.03
Comp. density [TMACs/s/mm^2]	0.58	18	4.5	50	5000
Vector size	256	100	56	148	300
Precision [bits]	8	5	5.1+	5.1+	5.1+
Latency/speed [ns]	2000/1.42	15	< 0.1	< 0.1	< 0.05

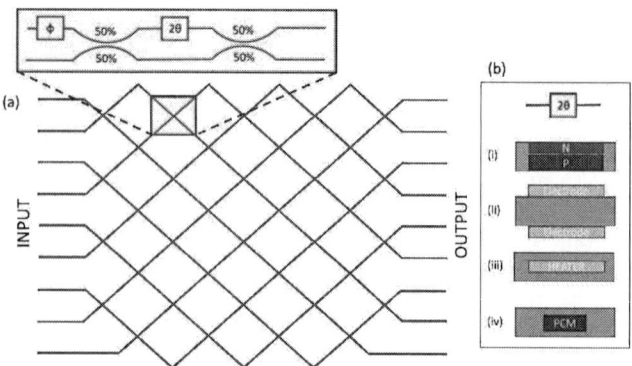

Fig. 3: (a) Photonic implementation of a network for 9x9 matrix multiplication based on Mach-Zenhder Interferometers (MZIs) [60]. Inset describes what each crossing consists of, i.e., a phase shifter (ϕ), then a 50/50 splitter, another phase shifter (2θ), and then a 50/50 combiner. (b) Top-view of phase shifter implementations in waveguides based on (i) thermo-optic effect, (ii) electro-optic effect, (iii) carrier plasma effect, and (iv) Phase Change Material (PCM)-induced shift. Blue lines where no crossing is present are optical waveguides.

a micro-heater, resulting in a change in the refractive index of the arm. In the second case, electrodes can establish an electric field (for electro-optic materials - not present in native Si platforms) that modifies the refractive index. In the third case, a p-n/p-i-n junction is used where carriers concentration in the depletion region is modified by an applied voltage, thus changing the refractive index by the plasma carrier effect [66]. All these approaches are of interest (depending on the platform available) and have enabled vector-matrix multiplication and PNNs for classification tasks [55].

However, in silicon platforms, one of the main limitations during the inference process is the need to dissipate energy to keep the values of the weights. This energy can account for up to 10 mW with a total π phase shift per MZI. Such a shift allows routing full signals from one output to the other of an MZI with two output ports (see inset of Fig. 3)(a) starting from a signal only coming from the former port.

In [55], the authors used a specific arrangement of mesh devices to perform matrix multiplication through singular value decomposition. This mapping is precisely defined between the elements of the matrix and the phase shifters MZIs in the mesh. Unlike conventional electronics like CPUs and GPUs, the energy consumed per calculation (Floating Point

Operations per Second (FLOP)) reduces to zero as the size of the mesh grows larger (assuming non-volatile weights). In contrast, conventional electronics require a fixed amount of energy per FLOP, and the overall energy consumption scales quadratically with the size of the problem, i.e., N^2 instead of just N for PNNs with N representing the mesh size. The authors in [55] also used this MZIs as weights to build a Feed-Forward Neural Network (FFNN) that could analyze vowels, achieving a simulated correctness of 90%, comparable to that of a digital computer which would reach 91.7%.

One solution to avoid constant power dissipation is setting the weights using waveguides integrating non-volatile materials such as PCMs, e.g., above the waveguide as in (iv) approach in Fig. 3)(b). Their response (change in the degree of crystalline to amorphous ratio) can be set using electrical or optical pulses [67]. Such materials are exciting also to implement STDP thanks to their very rapid response time (sub-ns) to stimuli, thus allowing to build of optical plastic synapses and spiking architectures based on ring resonators [54].

The recently started Horizon Europe NEUROPULS project investigates a series of approaches that leverage silicon photonics platforms with the addition of PCMs and III-V materials for building more efficient neuromorphic hardware.

An approach to improving CNN performance is to use an accelerator designed for matrix-vector products, such as a mesh of modulators programmed to perform a specific matrix multiplication. Singular value decomposition can factor the matrix, resulting in more efficient computation on the accelerator. This approach can also handle other matrix operations, including convolutions, and optimize hardware architecture for specific tasks, leading to significant training and inference times speed-ups [55].

Although optical components can benefit matrix operations, their size can limit the size of matrices that can be implemented. To overcome this limitation, alternative approaches can be explored. One possible solution is to employ pruning techniques that remove unnecessary connections and weights from the network, thereby reducing the overall size of the matrix. This makes it possible to implement larger systems using the available optical components.

Another option is to use block matrix decompositions, which involve dividing the matrix into smaller blocks that can be processed separately. This approach can enable the implementation of larger matrices using a smaller number of optical components. The smaller blocks can be computed

979-8-3503-4631-2/23 $31.00 © 2023 IEEE

independently and combined for the final result. This technique can also be combined with other optimization strategies, such as quantization and compression, further to reduce the size and complexity of the system.

The tensor-train approach proposed in [68] will be explored. This approach represents the matrix as a product of low-rank tensors, which can be processed more efficiently using optical components. The goal is to develop a scalable optical architecture capable of handling larger matrices and more complex neural networks using the abovementioned approaches. This will facilitate faster and more efficient training of neural networks using photonics.

The research will also explore RNN applications, including fully trainable RNNs as well as RCs. We will investigate the potential applications of RC in photonics for various applications, including nonlinear dispersion compensation of telecom signals, as demonstrated in [69]. The research aims to develop new techniques and approaches for utilizing photonics in Machine Learning (ML) and other fields, which could lead to significant advancements in the performance and efficiency of these systems.

To implement non-volatile optical weights, the proposed architecture will incorporate PCMs. Previous studies, such as [54] and [70], have explored these materials and shown a strong potential for neuromorphic systems. Incorporating non-volatile weights can significantly reduce power consumption compared to volatile weights, which require continuous driving or periodic refreshing. Including these materials in the proposed architecture will be crucial in developing low-power and high-performance systems for machine learning and other applications.

However, in addition to their non-volatility, PCMs have another advantage: their nonlinear dynamics. E.g., by exciting the material with pulses rather than continuous-wave excitation, the nonlinear behavior of the material enables other computing paradigms, such as SNNs. In such networks, the neurons communicate using brief pulses or spikes rather than continuously varying signals. This opens up the implementation of energy-efficient and highly parallel neural networks. To generate the spikes injected into the system, Advanced high-extinction ratio (ER > 8 dB) Q-switched spiking lasers can be used, which will be monolithically integrated into III-V materials on the same platform [71]. These hybrid III-V-on-Si spiking lasers are a scalable and cost-effective alternative to previous Q-switched lasers made purely from III-V materials.

III-V-on-Si spiking lasers will generate highly precise and controlled optical spikes, essential for many photonics and ML applications. These lasers offer several advantages, including high extinction ratios, low power consumption, and compatibility with standard silicon processing techniques.

V. RELIABILITY STUDIES AND CONCERNS

ANNs have an intrinsic error tolerance from an algorithmic point of view thanks to their redundant nature. However, hardware designs to deploy such algorithms must be analyzed to assess the impact of hardware restrictions or faulty manifestation on the network functional's behavior.

Due to manufacturing issues, hardware faults can occur randomly, provoked by neuron and synapse defects and imprecisions. Still, they can also be malicious, introduced by different kinds of attacks (i.e., laser beams fault injection or row hammer attack) [72], [73]. The authors of [73] have analyzed the misclassification rate of MLP based deep neural networks face to models derived from physical phenomena. Faulty-neuron behaviors have been injected randomly or deterministically, with injection scenarios considering single and multiple faulty neurons per layer. This is usually done during the function activation timeframe, which can be hundreds to thousands of cycles. Results indicate that in some cases, even a relatively small number of faulty neurons ($\approx 10\%$) can lead to a high risk of misclassification ($\approx 62\%$). As seen in section II, different activation functions have been studied for DNNs. They show that for a higher miss-classification rate (>50%), at least half of the neurons in a given hidden layer should be faulty, which is the case for sigmoid and tanh activation functions. In the case of ReLU, at least 3/4 of the neurons should be faulty to achieve the same miss-classification rate.

Extensive work has been dedicated in the last years to studying and evaluating AI hardware accelerators' errors and fault tolerance. An overview of fault tolerance techniques for feedforward neural networks is presented in [74]. In this paper, the authors review fault types, models, and measures used to evaluate performance and provide a taxonomy of the main techniques to enhance the intrinsic properties of some neural models based on the principles and mechanisms they exploit to achieve fault tolerance passively.

In [75], the authors present a study of the fault characterization and mitigation of Register-Transfer Level (RTL) model of NN accelerators by characterizing the vulnerability of NNs to application-level specifications, network topology, and activation functions, as well as architectural level specifications. In [76], authors present an experimental evaluation of the resilience of DNN systems (i.e., DNN software running on specialized accelerators) under Soft errors caused by high-energy particles.

An empirical study of DNNs resilience can be found in [77], where a fault injection framework named Ares, which can deal with fully connected and CNN-based DNN accelerators, is presented. It uses hardware fault models related to technology and environment variability, single event faults transients in memory elements, and algorithmic level faults models such as faults occurring in weights, activation, and hidden states. Fault injection is performed static, offline, before the inference process, and dynamically during inference execution.

The analysis of SNNs fault tolerance and reliability is a relatively newer field of research since their hardware implementations are much more recent than classical DNNs. Consensus has yet to be reached on the main applications of these networks. Moreover, the variety of signal-to-spike coding and training algorithms brings a specific heterogeneity in their characteristic which should be accounted for when their fault tolerance and reliability is discussed. One of the first works in this field, [78], estimates the accuracy of a FC SNN, capable of STDP learning designed with spintronic devices under the effect of process variability. Both the neuron and synapse behavior are strongly affected by process variability, and the accuracy drops by approximately 10% when assuming moderate variability compared to the ideal case.

Another interesting study is presented in [79], where a taxonomy of faults was defined for spiking neural networks. The accuracy of a hardware-implemented spiking neural network designed to perform STDP online training was analyzed under the assumption that (i) both learning and inference were performed on faulty hardware, (ii) only inference was performed on faulty hardware. This paper shows that performing learning directly on faulty hardware reduces the impact of faults on the network accuracy by an average of 15% and, in extreme cases, can reach up to 30%. Moreover, in [80], faults affecting the signal-to-spike conversion layer have the strongest impact on the network accuracy, with the synaptic stuck-at faults coming to a very close second. These faults strongly affect the learning process, which only exacerbates during inference.

On the other hand, faults, like delayed synapse activation or stuck lateral inhibition, have a marginal effect. The study presented in [81] takes a different approach. The fault tolerance study is performed on an SNN inference accelerator, a multi-layer SNN with supervised off-line learning, designed in VHDL and implemented on an FPGA. The fault injection experiments identify the parts of the design that need to be protected against faults and the inherently fault-tolerant parts. They have shown that the behavior under faults of the chosen type of SNN is similar to the behavior under faults of an ANN in that faults injected in the most significant bits of the synaptic values affect have a more substantial effect on network accuracy that when the faults are injected in the less critical bits, and also faults affecting the last layer (where the classification is performed) are more relevant than faults affecting the first layer (where there is higher computing redundancy). In addition, the authors have shown that their proposed SNN implementation is much more sensitive to faults injected in the routing of signals than in the synaptic weight or neuron computation.

Several studies have compared SNNs and their ANNs counterparts, mainly focusing on performance or power consumption rather than their relative fault tolerance. Therefore, a study was conducted to assess the fault resilience of both network types, assuming different quantizations of weights and neural computation and various training scenarios for the SNN. The study compared the fault tolerance of an MLP and an SNN with the same topology and bit precision. A 784 X 100 X 10 network was implemented to solve the MNIST classification problem [6]. The MLP uses a sigmoid activation function for all layers and has a base accuracy of 98% after 20 training epochs. The SNN used rate coding for signal-to-spike conversion and several training algorithms: (i) Shadow Training (ST) (with a base accuracy of 96%), (ii) BPTT (with a base accuracy of 97%), (iii) STDP (with a

base accuracy of 87%). The precision of synaptic weights was 16, 8, and 4 bits. Figure 4 illustrates selected results of the analysis, showing that the SNNs are more fault-tolerant than the MLP, and the SNN trained using BPTT is the most resilient of the three. This study also demonstrates how the SNN is trained in its fault resilience, with online training being more resilient than offline training. The last two columns in Figure 5 correspond to the SNN being trained by STDP, and the network accuracy degradation is computed when the presence of faults is assumed only at inference (AT) or during both training and inference (BT). These results show that the fault tolerance of the SNN depends on the training mechanism and that an SNN trained online is more resilient to faults than an SNN trained offline. The results suggest that SNNs have the potential to be more resilient to faults than ANNs, and further comparison between the two is necessary to consolidate these findings.

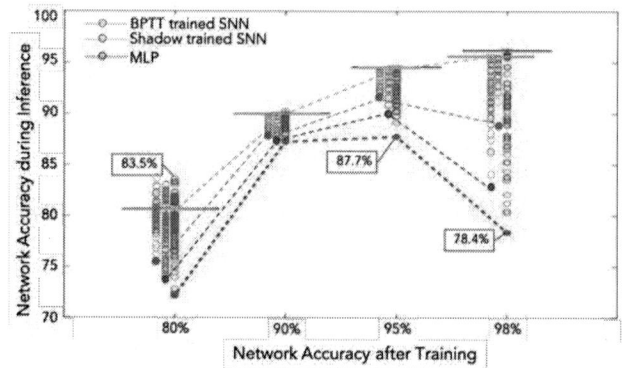

Fig. 4: Network accuracy under synaptic fault injection: a comparison between MLP and SNN. The synaptic weight precision is 8 bits, and 10^{-4} of the synaptic bits are considered faulty.

				Relative accuracy loss	
	ANN	ST SNN	BPTT SNN	STDP SNN	
				BT	AT
@Max accuracy	20,00%	14,58%	9,28%	3,07%	5,36%
95%	7,68%	6,32%	4,21%	3,17%	3,64%
90%	4,44%	4,00%	3,33%	3,21%	6,50%
80%	10,00%	7,50%	5,00%	3,33%	6,32%

Fig. 5: Maximum network accuracy degradation for different training scenarios under synaptic fault injection: a comparison between MLP and SNN. The synaptic weight precision is 8 bits, and 10^{-4} of the synaptic bits are considered faulty. AT = after training, BT = before training

VI. CONCLUSION

In this paper, the fundamentals of Convolutional Neural Network, Deep Neural Network, and Spiking Neural Network were introduced, and their digital counterparts were described using standard CMOS technologies. The discussion then focused on augmented silicon photonics improvements and the reliability aspects of Artificial Neural Network. The benefits and drawbacks of each technology were thoroughly analyzed, including its reliability. Our analysis suggests that the choice of technology for Artificial Neural Network design should depend on the specific application requirements and design constraints. While CMOS technology offers established fabrication processes and high reliability, it may have limitations in power consumption and scalability. Silicon photonics, on the other hand, provides high power efficiency and scalability, but their reliability is still under ongoing research. We hope this paper will serve as a valuable reference for researchers and practitioners in the field of Artificial Neural Network design as they explore the potential of these emerging technologies.

REFERENCES

[1] A. Marchisio et al., "Deep learning for edge computing: Current trends, cross-layer optimizations, and open research challenges," in *2019 IEEE Computer Society Annual Symposium on VLSI (ISVLSI)*. IEEE, 2019, pp. 553–559.

[2] M. Shafique et al., "An overview of next-generation architectures for machine learning: Roadmap, opportunities and challenges in the iot era," in *2018 Design, Automation & Test in Europe Conference & Exhibition (DATE)*. IEEE, 2018, pp. 827–832.

[3] S. Dutto et al., "Exploring deep learning for in-field fault detection in microprocessors," in *2021 Design, Automation & Test in Europe Conference & Exhibition (DATE)*, 2021, pp. 1456–1459.

[4] H. Sedjelmaci et al., "A lightweight anomaly detection technique for low-resource iot devices: A game-theoretic methodology," in *2016 IEEE International Conference on Communications (ICC)*, 2016, pp. 1–6.

[5] C.-W. Zhou et al., "A brief survey on deep neural networks," *arXiv preprint arXiv:1802.08882*, 2018.

[6] Y. LeCun et al., "Gradient-based learning applied to document recognition," in *Proceedings of the IEEE*, vol. 86, no. 11. IEEE, 1998, pp. 2278–2324.

[7] I. Goodfellow et al., "Deep learning," in *Deep learning*. MIT Press, 2016, pp. 1–800.

[8] C. D. Schuman et al., "Opportunities for neuromorphic computing algorithms and applications," *Nat Comput Sci*, vol. 2, no. 1, pp. 10–19, Jan. 2022, number: 1 Publisher: Nature Publishing Group. [Online]. Available: https://www.nature.com/articles/s43588-021-00184-y

[9] W. Maass, "Networks of spiking neurons: The third generation of neural network models," *Neural networks*, vol. 10, no. 9, 1997.

[10] B. Cramer et al., "The heidelberg spiking data sets for the systematic evaluation of spiking neural networks," *IEEE Transactions on Neural Networks and Learning Systems*, vol. 33, no. 7, pp. 2744–2757, 2022.

[11] A. Amirshahi and M. Hashemi, "Ecg classification algorithm based on stdp and r-stdp neural networks for real-time monitoring on ultra low-power personal wearable devices," *IEEE Transactions on Biomedical Circuits and Systems*, vol. 13, no. 6, pp. 1483–1493, 2019.

[12] A. Vallero et al., "Syra: Early system reliability analysis for cross-layer soft errors resilience in memory arrays of microprocessor systems," *IEEE Transactions on Computers*, vol. 68, no. 5, pp. 765–783, 2019.

[13] A. Ruospo et al., "A survey on deep learning resilience assessment methodologies," *Computer*, vol. 56, no. 2, pp. 57–66, 2023.

[14] C. D. James et al., "A historical survey of algorithms and hardware architectures for neural-inspired and neuromorphic computing applications," *Biologically Inspired Cognitive Architectures*, vol. 19, pp. 49–64, Jan. 2017. [Online]. Available: https://www.sciencedirect.com/science/article/pii/S2212683X16300561

[15] A. Krizhevsky et al., "Imagenet classification with deep convolutional neural networks," *Commun. ACM*, vol. 60, no. 6, pp. 84–90, may 2017. [Online]. Available: https://doi.org/10.1145/3065386

[16] D. E. Rumelhart et al., "Learning representations by back-propagating errors," *Nature*, vol. 323, no. 6088, pp. 533–536, 1986.

[17] G. E. Hinton and R. R. Salakhutdinov, "Reducing the dimensionality of data with neural networks," *Science*, vol. 313, no. 5786, pp. 504–507, 2006. [Online]. Available: https://www.science.org/doi/abs/10.1126/science.1127647

[18] F. Akopyan et al., "Truenorth: Design and tool flow of a 65 mw 1 million neuron programmable neurosynaptic chip," *IEEE Transactions on Computer-Aided Design of Integrated Circuits and Systems*, vol. 34, no. 10, 2015.

[19] S. Draghici, "On the capabilities of neural networks using limited precision weights," *Neural Networks*, vol. 15, no. 3, pp. 395–414, 2002. [Online]. Available: https://www.sciencedirect.com/science/article/pii/S0893608002000321

[20] P. A. Merolla et al., "A million spiking-neuron integrated circuit with a scalable communication network and interface," *Science*, vol. 345, no. 6197, pp. 668–673, 2014. [Online]. Available: https://www.science.org/doi/abs/10.1126/science.1254642

[21] A. L. Hodgkin and A. F. Huxley, "A quantitative description of membrane current and its application to conduction and excitation in nerve," *The Journal of Physiology*, vol. 117, no. 4, pp. 500–544, 1952.

[22] E. Izhikevich, "Simple model of spiking neurons," *IEEE Transactions on Neural Networks*, vol. 14, no. 6, pp. 1569–1572, 2003.

[23] C. Teeter et al., "Generalized leaky integrate-and-fire models classify multiple neuron types," *Nature Communications*, vol. 9, no. 1, p. 709, 2018.

[24] A. N. Burkitt, "A review of the integrate-and-fire neuron model: Ii. inhomogeneous synaptic input and network properties," *Biological Cybernetics*, vol. 95, no. 2, pp. 97–112, 2006.

[25] A. Carpegna et al., "Spiker: an fpga-optimized hardware accelerator for spiking neural networks," in *2022 IEEE Computer Society Annual Symposium on VLSI (ISVLSI)*, 2022, pp. 14–19.

[26] F. Corradi and G. Indiveri, "A neuromorphic event-based neural recording system for smart brain-machine-interfaces," *IEEE Transactions on Biomedical Circuits and Systems*, vol. 9, no. 5, pp. 699–709, 2015.

[27] G. V. der Sande et al., "Advances in photonic reservoir computing," *Nanophotonics*, vol. 6, no. 3, pp. 561–576, 2017. [Online]. Available: https://doi.org/10.1515/nanoph-2016-0132

[28] D. J. Enoka RM, "Rate coding and the control of muscle force," *Cold Spring Harb Perspect Med*, 2017.

[29] B. Petro et al., "Selection and optimization of temporal spike encoding methods for spiking neural networks," *IEEE Transactions on Neural Networks and Learning Systems*, vol. 31, no. 2, pp. 358–370, 2020.

[30] Z. Pan et al., "Neural population coding for effective temporal classification," in *2019 International Joint Conference on Neural Networks (IJCNN)*, 2019, pp. 1–8.

[31] L. Bottou, "Large-scale machine learning with stochastic gradient descent," in *Proceedings of COMPSTAT'2010*, Y. Lechevallier and G. Saporta, Eds. Heidelberg: Physica-Verlag HD, 2010, pp. 177–186.

[32] J. Duchi et al., "Adaptive subgradient methods for online learning and stochastic optimization," *J. Mach. Learn. Res.*, vol. 12, no. null, pp. 2121–2159, jul 2011.

[33] N. Srivastava et al., "Dropout: A simple way to prevent neural networks from overfitting," *J. Mach. Learn. Res.*, vol. 15, no. 1, pp. 1929–1958, jan 2014.

[34] R. Takahashi et al., "Ricap: Random image cropping and patching data augmentation for deep cnns," in *Proceedings of The 10th Asian Conference on Machine Learning*, ser. Proceedings of Machine Learning Research, J. Zhu and I. Takeuchi, Eds., vol. 95. PMLR, 14–16 Nov 2018, pp. 786–798. [Online]. Available: https://proceedings.mlr.press/v95/takahashi18a.html

[35] A. Fischer and C. Igel, "An introduction to restricted boltzmann machines," in *Progress in Pattern Recognition, Image Analysis, Computer Vision, and Applications*, L. Alvarez et al., Eds. Berlin, Heidelberg: Springer Berlin Heidelberg, 2012, pp. 14–36.

[36] S. J. Pan and Q. Yang, "A survey on transfer learning," *IEEE Transactions on Knowledge and Data Engineering*, vol. 22, pp. 1345–1359, 2010.

[37] V. Mnih et al., "Playing atari with deep reinforcement learning," 2013.

[38] P. M. Bi GQ, "Synaptic modifications in cultured hippocampal neurons: dependence on spike timing, synaptic strength, and postsynaptic cell type," *The Journal of Neuroscience*, vol. 18, no. 24, 1998.

[39] J. K. Eshraghian et al., "Training spiking neural networks using lessons from deep learning," *arXiv preprint arXiv:2109.12894*, 2021.

[40] S. B. Shrestha and G. Orchard, "SLAYER: Spike layer error reassignment in time," in *Advances in Neural Information Processing Systems 31*, S. Bengio et al., Eds. Curran Associates, Inc., 2018, pp. 1419–1428. [Online]. Available: http://papers.nips.cc/paper/7415-slayer-spike-layer-error-reassignment-in-time.pdf

979-8-3503-4631-2/23 $31.00 © 2023 IEEE

[41] T.-S. Chou et al., "Carlsim 4: An open source library for large scale, biologically detailed spiking neural network simulation using heterogeneous clusters," in *2018 International Joint Conference on Neural Networks (IJCNN)*, 2018, pp. 1–8.

[42] S. B. Furber et al., "The spinnaker project," *Proceedings of the IEEE*, vol. 102, no. 5, 2014.

[43] M. Davies et al., "Loihi: A neuromorphic manycore processor with on-chip learning," *IEEE Micro*, vol. 38, no. 1, 2018.

[44] J. Pei et al., "Towards artificial general intelligence with hybrid tianjic chip architecture," *Nature*, vol. 572, no. 7767, 2019.

[45] C. Frenkel et al., "A 0.086-mm^2 12.7-pj/sop 64k-synapse 256-neuron online-learning digital spiking neuromorphic processor in 28-nm cmos," *IEEE Transactions on Biomedical Circuits and Systems*, vol. 13, no. 1, 2019.

[46] A. Basu et al., "Spiking neural network integrated circuits: A review of trends and future directions," in *2022 IEEE Custom Integrated Circuits Conference (CICC)*, 2022.

[47] S. Li et al., "A fast and energy-efficient snn processor with adaptive clock/event-driven computation scheme and online learning," *IEEE transactions on circuits and systems. I, Regular papers*, vol. 68, no. 4, pp. 1543–1552, 2021.

[48] D. Neil and S.-C. Liu, "Minitaur, an event-driven fpga-based spiking network accelerator," *IEEE transactions on very large scale integration (VLSI) systems*, vol. 22, no. 12, pp. 2621–2628, 2014.

[49] Q. Wang et al., "Energy efficient parallel neuromorphic architectures with approximate arithmetic on fpga," *Neurocomputing (Amsterdam)*, vol. 221, pp. 146–158, 2017.

[50] D. Ma et al., "Darwin: A neuromorphic hardware co-processor based on spiking neural networks," *Journal of systems architecture*, vol. 77, pp. 43–51, 2017.

[51] Y. Paquot et al., "Optoelectronic Reservoir Computing," *Scientific Reports*, vol. 2, no. 1, p. 287, Feb. 2012. [Online]. Available: https://doi.org/10.1038/srep00287

[52] K. Vandoorne et al., "Experimental demonstration of reservoir computing on a silicon photonics chip," *Nat Commun*, vol. 5, no. 1, p. 3541, Mar. 2014, number: 1 Publisher: Nature Publishing Group. [Online]. Available: https://www.nature.com/articles/ncomms4541

[53] H.-T. Peng et al., "Neuromorphic Photonic Integrated Circuits," *IEEE J. Select. Topics Quantum Electron.*, vol. 24, no. 6, pp. 1–15, Nov. 2018. [Online]. Available: https://ieeexplore.ieee.org/document/8364605/

[54] J. Feldmann et al., "All-optical spiking neurosynaptic networks with self-learning capabilities," *Nature*, vol. 569, no. 7755, pp. 208–214, May 2019. [Online]. Available: http://www.nature.com/articles/s41586-019-1157-8

[55] Y. Shen et al., "Deep learning with coherent nanophotonic circuits," *Nature Photon*, vol. 11, no. 7, pp. 441–446, Jul. 2017. [Online]. Available: http://www.nature.com/articles/nphoton.2017.93

[56] M. A. Nahmias et al., "Photonic Multiply-Accumulate Operations for Neural Networks," *IEEE J. Select. Topics Quantum Electron.*, vol. 26, no. 1, pp. 1–18, Jan. 2020. [Online]. Available: https://ieeexplore.ieee.org/document/8844098/

[57] J. Feldmann et al., "Parallel convolutional processing using an integrated photonic tensor core," *Nature*, vol. 589, no. 7840, pp. 52–58, Jan. 2021. [Online]. Available: http://www.nature.com/articles/s41586-020-03070-1

[58] Y. Nahmias and Y. Loewenstein, "Neuromodulation of hebbian plasticity: relevance and mechanisms," *Current Opinion in Neurobiology*, vol. 69, pp. 150–159, 2021.

[59] M. Abdalla et al., "Minimum complexity integrated photonic architecture for delay-based reservoir computing," *Opt. Express*, vol. 31, no. 7, p. 11610, Mar. 2023. [Online]. Available: https://opg.optica.org/abstract.cfm?URI=oe-31-7-11610

[60] W. R. Clements et al., "Optimal design for universal multiport interferometers," *Optica*, vol. 3, no. 12, p. 1460, Dec. 2016. [Online]. Available: https://opg.optica.org/abstract.cfm?URI=optica-3-12-1460

[61] "In-Datacenter Performance Analysis of a Tensor Processing Unit | Proceedings of the 44th Annual International Symposium on Computer Architecture." [Online]. Available: https://dl.acm.org/doi/10.1145/3079856.3080246

[62] M. R. Mahmoodi and D. Strukov, "An ultra-low energy internally analog, externally digital vector-matrix multiplier based on NOR flash memory technology," in *Proceedings of the 55th Annual Design Automation Conference*. San Francisco California: ACM, Jun. 2018,

pp. 1–6. [Online]. Available: https://dl.acm.org/doi/10.1145/3195970.3195989

[63] M. A. Nahmias et al., "A Leaky Integrate-and-Fire Laser Neuron for Ultrafast Cognitive Computing," *IEEE Journal of Selected Topics in Quantum Electronics*, vol. 19, no. 5, pp. 1–12, Sep. 2013, conference Name: IEEE Journal of Selected Topics in Quantum Electronics.

[64] A. N. Tait et al., "Silicon Photonic Modulator Neuron," *Phys. Rev. Appl.*, vol. 11, no. 6, p. 064043, Jun. 2019, publisher: American Physical Society. [Online]. Available: https://link.aps.org/doi/10.1103/PhysRevApplied.11.064043

[65] N. Margalit et al., "Perspective on the future of silicon photonics and electronics," *Appl. Phys. Lett.*, vol. 118, no. 22, p. 220501, May 2021. [Online]. Available: https://aip.scitation.org/doi/10.1063/5.0050117

[66] W. Bogaerts et al., "Silicon microring resonators," *Laser & Photon. Rev.*, vol. 6, no. 1, pp. 47–73, Jan. 2012. [Online]. Available: https://onlinelibrary.wiley.com/doi/10.1002/lpor.201100017

[67] C. Rios et al., "Controlled switching of phase-change materials by evanescent-field coupling in integrated photonics [Invited]," *Opt. Mater. Express*, vol. 8, no. 9, p. 2455, Sep. 2018. [Online]. Available: https://opg.optica.org/abstract.cfm?URI=ome-8-9-2455

[68] X. Xiao et al., "Large-scale and energy-efficient tensorized optical neural networks on iii–v-on-silicon moscap platform," *APL Photonics*, vol. 6, no. 12, p. 126107, 2021. [Online]. Available: https://doi.org/10.1063/5.0070913

[69] S. Sackesyn et al., "Experimental realization of integrated photonic reservoir computing for nonlinear fiber distortion compensation," *Opt. Express, OE*, vol. 29, no. 20, pp. 30991–30997, Sep. 2021, publisher: Optica Publishing Group. [Online]. Available: https://opg.optica.org/oe/abstract.cfm?uri=oe-29-20-30991

[70] M. Miscuglio et al., "Artificial synapse with mnemonic functionality using gsst-based photonic integrated memory," in *2020 International Applied Computational Electromagnetics Society Symposium (ACES)*, 2020, pp. 1–3.

[71] K. Mekemeza-Ona et al., "All optical q-switched laser based spiking neuron," *Frontiers in Physics*, vol. 10, 2022. [Online]. Available: https://www.frontiersin.org/articles/10.3389/fphy.2022.1017714

[72] Y. Liu et al., "Fault injection attack on deep neural network," in *2017 IEEE/ACM International Conference on Computer-Aided Design (ICCAD)*, 2017, pp. 131–138.

[73] J. Breier et al., "Deeplaser: Practical fault attack on deep neural networks," 2018.

[74] C. Torres-Huitzil and B. Girau, "Fault and error tolerance in neural networks: A review," *IEEE Access*, vol. 5, pp. 17322–17341, 2017.

[75] B. Salami et al., "On the resilience of rtl nn accelerators: Fault characterization and mitigation," in *2018 30th International Symposium on Computer Architecture and High Performance Computing (SBAC-PAD)*, 2018, pp. 322–329.

[76] G. Li et al., "Understanding error propagation in deep learning neural network (dnn) accelerators and applications," in *SC17: International Conference for High Performance Computing, Networking, Storage and Analysis*, 2017, pp. 1–12.

[77] B. Reagen et al., "Ares: A framework for quantifying the resilience of deep neural networks," in *2018 55th ACM/ESDA/IEEE Design Automation Conference (DAC)*, 2018, pp. 1–6.

[78] E. I. Vatajelu and L. Anghel, "Fully-connected single-layer stt-mtj-based spiking neural network under process variability," in *2017 IEEE/ACM International Symposium on Nanoscale Architectures (NANOARCH)*, 2017, pp. 21–26.

[79] E.-I. Vatajelu et al., "Special session: Reliability of hardware-implemented spiking neural networks (snn)," in *2019 IEEE 37th VLSI Test Symposium (VTS)*, 2019, pp. 1–8.

[80] A. Bosio et al., "Rebooting computing: The challenges for test and reliability," in *2019 IEEE International Symposium on Defect and Fault Tolerance in VLSI and Nanotechnology Systems (DFT)*, 2019, pp. 8138–8143.

[81] T. Spyrou et al., "Reliability analysis of a spiking neural network hardware accelerator," in *2022 Design, Automation & Test in Europe Conference & Exhibition (DATE)*, 2022, pp. 370–375.

IP Session on Chiplet: Design, Assembly, and Test

Bapi Vinnakota
Lawrence Berkely Labs
Berkeley, CA, USA
bapi.vinnakota@ocproject.net

Jaber Derakhshandeh
IMEC, Kapeldreef 75
3001 Leuven, Belgium
jaber.derakhshandeh@imec.be

Eric Beyne
IMEC, Kapeldreef 75
3001 Leuven, Belgium
Eric Beyne @imec.be

Erik Jan Marinissen
IMEC, Kapeldreef 75
3001 Leuven, Belgium
Erik Jan Marinissen @imec.be

Sreejit Chakravarty
San Jose, CA, USA
sreejit_chakravarty@yahoo.com

Abstract—**This IP session will consist of three presentations. A summary of each presentation is given below.**

I. THE NEW OPEN CHIP-LET ECONOMY: BAPI VINNAKOTA

Multiple technological and business trends are driving a change to realizing semiconductor products as systems in package (SiP) that integrate multiple die, usually referred to as chip-lets, into a single package. Compared to monolithic SoCs, chip-let-based designs require an evolution in architecture, interfaces, design and manufacturing flows. Several large companies have already made the transition to chip-let-based designs with proprietary tools and technologies. Many have not and need the expertise to do so. A new and open chip-let economy to help companies adopt chip-let technologies is developing. It is based on and will require collaboration and standardization on multiple dimensions, ensuring that companies can interact in an open, efficient and scalable manner. This talk will profile the chip-let economy including its motivations, standards, participants and growth areas.

II. MULTI-DIE STACK ASSEMBLY AND INTERCONNECTS: JABER DERAKHSHANDEH, ERIC BEYNE, ERIK JAN MARINISSEN

The demand for computational performance in machine learning and artificial intelligence applications is growing rapidly [1], but CMOS technology struggles to keep up with this pace, leading to increased difficulty and cost. 3D integration technology offers a solution to this problem by assembling multiple dies in a single chip package. This technology offers several benefits, such as heterogeneous integration, improved yields, and reduced power dissipation due to shorter wire lengths [2].

This presentation discusses various aspects of 3D technologies, including different stacking processes such as die-to-die, die-to-wafer, and wafer-to-wafer. Electrical interconnects between stacked dies are formed using solder-based micro-bump pairs or Cu-Cu hybrid bonding, and interconnect faults caused by voiding in the solder joint, poor solder wetting, or particle entrapment can be effectively detected through electrical testing.

The resulting stacks can be 2.5D, 3D, or a combination of both, known as "5.5D," with through-silicon vias providing electrical access between the front- and back-side of the dies

in multi-tiered stacks. If a stack contains more than two tiers, some interconnects will connect to the backside of a die, and to solve this issue, wafers are thinned down, and through-silicon vias are created.

Multi-die stack assembly can be performed using either sequential or vertical collective methods. In sequential bonding, a full thermal compression bonding profile is applied for each stacked chip, while in the collective method, first dies are placed at room temperature and only after all dies are placed, a full thermal compression bonding profile is applied [3]. Finally, the presentation includes a roadmap for interconnect density and pitch for D2W and W2W bonding.

III. STANDARDIZING CHIP-LET INTERCONNECT TEST: SREEJIT CHAKRAVARTY

This presentation starts with a classification of various chip-let interconnect types and the emerging trends for such interconnects. Defect profiles for such technologies are presented, followed by a discussion of the test and repair problems that require a solution. To enable packaging of chip-lets from different vendors a standard for chip-let interconnect testing is required. The presentation will touch upon the evolving UCIe standard and an IEEE standardization disussion underway to fill gaps in existing test standard for chip-let interconnects.

IV. REFERENCES

1. Sri Samavedam et al., "Future Logic Scaling: Towards Atomic Channels and Deconstructed Chips", keynote address at IEEE International Electron Devices Meeting, December 2020, San Francisco, CA, USA, doi= 10.1109/IEDM13553.2020.9372023
2. Eric Beyne, "A view on the 3D technology landscape; Design and technology options for 3D systems-on-chip", https://www.imec-int.com/en/articles/view-3d-technology-landscape.
3. Jaber Derakhshandeh et al., "Low-Temperature Backside Damascene Processing on Temporary Carrier Wafer Targeting 7µm and 5µm Pitch Micro-Bumps for *N* Equal and Greater Than 2 Die-to-Wafer TCB Stacking," IEEE Electronic Components and Technology Conference (ECTC), July 2022, San Diego, CA, USA, doi= 10.1109/ECTC51906.2022.00179

979-8-3503-4631-2/23 $31.00 © 2023 IEEE

Effective and Efficient Testing of Large Numbers of Inter-Die Interconnects in Chiplet-Based Multi-Die Packages

Po-Yao Chuang[1,3]   Francesco Lorenzelli[1,4]   Sreejit Chakravarty[2]

Cheng-Wen Wu[3]   Georges Gielen[4]   Erik Jan Marinissen[1,5]

[1] IMEC	[2] Intel Corporation	[3] National Tsing-Hua Univ.	[4] KU Leuven	[5] TU/e
Kapeldreef 75	2200 Mission College Blvd.	101, Sec. 2, Kuang-Fu Road	Kasteelpark Arenberg 10	Den Dolech 2
3001 Leuven	Santa Clara, CA 95054-1549	Hsinchu 30013	3001 Leuven	5612 AZ Eindhoven
Belgium	U.S.A.	Taiwan	Belgium	the Netherlands

Abstract

Chiplet-based multi-die packages implement large numbers of inter-die interconnect bundles clustered in large micro-bump islands. These micro-bumps can be subject to manufacturing defects. The most common defect types are shorts and opens. Traditional interconnect automatic test pattern generation (I-ATPG) algorithms detect, for a given collection of interconnects, all shorts between any pair of interconnects, all open interconnects, and exclude any *aliasing*, independent from the interconnects' layout positions. Exploiting knowledge of their relative layout positions, we derive a new, improved I-ATPG algorithm. For a user-defined and scalable definition of *realistic* shorts, the new I-ATPG approach (1) increases the defect coverage significantly (in an example case, between 18% and 67%) by including realistic inter-bundle shorts between micro-bumps from adjacent bundles, and (2) reduces the overall test pattern count (and hence, the resulting test time) by 33% by providing test patterns for realistic shorts only.

1 Introduction

More than a decade ago, 3D chip stacking appeared for the first time on the Gartner Hype Cycle for Emerging Technologies [1] as an "Innovation Trigger". In the mean time, this technology went, nicely following Gartner's theory, through its "Peak of Inflated Expectations" and into a "Trough of Disillusionment". Today, under new names as *chiplet-based design*, *heterogeneous integration*, or *advanced packaging*, the technology is fully back in the saddle, equipped with adjusted expectations and several new useful standards [2, 3]. This time it is quickly gaining product ground, thanks to its still very compelling benefits. Instead of being manufactured as one monolithic die in a single technology node, multiple smaller chiplets can be made each in the technology which delivers the best price/performance for that part of the design. As yield increases superlinearly with decreasing chip area, the smaller chiplets increase the compound stack yield significantly, provided they are subjected to a "known-good die" pre-bond test prior to stack assembly [4]. And, in case chiplets are (completely or partly) stacked on top of each other, the final product has a reduced footprint, due to which wire lengths can drastically be reduced, leading to proportional performance improvements.

Intel's *Ponte Vecchio* product (see Figure 1) is a typical example of the type of 'monster chips' that can be put together with 3D chiplet assembly. Meant to serve as a GPU accelerator in data centers,

Ponte Vecchio contains a total of 47 chiplets that jointly employ five different technology nodes from both Intel and TSMC (including TSMC's 5nm node) with an aggregate count of over 100 billion transistors. It delivers a compute performance of more than 45 Tflops at FP32 precision and offers more than 2 TB/s of memory fabric bandwidth and over 2 TB/s of connectivity bandwidth, according to Intel [5].

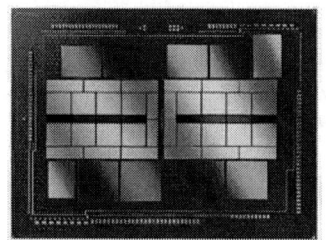
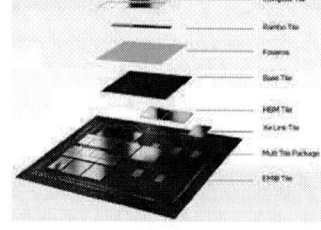

(a) Top view.   (b) Exploded view.

Figure 1: Intel's *Ponte Vecchio* product [5]. [Source: Intel Corp.]

Such chiplet-based products have a large number of inter-die interconnects; the number on *Ponte Vecchio* is >75,000. This number is expected to increase several times on future products [6]. The interconnects occur in *bundles*, collections of relatively small numbers (e.g., 32, 64, or 128) of functional interconnects with a common clock, which typically work at frequencies in the range of

---

* Reach the authors at po-yao.chuang.ext@imec.be, francesco.lorenzelli@imec.be, sreejit.chakravarty@intel.com, cww@ee.nthu.edu.tw, georges.gielen@kuleuven.be, and erik.jan.marinissen@imec.be.

979-8-3503-4631-2/23 $31.00 © 2023 IEEE

3-4 GHz [6]. The vast majority of the inter-die interconnects are synchronous, i.e., *registered* just before leaving the transmitting die as well as registered immediately after entering the receiving die. The large number of interconnect signals, their high-speed clocking requirements, and the fact that micro-bumps adhere best to their manufacturing specifications when in large, uniform arrays, result in the clustering of various bundles in different layout areas of the chiplets. Such a *micro-bump island* implements many functional interconnect bundles by means of one large, dense micro-bump matrix. The micro-bump densities are quickly increasing over time: from $400/mm^2$ (corresponding to an interconnect pitch of $50\mu m$) in the recent past, to $772/mm^2$ ($= 36\mu m$ pitch) on *Ponte Vecchio*, to $10,000/mm^2$ ($= 10\mu m$ pitch) on future products [6].

With such large numbers of inter-die interconnects with small feature sizes, it is unavoidable that occasionally manufacturing defects occur. The most common defects are shorts between two or more adjacent micro-bump pairs, followed by open interconnects. Intel reported that short defects (between data signals or between data and clock signals) occur almost $20\times$ more often than open defects [7]. Within one micro-bump island, shorts occur both within a single bundle, as well as between (adjacent) bundles.

This paper addresses interconnect automatic test pattern generation (I-ATPG) for effective, yet efficient testing of 3D interconnects. Traditional I-ATPG algorithms detects, for a given collection of interconnects, all shorts between any pair of interconnects, all open interconnects, and excludes any *aliasing*, independent from the interconnects' layout positions. However, as these layout positions are known to us anyway, we exploit this information in a new, improved I-ATPG algorithm that (1) increases the defect coverage to include *realistic* inter-bundle shorts between micro-bumps from adjacent bundles, and (2) reduces the overall test pattern count (and hence, the resulting test time) by providing test patterns for *realistic* shorts only – where the definition of *'realistic shorts'* is user-defined and scalable.

The remainder of this paper is organized as follows. In Section 2, we describe the traditional ATPG algorithms for point-to-point interconnects. In Section 3, we formulate the problem statement addressed in this paper. Section 4 presents the notion of *defect distance*, our way to let the user specify up to which distance between micro-bumps short defects are considered to be realistic. Section 5 addresses the shortcomings of traditional I-ATPG. In Section 6, we describe our new, proposed algorithm that exploits the concept of defect distance to generate a minimal set of test patterns, covering all intra- and inter-bundle shorts. Experimental results for two identical micro-bump islands, based on bundles with different aspect ratios, are shown in Section 7. Section 8 concludes this paper.

2 Related Prior Work

Throughout this paper, we use the following terminology for the test stimuli.

- *Test pattern.* A test pattern is a set of test stimuli that are simultaneously applied in parallel to all interconnects of a

micro-bump island. In Figure 2, test patterns are the columns in the set of stimuli. Others [8, 9] refer to test patterns as *parallel test vectors* (PTVs).

- *Code word.* A code word is the list of test stimuli that are subsequently applied to an individual interconnect. In Figure 2, code words are the rows in the set of stimuli. Others [8, 9] refer to code words as *sequential test vectors* (STVs).

Traditional digital ATPG methods for point-to-point interconnects are based on the Counting Sequence Algorithm by Kautz [10]. It uses simple binary counting to generate distinct code words, which is a requirement for detecting multiple-net shorts that are modeled as Wired-OR or Wired-AND faults [8]. The Modified Counting Sequence Algorithm by Goel and McMahon [11] excludes the all-zero and all-one code words from the counting sequence. This guarantees that each code word contains at least one '0' and one '1', which is a requirement for detecting single-net opens that are modeled as Stuck-at-Zero or Stuck-at-One [8]. *Aliasing* occurs if the response of a faulty net is equal to the response of another, fault-free net [12]. In this case, we cannot determine whether the fault-free net also suffers from the fault of the faulty net. Hence, aliasing reduces the diagnostic resolution and therefore, it is desirable to avoid it. The True/Complement Test Algorithm by Wagner [13] doubles the amount of test patterns of the original Counting Sequence Algorithm, by generating code words that consist of the concatenation of the original Counting Sequence code words and their inverted values. This leads to code words with an equal number of '0's and '1's, which prevents aliasing, as faults are guaranteed to introduce more zeros (Stuck-at-Zero, Wired-AND) or more ones (Stuck-at-One, Wire-OR) in the test responses.

For k interconnects, the Counting Sequence Algorithm requires $\lceil \log_2 k \rceil$ test patterns. The True/Complement Test Algorithm doubles that, and hence requires $2 \times \lceil \log_2 k \rceil$ test patterns. Figure 2 shows an example True/Complement test set for $k = 5$ interconnects with $2 \times \lceil \log_2 5 \rceil = 6$ test patterns. The code words are $(000; 111)$, $(001; 110)$, $(010; 101)$, $(011; 100)$, and $(100; 011)$, and they can arbitrarily be assigned to any of the five interconnects.

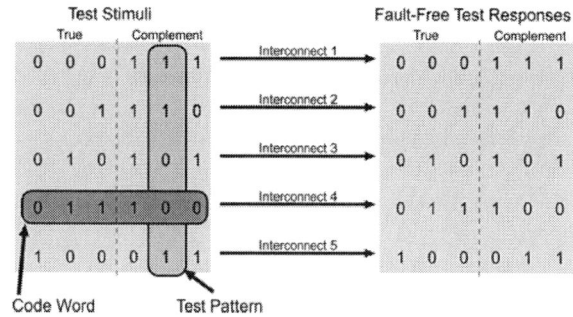

Figure 2: Terminology illustrated for an example True/Complement Test for five interconnects.

The True/Complement Test Algorithm is considered to be a very efficient test pattern generation algorithm, as the number of generated test patterns scales with the logarithm base two of the number of interconnects. For example, even for a million interconnects (i.e., $k=1,000,000$), we need only $2 \times \lceil \log_2 1,000,000 \rceil = 40$ test patterns for a test that covers *all* shorts between *any pair* of the million interconnects. Each of these 40 test patterns consists of 1,000,000 stimulus and 1,000,000 expected response bits, and hence still requires a significant time to load/unload if the test is applied from an external ATE; especially if a serial (one-bit in/out per clock cycle) test access mechanism is used [2] – but that is inherently due to the large problem instance at hand.

The I-ATPG algorithms described above are agnostic of the layout positions of the micro-bumps. They cover all shorts between any pair of interconnects, also if two micro-bumps are located so far from each other in the layout (for example: on opposite sides of a large micro-bump island), that it is rather unrealistic that they can be shorted due to a manufacturing defect while the same defect does *not* affect the many interconnects in between the shorted pair. Hence, even though these I-ATPG algorithms are generally considered to be efficient, it is possible to enhance their efficiency even further, by avoiding generating test patterns for unrealistic shorts. In this paper, we leave the precise definition of *'realistic'* and *'unrealistic'* shorts up to the user.

3   Problem Statement

In the sequel, we use the term *bump* for a 3D interconnect of which we know the layout position in a micro-bump island. A *bundle* is a (functionally coherent) set of bumps, e.g., a 64-bit bus. The *bounding box* for a bundle b is the minimal rectangle that encloses b.

Given: A set of bumps $R = \{r_1, r_2, \ldots, r_N\}$, arranged in a micro-bump island matrix of size $W \times H$, where bump r is located at coordinates $(x(r), y(r))$ relative to the island, with $0 \leq x(r) < W$ and $0 \leq y(r) < H$.
R is partitioned in a set of bundles $B = \{b_1, b_2, \ldots, b_M\}$, such that

- $\bigcup_{b \in B} b \subseteq R$,
- $\forall_{b_i, b_j \in B} (b_i \cap b_j = \emptyset)$, and
- $\forall_{b \in B} \forall_{r_i, r_j \in b} ((x(r_i) \neq x(r_j)) \vee (y(r_i) \neq y(r_j)))$.

For illustration purposes, Figure 3 shows the relatively small Example 1 of a micro-bump island consisting of $W \times H = 20 \times 8 = 160$ bumps. These bumps are partitioned over 10 bundles of different sizes, in the figure indicated by means of different bump colors. The sizes of the bounding boxes for the various bundles (in bump counts) are for each bundle indicated in its top-right bump.

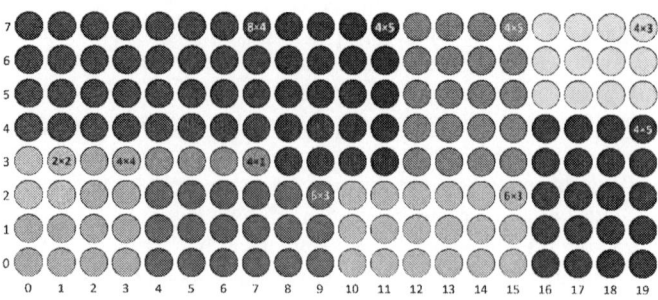

Figure 3: Example 1: a micro-bump island; constituting bundles are indicated by means of different bump colors.

Determine: For all bumps in R and a user-defined maximum defect distance (w, h) for realistic shorts, determine the test content, i.e., the test code words (in other words, the subsequent test pattern bits per bump) such that the test detects all realistic shorts, all opens, and excludes any (realistic) aliasing.

4   Defect Distance

It is clear that we have a need for a new I-ATPG approach which covers all realistic short defects, both within a bundle as well as between different bundles, but does not waste test patterns on unrealistic shorts. This first requires a definition of realistic short defects.

The *defect distance* is the diagonal of a user-defined rectangle, characterized by its *width* and *height* (w, h). For a given *victim* bump V, any bump that fits together with V in a (w, h) rectangle is a candidate for a *realistic* short defect with V.

Figure shows as example, for a user-defined maximum defect distance $(w, h) = (3, 2)$, a four-step procedure to determine the set of bumps (marked in red) with which a given victim bump V (marked in blue) can have realistic shorts. The bumps that remain gray after the completion of Step 4 are outside the defect distance and hence are considered to not have realistic shorts with V.

The maximum *defect distance* concept as defined above is flexible and powerful; it gives full user control over the notion of *realistic* shorts. In most cases, realistic shorts are limited to shorts between bumps that are directly adjacent to each other; this is achieved by setting $(w, h) = (2, 2)$. Obviously this strict definition for short defects implies a minimum number of test patterns and hence test time.

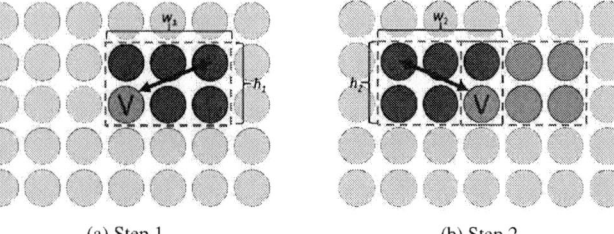
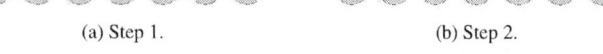

(a) Step 1.   (b) Step 2.   (c) Step 3.   (d) Step 4.

Figure 4: Four-step procedure to determine the potential aggressor bumps (red) for a specific victim bump (blue) for the user-defined max. defect distance $(w, h) = (3, 2)$.

979-8-3503-4631-2/23 $31.00 © 2023 IEEE

However, if there is a need to augment the defect distance, the user can easily do this. Note that the defect distance can be extended both in an x, y-symmetrical way (e.g., from $(w, h) = (2, 2)$ to $(3, 3)$), but also in an x, y-asymmetrical way (e.g., from $(w, h) = (2, 2)$ to $(3, 2)$). This can be useful, for example when the micro-bump pitch differs in the x- and y-directions. For example, in the first JEDEC Wide-I/O standard [14] (also known as "WIO1"), micro-bumps occur in pitches of 40μm horizontally and 50μm vertically [4].

5 Shortcomings of Traditional I-ATPG

The traditional I-ATPG approach covers *all* shorts between *all pairs* of the bump collection considered. On a micro-bump island consisting of multiple bundles, there are essentially two different ways in which we can apply this I-ATPG approach: (1) I-ATPG on the entire micro-bump island, and (2) I-ATPG per bundle.

The first approach inevitably covers short defects for bump pairs that are physically so distant from each other that a short defect between them is unrealistic, if other bump pairs in between are *not* shorted. Hence, the number of test patterns, which scales only logarithmically with the number of bumps, is larger than strictly necessary. Every test pattern provides test stimuli to all bumps and therefore takes a significant amount of time to load/unload from the ATE into the chip and vice versa. It is clear that this approach is wasting precious test time, while its set of covered shorts contains a large fraction of unrealistic defects.

Alternatively, we can run I-ATPG per bundle on the various bundles that together constitute the micro-bump island. This typically requires significantly less test patterns and hence test time. The test still covers unrealistic short defects, but as these are now confined to an individual bundle, there are far less of them. Within a bundle, I-ATPG generates unique code words for all bumps. However, it is problematic that the same number might be assigned to bumps in different adjacent bundles, as the I-ATPG is unaware of the bumps' layout locations. Consequently, we cannot guarantee that inter-bundle shorts are indeed covered by the test.

An example of such a case is shown in Figure 5, which builds on Example 1. The victim bump at coordinates $(0, 1)$ has two potential aggressors that share the same number and code word, one within the same bundle at coordinates $(0, 0)$ and one within a different bundle at coordinates $(0, 2)$.

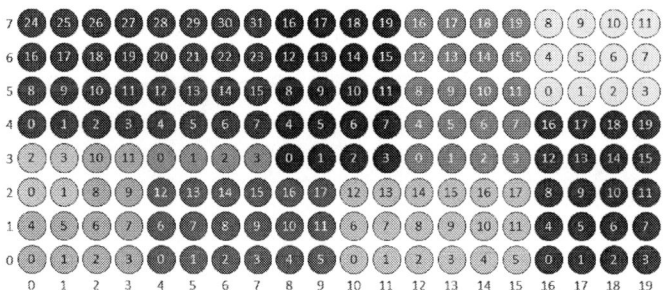

Figure 5: Example 1 after traditional I-ATPG per bundle and row-by-row code number assignments.

The coverage of all inter-bundle short defects can be guaranteed, if all bundles have the same size and bump code numbering is done for all bundles in the same way. To avoid duplicate code numbers in adjacent boundary bumps of different bundles within defect distance (w, h), we need a minimum bundle size of $(2w - 1, 2h - 1)$ to ensure that all inter-bundle shorts are covered.

Figure 6 shows an example. A micro-bump island of 20×8 bumps consists of 10 bundles of 5×4 bumps each, which are aligned both horizontally and vertically. Within each bundle, unique bump code numbers are assigned following a row-by-row counting sequence. In this example, not only all $(2, 2)$-, but even all $(3, 2)$-realistic inter-bundle shorts are covered, as that requires a minimum bundle size of $(5, 3)$.

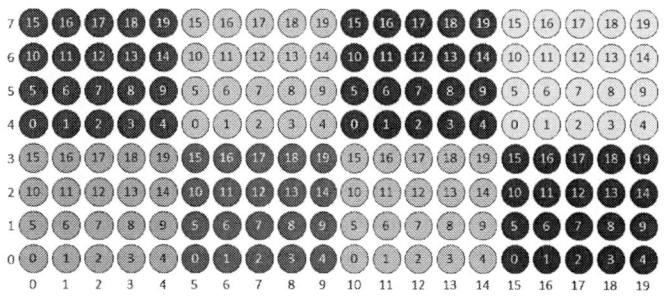

Figure 6: Example of per-bundle I-ATPG for which all $(3, 2)$-realistic inter-bundle shorts are covered.

6 Improved I-ATPG Algorithm

The pseudo-code of our new improved I-ATPG algorithm is given in Algorithm 1. The algorithm consists of two steps.

Algorithm 1 [3D Interconnect ATPG]

01: **input**: defect distance (w, h)
02: //Step 1: Bump Numbering
03: $bw = 2 \times \max(w, 2) - 1$
04: $bh = 2 \times \max(h, 2) - 1$
05: **for all** $r \in R$ **do** {
06:     $num(r) = (y(r) \bmod bh) \times bw + (x(r) \bmod bw)$
07: }
08: //Step 2: Number-to-Code and Pattern Generation
09: $\#pattern = \lceil \log_2(bw \times bh) \rceil$
10: **for all** $r \in R$ **do** {
11:     $code(r) = num(r)_2 \; ; \; \overline{num(r)_2}$
12: }

In Step 1 (Lines 02–07), first the bounding box size (bw, bh) for the set of realistic short aggressors is calculated (Lines 03–04) on the basis of the user-defined maximum defect distance (w, h). Then, the entire micro-bump island is partitioned in adjacent bounding boxes of size (bw, bh), while bundle boundaries are ignored. Next, per bounding box, all bumps of the entire micro-bump island are assigned a code number in a row-by-row fashion. For the micro-bump island of Example 1, the partitioning and numbering for defect distance $(2, 2)$ are shown in Figure 7.

979-8-3503-4631-2/23 $31.00 © 2023 IEEE

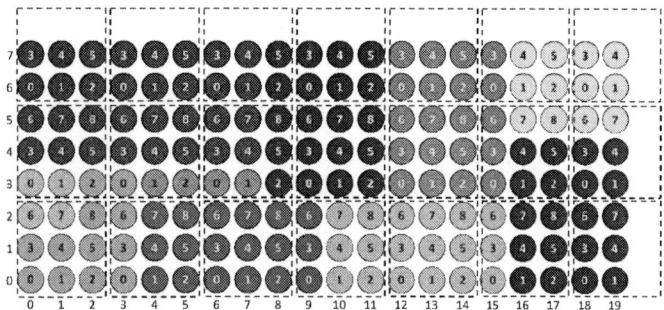

Figure 7: Example 1 with code number assignment for defect distance $(2, 2)$.

In Step 2 (Lines 08–12), based on the bump code numbers, the actual code words the test patterns are implicitly generated. For every bump r, the code word is the binary representation of its bump code number $num(r)$, stretched with leading zeros up till $\#pattern = \lceil \log_2(bw \times bh) \rceil$ binary digits, and its complement to formulate the corresponding code word $code(r)$.

Table 1 shows the number-to-code generation for Example 1. The code-word columns in this table implicitly define the eight test patterns.

$num(r)$	$code(r) = $ **True; Complement**
0	0000;1111
1	0001;1110
2	0010;1101
3	0011;1100
4	0100;1011
5	0101;1010
6	0110;1001
7	0111;1000
8	1000;0111

Table 1: Number-to-code generation for Example 1.

7 Experimental Results

In the preceding sections, Example 1 was used as a small running example to illustrate our approach. In this section, we feature a second example of a more real-life size, in order to determine the benefit range to be expected of the newly proposed 3D I-ATPG over the traditional I-ATPG.

Example 2: A micro-bump island consists of 2,496 interconnect bundles, each consisting of 64 functional interconnects. Typically, such bundles will also have a few special interconnects, such as a clock and a few spares. Sometimes, such special interconnects are shared between multiple bundles, which complicates their counting. As the number of special interconnects is small compared to the number of regular functional interconnects, we ignore the special interconnects in this example. Hence, the micro-bump island has a total number of $2,496 \times 64 = 159,744$ bumps, arranged in a 624×256 array.

We consider two different bundle layouts. They have been chosen such that they constitute the two extremes of a continuum with respect to their ratio of intra- versus inter-bundle shorts.

a. Each bundle consists of a rectangular layout, viz. a single row, i.e., 64×1 bumps. The total micro-bump island consists of 624 rows of 4 bundles each. This corresponds to a bump array of $(4 \times 64) \times (624 \times 1) = 256 \times 624 = 159,744$ bumps – see Figure 8(a).

b. Each bundle consists of a square layout, i.e., 8×8 bumps. The total micro-bump island consists of 78 rows of 32 bundles each. This corresponds to a bump array of $(32 \times 8) \times (78 \times 8) = 256 \times 624 = 159,744$ bumps – see Figure 8(b).

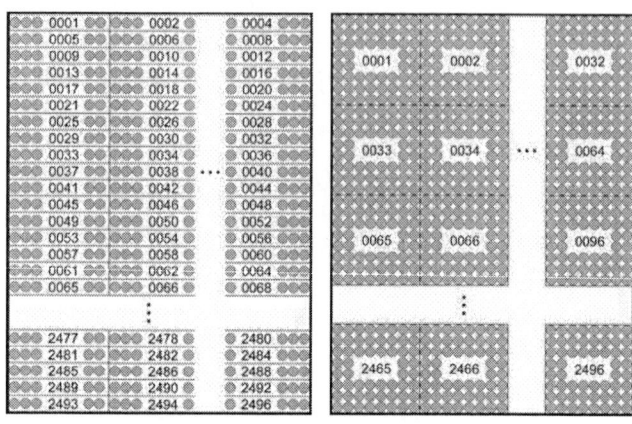

(a) Bundle = 64×1 bumps.   (b) Bundle = 8×8 bumps.

Figure 8: Layout configurations for Example 2.

The numerical results for both variants of this example are given in Table 2. In the first section of Table 2, bundle and bump numbers are given. It is clear that, although the aspect ratios of the two bundles considered are very different, the resulting micro-bump islands are identical in size and aspect ratio.

Parameter	Bundle = 64×1 Bumps	Bundle = 8×8 Bumps
Micro-Bump Island Layout Data		
#Bundles/matrix	$4 \times 624 = 2{,}496$ bundles	$32 \times 78 = 2{,}496$ bundles
Bump array	$(4 \times 64) \times (624 \times 1)$	$(32 \times 8) \times (78 \times 8)$
Total #bumps	$256 \times 624 = 159{,}744$	$256 \times 624 = 159{,}744$
1: Traditional I-ATPG for the Entire Micro-Bump Island		
#Patterns	$2 \times \lceil \log_2 159{,}744 \rceil = 36$ patterns	
All shorts	$12{,}758{,}992{,}896\ (= 100\%)\ \checkmark$	$12{,}758{,}992{,}896\ (= 100\%)\ \checkmark$
2: Traditional I-ATPG per Bundle		
#Patterns	$2 \times \lceil \log_2 64 \rceil = 12$ patterns	
All shorts		
– intra-bundle	$5{,}031{,}936\ (= 0.04\%)\ \checkmark$	$5{,}031{,}936\ (= 0.04\%)\ \checkmark$
– inter-bundle	$12{,}753{,}960{,}960\ (= 99.9\%)\ \times$	$12{,}753{,}960{,}960\ (= 99.9\%)\ \times$
All (2,2)-shorts		
– intra-bundle	$157{,}248\ (= 24.7\%)\ \checkmark$	$524{,}160\ (= 82.4\%)\ \checkmark$
– inter-bundle	$479{,}090\ (= 75.3\%)\ \times$	$112{,}178\ (= 17.6\%)\ \checkmark$
3: New 3D I-ATPG with Max. Defect Distance = (2,2)		
#Patterns	$2 \times \lceil \log_2 9 \rceil = 8$ patterns	
All (2,2)-shorts	$636{,}338\ (= 100\%)\ \checkmark$	$636{,}338\ (= 100\%)\ \checkmark$
– intra-bundle	$157{,}248\ (= 24.7\%)\ \checkmark$	$524{,}160\ (= 82.4\%)\ \checkmark$
– inter-bundle	$479{,}090\ (= 75.3\%)\ \checkmark$	$112{,}178\ (= 17.6\%)\ \checkmark$
4: New 3D I-ATPG with Max. Defect Distance = (3,2)		
#Patterns	$2 \times \lceil \log_2 15 \rceil = 8$ patterns	
All (3,2)-shorts	$953{,}952\ (= 100\%)\ \checkmark$	$953{,}952\ (= 100\%)\ \checkmark$
– intra-bundle	$312{,}000\ (= 32.7\%)\ \checkmark$	$778{,}752\ (= 81.6\%)\ \checkmark$
– inter-bundle	$641{,}952\ (= 67.3\%)\ \checkmark$	$175{,}200\ (= 18.4\%)\ \checkmark$

Table 2: Experimental results for Example 2.

The remainder of Table 2 consists of four sections:

1. Traditional I-ATPG for the entire micro-bump island

2. Traditional I-ATPG per bundle

3. New 3D I-ATPG with max. defect distance = $(2, 2)$

4. New 3D I-ATPG with max. defect distance = $(3, 2)$.

For each of these four cases, the table list the number of test patterns, as well as the number of shorts and whether their coverage is guaranteed (indicated by "✓") or not (indicated by "✗").

Traditional I-ATPG is layout-agnostic, and hence, its statistics are identical for the two different bundle layouts; it is only the bundle size that matters. Traditional I-ATPG on the entire micro-bump island covers all $C_2^{159,744} \geq 12.7B$ potential shorts between any pair of bumps with only 36 test patterns. If we apply traditional I-ATPG per bundle, then we reduce the test time to one third (i.e., 12 test patterns) and cover all ∼5M intra-bundle shorts. However, all inter-bundle shorts are not guaranteed because traditional I-ATPG has no fixed rule to indicate the bump code numbers assignment. Furthermore, bundles of functional bumps can be non-regularly distributed on the layout, depending on the chip designers.

The new 3D I-ATPG has been used for maximum defect distances $(2, 2)$ and $(3, 2)$, resulting in bounding box sizes of $(3, 3)$ respectively $(5, 3)$. This implies 9 resp. 15 unique bump code numbers and code words, in both cases resulting in 8 test patterns. These tests cover all 'realistic' (intra- and inter-bundle) shorts, but use different definitions for 'realistic', viz. all $(2, 2)$- respectively $(3, 2)$-realistic shorts. As both test cases use 8 test patterns and hence have identical test time, it appears to be best to select the test with the largest defect distance $(3, 2)$, as it offers ∼50% more coverage. Even if the 953,952-636,338 = 317,614 extra $(3, 2)$-only faults are less realistic, their coverage is for free!

8 Conclusion

In the context of chiplet-based heterogeneous designs, products with advanced multi-die packages start to appear that contain large numbers of inter-die interconnects bundles, implemented as even larger arrays ('islands') of micro-bumps. These interconnects incidentally suffer from short and/or open defects. Hence there is a need for effective and efficient manufacturing tests.

Traditional I-ATPG approaches such as based on the True/Complement version of the Counting Sequence Algorithm, detect, for a given set of interconnects, all shorts between any pair of interconnects, all open interconnects, and exclude any aliasing, in a bump-layout agnostic way. Its coverage also includes unlikely shorts between bumps that are rather far away from each other. By exploiting the relative locations of the various bumps, as known from the layout, we are able to avoid these unrealistic faults, and hence save on the test length and time.

In the newly proposed algorithm, we enable the user to constrain the faults under consideration, by specifying a maximum *defect distance* (w, h), containing bumps that are a candidate for a short defect. The algorithm then determines a bounding box size, partitions the micro-bump island according to this bounding box, assigns code numbers to the bump in each bounding box, and generates a code

word (and hence, implicitly, test patterns) on the basis of these code numbers.

Experimental results have shown both the effectiveness and the efficiency of the algorithm. Defect distance $(3, 2)$ covers ∼50% more short defects for the same test pattern count as defect distance $(2, 2)$ (which corresponds to shorts being restricted to adjacent bumps only). Both solutions reduce the test pattern count with 33% to 78% compared to the traditional I-ATPG approaches.

References

[1] Gartner Hype Cycle. https://www.gartner.com/en/research/methodologies/gartner-hype-cycle.

[2] IEEE Standards Association. *IEEE Std 1838^{TM}-2019, IEEE Standard for Test Access Architecture for Three-Dimensional Stacked Integrated Circuits*. IEEE, March 2020. doi:10.1109/IEEESTD.2020.9036129.

[3] UCIe. *Universal Chiplet Interconnect Express (UCIe)*. UCIe, February 2022.

[4] Erik Jan Marinissen et al. Direct Probing on Large-Array Fine-Pitch Micro-Bumps of a Wide-I/O Logic-Memory Interface. In *Proceedings IEEE International Test Conference (ITC)*, October 2014. doi:10.1109/TEST.2014.7035314.

[5] Ravi Mahajan. Advanced Packaging Architectures for Heterogeneous Integration. In *2019 IEEE PELS/PSMA Phoenix Workshop on Packaging and Integration in Power Delivery (PwrPack)*, October 2019.

[6] Sreejit Chakravarty. Special Session: A Call to Standardize Chip-let Interconnect Testing. In *Proceedings IEEE VLSI Test Symposium (VTS)*, April 2022. doi:10.1109/VTS52500.2021.9794149.

[7] Sreejit Chakravarty. 3D Interconnect Test Challenge. In *Proceedings IEEE European Test Symposium (ETS)*, May 2022.

[8] Najmi Jarwala and Chi W. Jau. A New Framework for Analyzing Test Generation and Diagnosis Algorithm for Wiring Interconnects. In *Proceedings IEEE International Test Conference (ITC)*, pages 63–70, October 1989.

[9] José T. de Sousa and Peter Y.K. Cheung. *Boundary-Scan Interconnect Diagnosis*. Kluwer Academic Publishers, Dordrecht, The Netherlands, 2001.

[10] William H. Kautz. Testing of Faults in Wiring Interconnects. *IEEE Transactions on Computers*, Vol. C-23(No. 4):358–363, April 1974.

[11] P. Goel and M.T. McMahon. Electronic Chip-In-Place Test. In *Proceedings IEEE International Test Conference (ITC)*, pages 83–90, October 1982.

[12] Henk D.L. Hollmann, Erik Jan Marinissen, and Bart Vermeulen. Optimal Interconnect ATPG under a Ground-Bounce Constraint. *Journal of Electronic Testing: Theory and Applications*, 21(1):17–31, February 2005. doi:10.1007/s10836-005-5284-9.

[13] Paul T. Wagner. Interconnect Testing with Boundary Scan. In *Proceedings IEEE International Test Conference (ITC)*, pages 52–57, October 1987.

[14] JEDEC. *Wide-I/O Single Data Rate (JEDEC Standard JESD229)*. JEDEC Solid State Technology Association, December 2011. http://www.jedec.org.

An Exploration of ATPG Methods for Redacted IP and Reconfigurable Hardware

Jackson Fugate, Greg Stitt, Naren Vikram Raj Masna,
Aritra Dasgupta, and Swarup Bhunia
University of Florida
Gainesville, FL, 32611
Email: {jacksonfugate, gstitt, nmasna, aritradasgupta, bhunias}@ufl.edu

Nij Dorairaj, David Kehlet
Intel Corporation
Email: {nij.dorairaj, david.kehlet}@intel.com

Abstract—Automated test-pattern generation (ATPG) is an important step of testing flows that is responsible for generating test values that expose faults in post-fabrication hardware. Previous work has introduced numerous ATPG methods that analyze application functionality to minimize the number of required tests. However, this existing work is misaligned with the emerging trend to use reconfigurable hardware, such as eFPGAs, to redact security-critical IP. When using reconfigurable hardware, application functionality is only known after serially loading a bitstream into a set of configuration flip-flops, which requires ATPG to do more general tests of the reconfigurable hardware as opposed to the targeted application. This more general testing results in prohibitively slow testing times that are on average 14.6× longer than the original design. In this paper, we explore novel ATPG and test methods for reconfigurable hardware to maximize stuck-at fault coverage, while minimizing testing time. We show significantly improved testing times that are on average 1.9× slower than the unredacted designs, without requiring any knowledge of the original application.

Keywords—Hardware IP protection, Reverse Engineering, Piracy, Confidentiality, ATPG, IP redaction

I. INTRODUCTION

Integrating reconfigurable hardware into ASICs in the form of embedded field-programmable gate arrays (eFPGAs) [1], [2] is becoming increasingly common due to flexibility and security advantages. One increasingly important security advantage is the ability of reconfigurable hardware to protect a design's confidentiality against reverse engineering and piracy attacks by hiding (i.e., *redacting*) the original design [3], [4]. However, these advantages come at the cost of increased testing challenges. FPGAs commonly use built-in self-test (BIST) methods [5], [6], but such methods often require partial reconfiguration or other specialized functionality, which may not be necessary for many use cases of embedded reconfigurable hardware, especially ones that do not use regular fabric-like architectures. An alternative is automatic test-pattern generation (ATPG)—a process for determining a minimal set of tests that expose faults—which is especially challenging for reconfigurable hardware, despite numerous effective strategies for fixed-logic ASICs [7]–[9].

The primary limitation of current ATPG methods is the requirement for analyzing the application functionality to determine test patterns. With reconfigurable hardware, the application's functionality is not revealed by the circuit's structure, and is instead specified at runtime by shifting a set of bits (i.e., a *bitstream*) into a serially loaded set of configuration flip-flops. Although a bitstream could potentially be provided to perform ATPG for reconfigurable hardware, such an approach is not ideal since is does not test the entire fault space. More importantly, for redacted IP, the bitstream cannot be used since this would reveal the original design. In this paper, we explore ATPG strategies and test methods to automatically generate and load bitstreams that achieve a desired fault coverage while minimizing testing time.

Figure 1 provides a comparison of traditional ATPG test methods with the proposed extensions required for redacted IP and/or reconfigurable hardware. In traditional methods (Figure 1(a)), the original design is first passed to *scan-chain insertion*, which creates a shift register out of the design flips-flops to enable loading and monitoring of values during testing. Scan-chain insertion outputs a test-ready design that is then passed to ATPG. ATPG analyzes the design to determine a series of scan-chain contents and primary inputs (collectively referred to as *test patterns)*. Finally, a *test method* applies the tests to the design by loading in the scan-chain contents for a test, then applying the primary inputs. The test method then scans out the contents of the scan chain, while also saving values of primary outputs. The test method then compares these values with known correct values provided by the ATPG tool. This test method then repeats for every test provided by ATPG, with the number of tests determined by the desired level of fault coverage.

By contrast, Figure 1(b) illustrates the proposed extensions for performing ATPG with reconfigurable hardware. First, in the case of IP redaction, the original design is passed to a redaction tool that replaces the application's logic with reconfigurable hardware. Alternatively, in the case of reconfigurable hardware without explicit IP redaction, the flow starts with the reconfigurable hardware. The flow then runs scan-chain insertion on the reconfigurable hardware to create a test-ready design. Next, like the traditional approach, ATPG produces test patterns, but in addition to producing scan-chain contents

---

DISTRIBUTION STATEMENT A. Approved for public release: distribution is unlimited.

979-8-3503-4631-2/23 $31.00 © 2023 IEEE

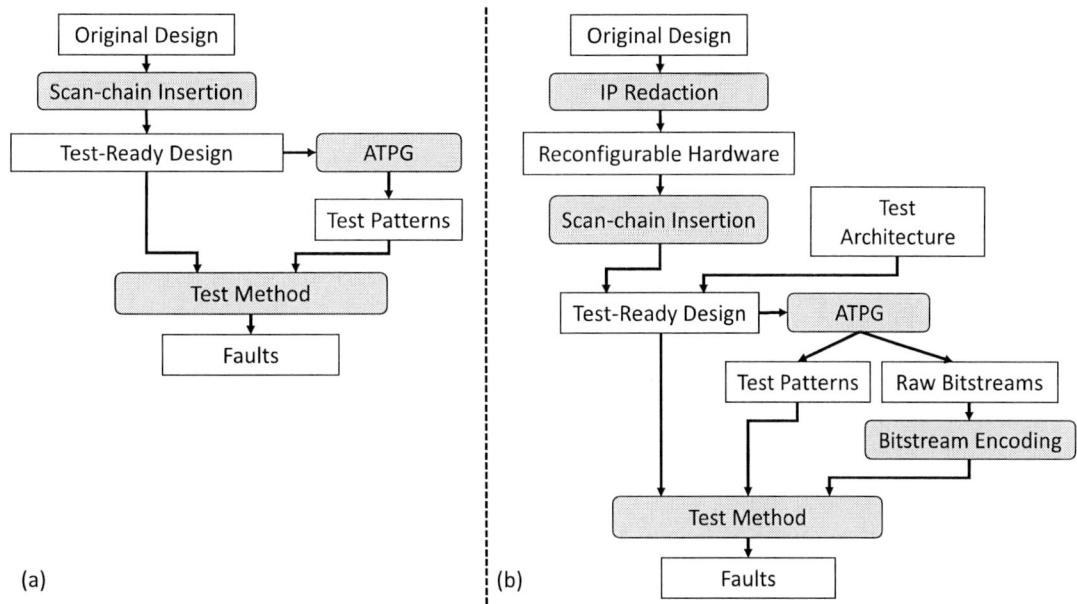

Fig. 1: A comparison of (a) a normal ATPG flow and (b) the proposed ATPG flow for redacted IP and reconfigurable hardware.

and primary inputs, ATPG now also produces bitstreams that are necessary to test the functionality of the reconfigurable hardware. The ATPG-provided bitstreams are in a raw format, which then must be encoded according to whatever format is expected by the reconfigurable hardware.

One other difference is that the flow extends the test-ready design with a *test architecture*. This architecture is necessary because unlike normal ASICs that simply load the scan chain serially, reconfigurable hardware must also load the bitstream, which contains additional configuration and error-checking logic. Finally, the test method leverages the test-ready design, the test architecture, the test patterns, and the bitstreams to provide the requested fault coverage.

In this paper, we present and evaluate four different ATPG test methods, using a baseline method that simply treats all configurable bits in reconfigurable hardware as primary inputs during ATPG. We evaluate the methods for the use case of redacted IP by comparing the testing times of the reconfigurable hardware with the testing times of the original design, with the goal of achieving 100% coverage of stuck-at faults. We show that the baseline method resulted in prohibitively slow testing times that were on average 14.6× longer than the original unredacted design. The proposed methods demonstrate significant improvements over the baseline, with testing times that averaged 1.9× slower than the original unredacted designs, all without requiring any knowledge of the original design.

The paper is organized as follows. Section II discusses architectural extensions that are required to perform testing of reconfigurable hardware. Section III discusses how we leverage and extend existing ATPG techniques, and how we combine ATPG with test methods to minimize testing time, while also discussing trade offs of each method. Section IV

presents experimental results comparing testing times of each method.

II. TEST ARCHITECTURE

Traditional ATPG assumes a set of primary inputs and scan chain for the design under test, which requires a simple test architecture that for each test simply loads the scan chain, applies the primary inputs, and then verifies primary outputs and scan-chain values. ATPG for reconfigurable hardware requires a more complicated test architecture because in addition to having to load a scan chain, the test architecture has to also configure the hardware with a bitstream.

Figure 2 illustrates a high-level overview of the envisioned test architecture that can be used for any of the proposed methods in Section III. Some of the methods require slight variations, which are discussed later.

The primary reason for requiring a different test architecture when using reconfigurable hardware is that ATPG now provides bitstreams in addition to test patterns. Exact functionality varies across test methods, but in general, each method will load a test pattern into the scan chain of the reconfigurable hardware, in addition to loading a bitstream. However, loading the bitstream will generally require additional logic depending on the exact reconfigurable architecture and application.

When ATPG provides bitstream values for testing, those values correspond to a *raw* bitstream that does not have any additional information other than the configuration bits. Most bitstreams for reconfigurable hardware typically include extra information such as bitstream length, checksum values, etc. that assist with configuration, while also enabling error checking. Although a custom configuration controller for ATPG could potentially be added, any reconfigurable hardware will already have a configuration controller, in which case it makes sense to reuse that controller as part of the test architecture.

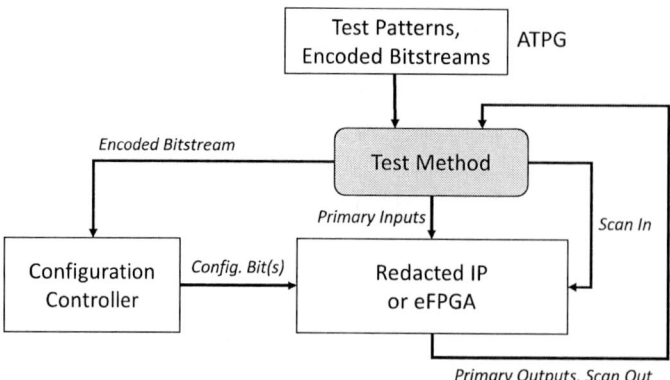

Fig. 2: The test architecture required for applying tests from ATPG to reconfigurable hardware. In addition to the normal scan chain that loads test values, testing reconfigurable hardware also requires a configuration controller.

Using an existing configuration controller will require the bitstream in the expected encoded format, which requires an additional step between ATPG and testing to encode every ATPG-provided raw bitstream.

In this paper, we assume the latency required for the fully encoded bitstream and the configuration controller generally have a negligible effect on total testing time. We therefore omit these times from the experiments, which allows us to evaluate ATPG methods independently from specific bitstream encodings.

III. ATPG AND TEST METHODS

In this section, we explore four strategies for extending existing ATPG methods to better handle reconfigurable hardware for redacted IP. Before using ATPG, a typical DFT flow uses scan-chain insertion [10] to create a shift register out of the design's flip-flops, while adding in various modes to control the flip-flop contents. ATPG then applies various algorithms [11], [12] to determine a set of test patterns. To test a design, each generated test is loaded into the scan chain and applied to primary inputs. The primary outputs and scan-chain contents are then compared with an expected outcome to detect faults.

A. Method 1: Basic

In Method 1 (referred to as the *basic* method), we extend ATPG by treating the contents of every configuration flip-flop as an input. With this approach, we can leverage existing ATPG tools to determine scan-chain contents, bitstream contents, in addition to primary-input values for every test.

Figure 3 illustrates a timing diagram of this behavior. Initially, the basic test method first loads the scan-chain contents for test 0. Upon completion, the method leverages the configuration controller within the test architecture to serially load the bitstream for test 0. Next, the method applies primary inputs and then checks all outputs, which would require two stages. The method first samples the primary outputs and then shifts out the contents of the scan chain. Finally, the method compares the outputs and scan-chain contents with correct values to identify faults.

This process then repeats for every ATPG-provided test, but with one difference after test 0. Since there is no cost for shifting in the scan-chain contents for test n while reading out the contents for test $n-1$, the method overlaps the scan in and scan out of consecutive tests.

The test architecture required for the basic method would generally be very similar to Figure 2. DFT tools provide the scan-chain interface, and the configuration controller provides a mechanism for the test method to shift in bitstreams.

The primary advantage of the basic method is that it is effective at minimizing the number of total tests to achieve coverage, since it directly uses tests provided by existing ATPG tools. However, it suffers from two limitations. First, ATPG normally uses a number of inputs that is much smaller than the number of configuration flip-flops, which could add anywhere from 10k to millions of additional inputs. Second, even if ATPG is able to produce effective bitstream contents, the test method must shift those contents into the configuration flip-flops for *every* generated test, which can require prohibitively long testing times.

B. Method 2: Overlapped

In Method 2 (the *overlapped* method), we reduce the primary test-time bottleneck of the basic method, which is the lengthy loading times of the scan chain and bitstream. One conceptually simple improvement is to modify the test architecture to enable simultaneous loading of the scan chain and bitstream. This improvement would generally require minimal overhead in the test architecture because the architecture already provides mechanisms to load both resources. The architecture would just need extra control logic, and possibly increased I/O requirements to load both resources in parallel.

Figure 4 illustrates a timing diagram of this overlapped method. Initially, the test architecture loads the scan chain and bitstream for test 0 at the same time, where the bitstream will generally take longer than the scan chain. After loading these resources, the remainder of the method is the same as the basic approach, with the scan out of test 0 and scan in of test 1 occurring simultaneously.

979-8-3503-4631-2/23 $31.00 © 2023 IEEE

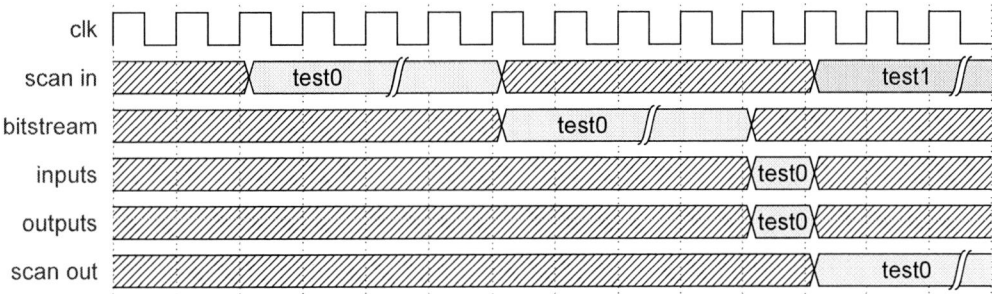

Fig. 3: A timeline of method 1 (*basic*). First, the test-application process loads the scan-chain contents and the bitstream for test 0. It then applies inputs and checks outputs. It then repeats this process for test 1 by first loading the chain, which simultaneously scans out the test results from test 0.

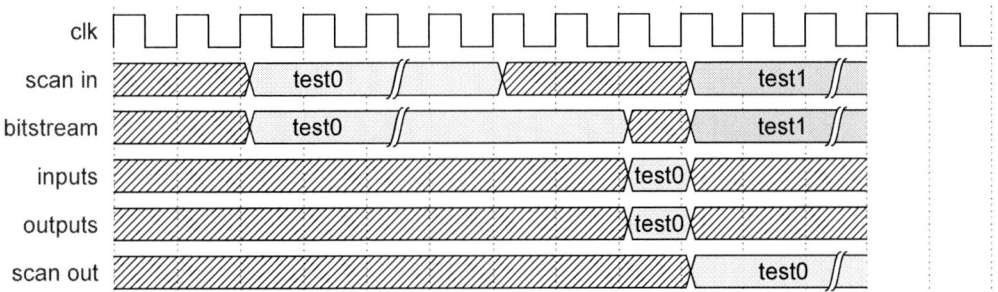

Fig. 4: A timeline of method 2 (*overlapped*). This method reduces testing times of the basic method by loading the bitstream and scan chain simultaneously.

C. Method 3: Shift

In Method 3 (the *shift* method), we again provide the configuration flip-flops as inputs to ATPG, but instead of using these patterns directly and shifting the entire bitstream for every pattern, we shift the bitstream by some fixed amount for each test. The motivation for this method is that shifting in an entire bitstream for every pattern can make testing time prohibitively long, especially if ATPG requires a large number of tests to achieve sufficient coverage.

Figure 5 illustrates a timing diagram of this method. Like the overlapped method, the shift method also loads in the bitstream and scan chain simultaneously, which makes test 0 identical to the previous method. However, for test 1, the shift method does not load in an entire new bitstream. Instead, it shifts the configuration chain by an amount equal to the scan chain, while partially loading in the ATPG-provided bitstream for test 1. By shifting the configuration chain by the length of the scan chain, the shift method completely hides the bottleneck for loading a bitstream. However, this method comes at the cost of not always using bitstreams determined by ATPG tools. As a result, the method essentially sacrifices coverage per test to reduce the time to apply each test. As long as the improvement in time per test outweighs the increase in number of required tests, then this method will reduce testing time.

To make the shifting method work with TetraMAX, we developed scripts that insert additional tests in between the normal ATPG-generated tests that provide the intermediate values of the configuration flip-flops after each shift. For each inserted test, we constrain the bitstream contents to ensure that TetraMAX generates appropriate scan-chain contents and primary inputs for the partially shifted bitstream.

One disadvantage of the shift method is the requirement for a more specialized test architecture. Specifically, the configuration controller would require additional control signals to allow the test method to halt the configuration of the bitstream as soon as the scan-chain contents were loaded. We do not expect this modification to create a significant resource overhead, but it will add some complexity.

In addition, a more significant disadvantage is that by shifting the bitstream by an amount equal to the length of the scan chain, the number of configuration flips-flops must be a multiple of the length of the scan chain. To ensure this requirement, we manually add "dummy" flip-flops, which can create a significant area overhead when the bitstream length is not close to a multiple of the scan chain length.

D. Method 4: Overlapped Parallel

In Method 4 (the *overlapped parallel* method), we extend the overlapped method with parallel loading for the bitstream. The motivation for this method is that the loading of the bitstream is generally the biggest testing bottleneck. By loading multiple bits each cycle, and overlapping that loading with the loading of the scan chain, we can significantly reduce times for each test.

Figure 6 illustrates a timing diagram of this method. The timing looks largely similar to the overlapped method, but now

979-8-3503-4631-2/23 $31.00 © 2023 IEEE

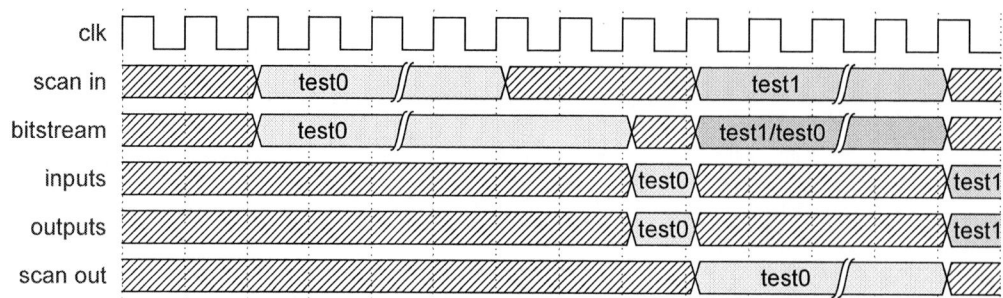

Fig. 5: A timeline of method 3 (*shift*). This method overlaps the loading of the scan chain and bitstream, but instead of loading in the entire new bitstream, it stops after the scan chain is loaded and treats the current state of the configuration bits as a new bitstream. The motivation for this approach is that loading the bitstream usually dominates testing time. By stopping the bitstream loading once the scan chain has been loaded, this approach reduces time for each individual test at the cost of an increase in the total number of tests.

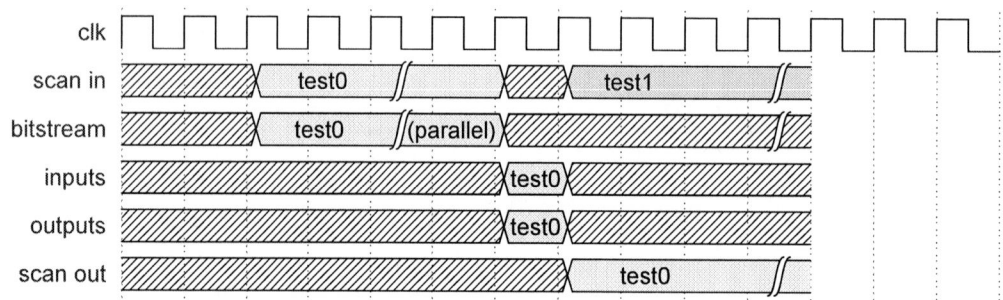

Fig. 6: A timeline of method 4 (*overlapped parallel*). This method reduces testing times by loading the bitstream in parallel. Potentially any number of bits can be loaded every cycle.

the loading time of the bitstream is reduced via parallel loading, which significantly reduces the time to apply each test. Although arbitrary amounts of parallelism could potentially be leveraged, it generally does not make sense to reduce the bitstream loading times beyond the time required for the scan chain because the total time to apply a test would be identical.

Unlike the previous methods, the overlapped parallel method would generally require more significant changes to the testing architecture, and possibly the reconfigurable hardware itself. Reconfigurable hardware typically loads a bitstream serially into the configuration chain. To load multiple bits in parallel, two options are possible. First, the configuration chain can be modified to shift by n bits every cycle, while also using n inputs for the new bits. While this modification is conceptually simple, ASICs often use highly optimized cells for scan chains that only shift by a single flip-flop. As a result, this method might have a significant area overhead. Alternatively, a likely better approach is to divide the original bitstream into separate smaller bitstreams that get loaded into multiple configuration chains that shift by the usual one flip-flop per cycle. While likely minimizing overhead, this approach would require a considerably different configuration controller that was capable of partitioning a bitstream into multiple configuration chains. Alternatively, the tool that generates the bitstream must have knowledge of the testing architecture to generate each individual bitstream. While none of these

issues are significant technical challenges, they pose practical challenges due to non-standard tool requirements.

This specialized test architecture is the primary disadvantage of the overlapped parallel method. Such an approach would rarely be used for an entire ASIC due to the increased I/O requirements. However, for any SoC with reconfigurable hardware, there will already be some amount of configuration control within the design, where increasing the complexity of that configuration logic for parallel testing could have acceptable overheads for many use cases.

IV. EXPERIMENTAL RESULTS

This section presents experiments comparing the testing times for the four methods from Section III. We evaluated the use case of redacted IP for five designs from the MIT-CEP benchmarks [13], and one (s27) from ISCAS89 benchmarks [14], which we included to evaluate one small example. The examples cover a range of cell counts and flip-flops: s27 (19 cells, 3 flops), FIR (1249 cells, 448 flops), SHA256 (5k cells, 1k flops), RSA (25k cells, 8k flops), AES192 (180k cells, 9.6k flops), GPS (181k cells, 10k flops), For each experiment, we passed the original code through our internal IP-redaction tool to replace the original design with reconfigurable hardware. This tool replaced all combinational logic with lookup tables that are configured via serially loaded flip-flops. We then used Synopsys Design Compiler to synthesize the designs, targeting the LEDA 250nm CMOS library. For ATPG, we

979-8-3503-4631-2/23 $31.00 © 2023 IEEE   242

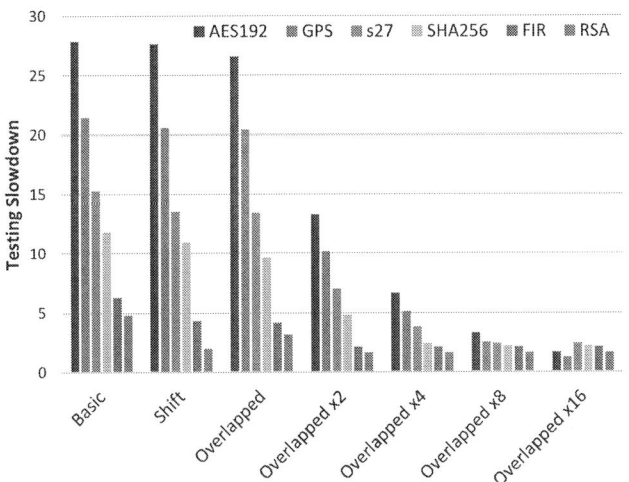

Fig. 7: A comparison of the testing slowdown, compared to an unredacted design, for each of the proposed ATPG methods.

used TetraMAX to generate flip-flop contents for the bitstream, scan-chain contents for design flip-flops, and primary inputs. For each example, we configured TetraMAX to run until achieving 100% coverage of stuck-at faults.

Because the testing times can vary significantly across different-sized examples, we normalized the results by comparing the reconfigurable hardware testing times with the testing time required by original design before redaction. The metric we use for this comparison is *testing slowdown*, which is the testing time for the reconfigurable hardware divided by the testing time for the original design. In the ideal case, the slowdown would be 1, meaning that the there is no overhead in testing the reconfigurable versions of each design.

The experiments include four different overlapped-parallel approaches, ranging from 2 bits per cycle (Overlapped x2) to 16 bits per cycle (Overlapped x16).

Figure 7 shows the testing slowdown of each proposed approach. The basic approach, as expected, performed the worst across all examples, with testing slowdowns ranging from $4.8\times$ (for RSA) to $27.8\times$ (for AES192), and averaging $14.6\times$. Despite lengthy testing times, the basic approach still might be useful for situations requiring a low-overhead test architecture. However, for many use cases, the testing slowdown might be prohibitive.

On average, the shift method achieved a modest 9.6% improvement over the basic approach, with an average testing slowdown $13.2\times$. The shift method provided the largest improvement for RSA, which reduced testing times by $2.4\times$. In general, the shift method did not see significant improvements, because despite much faster times for each individual test, the shifting required an average pattern increase of $8\times$ compared to the other methods, and a $12.6\times$ increase compared to the

original, unredacted design. The shift method achieves these modest testing time improvements at the cost of an average flip-flop overhead of 12.8%, with significant overheads of 32% for FIR and RSA.

The overlapped approach achieved an average 11.5% improvement compared to the basic approach due to the scan chains being much smaller than the bitstreams. Parallel loading achieved significant improvements, with testing times achieving an average improvement of 87% for 16 bits, which corresponded to a testing slowdown of $1.9\times$ compared to the original design. FIR and RSA saw no improvements after loading 2 bits in parallel. SHA256 and s27 saw no improvement after loading 8 bits in parallel. In general, higher amounts of parallelism only provide benefits for designs that have a large number of configuration flip-flops compared to design flip-flops.

One disadvantage of all the methods is that TetraMAX was unable to achieve 100% coverage. The original, unredacted designs all achieved 100% coverage. For the shift method, TetraMAX achieved 98.0% coverage on average, and the basic, overlapped, and parallel methods averaged 98.5% coverage.

CONCLUSIONS

In this paper, we explored ATPG methods for reconfigurable hardware, which we evaluated for the use case of redacting security-critical IP for protection against confidentiality attacks (e.g., reverse engineering, piracy). We showed that straight-forward mappings onto existing ATPG methods provide potentially prohibitive testing-time slowdowns averaging $14.6\times$ compared to unredacted IP. Using the methods we introduced, we demonstrated significant improvements in testing times that reduced this average slowdown to $1.9\times$.

For future work, we plan to investigate additional test methods. For example, scan-chain loading can be potentially parallelized in a similar way as bitstream loading. Alternatively, the shift method can be combined with parallel loading of the bitstream and/or the scan chain. An additional area of future work will be evaluating the overhead of the test architecture for specific applications and use case. Such analysis was outside the scope of this initial study due to the architecture being highly dependent on the bitstream encoding, the error checking done by the configuration controller, the test method, and each individual application.

ACKNOWLEDGMENTS

This work is supported by the National Security Technology Accelerator (NSTXL) Consortium contract number N00164-19-9-G007b.

REFERENCES

[1] P. Mohan, O. Atli, O. Kibar, M. Zackriya, L. Pileggi, and K. Mai, "Top-down physical design of soft embedded fpga fabrics," in *The 2021 ACM/SIGDA International Symposium on Field-Programmable Gate Arrays*, ser. FPGA '21. New York, NY, USA: Association for Computing Machinery, 2021, p. 1–10. [Online]. Available: https://doi.org/10.1145/3431920.3439297

Intel, the Intel logo, and other Intel marks are trademarks of Intel Corporation or its subsidiaries. *Other names and brands may be claimed as the property of others. LEGAL DISCLAIMER: No computer system can be absolutely secure.

979-8-3503-4631-2/23 $31.00 © 2023 IEEE

[2] T. Hotfilter, F. Kreß, F. Kempf, J. Becker, J. M. de Haro, D. Jiménez-González, M. Moretó, C. Alvarez, J. Labarta, and I. Baili, "Towards reconfigurable accelerators in hpc: Designing a multipurpose efpga tile for heterogeneous socs," in *Proceedings of the 2022 Conference amp; Exhibition on Design, Automation amp; Test in Europe*, ser. DATE '22. Leuven, BEL: European Design and Automation Association, 2022, p. 628–631.

[3] C. M. Tomajoli, L. Collini, J. Bhandari, A. K. T. Moosa, B. Tan, X. Tang, P.-E. Gaillardon, R. Karri, and C. Pilato, "Alice: An automatic design flow for efpga redaction," in *Proceedings of the 59th ACM/IEEE Design Automation Conference*, ser. DAC '22. New York, NY, USA: Association for Computing Machinery, 2022, p. 781–786. [Online]. Available: https://doi.org/10.1145/3489517.3530543

[4] P. Mohan, O. Atli, J. Sweeney, O. Kibar, L. Pileggi, and K. Mai, "Hardware redaction via designer-directed fine-grained efpga insertion," in *2021 Design, Automation Test in Europe Conference Exhibition (DATE)*, 2021, pp. 1186–1191.

[5] M. Abramovici and C. Stroud, "Bist-based test and diagnosis of fpga logic blocks," *IEEE Transactions on Very Large Scale Integration (VLSI) Systems*, vol. 9, no. 1, pp. 159–172, 2001.

[6] C. Stroud, J. Nall, M. Lashinsky, and M. Abramovici, "Bist-based diagnosis of fpga interconnect," in *Proceedings. International Test Conference*, 2002, pp. 618–627.

[7] R. P. Lajaunie and M. S. Hsiao, "An effective and efficient atpg-based combinational equivalence checker," in *Proceedings of the 15th ACM Great Lakes Symposium on VLSI*, ser. GLSVLSI '05. New York, NY, USA: Association for Computing Machinery, 2005, p. 248–253. [Online]. Available: https://doi.org/10.1145/1057661.1057722

[8] J. E. Nelson, J. G. Brown, R. Desineni, and R. D. Blanton, "Multiple-detect atpg based on physical neighborhoods," in *Proceedings of the 43rd Annual Design Automation Conference*, ser. DAC '06. New York, NY, USA: Association for Computing Machinery, 2006, p. 1099–1102. [Online]. Available: https://doi.org/10.1145/1146909.1147186

[9] F. Fummi, C. Marconcini, and G. Pravadelli, "An efsm-based approach for functional atpg," in *Proceedings of the 15th ACM Great Lakes Symposium on VLSI*, ser. GLSVLSI '05. New York, NY, USA: Association for Computing Machinery, 2005, p. 197–200. [Online]. Available: https://doi.org/10.1145/1057661.1057709

[10] S. Makar, "A layout-based approach for ordering scan chain flip-flops," in *Proceedings International Test Conference 1998 (IEEE Cat. No.98CH36270)*, 1998, pp. 341–347.

[11] Z. Hegedűs, "Efficiency test of automatic test pattern generation methods," *Periodica Polytechnica Electrical Engineering*, vol. 36, no. 3-4, p. 309–318, 1992. [Online]. Available: https://pp.bme.hu/ee/article/view/4529

[12] D. Duvalsaint, Z. Liu, A. Ravikumar, and R. D. Blanton, "Characterization of locked sequential circuits via atpg," in *2019 IEEE International Test Conference in Asia (ITC-Asia)*. IEEE, 2019, pp. 97–102.

[13] "Common evaluation platform," https://github.com/mit-ll/CEP.

[14] F. Brglez, D. Bryan, and K. Kozminski, "Combinational profiles of sequential benchmark circuits," in *1989 IEEE International Symposium on Circuits and Systems (ISCAS)*, 1989, pp. 1929–1934 vol.3.

Compact Set of Functional Broadside Tests with Fault Detection on Primary Outputs

Irith Pomeranz

School of Electrical and Computer Engineering
Purdue University
West Lafayette, IN 47907, U.S.A.
E-mail: pomeranz@ecn.purdue.edu

Abstract—When a broadside (launch-on-capture) test is extracted from a functional test sequence, the test addresses the need to avoid overtesting the circuit by exercising it under functional operation conditions. A broadside test t is extracted from a functional test sequence A by duplicating two or more functional capture cycles from A between the scan-in and scan-out operations of t. The functional capture cycles in t detect a fault f if it is activated and propagated to an observable output. This is typically a next-state variable whose value is observed during the scan-out operation of t. This article considers the added requirement that an extracted broadside test would propagate fault effects to the primary outputs. This reduces overtesting further by eliminating the detection of effects that cannot be propagated to the primary outputs during functional operation. The article makes several observations about the use of sequential fault simulation for the efficient extraction of a compact set of broadside tests that detect transition faults on the primary outputs. Experimental results for transition faults in benchmark circuits demonstrate an improved fault coverage and a reduced number of clock cycles for test application.

I. INTRODUCTION

Overtesting, where a fault-free circuit appears to be faulty under a scan-based test set for delay faults, was observed in [1]-[3]. Overtesting results from the activation of long paths that cannot be activated during functional operation [1], as well as excessive switching activity [2]-[3]. Both effects cause the circuit to be slower, resulting in the appearance of a delay fault in a circuit that will operate correctly during functional operation.

Broadside (launch-on-capture) tests address the need to avoid overtesting a circuit when they exercise it under functional operation conditions. Single-cycle scan-based tests that create functional operation conditions were defined in [4]. The definition was extended to two-cycle broadside tests in [5]. Two procedures for generating broadside tests that create functional operation conditions were described in [5]. One of these procedures extracts broadside tests from a functional test sequence. A similar extraction process was also used in [6]-[7]. Multicycle broadside tests are extracted from functional test sequences in [8].

A broadside test t is extracted from a functional test sequence A by duplicating two or more functional capture cycles from A between the scan-in and scan-out operations of t. This is illustrated by Figure 1. The sequence $A = a_0 \ a_1 \ ...$

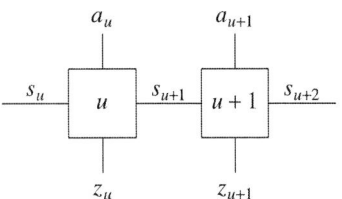

Fig. 1. Functional test sequence.

a_{L-1} consists of L functional capture cycles, of which two are shown in Figure 1. The sequence takes the circuit through the sequence of states $s_0 s_1 ... s_L$, where s_0 is the initial state of the circuit for functional operation. The extracted broadside test $t(u, u + 1) = \langle s_u, a_u, a_{u+1} \rangle$ consists of the two functional capture cycles shown in Figure 1. It starts with a scan-in operation that brings the circuit to state s_u. This is followed by two functional capture cycles with primary input vectors a_u and a_{u+1}. The test ends with a scan-out operation. The expected fault-free scan-out state is s_{u+2}, and the expected fault-free primary output vector is z_{u+1}.

In general, the test $t(u, u + 1)$ detects a fault f if it is activated and propagated to an observable output. This is typically a next-state variable whose value is observed during the scan-out operation of $t(u, u + 1)$.

This article considers the added requirement that a broadside test extracted from a functional test sequence would propagate fault effects to the primary outputs. The benefit is the potential to reduce overtesting further by eliminating the detection of effects that cannot be propagated to the primary outputs during functional operation. The tests are referred to as primary output extracted broadside, or $POEXB$ tests. The main limitations of using $POEXB$ tests are: (1) the increased computational cost for test extraction, and (2) the need for multicycle tests with large numbers of functional capture cycles to propagate fault effects to the primary outputs.

Considering the computational cost, the extraction process of $POEXB$ tests is performed using sequential fault simulation. Specifically, given a functional test sequence A, sequential fault simulation of A is useful in identifying clock cycles from which $POEXB$ tests can be extracted efficiently. Without sequential fault simulation, an excessive number of tests would have to be extracted and simulated as scan-based

979-8-3503-4631-2/23 $31.00 © 2023 IEEE

tests. The computational effort of sequential fault simulation is assumed to be acceptable. The use of sequential fault simulation for the efficient extraction of $POEXB$ tests is explained in Section II.

Considering the need for multicycle tests, unconstrained multicycle scan-based tests were considered earlier in [9]-[16], and were shown to contribute to test compaction. The same effect exists with extracted broadside and $POEXB$ tests. Thus, although the number of functional capture cycles in a test is increased, the effect on the total number of clock cycles required for applying the test set is mitigated by the fact that each test can potentially detect more faults.

The article observes that a single pass of sequential fault simulation of a sequence A does not identify all the $POEXB$ tests that can be extracted from A. It thus has a limited ability to support test compaction. The article describes a procedure that, using several passes of sequential fault simulation considering sequences of decreasing length, is able to extract from A all the shortest $POEXB$ tests for a transition fault f (a shortest $POEXB$ test for f uses a clock cycle where f is detected by A, and has the smallest number of functional capture cycles guaranteed to detect f). This procedure is used in a test compaction procedure that extracts a compact set of $POEXB$ tests from a functional test sequence. Of all the tests extracted for a fault, the test compaction procedure selects the one that detects the largest number of additional faults.

The approach to test compaction is similar to dynamic test compaction [17]-[24] in that it attempts to ensure that every additional test would detect one target fault, and as many additional faults as possible. This implies that certain faults would have to be targeted several times before they are detected. However, whereas dynamic test compaction procedures typically use unspecified values to detect more faults, the test compaction procedure described in this article considers $POEXB$ tests that are fully-specified. Multicycle tests with more functional capture cycles are used by the procedure when this helps achieve test compaction. In addition to test compaction, a by-product of searching for all the shortest tests that can be extracted from A is an improved transition fault coverage compared with a procedure that uses a single pass of sequential fault simulation.

The possibility of propagating fault effects to the primary outputs was considered in [25] as part of a test generation procedure for broadside tests that create functional operation conditions. However, the analysis in [25] does not consider faulty circuits. Therefore, it may not match the detection conditions of a specific fault or fault model. In addition, the tests computed in [25] are two-cycle tests, and fault detection occurs during scan-out operations. As a result, it is still possible to detect faults whose fault effects cannot be propagated to the primary outputs using the tests from [25].

It should be noted that fault detection on the primary outputs does not affect the test application process. Every test starts with a scan-in operation and ends with a scan-out operation that is overlapped with the scan-in operation of the next test. For $POEXB$ tests, values that appear on the scan outputs are ignored for fault detection. Output compaction logic can be used for capturing the primary output vectors at-speed.

The article is organized as follows. Section II provides the motivation for using sequential fault simulation for the extraction of $POEXB$ tests. Section III describes a procedure that uses a single pass of sequential fault simulation to extract $POEXB$ tests. Section IV discusses the use of several passes of sequential fault simulation for the extraction of $POEXB$ tests. Section V describes a test compaction procedure that extracts a compact set of $POEXB$ tests for transition faults from a functional test sequence. Section VI presents experimental results. Section VII concludes the article.

II. MOTIVATION FOR SEQUENTIAL FAULT SIMULATION

Using only logic simulation, without sequential fault simulation, all the $POEXB$ tests can be extracted from a sequence $A = a_0 a_1 ... a_{L-1}$ by considering every two clock cycles u_b and u_d such that $u_d - u_b \geq 1$. With clock cycle u_b defining the first clock cycle of the test, the scan-in state of the test is s_{u_b}. The last clock cycle of the test is defined by clock cycle u_d. The resulting test is $t(u_b, u_d) = \langle s_{u_b}, a_{u_b}, ..., a_{u_d} \rangle$, and it duplicates clock cycles u_b to u_d of A. A fault is not detected by the first clock cycle of $t(u_b, u_d)$ even if sequential fault simulation shows that it is detected at clock cycle u_b of A. This is because a transition fault requires two clock cycles for fault activation. Regardless of the relationship to A, the test $t(u_b, u_d)$ may detect a transition fault.

Thus, it is possible to obtain a set $U = \{ t(u_b, u_d) : 0 \leq u_b < L-1, u_b+1 \leq u_d < L \}$ of all the $POEXB$ tests based on A. Fault simulation of the tests in U as scan-based tests can be used for identifying $POEXB$ tests that detect transition faults. During fault simulation of U, a fault is marked as detected only if its fault effects reach a primary output.

However, considering every pair of clock cycles u_b and u_d such that $u_d - u_b \geq 1$ results in $O(L^2)$ tests that need to be simulated to determine whether they detect any target faults. The tests are multicycle tests with a number of functional capture cycles between two and L. This yields $O(L^3)$ functional capture cycles that need to be simulated. With the experimental setup in Section VI, U would contain approximately $4 \cdot 10^7$ tests with a total of approximately $9 \cdot 10^8$ functional capture cycles. Sequential fault simulation can be used for reducing these numbers by focusing on pairs of clock cycles that produce effective $POEXB$ tests. This is discussed in Sections III and IV.

III. SINGLE PASS OF SEQUENTIAL FAULT SIMULATION

This section describes a procedure that uses a single pass of sequential fault simulation to extract $POEXB$ tests for transition faults from a functional test sequence $A = a_0 a_1 ... a_{L-1}$.

For a transition fault f, sequential fault simulation of A yields a sequence of fault-free states $s_0 s_1 ... s_L$, and a sequence of faulty states $\hat{s}_0 \hat{s}_1 ... \hat{s}_L$. It also yields a sequence of fault-free primary output vectors $z_0 z_1 ... z_{L-1}$, and a sequence of faulty primary output vectors $\hat{z}_0 \hat{z}_1 ... \hat{z}_{L-1}$.

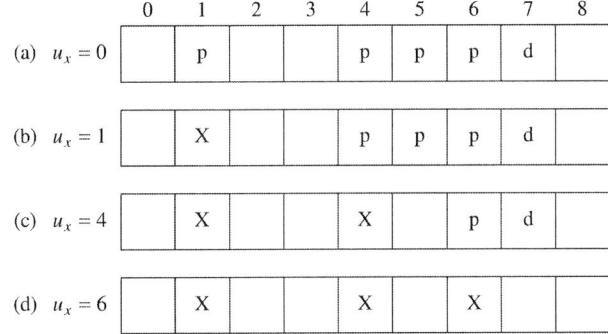

Fig. 2. Sequential fault simulation.

Two types of clock cycles are important for the extraction of $POEXB$ tests. Clock cycle u is referred to as a pseudo-detection clock cycle if $\hat{s}_{u+1} \neq s_{u+1}$. Clock cycle u is referred to as a detection clock cycle if $\hat{z}_u \neq z_u$. In both cases, fault effects appear in clock cycle u, and reach an observable output, either a next-state variable in the case of pseudo-detection, or a primary output in the case of detection. A clock cycle can be both a pseudo-detection and a detection clock cycle.

Figure 2(a) shows several clock cycles of a test sequence A when it is simulated under a transition fault f. Pseudo-detection clock cycles are marked with a p, and detection clock cycles are marked with a d. A transition fault requires at least two clock cycles for pseudo-detection or detection. The fault f is pseudo-detected at clock cycle 1. This implies that clock cycles 0 and 1 together pseudo-detect the fault, resulting in $\hat{s}_2 \neq s_2$. Without the requirement to detect the fault on a primary output, the scan out operation of the test $t(0,1) = \langle s_0, a_0, a_1 \rangle$ detects the fault. However, the fault is not detected by A on a primary output after being pseudo-detected at clock cycle 1, and before its fault effects disappear at clock cycle 2. Therefore, the pseudo-detection at clock cycle 1 does not yield a $POEXB$ test.

The transition fault f is pseudo-detected again starting at clock cycle 4, and detected on a primary output at clock cycle 7. A $POEXB$ test needs to use at least clock cycles 6 and 7 for activating and detecting the fault. However, based on Figure 2(a), the fault is also pseudo-detected at clock cycle 5. Whereas a $POEXB$ test $t(6,7) = \langle s_6, a_6, a_7 \rangle$ would use s_6 as a scan-in state for both the fault-free and faulty circuits, the sequence in Figure 2(a) brings the fault-free and faulty circuits to states s_6 and $\hat{s}_6$, respectively, such that $\hat{s}_6 \neq s_6$. Therefore, it is not guaranteed that $t(6,7)$ would detect the fault. The same issue exists for $t(5,7)$. Considering $t(4,7)$, the test would use s_4 as a scan-in state, and the fault would not be pseudo-detected by the first clock cycle of the test. This is because a transition fault requires two clock cycles for activation. Therefore, when clock cycle 4 of A is duplicated in $t(4,7)$, it creates a different next-state for the faulty circuit than clock cycle u of A. Consequently, the fault is not guaranteed to be detected by this test. The pseudo-detection that occurs at clock cycle 4 of A can occur when clock cycle 4 is duplicated

in the test $t(3,7) = \langle s_3, a_3, a_4, a_5, a_6, a_7 \rangle$. Considering its scan-in state s_3, the sequence A in Figure 2(a) takes the fault-free and faulty circuits to the same state, s_3, at clock cycle 3. This can be seen from the fact that clock cycle 2 is not a pseudo-detection clock cycle, i.e., $\hat{s}_3 = s_3$. The conclusion is that $t(3,7)$ takes the fault-free and faulty circuits through the same states as clock cycles 3 to 7 of A, and it is guaranteed to detect f. This is a $POEXB$ test for f.

Procedure 1 considers the set F of transition faults, and a functional test sequence A, for the extraction of $POEXB$ tests as illustrated by Figure 2(a). The procedure considers the faults from F one by one. For every fault it performs sequential fault simulation of f until it is detected on a primary output for the first time. If f is detected on a primary output at clock cycle u_d, the procedure extracts a $POEXB$ test $t(u_b, u_d)$ for f. Fault simulation with fault dropping of F under $t(u_b, u_d)$ removes from consideration faults that are detected by $t(u_b, u_d)$. The test is added to a test set T_1.

Procedure 1: Extraction of $POEXB$ tests using a single pass of sequential fault simulation

1) Assign $T_1 = \emptyset$.
2) For every fault $f \in F$:
   a) Simulate f under A until it is detected on a primary output.
   b) If f is detected by A on a primary output at clock cycle u_d:
      i) Find a clock cycle $0 \leq u_b < u_d$ such that the following conditions are satisfied: (1) either $u_b = 0$, or f is not pseudo-detected at clock cycle $u_b - 1$; (2) f is not pseudo-detected at clock cycle u_b; and (3) f is pseudo-detected at clock cycles $u_b < u < u_d$.
      ii) Define the test $t(u_b, u_d) = \langle s_{u_b}, a_{u_b}, ..., a_{u_d} \rangle$.
      iii) Perform fault simulation with fault dropping of F under $t(u_b, u_d)$, and add $t(u_b, u_d)$ to T_1.

Procedure 1 performs sequential fault simulation for at most $|F|$ faults. In effect, fault dropping reduces significantly the number of faults for which sequential fault simulation is required. For example, using the setup described in Section VI, benchmark circuit $s1423$ has 2,218 transition faults that can be detected by $POEXB$ tests. Procedure 1 performs sequential fault simulation for 138 of these faults.

IV. EXTRACTION OF $POEXB$ TESTS

This section discusses the use of several passes of sequential fault simulation for the extraction of $POEXB$ tests. The goal in this section is to extract from A all the shortest $POEXB$ tests that detect a transition fault f. A shortest $POEXB$ test for f uses a clock cycle u_d where f is detected by A on a primary output, and has the smallest number of functional capture cycles that is guaranteed to detect f. The test $t(3,7)$ based on Figure 2(a) is an example of a shortest test that uses clock cycle $u_d = 7$ for detecting the fault. The set of all the shortest $POEXB$ tests for a fault f is denoted by $T(f)$. The sets $T(f)$ will be used in Section V for test compaction.

979-8-3503-4631-2/23 $31.00 © 2023 IEEE

Initially, $T(f) = \emptyset$. The procedure for deriving $T(f)$ starts with sequential fault simulation of f under the entire test sequence A. The procedure uses the results of sequential fault simulation of A to derive tests as illustrated by Figure 2(a).

To obtain additional tests, the following observation is used. Let u_x be the first pseudo-detection clock cycle of f, i.e., the first clock cycle with $\hat{s}_{u_x+1} \neq s_{u_x+1}$. Suppose that sequential fault simulation is carried out starting in clock cycle u_x. This corresponds to a scan-in operation that brings the circuit to state s_{u_x} without pseudo-detecting the fault. Sequential fault simulation starting from clock cycle u_x will not change the states that the fault-free circuit traverses compared with the case where fault simulation starts at clock cycle 0. However, with $\hat{s}_{u_x} = s_{u_x}$, it will change the states that the faulty circuit traverses. This will change the clock cycles where f is pseudo-detected and detected. As a result it may be possible to extract additional $POEXB$ tests from A starting from clock cycle u_x. The tests will be $POEXB$ tests since the sequence starting from s_{u_x} at clock cycle u_x is also a functional test sequence. The differences occur only in the faulty circuit.

The process of suppressing the first pseudo-detection clock cycle and computing new states can be repeated until no pseudo-detection clock cycles remain.

Figure 2 illustrates this computation. With $u_x = 0$, the sequence from Figure 2(a) is obtained, and the test $t(3,7) = \langle s_3, a_3, a_4, a_5, a_6, a_7 \rangle$ is added to $T(f)$. With $u_x = 1$, the pseudo-detection at clock cycle 1 is suppressed by simulating the sequence again starting in clock cycle 1. With clock cycle 1 considered as the first clock cycle of A, the transition fault is not pseudo-detected at clock cycle 1. The result is shown in Figure 2(b). In this case, no new tests are obtained.

The next pseudo-detection clock cycle to be suppressed is $u_x = 4$. The result of sequential fault simulation starting at clock cycle 4 is shown in Figure 2(c). In this case, a new test can be extracted, $t(5,7) = \langle s_5, a_5, a_6, a_7 \rangle$.

Finally, suppressing the pseudo-detection at clock cycle $u_x = 6$ results in a sequence that does not detect the fault, and does not have additional pseudo-detection clock cycles, as shown in Figure 2(d). The procedure terminates with two tests in $T(f)$.

The procedure is summarized next.

Procedure $Extract - Tests(A, f)$:

1) Assign $T(f) = \emptyset$ and $u_x = 0$.
2) Simulate f under A starting from clock cycle u_x (f is not pseudo-detected at clock cycle u_x).
3) For every clock cycle $u_d > u_x$ where f is detected by A on a primary output:
   a) Find a clock cycle $u_x \leq u_b < u_d$ such that the following conditions are satisfied: (1) either $u_b = u_x$, or f is not pseudo-detected at clock cycle $u_b - 1$; (2) f is not pseudo-detected at clock cycle u_b; and (3) f is pseudo-detected at clock cycles $u_b < u < u_d$.
   b) Define the test $t(u_b, u_d) = \langle s_{u_b}, a_{u_b}, ..., a_{u_d} \rangle$.
   c) Add $t(u_b, u_d)$ to $T(f)$.

4) Assign $u_x = u_x + 1$. As long as f is not pseudo-detected at clock cycle u_x, and $u_x < L$, assign $u_x = u_x + 1$.
5) If $u_x < L$, go to Step 2.

Procedure $Extract-Tests$ performs sequential fault simulation for a fault f at most L times, where L is the length of A. In effect, the number of times a fault is pseudo-detected by A is typically significantly lower, and sequential fault simulation is needed a significantly lower number of times. In addition, with $0 \leq u_x < L$, sequential fault simulation is carried out for $L - u_x \leq L$ clock cycles. This number decreases as u_x is increased.

Another consideration related to Procedure $Extract-Tests$ is the number of tests in $T(f)$. An example based on benchmark circuit $b04$ is given next using the setup described in Section VI. The first three faults in the set of transition faults for $b04$ have 8776, 8795 and 2343 tests in $T(f)$. These large numbers of tests are not needed for test compaction. To limit the computational effort of Procedure $Extract - Tests$, the procedure is run with a constant limit on the number of tests in $T(f)$. The limit is denoted by τ. When the number of tests in $T(f)$ reaches τ, the procedure terminates for the fault.

V. TEST COMPACTION PROCEDURE

Similar to Procedure 1, the test compaction procedure described in this section considers the target faults in F one by one. The procedure achieves test compaction by using procedure $Extract-Test$ to extract all the shortest $POEXB$ tests for a fault, and then selecting the test that detects the largest number of faults.

The compact set of $POEXB$ tests is denoted by T_2. Initially, $T_2 = \emptyset$. For a fault $f \in F$, the procedure calls Procedure $Extract-Tests$ to obtain the set of tests $T(f)$ that consists of all the shortest $POEXB$ tests for f. The procedure simulates F under every test in $T(f)$. For a test $t(u_b, u_d)$, it finds a number of detected faults denoted by $n_{det}(u_b, u_d)$. Of all the tests in $T(f)$, the procedure selects the one with the largest value of $n_{det}(u_b, u_d)$. It adds the test to T_2, and removes the faults it detects from F.

The test compaction procedure is described by Procedure 2.

Procedure 2: Test compaction procedure

1) Assign $T_2 = \emptyset$.
2) For every fault $f \in F$:
   a) Call Procedure $Extract - Tests(A, f)$ to find the set of tests $T(f)$ that consists of all the shortest $POEXB$ tests for f.
   b) If $T(f) \neq \emptyset$:
      i) For every test $t(u_b, u_d) \in T(f)$, simulate F under $t(u_b, u_d)$, and find the number of detected faults, $n_{det}(u_b, u_d)$.
      ii) Select the test $t(u_b, u_d) \in T(f)$ with the largest value of $n_{det}(u_b, u_d)$.
      iii) Perform fault simulation with fault dropping of F under $t(u_b, u_d)$, and add $t(u_b, u_d)$ to T_2.

Similar to Procedure 1, Procedure 2 considers at most $|F|$ faults. For each fault it may call Procedure $Extract - Tests$

979-8-3503-4631-2/23 $31.00 © 2023 IEEE

and perform sequential fault simulation up to L times. In effect, fault dropping reduces significantly the number of faults for which Procedure 2 needs to call Procedure $Extract - Tests$. For example, with the setup described in Section VI, benchmark circuit $s1423$ has 2,218 transition faults that can be detected by $POEXB$ tests. Procedure 2 calls Procedure $Extract - Tests$ for 89 of these faults. This is lower than the 138 faults considered by Procedure 1 since Procedure 2 detects more faults with every test it adds to T_2. In addition, as discussed earlier, Procedure $Extract - Tests$ performs sequential fault simulation fewer than L times with shorter and shorter sequences.

VI. EXPERIMENTAL RESULTS

The results of Procedures 1 and 2 for transition faults in benchmark circuits are presented in this section.

$POEXB$ tests are extracted from a pool of 20 functional test sequences each of length 2,000. The pool is generated by a low-complexity sequential test generation procedure as in [8]. This pool is sufficient for achieving a high coverage of transition faults that are functionally-detectable. Experimental results indicate that the pool is preferred over a single sequence of length 40,000 considering the transition fault coverage.

Only benchmark circuits for which the pool achieves a transition fault coverage of 50% or higher are considered. This excludes benchmark circuits with high levels of sequential redundancy.

Procedure $Extract - Tests$ is run with a limit of $\tau = 200$ on the number of tests in $T(f)$. This limit was selected based on the fact that many faults have fewer than 200 tests in $T(f)$, and even when the number of tests exceeds 200, the selected test is typically one of the first 200.

After Procedure 1 or 2 is applied to produce the test set T_1 or T_2, the test set is compacted by applying forward-looking reverse order fault simulation to remove unnecessary tests. This is similar to reverse order fault simulation, but with a provision to identify tests that are not necessary without simulating them again in the reverse pass. Before applying forward-looking reverse order fault simulation, the test set is ordered such that a test with a larger number of functional capture cycles appears earlier. This order is based on the observation that tests with more functional capture cycles tend to detect more faults, and placing them earlier in the test set allows other tests to be removed.

Fault simulation of $POEXB$ tests captures fault detections only on the primary outputs. A fault is not considered to be detected if it can be detected only by observing scan chain outputs. The test application process of scan-based tests is not modified except to ignore scan-out values.

To provide a reference point for the transition fault coverage achieved by $POEXB$ tests, and the number of clock cycles required for test application, two-cycle broadside tests are extracted from the same 20 sequences without the added requirement to detect the faults only on the primary outputs. In this case, a fault is marked as detected if its fault effects appear on the primary outputs or next-state variables in the second

TABLE I
EXTRACTED BROADSIDE TESTS

| circuit | trans | extracted | | | conventional | |
		tests	cycles	f.c.	tests	f.c.
usb_phy	2418	106	10698	58.065	55	93.05
simple_spi	3820	151	20214	68.534	82	94.32
i2c	4290	164	21448	78.881	113	94.64
b20	38494	855	424574	77.791	603	89.08
b14	17180	437	109060	73.946	443	88.46
spi	11968	853	197272	85.528	811	99.18
systemcdes	12070	210	40510	99.660	91	99.68
des_area	18754	305	39778	100.000	130	100.00
b15	35404	764	343483	61.953	757	94.34
wb_dma	16556	206	108673	61.398	205	98.96
b21	38088	818	406222	72.585	-	-
s5378	10590	196	35655	71.445	179	92.06
systemcaes	39744	369	248638	77.169	196	94.38
s1423	2846	122	9346	81.377	71	88.55
sasc	3064	111	13326	80.320	45	94.68
b04	2284	93	6390	84.413	52	90.11
s35932	71864	151	262958	87.211	32	87.21

clock cycle of a test. Thus, the detection conditions are the same as for conventional scan-based tests. The resulting test set is denoted by T_{EXB}.

Column $trans$ of Table I shows the number of transition faults. Column $extracted$ of Table I shows the following information for T_{EXB}. Subcolumn tests shows the number of tests. Subcolumn $cycles$ shows the number of clock cycles required for applying T_{EXB}, including both functional capture and scan shift cycles. Subcolumn $f.c.$ shows its transition fault coverage.

Column $conventional$ of Table I shows the number of tests and transition fault coverage for a conventional (non-functional) two-cycle broadside test set.

The results of Procedures 1 and 2 are included in Table II. For the circuits in the first part of Table II, the transition fault coverage of T_2 is the same as that of T_1. For the circuits in the second part of Table II, the fault coverage of T_2 is higher than that of T_1. The circuits in each table are arranged from low to high transition fault coverage achieved by T_1. The two rows for every circuit show the results of Procedures 1 and 2.

After the name of the circuit, column sv shows the number of state variables (equal to the number of flip-flops). Column pi shows the number of primary inputs. Column pr has the index of the procedure applied. Column $tests$ shows the number of tests. Columns max and ave show the maximum and average number of functional capture cycles in a test. Column $cycles$ shows the number of clock cycles required for applying the test set, including both functional capture and scan shift cycles. Column $f.c.$ shows the transition fault coverage. Column $rtime$ shows the runtime in seconds on a Linux machine with a 3-GHz processor. The implementation of the procedure does not use efficient fault simulation procedures. Therefore, the runtime is higher than necessary. Column $ntime$ shows the runtime divided by the runtime of Procedure 1. This is referred to as the normalized runtime. Procedure 1 considers the first test that detects every fault. Procedure 2 performs a further optimization to compact the test set. The normalized runtime shows the increase in runtime that results from this optimization.

979-8-3503-4631-2/23 $31.00 © 2023 IEEE

TABLE II
EXPERIMENTAL RESULTS

circuit	sv	pi	pr	tests	max	ave	cycles	f.c.	rtime	ntime
usb_phy	98	14	1	54	101	23.61	6665	48.428	456.13	1.00
usb_phy	98	14	2	47	245	27.45	5994	48.428	3700.65	8.11
simple_spi	131	15	1	62	53	11.56	8970	50.366	1642.53	1.00
simple_spi	131	15	2	40	336	33.58	6714	50.366	5903.47	3.59
i2c	128	17	1	70	635	29.56	11157	60.769	1259.31	1.00
i2c	128	17	2	63	635	29.30	10038	60.769	13999.14	11.12
b20	494	33	1	426	227	18.84	218964	67.979	761668.19	1.00
b20	494	33	2	381	266	20.19	196401	67.979	2868443.00	3.77
b14	247	33	1	243	81	13.93	63653	69.267	54972.74	1.00
b14	247	33	2	193	124	18.03	51398	69.267	442672.97	8.05
spi	229	45	1	349	1794	20.89	87442	83.723	41844.13	1.00
spi	229	45	2	287	1794	28.59	74157	83.723	89149.67	2.13
systemcdes	190	130	1	179	19	4.33	34975	99.660	363.88	1.00
systemcdes	190	130	2	80	19	5.46	15827	99.660	8920.79	24.52
des_area	128	239	1	301	5	3.07	39580	100.000	1915.41	1.00
des_area	128	239	2	212	5	3.25	27952	100.000	53733.07	28.05
b15	447	36	1	409	591	16.09	189852	40.786	1015384.25	1.00
b15	447	36	2	365	527	18.91	170504	40.863	1860295.88	1.83
wb_dma	523	215	1	54	1015	233.11	41353	56.620	139776.80	1.00
wb_dma	523	215	2	46	1592	278.35	37385	56.632	212381.16	1.52
b21	494	33	1	413	142	19.11	212409	63.222	708381.19	1.00
b21	494	33	2	377	266	19.61	194125	63.230	1924544.50	2.72
s5378	179	35	1	68	614	27.03	14189	66.110	16938.67	1.00
s5378	179	35	2	46	968	77.00	11955	66.459	132306.23	7.81
systemcaes	670	258	1	42	261	49.86	30904	75.848	64153.40	1.00
systemcaes	670	258	2	36	292	62.83	27052	75.850	205655.50	3.21
s1423	74	17	1	78	356	34.54	8540	77.899	260.63	1.00
s1423	74	17	2	72	732	44.00	8570	77.934	1333.62	5.12
sasc	117	15	1	63	678	39.19	9957	78.590	368.78	1.00
sasc	117	15	2	51	250	43.35	8295	78.623	2353.76	6.38
b04	66	12	1	72	26	5.40	5207	79.685	164.23	1.00
b04	66	12	2	53	16	5.62	3862	79.729	1548.82	9.43
s35932	1728	35	1	87	23	7.66	152730	87.177	31154.85	1.00
s35932	1728	35	2	57	25	8.86	100729	87.211	294399.16	9.45

The following points can be seen from Tables I and II. The $POEXB$ tests in Table II have a lower transition fault coverage than the extracted broadside tests in Table I. This indicates that significant overtesting can occur when using scan-out operations for observing fault effects. Nevertheless, the fault coverage in Table II is still significant.

The total number of clock cycles required for T_{EXB} is higher than for T_2 partly because of the higher fault coverage, but mostly because of the use of two-cycle tests for T_{EXB}. The use of multicycle tests contributes significantly to test compaction in the case of T_2.

The use of Procedure 2 instead of Procedure 1 has two effects. The first effect is a reduction in the number of clock cycles required for applying the test set set T_2 compared with that of the test set T_1. The second effect is related to the transition fault coverage. There are several circuits, included in the second part of Table II, where the transition fault coverage of T_2 is higher than that of T_1. Although the difference in the fault coverage is small, it typically goes together with a reduced number of clock cycles.

Both effects result from the fact that Procedure 2 searches for $POEXB$ tests that are not visible from a single pass of sequential fault simulation. This allows it to select tests with more functional capture cycles, that detect more faults, requiring fewer tests. The same search also leads to the increased fault coverage of T_2.

The normalized runtime does not grow with the size of the circuit. The main contribution to the increased normalized runtime of Procedure 2 is the number of times sequential fault simulation is carried out by Procedure $Extract-Tests$. The increase is similar for circuits of different sizes. Such an increase is typical of dynamic test compaction that needs to target certain faults several times before they are detected.

VII. CONCLUDING REMARKS

This article considered the extraction of broadside tests from functional test sequences such that a test would propagate fault effects to the primary outputs. This reduces overtesting by eliminating the detection of effects that cannot be propagated to the primary outputs during functional operation. The article observed that the test extraction process benefits from sequential fault simulation. A test compaction procedure applied sequential fault simulation repeatedly to a shorter and shorter sequence to obtain additional tests. Considering transition faults one by one, the test compaction procedure selected a test for every fault that detects the largest number of additional faults. Experimental results for transition faults in benchmark circuits demonstrated the effectiveness of the test compaction procedure in producing more compact test sets that sometimes also have an increased transition fault coverage.

REFERENCES

[1] J. Rearick, "Too Much Delay Fault Coverage is a Bad Thing", in Proc. Intl. Test Conf., 2001, pp. 624-633.

979-8-3503-4631-2/23 $31.00 © 2023 IEEE

[2] J. Saxena, K. M. Butler, V. B. Jayaram, S. Kundu, N. V. Arvind, P. Sreeprakash and M. Hachinger, "A Case Study of IR-Drop in Structured At-Speed Testing", in Proc. Intl. Test Conf., 2003, pp. 1098-1104.

[3] S. Sde-Paz and E. Salomon, "Frequency and Power Correlation between At-Speed Scan and Functional Tests", in Proc. Intl. Test Conf., 2008, Paper 13.3, pp. 1-9.

[4] I. Pomeranz, "On the Generation of Scan-Based Test Sets with Reachable States for Testing under Functional Operation Conditions", in Proc. Design Autom. Conf., 2004, pp. 928-933.

[5] I. Pomeranz and S. M. Reddy, "Generation of Functional Broadside Tests for Transition Faults", in IEEE Trans. on Computer-Aided Design, Oct. 2006, pp. 2207-2218.

[6] M. Valka, A. Bosio, L. Dilillo, P. Girard, S. Pravossoudovitch, A. Virazel, E. Sanchez, M. De Carvalho and M. Sonza Reorda "A Functional Power Evaluation Flow for Defining Test Power Limits during At-Speed Delay Testing", in Proc. IEEE European Test Symp., 2011, pp. 153-158.

[7] A. Touati, A. Bosio, L. Dilillo, P. Girard, A. Virazel, P. Bernardi and M. Sonza Reorda "Exploring the Impact of Functional Test Programs Re-used for Power-aware Testing", in Proc. Design, Automation & Test in Europe Conf., 2015, pp. 1277-1280.

[8] I. Pomeranz, "A Static Test Compaction Procedure for Large Pools of Multicycle Functional Tests", in IET Computers & Digital Techniques, Vol. 12, No. 5, Sep. 2018, pp. 233-240.

[9] S. Y. Lee and K. K. Saluja, "Test Application Time Reduction for Sequential Circuits with Scan", in IEEE Trans. on Computer-Aided Design, Sept. 1995, pp. 1128-1140.

[10] I. Pomeranz and S. M. Reddy, "Static Test Compaction for Scan-Based Designs to Reduce Test Application Time", in Proc. Asian Test Symp., 1998, pp. 198-203.

[11] X. Lin and R. Thompson, "Test Generation for Designs with Multiple Clocks", in Proc. Design Autom. Conf., 2003, pp. 662-667.

[12] G. Bhargava, D. Meehl and J. Sage, "Achieving Serendipitous N-Detect Mark-Offs in Multi-Capture-Clock Scan Patterns", in Proc. Intl. Test Conf, 2007, Paper 30.2.

[13] I. Pomeranz, "A Multi-Cycle Test Set Based on a Two-Cycle Test Set with Constant Primary Input Vectors", in IEEE Trans. on Computer-Aided Design, July 2015, pp. 1124-1132.

[14] S. Wang, H. T. Al-Awadhi, S. Hamada, Y. Higami, H. Takahashi, H. Iwata and J. Matsushima, "Structure-Based Methods for Selecting Fault-Detection-Strengthened FF under Multi-Cycle Test with Sequential Observation", in Proc. Asian Test Symp., 2016, pp. 209-214.

[15] I. Pomeranz, "$LFSR$-Based Generation of Multicycle Tests", in IEEE Trans. on Computer-Aided Design, March 2017, pp. 503-507.

[16] T. McLaurin and I. P. Lawrence, "Improving Power, Performance and Area with Test: A Case Study", in Proc. Intl. Test Conf., 2018, pp. 1-10.

[17] P. Goel and B. C. Rosales, "Test Generation and Dynamic Compaction of Tests", in Proc. Test Conf., 1979, pp. 189-192.

[18] I. Pomeranz, L. N. Reddy and S. M. Reddy, "COMPACTEST: A Method to Generate Compact Test Sets for Combinational Circuits", in Proc. Intl. Test Conf., 1991, pp. 194-203.

[19] J.-S. Chang and C.-S. Lin, "Test Set Compaction for Combinational Circuits", in Proc. Asian Test Symp., 1992, pp. 20-25.

[20] Y. Matsunaga, "MINT -An Exact Algorithm for Finding Minimum Test Sets", in IEICE Trans. Fundamentals., vol. E76-A, No. 10, Oct. 1993, pp. 1652-1658.

[21] S. Kajihara, I. Pomeranz, K. Kinoshita and S. M. Reddy, "Cost-Effective Generation of Minimal Test Sets for Stuck-at Faults in Combinational Logic Circuits", in IEEE Trans. on Computer-Aided Design, Dec. 1995, pp. 1496-1504.

[22] I. Hamzaoglu and J. H. Patel, "Test Set Compaction Algorithms for Combinational Circuits", in Proc. Intl. Conf. on Computer-Aided Design, 1998, pp. 283-289.

[23] S. Alampally, R. T. Venkatesh, P. Shanmugasundaram, R. A. Parekhji and V. D. Agrawal, "An Efficient Test Data Reduction Technique through Dynamic Pattern Mixing Across Multiple Fault Models", in Proc. VLSI Test Symp., 2011, pp. 285-290.

[24] D. Xiang, J. Li, K. Chakrabarty and X. Lin, "Test Compaction for Small-Delay Defects Using an Effective Path Selection Scheme", in ACM Trans. on Design Automation, July 2013, Vol. 18, No. 3, Article 44, pp. 1-23.

[25] I. Pomeranz and S. M. Reddy, "On Functional Broadside Tests with Functional Propagation Conditions", in IEEE Trans. on VLSI Systems, June 2011, pp. 1094-1098.

Innovation Practices Track: Testability and Dependability of AI Hardware and Autonomous Systems

Fei Su
Intel Corporation
U.S.
fei.su@intel.com

Eric Zhang
Untether AI
Canada
ericz@untether.ai

Arjun Chaudhuri
nVidia
U.S.
achaudhuri@nvidia.com

Michael Paulitsch
Intel Corporation
Germany
michael.paulitsch@intel.com

I. INTRODUCTION (FEI SU)

Testability and dependability (e.g., safety) of AI hardware (e.g., GPU, AI accelerators) and AI-based autonomous systems has been emerging as an important R&D topic in order to address increasing resiliency. In this session we will invite the industry experts to discuss the various aspects of this new field.

II. CHALLENGES AND CONSIDERATIONS FOR ENSURING SAFE DEPLOYMENT OF AUTONOMOUS VEHICLES FROM A CHIP-LEVEL PERSPECTIVE (ERIC ZHANG)

Safety is the top priority in autonomous vehicles (AVs). This talk will explore the key challenges of safely deploying AVs from a chip-level perspective. It will begin by visiting the idea of "how safe is safe enough" and the design considerations needed to ensure safety in AI-based autonomous driving systems. We will focus on some critical design considerations for enhancing safety beyond ISO26262, such as preventive maintenance and fault prediction.

Furthermore, we will discuss the possibility of building a safety case for certifiable AI-based safety systems in AVs using a data-driven approach. With the increasing complexity of AV systems, establishing a robust safety case is essential to demonstrate their safety and reliability. By collecting and analyzing data at different hierarchy, safety engineers can build a strong case that provides evidence of the safety performance of these systems.

Overall, this talk will provide insights into the challenges and considerations for ensuring safety in AVs from a chip-level perspective, as well as the importance of building a data-driven safety case for certifiable AI-based safety systems in AVs.

III. AWPERA: ATPG-AWARE OBSERVATION POINT INSERTION USING GRAPH CONVOLUTIONAL NETWORKS (ARJUN CHAUDHURI)

We present AWPERA, a framework for ATPG-aware Observation Point Insertion (OPI) using Graph Convolutional Networks (GCNs), with the motivation of achieving – 1) better quality Observation Points (OPs) than commercial EDA tools leading to a lower pattern count and improved testability for the mitigation of Silent Data Corruption (SDC) errors; 2) faster turnaround time to generate the observation points. We employ GCNs to learn the topology of a logic circuit along with the features that influence circuit testability. The graph structure of the logic circuit is used to train two GCN-based deep learning models – the first model predicts signal probabilities at different nets and the second model uses these signal probabilities along with other circuit-based structural features to predict the

reduction in test pattern count when OPs are inserted at different locations in the design. The structural features include gate type, gate logic, reconvergent-fanouts, and testability metrics like SCOAP. We further present a dynamic OPI flow that captures the incremental changes in the observability of the faults after each OP is inserted. The proposed iterative OPI method considers the updated testability landscape of the netlist, after prior OPs have been inserted, before proceeding with the insertion of the next OP. The dynamic OPI flow can identify better OPs than the static OPI flow which is agnostic to the changing dynamics of circuit testability as OPs are inserted sequentially. Evaluation across multiple netlists shows a mean reduction of 5.74% in the test pattern count with a test point budget of 50 OPs.

IV. APPLICATION-DRIVEN TARGET SETTING, FAULT IMPACT ON MACHINE LEARNING RESILIENCY (MICHAEL PAULITSCH)

Chip area is growing significantly in the next years combined with production node downscaling associated effects, pushing requirements to detect and tolerate faults to new levels. Setting balanced error resiliency targets during chip design that trade-off between an integrated circuit's resiliency and performance and cost is essential in such an environment. A good understanding of chip targets related to application requirements, but are also effective overall considering hardware, system and application software approaches and its capabilities to mitigate errors.

We present a novel multi-level fault impact analysis framework (HDFIT), which addresses impact of hardware errors of modern computing accelerators like systolic arrays to different applications. Core to this tool is the efficient multi-level simulation approach considering hardware error impact of key accelerator elements like systolic arrays, combined with software stacks (libraries like Intel's oneAPI approach) and application software. We apply this to exemplary AI/ML models to show the efficiency of the approach and the impact of errors on applications. Results show that most hardware errors (>90 %) don't have measurable application-level impact for selected AI/ML models considering real hardware implementations. Such results and tooling capability become an essential element of the overall reliability analysis and an effective way to also validate any hardware resiliency feature changes early in the design cycle.

To illustrate automated resiliency approaches at higher levels (AI/ML model) we also show the effectiveness of automated application-level monitors to hardware errors for AI/ML approaches using Intel's OpenVino toolkit (resiliency feature 'range supervision').

979-8-3503-4631-2/23 $31.00 © 2023 IEEE

Special Session: CAD for Hardware Security -
Promising Directions for Automation of Security Assurance

Sohrab Aftabjahani
Data Center Platform
Engineering and Architecture
Intel Corporation
Hillsboro, OR USA
sohrab.aftabjahani@intel.com

Mark Tehranipoor
Department of Electrical and
Computer Engineering
University of Florida
Gainesville, FL USA
tehranipoor@ece.ufl.edu

Farimah Farahmandi
Department of Electrical and
Computer Engineering
University of Florida
Gainesville, FL USA
farimah@ece.ufl.edu

Bulbul Ahmed
Department of Electrical and
Computer Engineering
University of Florida
Gainesville, FL USA
ahmed.b@ufl.edu

Ryan Kastner
Department of Computer Science
and Engineering
UC San Diego
San Diego, CA USA
kastner@ucsd.edu

Francesco Restuccia
Department of Computer Science
and Engineering
UC San Diego
San Diego, CA USA
frestuccia@eng.ucsd.edu

Andres Meza
Department of Computer Science
and Engineering
UC San Diego
San Diego, CA USA
anmeza@ucsd.edu

Kaki Ryan
Department of Computer Science
and Engineering
UC San Diego
San Diego, CA USA
kakiryan@ucsd.edu

Nicole Fern
Riscure Inc.
San Francisco, CA USA
fern@riscure.com

Jasper van Woudenberg
Riscure Inc.
San Francisco, CA USA
vanwoudenberg@riscure.com

Rajesh Velegalati
Riscure Inc.
San Francisco, CA USA
velegalati@riscure.com

Cees-Bart Breunesse
Riscure Inc.
San Francisco, CA USA
breunesse@riscure.com

Cynthia Sturton
Department of Computer Science
University of North Carolina
Chapel Hill, NC USA
csturton@cs.unc.edu

Calvin Deutschbein
Department of Computing and
Data Science
Willamette University
Salem, OR, USA
ckdeutschbein@willamette.edu

Abstract— Hardware security creates a hardware-based security foundation for secure and reliable operation of systems and applications used in our modern life. The presence of design for security, security assurance, and general security design life cycle practices in product life cycle of many large semiconductor design and manufacturing companies these days indicates that the importance of hardware security has been very well observed in industry. However, the high cost, time, and effort for building security into designs and assuring their security - due to using many manual processes - is still an important obstacle for economy of secure product development. This paper presents several promising directions for automation of design for security and security assurance practices to reduce the overall time and cost of secure product development. First, we present security verification challenges of SoCs, possible vulnerabilities that could be introduced inadvertently by tools mapping a design model in one level of abstraction to its lower level, and our solution to the problem by automatically mapping security properties from one level to its lower level incorporating techniques for extension and expansion of the properties. Then, we discuss the foundation necessary for further automation of formal security analysis of a design by incorporating threat model and common security vulnerabilities into an intermediate representation of a hardware model to be used to automatically determine if there is a chance for direct or indirect flow of information to compromise confidentiality or integrity of security assets. Finally, we discuss a pre-silicon-based framework for practical and time-and-cost effective power-side channel leakage analysis, root-causing the side-channel leakage by using the automatically generated leakage profile of circuit nodes, providing insight to mitigate the side-channel leakage by addressing the high leakage nodes, and assuring the effectiveness of the mitigation by reprofiling the leakage to prove its acceptable level of elimination. We hope that sharing these efforts and ideas with the security research community can accelerate the evolution of security-aware CAD tools targeted to design for security and security assurance to enrich the ecosystem to have tools from multiple vendors with more capabilities and higher performance.

Keywords—CAD for Security, Hardware Security, Security Verification, Security Assurance, Property Generation, Property Extension, Security Property Expansion, Security Property Translation, Formal Security Analysis, Power Side-Channel Leakage, Power Side-Channel Analysis, Design for Security.

I. INTRODUCTION

While new hardware security vulnerabilities are being identified every day through vulnerability/security research efforts of security community from academia and industry [1,2,3], fixing hardware security issues are expensive and sometimes even not possible. Moreover, design for security and security assurance still rely heavily on manual processes, which makes secure product development costly, slow, and limited to the best effort possible within budget and execution timeline. Because manual security analysis of large and complex systems is error-prone and sometimes even not possible with sufficient

---

The work in section II (III) was supported in part by Synopsys and DARPA under grant number HR0011-20-9-0043 (HR001119S0035-05) and SRC Task 3194.001. The views, opinions, and/or findings expressed are those of the author(s) and should not be interpreted as representing the official views or policies of the Department of Defense or the U.S. Government. We would like to acknowledge Peter Grossmann, Alex Dich, Kate Gillis, Chris Quinkert, Shreyas Sen, Archisman Ghosh, Patrick Schaumont, and Pantea Kiaei, for their valuable contributions with the research work of section IV.

979-8-3503-4631-2/23 $31.00 © 2023 IEEE

coverage, there is a need for scalable, affordable, accurate, and automated tools for design for security and security assurance. These tools can bring security execution efficiency to teams of security experts and even can help reduce the size of such high-cost teams.

In 2015, a group of security experts from industry and academia was formed (some are among the authors above) to work on closing this gap and energizing this movement for the first time to realize an ecosystem of security-aware CAD tools for Design for Security and Security Assurance. Since security has various aspects based on mitigation of different types of vulnerabilities and attacks, usually specialized tools are required for analysis of various aspect of security, which can be thought of plugins for a general framework for security analysis to guide design for security. This group did their best attempt to bring influence on national and commercial investment decisions to support CAD for security research [3-5]. They publicized their vision regarding the future need for CAD for Security in Hardware Security and Electronic Design Automation (EDA), and Test and Validation conferences (including HOST, DAC, VTS, GLVSLSI, IVSW [6] and MTV [7]) by organizing panels, tutorial sessions, keynotes, and invited talks.

Trust-Hub [8], sponsored by National Science Foundation (NSF), was used as the main venue to collect the list of CAD for Security solutions [9] for vulnerability analysis with respect to various types of attacks (based on Taxonomy of Physical Attacks [10]) and guidance for design for security. The solutions as different pieces of puzzle can be put together to form a general framework for design for security and security assurance. Trust-Hub was also used to promote collaboration among researchers and attract prospective EDA investors from government and industry. The online catalog of the solutions from many partners from academia and industry - addressing various aspects of design for security and security verification - with more than 100 entries was modernized in 2021 and was made available to support realizing building blocks of the planned "General Framework for Design for Security and Security Assurance". CAD for Assurance [11] formed later to realize a similar vision.

This vision led to bringing many EDA and security companies (Synopsys, Siemens EDA Business, Cadence, ANSYS, Tortuga Logic, ..., Intel, AMD, IBM, TI, NXP, Analog Devices, ...), academic research institutions (UoF, UCSD, UT-Austin, UT-Dallas, GaTech, ...), and government (DARPA, AFRL, Navy, NSF, ...) to get involved in this collaborative research to form the general framework and realize an ecosystem of security-aware tools and/or to support this mission by funding directly or indirectly (such as through Semiconductor Research Corporation) the research and development effort.

The rest of the paper is organized as follows. Sections II, III, and IV focus on the invited works focusing on CAD for security assurance from prominent academic and industry security researchers. Section II discusses a solution for detection of vulnerabilities possibly introduced by high- and logic- level synthesis. Section III introduces an intermediate representation for hardware security verification. Section IV presents a pre-silicon framework to detect, root-cause, and mitigate power side-channel leakage. Section V concludes the paper.

II. SoC Security Verification: Challenges and Solutions

A. Introduction

System-on-Chips (SoCs) are becoming increasingly complex as they integrate various intellectual property (IP) components to utilize their functionalities. As a result, they are used in various devices, including highly security-critical ones such as in military, medical, and financial transaction devices. Thus, it is crucial to secure SoCs against a wide range of threats to secure their operation and protect their sensitive information. However, SoC development involves multiple parties and stages. This increases the likelihood of security vulnerabilities. While existing verification techniques, such as simulation-based verification [13], formal verification, fuzz testing [14], and machine learning approaches, have been used to verify SoC security, each has its own limitations. Security property-based verification is a commonly used method that relies on manually developing security properties for each abstraction level, which is error-prone and significantly increases verification time. In addition, this approach cannot detect vulnerabilities that arise during translations between abstraction levels [12]. To overcome these challenges, a new methodology has been proposed that automatically translates security properties from higher levels to lower levels of abstraction and extends and expands the set of properties to detect new vulnerabilities. This methodology enables the reuse of verification efforts and ensures security at every level of abstraction.

B. Security Verficiation Challenges

In this section, we will examine some of the key challenges associated with modern SOC security verification, specifically those associated with pre-silicon stage.

Lack of standards: SoC security verification lacks a widely accepted methodology, which may lead to ad hoc approaches by verification engineers.

Complex design and verification environments: As SoCs become more complex with multiple components and IPs integrated from various vendors, developing effective security verification techniques is increasingly challenging.

Dynamic Nature of threats: Ensuring the security of SoCs against present and future threats is a challenging task due to the dynamic nature of security threats and the continuous discovery of new vulnerabilities.

Time and Cost Constraints: Due to tight design schedules associated with SoCs, as well as the high cost and risk of design errors, security verification must be completed on a timely basis and within an appropriate budget.

Lack of Expertise: Due to the fact that it requires highly specialized skills and knowledge in order to verify the security of SoCs, there is a shortage of specialists in this field, making it challenging to evaluate the security of SoCs [15].

Lack of Security-aware EDA tools: EDA tools may introduce hardware vulnerabilities during design optimization for better performance, area, and power consumption. The two categories of the mappings -HLL to RTL and RTL to Gate Level (GL)- and their possibly introduced vulnerabilities are discussed below.

979-8-3503-4631-2/23 $31.00 © 2023 IEEE

1. C/C++/SystemC to RTL: High-level synthesis (HLS) boosts SoC productivity via high-level language-based logic creation but may pose security threats due to insufficient security considerations during HLL-to-RTL conversion.

Timing side-channel vulnerabilities:
HLS can introduce block-level handshaking signals, creating timing side-channel vulnerabilities, such as a ready signal that leaks the secret key [16].

Information Leakage vulnerabilities: HLS optimization primarily aim to enhance power, area, and performance, which can inadvertently create insecure pipeline registers and lead to vulnerabilities in information leakage. For instance, flushing intermediate values can result in information leakage vulnerabilities [17].

Bus Width Mismatch: Security-critical designs may encounter potential data loss or leakage when different data types in C/C++ are implemented to RTL buses with different widths.

Side-channel Leakage: HLS tools optimize RTL power consumption without considering power side-channel leakage. This can lead to generated RTLs being vulnerable to power side-channel leakage attacks and weakening side-channel countermeasures in C/C++ [18].

2. RTL to Gate level: RTL-to-gate-level netlist transformation can introduce hardware vulnerabilities due to design optimization without security considerations. DFT structures may also create security vulnerabilities [19], as discussed below.

State encoding vulnerability: RTL is transformed into a gate-level netlist during synthesis, comprising control and data logic. Flip-flops encode design states in the control logic. Nahiyan et al. [20] demonstrated that some state encodings can be exploited to leak information via fault injection violating certain flip-flops' setup times.

Dangerous don't-cares states: Synthesis tools may introduce additional don't-care states during optimization of RTL designs, which are not related to the design's functionality, but can be exploited by fault-injection attacks to access the protected states, leading to the leakage of sensitive data [21].

Side-channel: DFT accelerates design testing but can compromise security by increasing the controllability and observability of internal states, leading to sensitive information leaks. A non-security-aware DFT can enable attackers to infer sensitive information by analyzing internal operation results [22].

Unwanted access point: DFT may grant unwanted access to critical design registers if not inserted securely, such as a flag

register used to identify user access rights to specific memory. Insertion into the scan chain can expose it to attackers, creating security risks [19].

C. Motivating Example

The scenario presented in Fig. 1 highlights security threats that can arise during high-level synthesis of C/C++ designs to the RTL. During the high-level synthesis of a C++ implementation of AES encryption module using the default

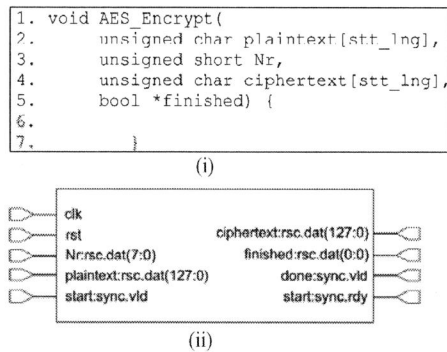

Fig. 1: (i) C++ function of AES implementation, (ii) Primary IO ports of HLS-generated RTL implementation of AES.

configuration of Catapult [23], three handshaking signals were introduced that can potentially create vulnerabilities such as information leakage, confidentiality violation, integrity violation, and side-channel leakage. Verification engineers need to extend their verification methodologies to cover these vulnerabilities at the RTL level. The assumption that a design is secure if no vulnerabilities are found in HLL implementation may not be true, as the introduction of handshaking ports in the RTL design may create new potential attack surfaces. Therefore, there is a need to extend and expand the security properties to ensure security at the lower abstraction levels.

D. Solution

Among various security verification frameworks, AutoMap [24] offers a promising solution for verifying the security of SoCs by employing a multi-faceted approach that includes security property mapping, extension, and expansion. AutoMap is depicted at high level in Fig. 2, and different components of the framework are discussed in detail in this section, including the three distinct modules: mapping, extension, and expansion of security properties.

1) Mapping of Security Properties
This module involves translating security properties from a higher abstraction layer to a lower abstraction level. Name mapping, timing extraction, and property mapping are the primary tasks of this module.

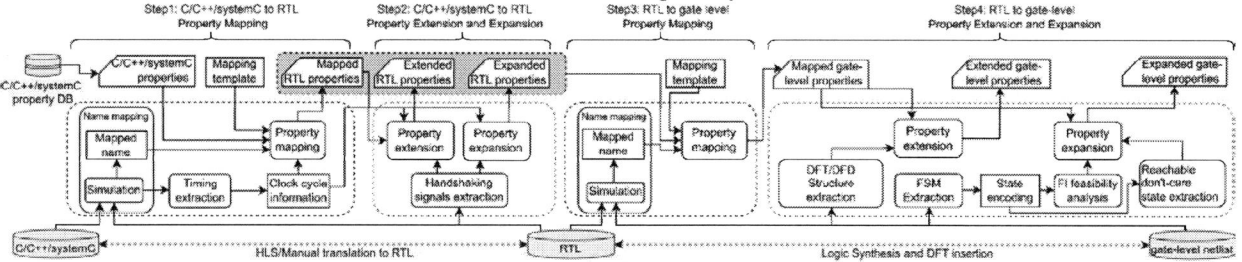

Fig. 2: Overview of the proposed framework.

Name mapping: This module identifies equivalent signals between the higher and lower abstraction levels, particularly from C/C++ to RTL, where signal names can change significantly. Random test patterns can be used to analyze the behavior of signals, registers, or ports to find their equivalent counterpart. For RTL to gate-level mapping, simple string-matching techniques can be used since signal names often stay the same or change slightly.

Timing extraction: When translating a design from C/C++ to RTL, signal relationships in C/C++ are event-based, lacking cycle-accurate information. The timing extraction module extracts clock cycle information between signals. For example, $a \rightarrow next(b)$ expresses that the occurrence of a precedes b without specifying cycle details. This module obtains these details by analyzing RTL design behavior and simulation outcomes.

Property mapping: The property mapping module completes the security property mapping process by utilizing the mapped name of the signals, extracted timing information, and the mapping template. The mapping template comprises the property tag and property mapping rule, which guide the property mapping process. For example, Table I illustrates an example of property mapping, in which "in_1" is mapped to "req", "out_1" to "grant", and "next" to a 1 clock-cycle delay. This step is essential in ensuring that the security properties are accurately mapped between different abstraction levels of the system.

TABLE I.   EXAMPLE MAPPING OF SECURITY PROPERTIES BETWEEN C++ AND RTL

Property in C/C++	Property in RTL
In_1 => Next(out_1)	@ (posedge clk) req |-> grant

2) Security Property Extension

This module identifies potential attack surfaces introduced during design transformation between abstraction levels, such as handshaking signals during HLL to RTL translation and DFT structures during RTL to gate-level translation. It then extends mapped security properties to these surfaces to verify if they cause vulnerabilities. No new properties are created during property extension, but rather, mapped properties are redefined to integrate the additional attack surfaces to identify potential vulnerabilities during verification. Table II shows an example of security property extension from RTL to the gate-level.

TABLE II.   EXAMPLE OF PROPERTY EXTENSION

RTL Property	Extended Gate-level Property
check_spv -create -from key -to_all {in_1 out_1};	check_spv -create -from key -to_all {in_1 out_1 test_si test_so};

3) Security Property Expansion

Security property expansion creates new security properties to maintain security requirements during design transformations to lower abstraction levels. Such transformations may introduce new attack surfaces, which can be exploited to violate the original requirements. For example, verifying access control properties at RTL may be insufficient, as don't-care vulnerabilities at the gate level could allow unauthorized access. Thus, new security properties must be created at the later

abstraction levels to verify the same security requirements. Fig. 3 shows how don't-care states in the AES encryption module at the gate-level abstraction could allow attackers to access protected states. Table III provides an example of security property expansion to ensure that the design does not violate access control requirements.

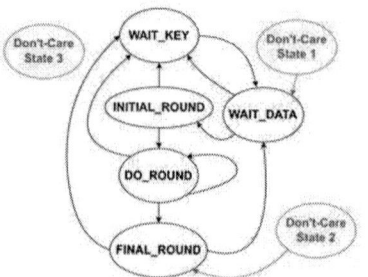

Fig. 3: Control logic of AES encryption.

TABLE III.   EXAMPLE OF PROPERTY EXPANSION

RTL Property	Expanded Gate-level Property
FINAL_ROUND can only be accessed from DO_ROUND	Don't-care state 2 cannot be accessed by fault-injection attack in state FFs.

E. Results

In a case study on AES encryption module, we demonstrate the effectiveness of AutoMap. The FSM for this module is shown in Fig. 3, with the original states and their encodings listed as WAIT_KEY (001), WAIT_DATA (010), INITIAL_ROUND (011), DO_ROUND (100), and FINAL_ROUND (000). When synthesizing the RTL with Synopsys Design Compiler [25], three additional don't-care states were introduced and marked in red. Starting from C/C++ abstraction, we verify the design of the AES encryption module at RTL and gate-level netlist, mapping and extending the security properties along the way. The verification starts at the C/C++ implementation with the following two properties.

Property 1: *finished signal should be high only at the final round.*

This property is used to verify that pre-encrypted intermediate data is not leaked, which can lead to potential key-leakage through side-channel analysis.

Property 2: *Key should not flow to Nr, plaintext, ciphertext, and finished signal (Peripheral ports).*

This property checks if the key is observed from the peripheral ports. Table IV lists all the mapped, extended, and expanded properties used in verification, along with newly identified vulnerabilities during the translation from C/C++ to RTL to gate-level. The extended RTL property P6 identified the information leakage vulnerability ILL(HSS) through handshaking signals, while the base properties in C/C++ did not find any vulnerabilities. At the RTL-level, the base property P7 identified the information leakage vulnerability IL(PIO). However, at the gate-level, the expanded properties uncovered four more vulnerabilities: IL(DFT), FSE, DCS, and IL(DFT), which respectively represent information leakage through DFT ports, FSM state encoding vulnerability, and the don't-care vulnerability.

979-8-3503-4631-2/23 $31.00 © 2023 IEEE   256

TABLE IV.   RESULT OF SECURITY PROPERTY MAPPING, EXTENSTION, AND EXPANSION.

Abst.	#	Property	Type	Vuln.
C/C++	P1	"finished" should only be asserted at the final round	Base	-
	P2	The cryptographic key must not be observable from Nr, plaintext, finished, and ciphertext	Base	-
RTL	P3	"finished" should only be asserted at FINAL_ROUND	Mapped (P1)	-
	P4	"finished" should only be asserted 10 cycles after the INITIAL_ROUND	Expanded (P1)	-
	P5	The cryptographic key register must not be observable from Nr_rsc_dat, plaintext_rsc_dat	Mapped (P2)	-
	P6	The cryptographic key register must not be observable from Nr_rsc_dat, plaintext_rsc_dat, finished_rsc_dat, done_sync_vld, start_sync_rdy, start_sync_vld	Extended (P2)	IL(HSS)
	P7	The state registers must not be observable or controllable from data_stable, key_ready, finished, and round_type_sel	Base	IL(PIO)
	P8	Only DO_ROUND can access FINAL_ROUND	Base	-
	P9	DO_ROUND must occur before FINAL_ROUND is occurred	Base	-
Gate	P10	The cryptographic key must not be observable from pe Nr_rsc_dat, plaintext_rsc_dat, finished_rsc_dat, done_sync_vld, start_sync_rdy, start_sync_vld test_si test_so	Extended (P6)	IL(DFT)
	P11	The FSM state encoding cannot not be susceptible to fault-injection attack to access FINAL_ROUND	Expanded(P3, P9)	FSE
	P12	FINAL_ROUND cannot be accessed from don't-care states with encoding 111, 110, and 101	Expanded(P3, P9)	DCS

F. Conclusion And Future Direction

In summary, the proposed framework enables efficient security verification of SoCs across different abstraction levels. The approach involves starting with an initial set of security properties and extending and expanding them throughout the design flow to identify vulnerabilities. However, this poses challenges, such as ensuring that the initial set of properties covers all relevant security aspects, as any gaps between these properties and the security goal will propagate across abstraction levels and identifies all attack surfaces during property extension and expansion. Despite these challenges, the proposed methodology shows promise and can be enhanced by incorporating machine learning and applied to newer design versions. It is important not to overlook security vulnerabilities due to design optimization and new features introduced in newer versions, which is a non-trivial task.

III. AN INTERMEDIATE REPRESENTATION FOR HARDWARE SECURITY VERIFICATION

Hardware security verification can uncover critical weaknesses and vulnerabilities but requires simulation, emulation, formal verification, code review, post-silicon penetration testing, and significant manual analysis, limiting scalability, effectiveness, and reproducibility [35]. The verification process is ad hoc, making it challenging to assess its efficacy and difficult to replicate. For example, when performing a code review, a verification engineer proceeds based on their experience. While the result may be documented, the complete verification process - the steps taken to uncover vulnerabilities and the process by which the hardware is examined - is often vaguely described, if at all.

Significant work is required to enable more effective and efficient hardware security verification. *Verification engineers need tools that provide automated, quantitative, and actionable insight into their hardware's security posture.* A hardware security framework that aids verification efforts and helps document the verification process would be an invaluable step toward automating hardware security verification.

Developing a standardized and extensible analysis framework is critical to improving hardware security verification. Open-source frameworks have aided programming language, compiler, and architecture research and development [28,33]. While there is a trend towards open-source hardware design tools [26,27,29-32], these do not consider security as a first-class citizen. Open-source hardware security design tools are required to push hardware security verification forward.

The Intermediate Representation (IR) is critical in developing any hardware security verification framework. The IR should be independent, e.g., it can interact with different languages (SystemVerilog, VHDL, Chisel) and tools (formal methods, simulation, emulation). And it should allow for various forms of security analysis and verification, including formal methods, simulation, and emulation. And it should provide evidence of the verification effort – summaries and statistics of the types of analysis performed, findings, and corrective actions.

We discuss the essential elements of a hardware security IR. We identify noninterference analysis using hardware information flow tracking (IFT) [39] as a critical aspect of a hardware security IR. We propose some initial ideas on adding IFT to existing hardware IRs. But first, we discuss the hardware security verification process and point out some deficiencies that motivate the requirement for the hardware security IR.

A. Hardware Security Verification

Security verification is an increasingly important part of the hardware design but requires significant manual effort [35]. Hardware security verification engineers must have deep knowledge of the hardware under verification, understanding how it behaves and if it correctly adheres to the desired security goals under different threat models. Thus, hardware security verification is a complex and costly undertaking.

The hardware security verification process involves the following steps [35,37,38]: 1) Create Threat Model, 2) Identify Assets, 3) Articulate Common Weaknesses, 4) Define Security Requirements, 5) Specify Security Properties, and 6) Verify Security Properties.

The first five steps are left to the verification engineer. Unfortunately, only a few tools are available to develop the threat model, identify assets, and map them to weaknesses, requirements, and properties. Identifying assets often requires deep insights into the hardware and its security. Once identified, the verification engineer must understand how those assets are accessed, how asset information moves throughout the hardware, and the conditions under which assets can be read,

979-8-3503-4631-2/23 $31.00 © 2023 IEEE

written, and reset. This is very challenging; we need tools to aid and automate this process.

Once formal security properties exist, a verification engineer can perform Step 6 using tools like Cycuity Radix, Cadence Security Path Verification (SPV), and Siemens Questa Secure Check (SC). Property-driven hardware security describes generating properties and using verification tools to assess the hardware's security [36]. Unfortunately, developing properties is challenging and primarily a manual process. Automate security property generation makes it easier for the verification engineer to create and refine security properties are crucial to advancing hardware security verification [35, 48-50].

An important aspect of hardware security verification is assessing potential asset weaknesses. Is the asset value properly initialized? Where does the asset information flow after it is initialized? Who can read the asset value? Who can write it? Who can reset it? Under what conditions? Is the asset fully reset? The verification engineer should be able to use hardware security verification tools to understand these questions. A hardware security IR must help answer these questions.

B. Intermediate Representation (IR)

IRs are essential for any compiler, and hardware design tools are no different. An IR consists of a frontend that interfaces with input programming languages, transformations that optimize the compilation/synthesis process, and a backend to emit lowered code. Standardizing IRs allows for better community support and feature development. There are many hardware IRs, which we enumerate in the following. None explicitly focuses on security verification.

LLVM is a modular and reusable compiler and toolchain [33]. LLVM is a popular IR used for various purposes, including domain-specific compilers [28], hardware design [40], and processor security validation [51].

Yosys is an open-source framework for RTL synthesis. It includes optimization and technology mapping. Yosys uses the RTLIL [31] to internally represent the RTL netlist. SymbiYosys integrates formal verification techniques into the analysis.

FIRRTL [32] is an abstraction built initially to translate Chisel to a digital circuit. FIRRTL uses an abstract syntax tree to represent the circuit at three levels. It primarily targets input from Chisel but supports Verilog with help from Yosys.

LLHD [27] is a multilevel IR for circuit design. It supports behavioral and structural representations and aims to facilitate synthesis, formal verification, simulation, and testing. LLHD uses a static single assignment IR.

CoreIR [26] provides first-class support for generators by providing an API for implementing Foreign Generator Interfaces. CoreIR focuses on functional verification and integrates with the COSA model checker [29].

PyRTL [30] is a Python-based hardware design framework. PyRTL performs elaboration through execution to generate hardware as the PyRTL file executes. It uses a well-defined set of composable primitives to allow users to specify synthesizable hardware easily.

The previous IRs have standard features relevant to hardware design. They use IR data structures like abstract syntax tree (AST) and static single assignment (SSA). They can parse hardware specification languages like Verilog, Chisel, and VHDL. The IRs primarily focus on synthesis. They interface with EDA tools for physical synthesis, formal verification, simulation, and testing. Some IRs support functional verification and interface with formal methods.

None of these IRs explicitly consider hardware security verification. Security verification requires unique analysis tools; thus, hardware IRs must be enhanced and potentially rethought to consider security. Incorporating noninterference analysis [41] into the IR is critical for security analysis.

C. Hardware Security IR Requirements

Consider the verification related to the security of the key used to encrypt and decrypt data stored in the one-time programmable memory [47]. The hardware security verification engineer wants to know where the key information can flow and ensure it never flows to a publicly viewable memory location. They also want to know when the key flows, as these present good opportunities for power side-channel and fault injection attacks and mitigations. The hardware security IR should provide the ability to perform this verification analysis.

The IR should document the analysis performed by the verification engineer. It should allow the designer to query the IR for security-related questions. It should record those queries and associated analyses. It should summarize how the verification engineer verified potential weaknesses and assessed and documented potential vulnerabilities.

The core components of a hardware security IR are: 1) A **frontend** to accept hardware design specifications (SystemVerilog, VHDL, Chisel), other hardware IRs (FIRRTL, LLHD, CoreIR), and security properties (SVA, PSL, Radix rules). 2) **Transforms and analysis** to understand flow relationships and connectivity throughout the hardware. 3) **Verification results** include formal proofs, simulation results, and testing coverage.

The hardware security IR requires a frontend to parse the hardware specifications and hardware security property languages. A property provides information about assets, access control requirements, and relevant CWEs.

A hardware security IR provides transforms and analysis to understand security weaknesses. Understanding explicit and control flows is critical to hardware security verification and is a core part of any hardware security IR. Explicit flows occur from an assignment operation. The variable on the left-hand side of the statement explicitly receives information from variables on the right-hand side. These are direct flows and often must be constrained as described by the security properties. Assignments are often guarded with predicate conditions under which they will be executed. Information about the predicate variables is transmitted in these assignments as an implicit control flow.

A hardware security IR would model explicit and implicit into a directed graph with explicit and control flow edges between hardware variables. The nodes and edges should be

979-8-3503-4631-2/23 $31.00 © 2023 IEEE

annotated with line numbers and module names to help locate assets and debug security weaknesses.

The IR can be traversed using standard graph algorithms to derive metrics and security assessments. For example, the number of edges on the shortest path from an asset to a publicly viewable memory location is important for security analysis. If that path is short, it is likely more vulnerable than a path requiring many activation steps. The predicates related to the control flows should be succinctly summarized, as they are often the focus of security verification.

A hardware security IR must efficiently and succinctly express and assess noninterference relationships related to the assets. Many security properties related to integrity, confidentiality, and availability require modeling noninterference [41]. Noninterference can be modeled using information flow tracking (IFT), which adds security labels that encode how information flows. Noninterference states that changes in HIGH inputs shall not result in changes in LOW outputs. For example, a confidentiality security property sets the label of a sensitive asset (e.g., secret key) to HIGH. The noninterference analysis would determine how this sensitive information moves through the hardware. That is, LOW objects can learn nothing about sensitive key information. Noninterference is expressible in property-driven hardware security tools, e.g., using the $=/=>$ operator from Radix tools and via source/sink arguments in Security Path Verification (SPV) and Security Coverage (SC) tools.

The backend of a hardware security IR must provide evidence of the verification. This can be identifying relevant CWEs, design review concerns, documentation of completed analyses, testbench coverage metrics, and formal proofs. These reports should be emitted in a manner that can be documented and turned into certification reports.

D. Hardware Security IR Characteristics

A hardware security IR should build upon and integrate with existing open-source hardware design tools [26,27,30-32,40]. Many tools use similar internal IRs, typically variants of an AST or SSA. A hardware security IR should model behavioral and structural dependencies, which is natural upon AST and SSA IRs. A hardware security IR should refrain from reinventing the wheel and reuse IRs and associated analysis techniques.

The hardware security IR should leverage existing verification methods. Many strong open-source and commercial verification tools exist that can and should be used for security, e.g., assertion-based functional verification tools are valuable even though they are less expressive than IFT properties [32]. The IR should allow formal analysis for complete local analysis. The IR should be amenable to simulation and emulation, which quickly becomes necessary when the verification considers security concerns spread across the hardware.

The hardware security IR should enable security-specific queries about the hardware. It should allow for selecting different assets and understanding how information flows from those assets flows in different testbenches. It should allow the verification engineer to query the assets' characteristics and derive metrics to quantify the severity of any weaknesses.

The IR should allow static security analysis related to reachability, cone of influence, flow conditions, and the paths between assets and security boundaries. It should provide statistics on those paths (edges/nodes, cumulative conditions, proximity to boundaries). These static analyses should allow different tools to interact and extract the security information.

The IR should allow for timing information flow analysis. Of particular interest are timing flows that play a crucial role in Spectre, Meltdown, cache side channels, and other microarchitectural attacks [44, 45]. Timing flows consist of scenarios where the time in which a value is written conveys information [43]. These are strongly related to control flow conditions and are examples of implicit flow [46].

The IR should support simulation-based analysis that extracts security-relevant information related to an execution of a functional testbench or set of testbenches. The IR can be annotated with data derived from the simulation. The IR should capture information relevant to functional simulation and data from the IFT security labels. The security label values summarize noninterference related to the assets, which is crucial to design properties pertaining to confidentiality, integrity, and integrity. Important simulation data includes the number of times a statement is executed and how many times a statement is executed with a specific security label.

The IR should allow visualizations that help security verification engineers understand the design's security posture more easily. For example, one often wishes to mark assets and understand where the information related to that asset could go and under what conditions. An IR that provides visualization of those information flows with a summarization of conditions and other relevant information related to the flows would be extremely valuable for security verification. This gives unique visual insights into the design's security and provides verification engineers better intuition to assess weaknesses, understand vulnerabilities, and derive security properties.

IV. HARNESSING PRE-SILICON SIMULATION TO DETECT AND ROOT-CAUSE SIDE-CHANNEL LEAKAGE

A. Introduction

Side channel analysis and fault injection (FI) are a solid part of the relevant attack space for any chip that can be physically reached by an attacker. These hardware attacks allow key extraction from cryptographic implementations, and fault injection specifically allows complete takeover of devices. Current commercial ASIC design tools do not offer integrated analysis to verify resistance against Side-Channel Analysis (SCA) attacks at design time. The effect is that countermeasure design is a manual and error-prone process; multiple tapeouts may be required, and even for experts it is nontrivial to pinpoint and mitigate sources of vulnerability.

Therefore, the research goal is to increase the ability for non-security-expert designers to create a resistant chip without having to perform multiple chip fabrications. This results in a significant increase in security, at significant cost savings.

B. Proposed Pre-Silicon SCA Framework

SCA vulnerability detection and countermeasure insertion can be performed at RTL and at gate level abstractions. The methodology, Side Channel Analysis Testbench Emulator (SCATE) is shown in Fig. 4. SCATE integrates easily with a traditional ASIC design flow.

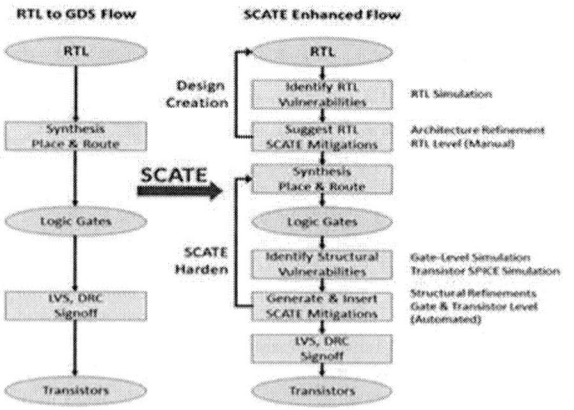

Fig. 4: Proposed SCATE framework.

SCATE starts with power trace generation using RTL or gate level circuit modeling, using industry-standard EDA software. In our experiments on various AES cores, power trace generation at the block level can be performed between 0h29m and 12h56m depending on core gate count and RTL/gate level modeling. SCATE vulnerability detection analyzes the generated power traces, by using Architecture Correlation Analysis (ACA) [52] to identify and rank the cells/gates of a design according to their contribution to side-channel leakage. There are three steps to this detection:

1. A power analysis that statistically tests when the side channel leakage occurs in time.
2. A signal analysis that statistically tests which signals leak at which time, and ranks them by T-score.
3. A ranking improvement that uses design and simulation data of the power consumption of the signals.

Once the leaky gates are identified, a designer can be presented with a hierarchical view of the leakage, from most vulnerable blocks down to individual signals. This allows localization of leakage by the designer, enabling iterative countermeasure addition to observe incremental decreases of side channel leakage.

C. RTL-Level Countermeasure Insertion Case Study- Masking

In order to validate our approach, we implemented a classic Boolean mask AES 128 cipher, called CAES. We then generated 1000 input vectors to perform a non-specific Welch t-test. Group 0 of the test vectors is generated with all random key/input, group 1 has inputs and keys computed to create a fully zero S-box output in round 5.

The netlist for gate-level simulation is ~8K gates in Skywater 130nm technology and runs on a 10MHz clock. We use Cadence Joules to generate one power simulation value every clock cycle. This simulation dataset is acquired in ~2-3 hours with a single CPU core/license. First, we create a simulation set where the masks are disabled, and as expected,

we obtain significant leakage. Next, we produce simulations where the masks are random, as they would be in a production environment. Unexpectedly, we still observe 29650 leaky signals; T-scores over five are considered leaky. Fig. 5 shows the Welch-T leakage test over time and the distribution of the amount of leakage for each of the signals in the design.

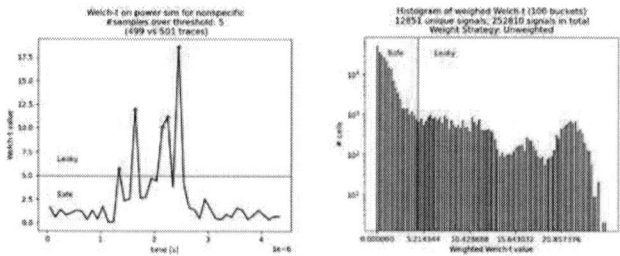

Fig. 5: Welch T-test value vs. simulation time and histogram for signals.

Review of the RTL shows no indication of errors, but further analysis points at the synthesis tool introducing leakage. The security of Boolean masking relies crucially on the orders of masking operations. For example, if the order of operations in *maskA_val1 ^ maskB_val2 ^ maskA* is re-arranged, the unmasked value of *val1* may leak. We instruct the synthesis tooling to not rearrange the logic in the sensitive parts of the design: the Canright S-boxes and some of the round key logic. After adding this instruction, we find that the first-order leakage disappears, in both the power simulation data and the signals (1 leaky signal left). This case study demonstrates the leakage may not be found in RTL, but can be introduced in later stages of a design. Therefore, leakage detection is needed at each design stage in order to produce a secure ASIC.

D. SoC-Level Analysis

SCATE is not limited only to block-level analysis. In [53], ACA is applied to a System-on-Chip (SoC) containing a RISC-V core along with an AES-128 hardware accelerator. Leakage is detected outside the AES engine in CPU registers and the CPU pipeline. The RISC-V core reads a secret key and plaintext from memory then transfers the data to the AES accelerator. Software is also responsible for writing to the appropriate control registers in the AES peripheral and transferring the ciphertext from the accelerator back to memory upon completion of the encryption operation. The leakage detected in the CPU indicates that the key transfer operation needs to be protected in addition to any leakage in the AES accelerator itself. These results highlight the importance of looking at the contribution of microarchitecture details towards side-channel leakage. This is often ignored in the current workflow where hardware designers of cryptographic accelerators are mainly concerned with side-channel leakage in the accelerator itself, and software designers rarely perform analysis below the assembly level of abstraction.

E. Conclusion

Our framework demonstrates that RTL and gate-level power trace generation, used for performing SCA analysis, can be readily integrated into traditional ASIC design flows. At the block level, we have shown gate-level power trace generation turnaround times of a few hours for 256 traces and scaling trends indicating that ~1000 traces can be generated overnight. Our

979-8-3503-4631-2/23 $31.00 © 2023 IEEE

research also shows that there are no technical barriers to generating thousands of traces in parallel in a matter of minutes if cost barriers to hardware and software access are removed.

We have developed software for post-processing power trace data and circuit activity data to detect information leakage on a gate level model of a circuit. We have shown that leakage in a circuits' model remains largely stable as that model is refined in synthesis and place and route. However, we also show refinement can introduce leakage: we demonstrate our software's ability to detect leakage in masked AES implementations and detect improvements in information leakage after errors are corrected.

F. Future Work

With the pre-silicon SCA concept demonstrated, there is a variety of future research directions:

- Validate the approach from design to silicon by conducting silicon proof experiments.
- Augment the simulated SCA analysis test suite with alternate methods like electromagnetic simulations, deep learning, and other post-silicon SCA technologies.
- Test our framework on ECC, RSA, post-quantum crypto, lightweight crypto, FPGAs, and non-crypto targets like neural networks, CPUs etc.

V. SUMMARY

We went through the story of the CAD for Security journey that some of the authors were involved in to bring their vision to reality and make life easier for security design and assurance engineers and enable more economic design and validation of high-quality secure products.

We introduced few promising directions to improve security assurance quality and scalability for the next generation of security assurance tools. First, automapping security properties allows economic verification at various levels of abstraction without losing accuracy and even with increased accuracy due to considering the added circuitries as a design model goes through various stages of design process from HLL to RTL and to GL. Second, intermediate representation of hardware models can be enhanced with threat models and typical vulnerability to create security-enhanced models capable of analysis of direct and indirect information flow for security verification. Third, the proposed pre-silicon power side-channel analysis framework makes it possible to mitigate side-channel leakage prior to Silicon tapeout, which leads to enormous cost-and-time saving of a product by using a ranked list of leakage nodes to mask the highest leakage nodes.

We hope this paper incentivize interested reader to invest in CAD for security research area to improve the next generation of security design and assurance tools as well as security-aware (whether backend/frontend or pre-/post- silicon) design, verification and testing tools and join the effort to take part in realizing an ecosystem of security-aware CAD tools to make development of secure products more affordable.

REFERENCES

[1] S. Bhunia and M. Tehranipoor, *Hardware Security - A Hands on Learning Approach*, S. Bhunia and M. Tehranipoor, Eds. Morgan Kaufmann, 2019

[2] M. Tehranipoor and C. Wang, *Introduction to Hardware Security and Trust*, Springer Publishing Company, Incorporated, 2011.

[3] R. Cammarota, S. Aftabjahani, et al., Security and Privacy Chapter, *Semiconductor Research Opportunities – An Industry Vision and Guide*, Semiconductor Research Corporation and Semiconductor Research Association, 2017, pp. 45-51. [Online]. Available: Semiconductor Research Opportunities – An Industry Vision and Guide

[4] M. Tehranipoor, R. Cammarota, S. Aftabjahani, et al., "Chapter 3: Microlectronics Secuity and Trust - Grand Challenges", *TAME: Trusted and Assured MicroElectronics Working Group Report*, Dec., 2019. [Online]. https://dforte.ece.ufl.edu/wp-content/uploads/sites/65/2020/08/TAME-Report-FINAL.pdf

[5] D. Gardner, P. Ramrakhani, S Jeloka, P. Song, C. Vishik, S. Aftabjahani, R. Cammarota, M. Chen, A. Xhafa, J. Oakley, and D. Yeh, Research Needs: Trustworthy and Secure Semiconductors and Systems (T3S), Semiconductor Research Corporation, 2019. [Online]. Available: https://www.src.org/program/grc/t3s/research-needs/2019/2019-t3s.pdf .

[6] M. Abadir and S. Aftabjahani, "An Overview of the International Microprocessor/ SoC Test, Security and Validation (MTV)Workshop," in *2019 IEEE International Test Conference (ITC)*, Washington, DC, USA, 2019, pp. 1-2.

[7] M. Abadir and S. Aftabjahani, "An Overview of the International Verification and Security Workshop (IVSW)," in *2019 IEEE International Test Conference (ITC)*, Washington, DC, USA, 2019, pp. 1-2.

[8] Welcome to Trust-Hub, *Trust-Hub*. Accessed on: Apr. 08, 2021. [Online]. Available URL: https://www.trust-hub.org

[9] "CAD/IP for Security," *Trust-Hub*. Accessed on: Apr. 08, 2021. [Online]. Available URL: https://www.trust-hub.org/#/cad-ip-sec/cad-solutions

[10] "The Vulnerability Database," *Trust-Hub*. Accessed on: Apr. 08, 2021. [Online]. Available URL: https://www.trust-hub.org/#/vulnerability-db/physical-vulnerabilities

[11] CAD for Assurance, https://cadforassurance.org

[12] F. Farahmandi, Y. Huang, P. Mishra, "SoC Security Verification Challenges," *System-on-Chip Security: Validation and Verification*, 2020, pp. 15-35.

[13] W. Chen, S. Ray, J. Bhadra, M. Abadir and L. C. Wang, "Challenges and Trends in Modern SoC Design Verification," in *IEEE Design & Test*, vol. 34, no. 5, pp. 7-22, Oct. 2017.

[14] K. Z. Azar, M. M. Hossain, A. Vafaei, H.A. Shaikh, N. N. Mondol, F. Rahman, M. Tehranipoor, F. Farahmandi, "Fuzz, Penetration, and AI Testing for SoC Security Verification: Challenges and Solutions," Cryptology ePrint Archive, 2022. [Online]. Available: https://eprint.iacr.org/2022/394.pdf .

[15] F. Farahmandi, Y. Huang, and P. Mishra, *System-on-Chip Security*. Springer, 2020.

[16] Z. Jiang, S. Dai, G. E. Suh and Z. Zhang, "High-Level Synthesis with Timing-Sensitive Information Flow Enforcement," *2018 IEEE/ACM International Conference on Computer-Aided Design (ICCAD)*, San Diego, CA, USA, 2018, pp. 1-8;

[17] N. Pundir, S. Aftabjahani, R. Cammarota, M. Tehranipoor, and F. Farahmandi, "Analyzing security vulnerabilities induced by high-level synthesis," *ACM Journal on Emerging Technologies in Computing Systems (JETC) 18.3*, 2022, pp. 1-22.

[18] L. Zhang, D. Mu, W. Hu, Y. Tai, J. Blackstone and R. Kastner, "Memory-Based High-Level Synthesis Optimizations Security Exploration on the Power Side-Channel," in *IEEE Transactions on Computer-Aided Design of Integrated Circuits and Systems*, vol. 39, no. 10, pp. 2124-2137, Oct. 2020.

[19] D. Hely, M. . -L. Flottes, F. Bancel, B. Rouzeyre, N. Berard and M. Renovell, "Scan design and secure chip [secure IC testing]," Proceedings. 10th IEEE International On-Line Testing Symposium, Funchal, Portugal, 2004, pp. 219-224.

[20] A. Nahiyan, F. Farahmandi, P. Mishra, D. Forte and M. Tehranipoor, "Security-Aware FSM Design Flow for Identifying and Mitigating Vulnerabilities to Fault Attacks," in *IEEE Transactions on Computer-Aided Design of Integrated Circuits and Systems*, vol. 38, no. 6, pp. 1003-1016, June 2019.

[21] A. Nahiyan, K. Xiao, K. Yang, Y. Jin, D. Forte and M. Tehranipoor, "AVFSM: A framework for identifying and mitigating vulnerabilities in FSMs," 2016 53nd ACM/EDAC/IEEE Design Automation Conference (DAC), Austin, TX, USA, 2016.

[22] Bo Yang, Kaijie Wu and Ramesh Karri, "Scan based side channel attack on dedicated hardware implementations of Data Encryption Standard," *2004 International Conferce on Test*, Charlotte, NC, USA, 2004, pp. 339-344.

[23] https://eda.sw.siemens.com/en-US/ic/catapult-high-level-synthesis/

[24] B. Ahmed, F. Rahman, N. Hooten, F. Farahmandi and M. Tehranipoor, "AutoMap: Automated Mapping of Security Properties Between Different Levels of Abstraction in Design Flow," *2021 IEEE/ACM International Conference On Computer Aided Design (ICCAD)*, Munich, Germany, 2021, pp. 1-9.

[25] https://www.synopsys.com.

[26] R. Daly, L. Truong and P. Hanrahan, "Invoking and linking generators from multiple hardware languages using coreir," in *Proceedings of the 1st Workshop on Open-Source EDA Technology*, 2018.

[27] F. Schuiki, A. Kurth, T. Grosser, and L. Benini, "LLHD: A multi-level intermediate representation for hardware description languages," in *Proceedings of the 41st ACM SIGPLAN Conference on Programming Language Design and Implementation*, pp. 258-271. 2020.

[28] C. Lattner et al., "MLIR: Scaling Compiler Infrastructure for Domain Specific Computation," *2021 IEEE/ACM International Symposium on Code Generation and Optimization (CGO)*, Seoul, Korea (South), 2021, pp. 2-14.

[29] C. Mattarei, M. Mann, C. Barrett, R. G. Daly, D. Huff and P. Hanrahan, "CoSA: Integrated Verification for Agile Hardware Design," *2018 Formal Methods in Computer Aided Design (FMCAD)*, Austin, TX, USA, 2018, pp. 1-5.

[30] D. Dangwal, G. Tzimpragos and T. Sherwood, "Agile Hardware Development and Instrumentation With PyRTL," in *IEEE Micro*, vol. 40, no. 4, pp. 76-84, 1 July-Aug. 2020.

[31] C. Wolf, J. Glaser, J. Kepler, "Yosys-a free Verilog synthesis suite," in *Proceedings of the 21st Austrian Workshop on Microelectronics (Austrochip)*, p. 97, 2013.

[32] A. Izraelevitz, J. Koenig, P. Li, R. Lin, A. Wang, A. Magyar, D. Kim, C. Schmidt, C. Markley, J. Lawson, and J. Bachrach, "Reusability is FIRRTL ground: Hardware construction languages, compiler frameworks, and transformations", in *2017 IEEE/ACM International Conference on Computer-Aided Design (ICCAD)*, Irvine, CA, USA, 2017, pp. 209-216.

[33] C. Lattner and V. Adve, "LLVM: a compilation framework for lifelong program analysis & transformation," *International Symposium on Code Generation and Optimization*, 2004. CGO 2004., San Jose, CA, USA, 2004, pp. 75-86.

[34] S. Ray, N. Ghosh, R. J. Masti, A. Kanuparthi and J. M. Fung, "INVITED: Formal Verification of Security Critical Hardware-Firmware Interactions in Commercial SoCs," 2019 56th ACM/IEEE Design Automation Conference (DAC), Las Vegas, NV, USA, 2019, pp. 1-4.

[35] R. Kastner, F. Restuccia, A. Meza, S. Ray, J. M. Fung, and C. Sturton. "Automating hardware security property generation," in *Proceedings of the 59th ACM/IEEE Design Automation Conference*, pp. 1384-1387. 2022.

[36] W. Hu, A. Althoff, A. Ardeshiricham and R. Kastner, "Towards Property Driven Hardware Security," *2016 17th International Workshop on Microprocessor and SOC Test and Verification (MTV)*, Austin, TX, USA, 2016, pp. 51-56.

[37] H. Khattri, N. K. V. Mangipudi and S. Mandujano, "HSDL: A Security Development Lifecycle for hardware technologies," *2012 IEEE International Symposium on Hardware-Oriented Security and Trust*, San Francisco, CA, USA, 2012, pp. 116-121.

[38] S. Aftabjahani, R. Kastner, M. Tehranipoor, F. Farahmandi, J. Oberg, A. Nordstrom, N. Fern, and A. Althoff, "Special Session: CAD for Hardware Security-Automation is Key to Adoption of Solutions," *2021 IEEE 39th VLSI Test Symposium (VTS)*, San Diego, CA, USA, 2021, pp. 1-10.

[39] W. Hu, A. Ardeshiricham, and R. Kastner, "Hardware information flow tracking," *ACM Computing Surveys (CSUR)*, vol. 54, no. 4: pp. 1-39, 2021.

[40] The CIRCT community, CIRCT: Circuit IR Compilers and Tools, 2020. [Online]. Available: https://github.com/llvm/circt

[41] J. A. Goguen and J. Meseguer, "Security Policies and Security Models," *1982 IEEE Symposium on Security and Privacy*, Oakland, CA, USA, 1982, pp. 11-11.

[42] F. Restuccia, A. Meza and R. Kastner, "Aker: A Design and Verification Framework for Safe and Secure SoC Access Control," *2021 IEEE/ACM International Conference On Computer Aided Design (ICCAD)*, Munich, Germany, 2021, pp. 1-9.

[43] J. Oberg, S. Meiklejohn, T. Sherwood and R. Kastner, "Leveraging Gate-Level Properties to Identify Hardware Timing Channels," in *IEEE Transactions on Computer-Aided Design of Integrated Circuits and Systems*, vol. 33, no. 9, pp. 1288-1301, Sept. 2014.

[44] M. Lipp, M. Schwarz, D. Gruss, T. Prescher, W. Haas, J. Horn, A. Fogh, J. Horn, S. Mangard, P. Kocher, D. Genkin, Y. Yarom and M. Hamburg, "Meltdown: Reading kernel memory from user space," *Communications of the ACM*, vol. 63, no. 6, pp.46-56, 2020.

[45] P. Kocher, J. Horn, A. Fogh, D. Genkin, D. Gruss, W. Haas, M. Hamburg, M. Lipp, S. Mangard, T. Prescher and Y. Yarom, "Spectre attacks: Exploiting speculative execution, " *Communications of the ACM*, vol. 63, no. 7, pp. 93-101, 2020.

[46] A. Ardeshiricham, W. Hu and R. Kastner, "Clepsydra: Modeling timing flows in hardware designs," *2017 IEEE/ACM International Conference on Computer-Aided Design (ICCAD)*, Irvine, CA, USA, 2017, pp. 147-154.

[47] A. Meza, F. Restuccia, J. Oberg, D. Rizzo, and R. Kastner, "Security Verification of the OpenTitan Hardware Root of Trust", *IEEE Security & Privacy*.

[48] C. Deutschbein, A. Meza, F. Restuccia, R. Kastner, and C. Sturton, "Isadora: automated information-flow property generation for hardware security verification," *Journal of Cryptographic Engineering*, pp. 1-17, 2022.

[49] R. Zhang and C. Sturton, "Transys: Leveraging Common Security Properties Across Hardware Designs," *2020 IEEE Symposium on Security and Privacy (SP)*, San Francisco, CA, USA, 2020, pp. 1713-1727.

[50] R. Zhang, N. Stanley, C. Griggs, A. Chi, and C. Sturton, "Identifying security critical properties for the dynamic verification of a processor," *ACM SIGARCH Computer Architecture News*, vol. 45, no. 1, pp. 541-554, 2017.

[51] R. Zhang, C. Deutschbein, P. Huang and C. Sturton, "End-to-End Automated Exploit Generation for Validating the Security of Processor Designs," *2018 51st Annual IEEE/ACM International Symposium on Microarchitecture (MICRO)*, Fukuoka, Japan, 2018, pp. 815-827.

[52] Y. Yao, T. Kathuria, B. Ege and P. Schaumont, "Architecture Correlation Analysis (ACA): Identifying the Source of Side-channel Leakage at Gate-level," *2020 IEEE International Symposium on Hardware Oriented Security and Trust (HOST)*, San Jose, CA, USA, 2020, pp. 188-196.

[53] P. Kiaei and P. Schaumont, "SoC Root Canal! Root Cause Analysis of Power Side-Channel Leakage in System-on-Chip Designs", in *IACR Transactions on Cryptographic Hardware and Embedded Systems (TCHES)*, vol. 2022, no. 4, pp. 751–773, Aug. 2022.

979-8-3503-4631-2/23 $31.00 © 2023 IEEE

Session Title: *"Unified Approaches for Silicon Debug"*,

Moderator & Organizer: Sankaran Menon, Intel Corp.

Talk 1: *"Scan and Embedded Memory State Extraction for Silicon Debug, Validation, and Analysis "*,
Authors: Mike Ricchetti, Yervant Zorian, Synopsys

Abstract:

Continued growth in the complexity of silicon and systems, in addition to new operating system functionality and software applications, are contributing factors to the challenges of first-silicon debug, product ramp-up, and system validation. Silicon debug is time-consuming and is often in the in the critical path of new silicon bring-up, and it gates TTV (Time-to-Volume). Comprehensive solutions for silicon debug will improve productivity and reduce costs, helping to accelerate meeting TTV and TTM milestones It can also provide support for Silicon Lifecycle Management (SLM) and analytics.

Silicon debug relies on the visibility and observability of hardware states that are the suspected cause of system bugs. In addition, debug-specific circuitry may be added for advance debugging. This presentation will describe an architecture and design automation for system-level state extraction. This methodology is aligned with the proposed IEEE standard P2929 for state extraction of scan and embedded memories.

Talk 2: "Use-cases and ROI for unified verification, test and silicon debug sequences using Design-for-Verification"
Authors: David Akselrod, Francis Cheng, Kevin Zhou, Yuan Zhang, AMD

Abstract:

The area of Design-for-Verification (DfV) is gaining focus from the chipmaker industry, especially in the era of ever-increasing complexity of multi-die chip design. The ability to address numerous pain-points throughout chip design cycle in virtually each functional area has turned attention to what DfV techniques offer not only to the core areas of Verification and Emulation, but also to post-silicon stages of Test, Validation and Diagnostics. In addition to covering DfV application areas, this presentation will explore the Return-on-Investment analysis of applying DfV in modern chip design.

Talk 3: *"Platform-level observability of SoC Scan and Array Functional states for System and In-Field Debug"*
Authors: Sankaran Menon, Rolf Kuehnis and Rakesh Kandula, Intel Corporation

Abstract:

Comprehensive and Efficient SoC debug is extremely important for time-to-market and for faster throughput of SoCs. Traditional platform debug involves opening the chassis and connecting cables to the SoC for debug, which is time-consuming and cumbersome. Platform-level observability of SoC Scan and Array functional states mitigates the need to open the chassis, thus saving tremendous amount of time and effort. This is done using the available closed-chassis interfaces such as USB Type-C, which is becoming the choicest interface by most OEMs/ODMs, making this the ubiquitous connector/receptacle.

This paper provides a debug architecture framework for capturing the scan and array functional state extraction from the SoC, aligning with the IEEE P2929 proposed standard for Scan and Array system-level extraction for functional validation and debug. The extracted debug information is sent via the platform-level USB Type-C closed chassis interface to enable platform-level debug. We also show that this debug information can also be used for in-field debug as well as for Silicon Lifecycle Management (SLM).

Author index

AFTABJAHANI, Sohrab

- Special Session: CAD for Hardware Security - Promising Directions for Automation of Security Assurance

AHMADILIVANI, Mohammad Hasan

- Special Session: Approximation and Fault Resiliency of DNN Accelerators

AHMED, Bulbul

- Special Session: CAD for Hardware Security - Promising Directions for Automation of Security Assurance

AHMED, Soyed Tuhin

- A Low Overhead Checksum Technique for Error Correction in Memristive Crossbar for Deep Learning Applications

AKSELROD, David

- Unified Approaches for Silicon Debug

AKSHAY, Jaiswal

- Targeted Custom High-Voltage Stress Patterns on Automotive Designs

ALEXANDRESCU, Dan

- Innovation Practices Track: VLSI Functional Safety

ALRAHIS, Lilas

- Graph Neural Networks for Hardware Vulnerability Analysis - Can you Trust your GNN?

AMANOR-BOADU, Judy

- Pre and post silicon server platform transient performance using trans-inductor voltage regulator

AMROUCH, Hussam

- Reliable Brain-inspired AI Accelerators using Classical and Emerging Memories

ANGHEL, Lorena

- Hardware design and Reliability Mitigation of Binary Bayesian Reasoning

ANGIONE, Francesco

- A guided debugger-based fault injection methodology for assessing functional test programs

APPELLO, davide

- A guided debugger-based fault injection methodology for assessing functional test programs

ARORA, Manish

- Allocating Physically Aware Embedded Memory Test & Repair Processor using Floorplan Info at the RTL Design Level

AVHAD, Gurumurti

- Auxiliary State Machine Controlled Autonomous Design Verification Framework

AZIZ, Ahmedullah

- Reliable Brain-inspired AI Accelerators using Classical and Emerging Memories

BAHRAMI, Javad

- Special Session: Security Verification & Testing for SR-Latch TRNGs

BANERJEE, Sanmitra

- Special Session: Using Graph Neural Networks for Tier-Level Fault Localization in Monolithic 3D Ics

BARBARESCHI, Mario

- Special Session: Approximation and Fault Resiliency of DNN Accelerators

BARONE, Salvatore

- Special Session: Approximation and Fault Resiliency of DNN Accelerators

BAZAZ, Rishik

- Pre and post silicon server platform transient performance using trans-inductor voltage regulator

BERNARDI, Paolo

- A guided debugger-based fault injection methodology for assessing functional test programs

BERTANI, claudia

- A guided debugger-based fault injection methodology for assessing functional test programs

BEYNE, Eric

- IP Session on Chiplet: Design, Assembly and Test

BHUNIA, Swarup

- An Exploration of ATPG Methods for Redacted IP and Reconfigurable Hardware

BIENSTMAN, Peter

- Special Session: Neuromorphic hardware design and reliability from traditional CMOS to emerging technologies

BORZA, Mike

- CAPEC: A Cellular Automata Guided FSM-based IP Authentication Scheme

BOSIO, Alberto

- Special Session: Approximation and Fault Resiliency of DNN Accelerators
- Special Session: Neuromorphic hardware design and reliability from traditional CMOS to emerging technologies

BREUNESSE, Cees-Bart

- Special Session: CAD for Hardware Security - Promising Directions for Automation of Security Assurance

CARPEGNA, Alessio

- Special Session: Neuromorphic hardware design and reliability from traditional CMOS to emerging technologies

CETIN, Ahmet Enis

- Hybrid Binary Neural Networks: A Tutorial Review

CHAKRABARTY, Krishnendu

- Functional Test Generation for AI Accelerators using Bayesian Optimization
- Special Session: Using Graph Neural Networks for Tier-Level Fault Localization in Monolithic 3D Ics

CHAKRAVARTY, SREEJIT

- Effective and Efficient Testing of Large Numbers of Inter-Die Interconnects in Chiplet-Based Multi-Die Packages

CHAKRAVARTY, Sreejit

- Silent Data Errors: Sources, Detection, and Modeling
- IP Session on Chiplet: Design, Assembly and Test

CHANG, Shuo-Wen

- Outlier Detection for Analog Tests Using Deep Learning Techniques

CHAO, Mango

- Outlier Detection for Analog Tests Using Deep Learning Techniques

- Test Generation for Defect-Based Faults of Scan Flip-Flops

CHARBONNIER, Benoit

- Special Session: Neuromorphic hardware design and reliability from traditional CMOS to emerging technologies

CHAUDHURI, Arjun

- Functional Test Generation for AI Accelerators using Bayesian Optimization
- Special Session: Using Graph Neural Networks for Tier-Level Fault Localization in Monolithic 3D Ics
- Innovation Practices Track: Testability and Dependability of AI Hardware and Autonomous Systems

CHEN, Ching-Yuan

- Functional Test Generation for AI Accelerators using Bayesian Optimization

CHEN, Chun

- Vmin Prediction Using Nondestructive Stress Test

CHEN, Harry

- Vmin Prediction Using Nondestructive Stress Test

CHEN, Jian-Jia

- Reliable Brain-inspired AI Accelerators using Classical and Emerging Memories

CHU, Ying-Hua

- Outlier Detection for Analog Tests Using Deep Learning Techniques

CHUANG, Po-Yao

- Effective and Efficient Testing of Large Numbers of Inter-Die Interconnects in Chiplet-Based Multi-Die Packages

CHUDASAMA, Bhrugurajsinh

- Allocating Physically Aware Embedded Memory Test & Repair Processor using Floorplan Info at the RTL Design Level

CRON, Adam

- CAPEC: A Cellular Automata Guided FSM-based IP Authentication Scheme

DANESHTALAB, Masoud

- Special Session: Approximation and Fault Resiliency of DNN Accelerators

DANGER, Jean-Luc

- Special Session: Security Verification & Testing for SR-Latch TRNGs

DELLA TORCA, Salvatore

- Special Session: Approximation and Fault Resiliency of DNN Accelerators

DERAKHSHANDEH, Jaber

- IP Session on Chiplet: Design, Assembly and Test

DEUTSCHBEIN, Calvin

- Special Session: CAD for Hardware Security - Promising Directions for Automation of Security Assurance

DI CARLO, Stefano

- Special Session: Neuromorphic hardware design and reliability from traditional CMOS to emerging technologies

DI GRUTTOLA GIARDINO, Nicola

- A guided debugger-based fault injection methodology for assessing functional test programs

DORAIRAJ, Nij

- An Exploration of ATPG Methods for Redacted IP and Reconfigurable Hardware

EBRAHIMABADI, Mohammad

- Special Session: Security Verification & Testing for SR-Latch TRNGs

EHRENBERG, Heiko

- Refreshing the JTAG Family

FARAHMANDI, Farimah

- CAPEC: A Cellular Automata Guided FSM-based IP Authentication Scheme
- Special Session: CAD for Hardware Security - Promising Directions for Automation of Security Assurance

FERN, Nicole

- Special Session: CAD for Hardware Security - Promising Directions for Automation of Security Assurance

GAVARINI, Gabriele

- Special Session: Approximation and Fault Resiliency of DNN Accelerators

GHUKASYAN, Artur

- Overcoming Embedded Memory Test & Repair Challenges in the Gate-All-Around Era

GIELEN, Georges

- Effective and Efficient Testing of Large Numbers of Inter-Die Interconnects in Chiplet-Based Multi-Die Packages

GIZOPOULOS, Dimitris

- Silent Data Errors: Sources, Detection, and Modeling

GOPALSAMY, Subashini

- Fully Deterministic Storage Based Logic Built-In Self-Test

GUILLEY, Sylvain

- Special Session: Security Verification & Testing for SR-Latch TRNGs

GUPTA, Sandeep

- Design for testability (DFT) for RSFQ circuits

HARUTYUNYAN, Gurgen

- An Efficient External Memory Test Solution: Case Study for HPC Application
- Overcoming Embedded Memory Test & Repair Challenges in the Gate-All-Around Era

HE, Chen

- Innovation Practices Track: Innovation on Telemetry Monitoring

HEMARAM, Surendra

- A Low Overhead Checksum Technique for Error Correction in Memristive Crossbar for Deep Learning Applications

HILL, Ian

- Gerabaldi: A Temporal Simulator for Probabilistic IC Degradation and Failure Processes

HO, Shu-Yin

- Reliable Brain-inspired AI Accelerators using Classical and Emerging Memories

HSIEH, Cheng-Yun

- Diagnosis of Quantum Circuits in the NISQ Era

HUHN, Sebastian

- A Novel LBIST Signature Computation Method for Automotive Microcontrollers using a Digital Twin

HUNG, Shao-Chun

- Special Session: Using Graph Neural Networks for Tier-Level Fault Localization in Monolithic 3D Ics

HUNTER, Marc

- Innovation Practices Track: Innovation on Telemetry Monitoring

ISLAM, Md Mazharul

- Reliable Brain-inspired AI Accelerators using Classical and Emerging Memories

IVANOV, Andre

- Gerabaldi: A Temporal Simulator for Probabilistic IC Degradation and Failure Processes

JENIHHIN, Maksim

- Special Session: Approximation and Fault Resiliency of DNN Accelerators

JENNIFER, Kitchen

- Architectural Radiation Hardening of CMOS Power Management Circuits through Bias Tuning

JIN, Robert

- Silicon Lifecycle Management Challenges and Opportunities

Jan, Erik

- Effective and Efficient Testing of Large Numbers of Inter-Die Interconnects in Chiplet-Based Multi-Die Packages

Jia-Wei, Eric

- Vmin Prediction Using Nondestructive Stress Test

KALLURU, Venkata

- An Exploration of ATPG Methods for Redacted IP and Reconfigurable Hardware

KARIMI, Naghmeh

- Special Session: Security Verification & Testing for SR-Latch TRNGs

KASTNER, Ryan

- Special Session: CAD for Hardware Security - Promising Directions for Automation of Security Assurance

KEHLET, David

- An Exploration of ATPG Methods for Redacted IP and Reconfigurable Hardware

KEIM, Martin

- Refreshing the JTAG Family

KIBRIA, Rasheed

- CAPEC: A Cellular Automata Guided FSM-based IP Authentication Scheme

KIM, Gooyoung

- Kernel Smoothing Technique Based on Multiple-Coordinate System for Screening Potential Failures in NAND Flash Memory

KIM, Jongmin

- Kernel Smoothing Technique Based on Multiple-Coordinate System for Screening Potential Failures in NAND Flash Memory

KLIMASCH, Leon

- A Novel LBIST Signature Computation Method for Automotive Microcontrollers using a Digital Twin

KOLI, Gauri

- Architectural Radiation Hardening of CMOS Power Management Circuits through Bias Tuning

KRISTOFOR, Dickson

- Targeted Custom High-Voltage Stress Patterns on Automotive Designs

KUMAR, Arun

- An Efficient External Memory Test Solution: Case Study for HPC Application

KUMAR, Vaibhav

- Innovation Practices Track: VLSI Functional Safety

KUMAR, Vinay

- Allocating Physically Aware Embedded Memory Test & Repair Processor using Floorplan Info at the RTL Design Level

Kumar, Abhishek

- Predicting the Silent Data Error Prone Devices Using Machine Learning

LI, Chen-Hong

- Test Generation for Defect-Based Faults of Scan Flip-Flops

LI, Chien-Mo

- Diagnosis of Quantum Circuits in the NISQ Era
- Vmin Prediction Using Nondestructive Stress Test

LI, Leon

- Thwarting Reverse Engineering Attacks through Keyless Logic Obfuscation

LI, Mingye

- Design for testability (DFT) for RSFQ circuits

LI, Yen-Wei

- Diagnosis of Quantum Circuits in the NISQ Era

LI, Yu-Min

- Diagnosis of Quantum Circuits in the NISQ Era

LIAO, Jeng-Yu

- Vmin Prediction Using Nondestructive Stress Test

LIN, Chin-Kuan

- Outlier Detection for Analog Tests Using Deep Learning Techniques

LIN, Yunkun

- Design for testability (DFT) for RSFQ circuits

LORENZELLI, Francesco

- Effective and Efficient Testing of Large Numbers of Inter-Die Interconnects in Chiplet-Based Multi-Die Packages

LU, Cheng-Che

- Outlier Detection for Analog Tests Using Deep Learning Techniques

MAKRIS, Yiorgos

- Machine Learning-Based Adaptive Outlier Detection for Underkill Reduction in Analog/RF IC Testing

MARINISSEN, Erik Jan

- IP Session on Chiplet: Design, Assembly and Test

MASNA, Naren Vikram Raj

- An Exploration of ATPG Methods for Redacted IP and Reconfigurable Hardware

MAYAHINIA, Mahta

- A Low Overhead Checksum Technique for Error Correction in Memristive Crossbar for Deep Learning Applications

MENON, Sankaran

- Unified Approaches for Silicon Debug

MEZA, Andres

- Special Session: CAD for Hardware Security - Promising Directions for Automation of Security Assurance

MOON, Youngseon

- Kernel Smoothing Technique Based on Multiple-Coordinate System for Screening Potential Failures in NAND Flash Memory

MUKHERJEE, Nilanjan

- Silicon Lifecycle Management Challenges and Opportunities

MÜNCH, Christopher

- A Low Overhead Checksum Technique for Error Correction in Memristive Crossbar for Deep Learning Applications

NAHAR, Amit

- Machine Learning-Based Adaptive Outlier Detection for Underkill Reduction in Analog/RF IC Testing

NEETHIRAJAN, Deepika

- Machine Learning-Based Adaptive Outlier Detection for Underkill Reduction in Analog/RF IC Testing

NGUYEN, Liam

- Architectural Radiation Hardening of CMOS Power Management Circuits through Bias Tuning

NIEN, Yu-Teng

- Test Generation for Defect-Based Faults of Scan Flip-Flops

NIRANJAN, Vineeth

- Machine Learning-Based Adaptive Outlier Detection for Underkill Reduction in Analog/RF IC Testing

NITZAN, Meirav

- Innovation Practices Track: VLSI Functional Safety

OBILISETTY, Sashi

- Innovation Practices Track: Innovation on Telemetry Monitoring

ORAILOGLU, Alex

- Thwarting Reverse Engineering Attacks through Keyless Logic Obfuscation

OUYANG, Keqing

- An Efficient External Memory Test Solution: Case Study for HPC Application

PAN, Hongyi

- Hybrid Binary Neural Networks: A Tutorial Review

PANDE, Partha

- Fault Criticality-aware GNN Training on ReRAM-based Processing-in-Memory Systems

PAPADIMITRIOU, George

- Silent Data Errors: Sources, Detection, and Modeling

PAULITSCH, Michael

- Innovation Practices Track: Testability and Dependability of AI Hardware and Autonomous Systems

PAVANELLO, Fabio

- Special Session: Neuromorphic hardware design and reliability from traditional CMOS to emerging technologies

PENG, Minqiang

- An Efficient External Memory Test Solution: Case Study for HPC Application

POMERANZ, Irith

- Expanding a Pool of Functional Test Sequences to Support Test Compaction
- Fully Deterministic Storage Based Logic Built-In Self-Test
- Compact Set of Functional Broadside Tests with Fault Detection on Primary Outputs

PORTOLAN, Michele

- Refreshing the JTAG Family

QI, Kang
- An Efficient External Memory Test Solution: Case Study for HPC Application

RAHMAN, Fahim
- CAPEC: A Cellular Automata Guided FSM-based IP Authentication Scheme

RAHMAN, Mohammad
- CAPEC: A Cellular Automata Guided FSM-based IP Authentication Scheme

RAHMAN, Mridha Md Mashahedur
- CAPEC: A Cellular Automata Guided FSM-based IP Authentication Scheme

RAIK, Jaan
- Special Session: Approximation and Fault Resiliency of DNN Accelerators

RAMESH, Saidapet
- Targeted Custom High-Voltage Stress Patterns on Automotive Designs

REARICK, Jeff
- Refreshing the JTAG Family

REDDY, Bandi
- CAPEC: A Cellular Automata Guided FSM-based IP Authentication Scheme

RESTUCCIA, Francesco
- Special Session: CAD for Hardware Security - Promising Directions for Automation of Security Assurance

RICCHETTI, Mike
- Unified Approaches for Silicon Debug

RICE, Ritchie
- Pre and post silicon server platform transient performance using trans-inductor voltage regulator

ROBERT, Marchese
- Targeted Custom High-Voltage Stress Patterns on Automotive Designs

RUOSPO, Annachiara

- Special Session: Approximation and Fault Resiliency of DNN Accelerators

RYAN, Kaki

- Special Session: CAD for Hardware Security - Promising Directions for Automation of Security Assurance

SANCHEZ, Ernesto

- Special Session: Approximation and Fault Resiliency of DNN Accelerators

SAVINO, Alessandro

- Special Session: Neuromorphic hardware design and reliability from traditional CMOS to emerging technologies

SETHI, Ankush

- Innovation Practices Track: VLSI Functional Safety

SHAIK, Mohammad Ershad

- Predicting the Silent Data Error Prone Devices Using Machine Learning

SHANKARANARAYANAN, Bharath

- Allocating Physically Aware Embedded Memory Test & Repair Processor using Floorplan Info at the RTL Design Level

SHUMA, Azizi

- Pre and post silicon server platform transient performance using trans-inductor voltage regulator

SINANOGLU, Ozgur

- Graph Neural Networks for Hardware Vulnerability Analysis - Can you Trust your GNN?

SINGH, Adit

- Silent Data Errors: Sources, Detection, and Modeling

STITT, Greg

- An Exploration of ATPG Methods for Redacted IP and Reconfigurable Hardware

STURTON, Cynthia

- Special Session: CAD for Hardware Security - Promising Directions for Automation of Security Assurance

SU, Fei

- Innovation Practices Track: VLSI Functional Safety

- Innovation Practices Track: Innovation on Telemetry Monitoring
- Innovation Practices Track: Testability and Dependability of AI Hardware and Autonomous Systems

SUNNY, Thota

- Targeted Custom High-Voltage Stress Patterns on Automotive Designs

TAHERI, Mahdi

- Special Session: Approximation and Fault Resiliency of DNN Accelerators

TAHOORI, Mehdi

- Hardware design and Reliability Mitigation of Binary Bayesian Reasoning
- A Low Overhead Checksum Technique for Error Correction in Memristive Crossbar for Deep Learning
Applications

TALUKDAR, Jonti

- Functional Test Generation for AI Accelerators using Bayesian Optimization

TAMDEM, Horthense

- Pre and post silicon server platform transient performance using trans-inductor voltage regulator

TANCORRE, Vincenzo

- A guided debugger-based fault injection methodology for assessing functional test programs

TEHRANIPOOR, Mark

- CAPEC: A Cellular Automata Guided FSM-based IP Authentication Scheme
- Special Session: CAD for Hardware Security - Promising Directions for Automation of Security Assurance

THOMANN, Simon

- Reliable Brain-inspired AI Accelerators using Classical and Emerging Memories

TILLE, Daniel

- A Novel LBIST Signature Computation Method for Automotive Microcontrollers using a Digital Twin

TSHAGHARYAN, Grigor

- An Efficient External Memory Test Solution: Case Study for HPC Application
- Overcoming Embedded Memory Test & Repair Challenges in the Gate-All-Around Era

VAN VAERENBERGH, Thomas

- Special Session: Neuromorphic hardware design and reliability from traditional CMOS to emerging
technologies

VAN WOUDENBERG, Jasper

- Special Session: CAD for Hardware Security - Promising Directions for Automation of Security Assurance

VATAJELU, Elena Ioana

- Special Session: Neuromorphic hardware design and reliability from traditional CMOS to emerging technologies

VELEGALATI, Rajesh

- Special Session: CAD for Hardware Security - Promising Directions for Automation of Security Assurance

VINNAKOTA, Bapi

- IP Session on Chiplet: Design, Assembly and Test

WANG, Bin BW

- Allocating Physically Aware Embedded Memory Test & Repair Processor using Floorplan Info at the RTL Design Level

WANG, Isaac

- An Efficient External Memory Test Solution: Case Study for HPC Application

WANG, Yung-Jheng

- Test Generation for Defect-Based Faults of Scan Flip-Flops

WEBSTER, Dallas

- Machine Learning-Based Adaptive Outlier Detection for Underkill Reduction in Analog/RF IC Testing

WEI, Ming-Liang

- Reliable Brain-inspired AI Accelerators using Classical and Emerging Memories

WU, Cheng-Wen

- Effective and Efficient Testing of Large Numbers of Inter-Die Interconnects in Chiplet-Based Multi-Die Packages

WU, Kai-Chiang

- Outlier Detection for Analog Tests Using Deep Learning Techniques
- Test Generation for Defect-Based Faults of Scan Flip-Flops

WU, Pei-Yin

- Test Generation for Defect-Based Faults of Scan Flip-Flops

XANTHOPOULOS, Constantinos

- Machine Learning-Based Adaptive Outlier Detection for Underkill Reduction in Analog/RF IC Testing

YANG, Chia-Lin

- Reliable Brain-inspired AI Accelerators using Classical and Emerging Memories

YAYLA, Mikail

- Reliable Brain-inspired AI Accelerators using Classical and Emerging Memories

ZHANG, Eric

- Innovation Practices Track: Testability and Dependability of AI Hardware and Autonomous Systems

ZHU, Yunnong

- An Efficient External Memory Test Solution: Case Study for HPC Application

ZORIAN, Yervant

- Silicon Lifecycle Management Challenges and Opportunities
- Overcoming Embedded Memory Test & Repair Challenges in the Gate-All-Around Era

IEEE
445 Hoes Lane
Piscataway, NJ 08854-4141

ISBN 979-8-3503-4631-2